THE
NUCLEAR
RECEPTOR
FactsBook

Other books in the FactsBook Series:

Katherine A. Fitzgerald, Luke A.J. O'Neill, Andy Gearing and Robin E. Callard
The Cytokine FactsBook, 2nd edn plus Cytokine Webfacts

Steve Watson and Steve Arkinstall
The G-Protein Linked Receptor FactsBook

Shirley Ayad, Ray Boot-Handford, Martin J. Humphries, Karl E. Kadler
and C. Adrian Shuttleworth
The Extracellular Matrix FactsBook, 2nd edn

Grahame Hardie and Steven Hanks
The Protein Kinase FactsBook
The Protein Kinase FactsBook CD-Rom

Edward C. Conley
The Ion Channel FactsBook
I: Extracellular Ligand-Gated Channels

Edward C. Conley
The Ion Channel FactsBook
II: Intracellular Ligand-Gated Channels

Edward C. Conley and William J. Brammar
The Ion Channel FactsBook
IV: Voltage-Gated Channels

Kris Vaddi, Margaret Keller and Robert Newton
The Chemokine FactsBook

Marion E. Reid and Christine Lomas-Francis
The Blood Group Antigen FactsBook

A. Neil Barclay, Marion H. Brown, S.K. Alex Law, Andrew J. McKnight,
Michael G. Tomlinson and P. Anton van der Merwe
The Leucocyte Antigen FactsBook, 2nd edn

Robin Hesketh
The Oncogene and Tumour Suppressor Gene FactsBook, 2nd edn

Jeffrey K. Griffith and Clare E. Sansom
The Transporter FactsBook

Tak W. Mak, Josef Penninger, John Rader, Janet Rossant
and Mary Saunders
The Gene Knockout FactsBook

Bernard J. Morley and Mark J. Walport
The Complement FactsBook

Steven G.E. Marsh, Peter Parham and Linda Barber
The HLA FactsBook

Hans G. Drexler
The Leukemia-Lymphoma Cell Line FactsBook

Clare M. Isacke and Michael A. Horton
The Adhesion Molecule FactsBook, 2nd edn

Marie-Paule Lefranc and Gérard Lefranc
The Immunoglobulin FactsBook

Marie-Paule Lefranc and Gérard Lefranc
The T-Cell Receptor FactsBook

THE NUCLEAR RECEPTOR
FactsBook

Vincent Laudet[1] and
Hinrich Gronemeyer[2]

[1]*Laboratoire de Biologie Moleculaire et Cellulaire, Ecole Normale Supérieure de Lyon, Lyon, France.*
[2]*Institut de Genetique et de Biologie Moleculaire et Cellulaire, (CNRS, INSERM, ULP), Strasbourg, France.*

ACADEMIC PRESS
A Division of Harcourt, Inc.

San Diego San Francisco New York Boston
London Sydney Tokyo

Academic Press
A division of Harcourt, Inc.
Harcourt Place, 32 Jamestown Road, London NW1 7BY, UK
http://www.academicpress.com

Academic Press
A division of Harcourt, Inc.
525 B Street, Suite 1900, San Diego, California 92101-4495, USA
http://www.academicpress.com

ISBN 0-12-437735-1

A catalogue record for this book is available from the British Library

Library of Congress Catalog Card Number: 2001096534

Typeset by Mackreth Media Services, Hemel Hempstead, UK
Printed in Great Britain by CPI Bookcraft

01 02 03 04 05 BC 9 8 7 6 5 4 3 2 1

Contents

Section I THE INTRODUCTORY CHAPTERS

Section II THE NUCLEAR RECEPTORS

Contents

*Official nomenclature appears in parentheses.

Preface

Multicellular organisms require specific intercellular communication to organize properly the complex body plan during embryogenesis, and maintain the physiological properties and functions during the entire life. Signaling through nuclear receptors is a major signal transduction paradigm invented by metazoans to provide a plethora of intra- and intercellular communication networks. Nuclear receptors constitute a large family of ligand-inducible transcription factors that regulate gene-initiated programs at the basis of a multitude of (patho)physiological phenomena. They act by (1) responding directly to a large variety of hormonal and metabolic signals; (2) integrating diverse signaling pathways as nuclear receptors are themselves targets of post-translational modifications; and (3) regulating the activities of other major signalling cascades (called signal transduction crosstalk).

This book is intended to provide a comprehensive summary of the current knowledge of the mechanism(s) of action of this important class of transcription regulatory receptors which are of major importance for research in a large number of areas related to animal development and physiology, medicine, and pharmacological drug development. The introductory section is conceptual whilst the major part of the book is an in-depth discussion of each nuclear receptor that provides the reader with virtually all of the data presently available.

Compilation of this book was hard work as the field is vast and evolving rapidly. We are grateful to Pierre Chambon for his critical interest and suggestions. We apologize to all colleagues whose recent data are not considered in this book or whose work is, in spite of all efforts, mentioned or cited inadequately.

V.L. would like to thank deeply all the members of his laboratory for their patience, kindness and knowledge of nuclear receptors that was extremely useful during the preparation of this book. The help of Jacques Samarut, Frédéric Flamant, Pierre Jurdic, Bart Staels, Jamshed Tata, Barbara Demeneix, Gunther Schutz and Frances Sladek for information, criticisms and sharing of unpublished data is also acknowledged. Last but not least, thanks to Catherine Hänni and Paloma for the numerous days of absence and to Denise and Edmond Hänni for the old table and the fresh garden of La Bastide where this book was written. Work from V.L.'s laboratory is supported by the Centre National de la Recherche Scientifique, Ecole Normale Supérieure de Lyon, Ministère de l'Education Nationale, de la Recherche et de la Technologie, Association pour la Recherche sur le Cancer, Fondation pour la Recherche Médicale, Ligue Nationale contre le Cancer, and Région Rhône-Alpes. V.L. also thanks the Institut Universitaire de France for support.

H.G. thanks all the members of his laboratory for their patience, support and help while this book was being prepared. Work in the laboratory was supported by the Association for International Cancer Research, Association pour la Recherche sur le Cancer, Fondation pour la Recherche Médicale, Institut National de la Santé et de la Recherche Médicale, the Centre National de la Recherche Scientifique, the Hôpital Universitaire de Strasbourg and Bristol-Myers Squibb.

The authors hope that there are a minimum of omissions and inaccuracies and that these can be recitified in later editions. We would appreciate if such points were forwarded to the Editor, Nuclear Receptor FactsBook, Academic Press, 32 Jamestown Road, London NW1 7BY, UK.

Vincent Laudet

Hinrich Gronemeyer and co-workers.

Abbreviations

4-ABB	4-amino butyl benzoate
AD	activating domain
ADH	alcohol dehydrogenase
3-AEB	3-amino ethyl benzoate
AEV	avian erythroblastosis virus
AF	activation function
AHC	adrenal hypoplasia congenita
AIB	amplified in breast cancer
AML	acute myeloid leukemia
AP	activating protein
aP2	adipocyte fatty acid-binding protein 2
APC	adenomatous polyposis coli
APL	acute promyelocytic leukemia
ApoB	apolipoprotein B
AR	androgen receptor
ARC	activator-recruited cofactor
ARE	androgen response element
ARF	adipocyte regulatory factor
Arp	apoAI regulatory protein
ASC-2	activating signal cointegrator-2
ATF-2	activating transcription factor 2
BAR	bile acid receptor
bp	base pair
BCAR	breast cancer antiestrogen resistance
BMP	bone morphogenetic protein
BRL49653	rosiglitazone
CaMKIV	Ca^{2+}/calmodulin-dependent protein kinase IV
CAR	constitutively active receptor, constitutive androstane receptor
CARLA	coactivator-dependent receptor ligand assay
CBP	CREB-binding protein
CDK	cyclin-dependent kinase
C/EBP	CAAT/enhancer binding protein
CEF	chicken embryo fibroblasts
CGP52608	1-(3-allyl-4-oxothiazolidine-2-ylidene)-4-methylthio-semicarbazone
CIP	CBP-interacting protein
CMV	cytomegalovirus
CNS	central nervous system
CoRNR box	corepressor nuclear receptor box
COUP-TF	chicken ovalbumin upstream promoter transcription factor
COX-2	cyclooxygenase-2
9C-RA	9-*cis* retinoic acid
CRABP	cellular retinoic acid binding protein
CRALBP	cellular retinaldehyde binding protein

CRBPI	cellular retinal binding protein I
CREB	cyclic AMP response element binding factor
CREM	cyclic AMP response element modulator protein
CRH	corticotrophin-releasing protein
CRSP	cofactor required for Sp1 activation
CTE	C-terminal extension
CYP	cytochrome P450
3D	three-dimensional
DBD	DNA-binding domain
p,p'-DDE	1,1-dichloro-2,2-bis(*p*-chlorophenyl)ethylene
DES	diethylstilbestrol
DHEA	dihydroepiandrosterone
DHR	*Drosophila* hormone receptor
DHT	dihydrotestosterone
DIMIT	3,5-dimethyl-3'-isopropylthyronine
dpc	days post-coitum
DR	direct repeat
DRIP	vitamin D receptor interacting protein complement
DSF	dissatisfaction
DSS	dosage-sensitive sex reversal
20E	20-hydroxyecdysone
E2F	E2 factor
EB1089	1-25-dihyroxy-22,24-diene-24,27-trihomo vitamin D
EBV	Epstein-Barr virus
EC	embryonal carcinoma
EcR	ecdysone receptor
EcRE	ecdysone receptor response element
EGF	epidermal growth factor
ELP	embryonal LTR-binding protein
EMS	ethyl methanesulfonate
EMSA	electrophoretic mobility shift assay
ER	estrogen receptor, everted repeat
ERA-1	early retinoic acid-induced F9 teratocarcinoma stem cell gene1
ERE	estrogen response element
ERM	ets related member
ERR	estrogen receptor related receptor
ES	embryonic stem
EST	expressed sequence tag
FAAR	fatty acid-activated receptor
FGR	familial glucocorticoid resistance
FISH	fluorescent *in situ* hybridization
FPPS	farnesyl pyrophosphate synthase
FRET	fluorescence energy transfer
FSH	follicle-stimulating hormone
FTF	α-fetoprotein transcription factor
FTZ-F1	*fushi tarazu*–factor 1
FXR	farnesoid receptor
GAL	galactose
GCNF	germ cell nuclear factor

G-CSF	granulocyte colony-stimulating factor
GM-CSF	granulocyte–macrophage colony-stimulating factor
GFP	green fluorescent protein
GH	growth hormone
GHF	growth hormone-specific transcription factor
GnRH	gonadotrophin-releasing hormone
GR	glucocorticoid receptor
GRE	glucocorticoid response element
GRF	GCNF-related factor
GRIP	glucocorticoid receptor interacting protein
GST	glutathione S-transferase
HAT	histone acetyltransferase
HBV	hepatitis B virus
HDAC	histone deacetylase
HETE	hydroxyeicosatetraenoic acid
HHG	hypogonadotropic hypogonadism
HIV	human immunodeficiency virus
HMG	high motility group proteins
HNF4	hepatocyte nuclear factor 4
HODE	hydroxyoctadecadienoic acid
HPA	hypothalamic–pituitary–adrenal
hpf	hours post-fertilization
HPV	human papillomavirus
H-2RIIBP	H-2 region II binding protein
HRE	hormone receptor response element
Hsp	heat-shock protein
HSV	herpes simplex virus
IBABP	ileal bile acid binding protein
IGF	insulin-like growth factor
IGFBP	insulin-like growth factor binding protein
IL	interleukin
IR3	3 bp-spaced inverted repeats
IκB	inhibitor of NFκB
JAK	Janus kinase
JH	juvenile hormone
JNK	Jun N-terminal kinase
LBD	ligand-binding domain
LBP	ligand-binding pocket
LH	luteinizing hormone
LPL	lipoprotein lipase
LRH	liver receptor homologous protein
LT	leukotriene
LTR	long terminal repeat
LXR	liver X receptor
MAP	mitogen-activated protein
MAPK	mitogen-activated protein kinase
MCAD	medium-chain acyl coenzyme A dehydroenase
MED	mediator (multiprotein complex involved in transcription activation)

MHC	major histocompatibility complex
mHMG–CoAS	mitochondrial hydroxymethylglutaryl–CoA synthase
MIS	Müllerian inhibiting substance
MKK	mitogen-activated protein kinase
MLP	major late promoter
MMTV	mouse mammary tumor virus
MoMLV	Moloney murine leukaemia virus
MOZ	monocytic zinc finger
MR	mineralocorticoid receptor
NAT	negative regulator of activated transcription
NBFI-B	nerve growth factor-induced B
NBRE	NGFI-B response element
NCNF	neuronal cell nuclear factor
NCoR	nuclear receptor corepressor
NFAT	nuclear factor of activated cells
NFκB	nuclear factor κB (a transcription factor)
NHR	nuclear hormone receptor
NIDDM	non-insulin-dependent diabetes mellitus
NIX	neural interacting factor
NLS	nuclear localization signal
NMDA	N-methyl-D-aspartic acid
NMR	nuclear magnetic resonance
NPM	nuclear phosphoprotein nucleophosmin
NR	nuclear receptor
NRE	negative regulatory element
NRIF3	nuclear receptor interacting factor 3
NRRE	nuclear receptor responsive element
NSAID	non-steroidal anti-inflammatory drug
NSD-1	NR-binding SET-domain containing protein
NTCP	Na^+/taurocholate-cotransporting peptide
NUT	subunit of the yeast mediator complex
OCT	22-oxa-1α,25-dihydroxyvitamin D_3
Oct	octamer-binding protein
1,25(OH)$_2$-D$_3$	1,25-dihydroxyvitamin D_3
OHT	hydroxy-tamoxifen
OR	orphan receptor
p300	paralog of CBP
PAL	palindrome
pal	palindromic element
PAR	pregnane-activated receptor
PARP	poly(ADP-ribose) polymerase
PB	phenobarbital
PBP	PPAR-binding protein
PC2	positive coactivator 2
P/CAF	p300/CREB binding protein [CBP]-associated factor
P/CIP	p300/CBP/co-integrator-associated protein
PCN	pregnenolone 16-carbonitrile
PCR	polymerase chain reaction
PDB	protein database in Brookhaven

PEPCK	phosphoenolpyruvate carboxykinase
PGC	peroxisome proliferator-activated receptor gamma coactivator
PKA	protein kinase A
PKB	protein kinase B
PLTP	phospholipid transfer protein
PML	promyelocytic leukemia
PNMT	phenylethanolamine N-methyltransferase
PNR	photoreceptor-specific nuclear receptor
PNRC	proline-rich nuclear receptor coregulatory protein
POD	PML oncogenic domain
POMC	pro-opiomelanocortine
PPAR	peroxisome proliferator-activated receptor
PPARE	peroxisome proliferator-activated receptor response element
PPRE	peroxisome proliferator response element
PR	progesterone receptor
PRF	pre-finger region
PSA	prostate-specific antigen
PSU	pet mutant suppressor
PTH	parathyroid hormone
PTHrP	parathyroid hormone-related peptide
PXR	pregnane X receptor
RA	retinoic acid
RAC3	receptor-associated coactivator 3
RACE	rapid amplification of cDNA ends
RAP250	nuclear receptor activating protein 250
RAR	retinoic acid receptor
RARE	retinoic acid receptor response element
Rb	retinoblastoma
RE	response element
RevRE	Rev-erb response element
RIP	receptor interacting protein
RLD	rat liver-derived
RNR	retina-specific nuclear receptor
ROR	RAR-related orphan receptor
RORE	ROR response element
RT	reverse transcriptase
RTR	retinoid receptor-related testis-associated receptor
RU	Roussel–Uclaf compound (RU is followed by a number)
RXR	retinoid X receptor
RXRE	retinoid X receptor response element
SERM	selective estrogen receptor modulators
SET	homologous domain in Su(var)3-9, enhancer of zeste and trithorax
SF	steroidogenic factor
SFRE	SF-1 response element
SHP	short heterodimer partner
SMCC	Srb and mediator protein-containing complex
SMRT	silencing mediator of repressed transcription
SRA	steroid receptor RNA activator
StAR	steroidogenic acute regulatory protein

STAT	signal transducers and activators of transcription
SUG-1	suppressor of the GAL4 protein
Sun-CoR	small unique nuclear receptor corepressor
SVP	seven up
SXR	steroid and xenobiotic receptor
SWI	SWI ('switch') genes required for transcriptional activation
T3	3,5,3'-triiodo-L-thyronine
T4	3,5,3',5'-tetraiodo-L-thyronine
TAF	TATA-associated factors, TBP-associated factors
TAT	tyrosine aminotransferase
TBP	TATA-binding protein
TCF	T-cell-specific transcription factor
TCPOBOP	1,4-bis[2-(3,5-dichloropyridyloxy)]benzene
TFII	polymerase II associated transcription factor
TGF	transforming growth factor
TIF	transcription intermediary factor
TLX	vertebrate homologue of *Drosophila* tailless gene
TNF	tumor necrosis factor
TOR	T-cell orphan receptor
TPA	12-*O*-tetradecanoyl phorbol-13-acetate
TR	thyroid hormone receptor
TRAP	thyroid hormone receptor-associated protein
TRE	thyroid hormone response element
TRH	thyrotropin-releasing hormone
TRIAC	3'-tri-iodothyroacetic acid
TRP	tetratricopeptide
TSH	thyroid-stimulating hormone
TTNPB	(E)-4-[2-(5,6,7,8-tetrahydro-5,5,8,8-tetramethyl-2-naphthylenyl)-1-propenyl] benzoic acid
TTR	transthyretin hormone
TZD	thiazolidinedione
UCP	uncoupling protein
UR	ubiquitous receptor
USF	upstream regulatory factor
USP	ultraspiracle
UTR	untranslated region
VAD	vitamin A deficiency
VDR	vitamin D receptor
VDRE	vitamin D receptor response element
VEGF	vascular endothelial growth factor
VIP	vasoactive intestinal peptide
VLDL	very-low-density lipoprotein
VP	viral protein
WT1	Wilm's tumor gene
ZK	Schering AG compound (ZK followed by a number)
ZPA	zone of polarizing activity

Glossary

Ad-4BP
Ad4-binding protein; Ad4 is a *cis*-acting element of the bovine CYP11B1 gene.

CBP (=CREB binding protein)
Transcriptional coactivator (265 kDa) of CREB and of c-*Myb*. Only binds the phosphorylated form of CREB.

CDK (=cyclin-dependent kinase)
Family of kinases including cdc28, cdc2 and p34cdc2 that are only active when they form a complex with cyclins.

CREB
Cyclic AMP response element binding factor. Basic leucine zipper (bZip) transcription factor involved in activating genes through cAMP; binds to CRE element TGANNTCA. Phosphorylation by cAMP-dependent protein kinase (PKA) at serine-119 is required for interaction with DNA and phosphorylation at serine-133 allows CREB to interact with CBP (CREB-binding protein) leading to interaction with RNA polymerase II.

epidermal growth factor (EGF)
A mitogenic polypeptide (6 kDa) initially isolated from male mouse submaxillary gland. The name refers to the early bioassay, but EGF is active on a variety of cell types, especially but not exclusively epithelial. A family of similar growth factors are now recognized.

GCN5
Yeast GCN5 is one component of a putative adaptor complex that includes ADA2 and ADA3 and functionally connects DNA-bound transcriptional activators with general transcription factors.

HeLa cells
An established line of human epithelial cells derived from a cervical carcinoma (said to be from Henrietta Lacks).

IκB (IkappaB)
Protein that inhibits NFκB by binding to the p65 subunit. It is thought to prevent NFκB from entering the nucleus. Two forms have been identified IκBα (37 kDa) and IκBβ (43 kDa).

JAK (= Janus kinase)
Family of intracellular tyrosine kinases (120–140 kDa) that associate with cytokine receptors (particularly but not exclusively interferon receptors) and are involved in the signaling cascade. JAK is so-called either from Janus kinase (Janus was the gatekeeper of heaven) or 'just another kinase'. JAK has neither SH2 nor SH3 domains.

MAPK; MAP kinase (= mitogen-activated protein kinase; externally regulated kinase, ERK)
Serine-threonine kinases that are activated when quiescent cells are treated with mitogens, and that therefore potentially transmit the signal for entry into cell cycle. One target is trancription factor p62TCF. MAP kinase itself can be phosphorylated by MAP kinase kinase and this may in turn be controlled by raf-1.

NFκB (= NFkappaB)
A transcription factor (originally found to switch on transcription of genes for the kappa class of immunoglobulins in B-lymphocytes). It is involved in activating the transcription of more than 20 genes in a variety of cells and tissues. NFκB is found in the cytoplasm in an inactive form, bound to the protein IκB. A variety of stimuli, such as tumor necrosis factor, phorbol esters and bacterial lipopolysaccharide activate it, by releasing it from IκB, allowing it to enter the nucleus and bind to DNA. It has two subunits, p50 and p65, that bind DNA as a heterodimer. The dimerization and DNA binding activity are located in N-terminal regions of 300 amino acids that are similar to regions in the Rel and *dorsal* transcription factors.

Oct (=octamer binding protein)
Transcription factor that binds to the octamer motif. Examples: mammalian proteins Oct-1, Oct-2.

PKB (=protein kinase B); Akt
Product of the normal gene homologue of v-*akt*, the transforming oncogene of AKT8 virus. A serine/threonine kinase (58 kDa) with SH2 and PH domains, activated by PI3 kinase downstream of insulin and other growth factor receptors. AKT will phosphorylate GSK3 and is involved in stimulation of Ras and control of cell survival. Three members of the Akt/PKB family have been identified, Akt/PKBα, AKT2/PKBβ, AKT3/PKBγ.

PPAR (= peroxisome proliferator-activated receptors)
Member of the nuclear receptor family of ligand-activated transcription factors. PPARα stimulates β-oxidative degradation of fatty acids, PPARγ promotes lipid storage by regulating adipocyte differentiation. Are implicated in metabolic disorders predisposing to atherosclerosis and inflammation. PPARα-deficient mice show prolonged response to inflammatory stimuli. PPARα is activated by gemfibrozil and other fibrate drugs.

Protein kinase A
Cyclic AMP-dependent protein kinase.

Rel
Protein that acts as a transcription factor. It was first identified as the oncogene product of the lethal, avian retrovirus Rev-T. It has a N-terminal region of 300 amino acids that is similar to the N-terminal regions of NFκB subunits. Rel/NF-kappa B proteins are a small family of transcription factors which serve as pivotal regulators of immune, inflammatory and acute phase responses.

Rev-erb
Nuclear hormone receptor-related protein encoded on the opposite strand of the alpha-thyroid hormone receptor (TR) gene.

Smad proteins
Intracellular proteins that mediate signaling from receptors for extracellular TGFβ-related factors. Smad2 is essential for embryonic mesoderm formation and establishment of anterior–posterior patterning. Smad4 is important in gastrulation. Smads1 and 5 are activated (serine/threonine phosphorylated) by BMP receptors, Smad2 and 3 by activin and TGFβ receptors.

Spi-1
Proto-oncogene encoding a transcription factor (PU1) that binds to purine-rich sequences (PU boxes) expressed in hematopoietic cells.

SRB
The SRB complex that is part of the yeast RNA polymerase II holoenzyme.

STATs (= signal transducers and activators of transcription)
Contain SH2 domains that allow them to interact with phosphotyrosine residues in receptors, particularly cytokine-type receptors; they are then phosphorylated by JAKs, dimerize and translocate to the nucleus where they act as transcription factors. Many STATs are known; some are relatively receptor specific, others more promiscuous, so that a wide range of responses is possible with some STATs being activated by several different receptors, sometimes acting synergistically with other STATs.

SWI/SNF complex
The SWI/SNF complex remodels nucleosome structure in an ATP-dependent manner. In yeast the SWI/SNF chromatin remodeling complex is comprised of 11 tightly associated polypeptides (SWI1, SWI2, SWI3, SNF5, SNF6, SNF11, SWP82, SWP73, SWP59, SWP61, and SWP29). SWP59 and SWP61 are encoded by the ARP9 and ARP7 genes, respectively, which encode members of the actin-related protein (ARP) family.

VLDL (= very-low-density lipoprotein)
Plasma lipoproteins with density of 0.94–1.006 g/cm^3; made by the liver. Transport triacylglycerols to adipose tissue. Apoproteins B, C and E are found in VLDL. Protein content, about 10%, much lower than in very-high-density lipoprotein.

THE INTRODUCTORY CHAPTERS

1 General organization of nuclear receptors

NUCLEAR RECEPTORS REGULATE INTRACELLULAR AND INTERCELLULAR COMMUNICATION IN METAZOANS

Multicellular organisms require specific intercellular communication to organize the complex body plan properly during embryogenesis, and to maintain its properties and functions during the entire life span. While growth factors, neurotransmitters and peptide hormones bind to membrane receptors (thereby inducing the activity of intracellular kinase cascades or the JAK-STAT/Smad signaling pathways), other small, hydrophobic signaling molecules, such as steroid hormones, certain vitamins and metabolic intermediates, enter or are generated within the target cells and bind to cognate members of a large family of nuclear receptors (Fig. 1). Nuclear receptors (NRs) are of major importance for metazoan intercellular signaling, as they bring together different intracellular and extracellular signals to initiate and regulate gene expression programs. They act as transcription factors that: (1) respond directly through physical association with a large variety of hormonal and other regulatory, as well as metabolic signals; (2) integrate diverse signaling pathways as they are themselves targets of post-translational modifications; and (3) regulate the activities of other signaling cascades (commonly referred to as 'signal transduction cross-talk'). The genetic programs that they modulate affect virtually all aspects of the life of a multicellular organism, covering such diverse aspects as, for example, embryogenesis, homeostasis and reproduction, or cell growth and death. Their gene regulatory power and selectivity have prompted intense research into these key factors, which is now starting to decipher a complex network of molecular events accounting for their transcription regulatory capacity. The study of these molecular processes has also shed light on multiple general principles underlying transcription regulation, and it will be a challenge for the future to uncover the molecular rules that define selective NR-dependent spatial and temporal control of gene expression.

THE FAMILY AND ITS LIGANDS

To date, 65 different nuclear receptors have been identified throughout the animal kingdom ranking from nematodes to man (Fig. 2). They constitute a family of transcription factors that share a modular structure of 5–6 conserved domains encoding specific functions (Fig. 3)[1,2]. The most prominent distinction to other transcription factors is their capacity specifically to bind small hydrophobic molecules. These ligands constitute regulatory signals, which change the transcriptional activity of the corresponding nuclear receptor after binding. For some time a distinction was made between classical nuclear receptors with known ligands and so-called 'orphan' receptors – receptors without or with an unknown ligand. However, recent years have seen the identification of ligands for many of these orphan receptors, making this distinction rather superficial[3–6]. Moreover, the classification of nuclear receptors into 6–7 phylogenetic subfamilies with groups that comprise both orphan and non-orphan receptors further dismisses such a discrimination[7].

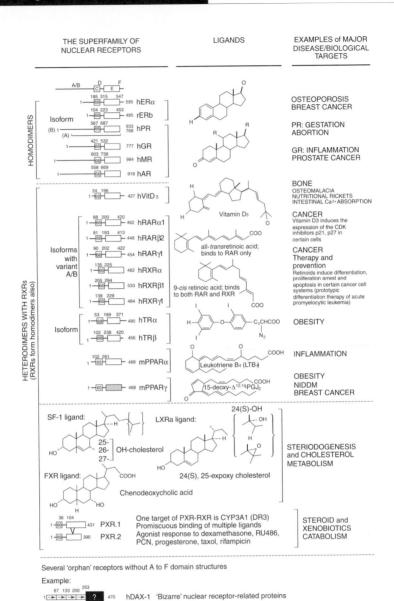

Figure 1 *Members of the superfamily of nuclear receptors. On the left the receptors are illustrated with their regions A to F (see text for a description); the structures of the corresponding ligand family/families are shown in the center. On the right the diseases in which these receptors and their cognate ligands (agonists, antagonists, etc.) are implicated (as prognostic factors, therapeutic or disease preventive targets) are given. At the bottom, receptors involved in steroidogenesis, cholesterol metabolism and xenobiotic catabolism are shown, as well as a member of the orphan receptors that display only conservation of the DNA- or ligand-binding domains. For further details, see the text and references cited therein.*

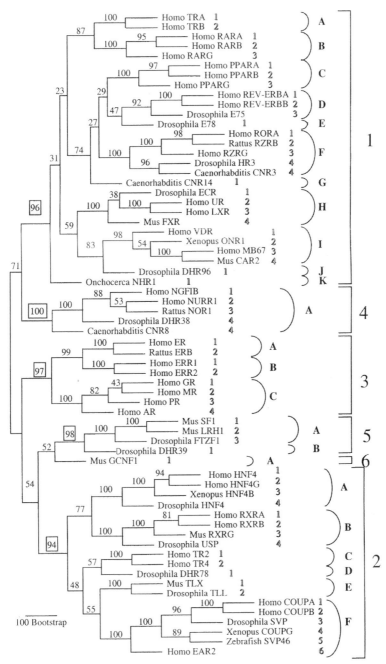

Figure 2 *Phylogenetic tree of 65 nuclear receptor genes in vertebrates, arthropods and nematodes. For a detailed description, see Nuclear Receptors Nomenclature Committee (1999)[7] and the regular updates at http://www.ens-lyon.fr/LBMC/laudet/nomenc.html.*

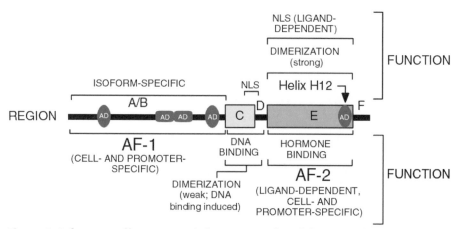

Figure 3 *Schematic illustration of the structural and functional organization of nuclear receptors. The evolutionary conserved regions C and E are indicated as boxes and a black bar represents the divergent regions A/B, D, and F. Note that region F may be absent in some receptors. Domain functions are depicted below and above the scheme. Most of these are derived from structure : functions studies of steroid, thyroid and retinoid receptors. In contrast to steroid hormone receptors, several members of the family (retinoid and thyroid hormone receptors) do not bind hsp90. Two transcription activation functions (AFs) have been described in several nuclear receptors, a constitutively active (if taken out of the context of the receptor) AF-1 in region A/B and a ligand-inducible AF-2 in region E. Within these activation functions, autonomous transactivation domains (ADs) have been defined in the estrogen (ER) and progesterone receptor (PR) N-terminal regions. They may even be specific to an isoform for a certain receptor (like PRB for the isoform B of PR). In the case of the estrogen, retinoid and thyroid hormone receptors, an autonomous activatin domain (AF-2 AD) has been detected at the C-terminal end of the ligand binding domain E. NLS, nuclear localization signal.*

The classification of nuclear receptors is made by virtue of the homology to other family members, with the DNA binding domain (region C; Fig. 3) and the ligand-binding domain (region E) having the highest evolutionary conservation. Table 1 lists all known nuclear receptors with their trivial names and the novel acronyms according to a unified nomenclature system[7]; the phylogenetic tree for this family is shown in Fig. 2. It has been proposed that nuclear receptors have evolved from an ancestral orphan 'receptor' through early diversification, and only later acquired ligand binding[8]. Nonetheless, as long as the identification of ligands for previous orphan receptors continues it cannot be formally excluded that all nuclear receptors may have cognate ligands. On the other hand, the concept that the nuclear receptor family has evolved from an ancestral orphan receptor and ligand binding has been acquired during evolution has found broad recognition. It is possible that a number of orphan receptors function exclusively as constitutive repressors or activators of transcription. Rev-Erb is an example of such a constitutive repressor, which recruits corepressors but lacks a functional activation function, AF-2 (see below for a description of corepressors and activation functions). For historical reasons, research focussed initially on the

Table 1 *Nomenclature for nuclear receptors and commonly used trivial names*

Subfamilies and group	Genes	Trivial names	Accession numbers[a]
1A	NR1A1	TRα, c-ErbA-1, THRA	M24748
	NR1A2	TRβ, c-ErbA-2, THRB	X04707
1B	NR1B1	RARα	X06538
	NR1B2	RARβ, HAP	Y00291
	NR1B3	RARγ, RARD	M57707
1C	NR1C1	PPARα	L02932
	NR1C2	PPARβ, NUC1, PPARδ, FAAR	L07592
	NR1C3	PPARγ	L40904
1D	NR1D1	Rev-erbα, EAR1, EAR1A	M24898
	NR1D2	Rev-erbβ, EAR1β, BD73, RVR, HZF2	L31785
	NR1D3	E75	X51548
1E	NR1E1	E78, DR-78	U01087
1F	NR1F1	RORα, RZRα	U04897
	NR1F2	RORβ, RZRβ	Y08639
	NR1F3	RORγ, TOR	U16997
	NR1F4	HR3, DHR3, MHR3, GHR3, CNR3,	M90806
		CHR3	U13075
1G	NR1G1	CNR14	U13074
1H	NR1H1	ECR	M74078
	NR1H2	UR, OR-1, NER1, RIP15, LXRβ	U07132
	NR1H3	RLD1, LXR, LXRα	U22662
	NR1H4	FXR, RIP14, HRR1	U09416
1I	NR1I1	VDR	J03258
	NR1I2	ONR1, PXR, SXR, BXR	X75163
	NR1I3	MB67, CAR1	Z30425
	NR1I4	CAR2, CARβ	AF00932
1J	NR1J1	DHR96	U36792
1K	NR1K1	NHR1	U19360
2A	NR2A1	HNF4	X76930
	NR2A2	HNF4G	Z49826
	NR2A3	HNF4B	Z49827
	NR2A4	DHNF4, HNF4D	U70874
2B	NR2B1	RXRA	X52773
	NR2B2	RXRB, H-2RIIBP, RCoR-1	M84820
	NR2B3	RXRG	X66225
	NR2B4	USP, Ultraspiracle, 2C1, CF1	X52591
2C	NR2C1	TR2, TR2-11	M29960
	NR2C2	TR4, TAK1	L27586
2D	NR2D1	DHR78	U36791
2E	NR2E1	TLL, TLX, XTLL	S72373
	NR2E2	TLL, Tailless	M34639
2F	NR2F1	COUP-TFI, COUP-TFA, EAR3, SVP44	X12795

Continued

Table 1 *Continued*

Subfamilies and group	Genes	Trivial names	Accession numbers[a]
	NR2F2	COUP-TFII, COUP-TFB, ARP1, SVP40	M64497
	NR2F3	SVP, COUP-TF	M28863
	NR2F4	COUP-TFIII, COUP-TFG	X63092
	NR2F5	SVP46	X70300
	NR2F6	EAR2	X12794
3A	NR3A1	ERα	X03635
	NR3A2	ERβ	U57439
3B	NR3B1	ERR1, ERRα	X51416
	NR3B2	ERR2, ERRβ	X51417
3C	NR3C1	GR	X03225
	NR3C2	MR	M16801
	NR3C3	PR	M15716
	NR3C4	AR	M20132
4A	NR4A1	NGFI-B, TR3, N10, NUR77, NAK1	L13740
	NR4A2	NURR1, NOT, RNR1, HZF-3, TINOR	X75918
	NR4A3	NOR1, MINOR	D38530
	NR4A4	DHR38, NGFI-B	U36762
		CNR8, C48D5	U13076
5A	NR5A1	SF1, ELP, FTZ-F1, AD4BP	D88155
	NR5A2	LRH1, xFF1rA, xFF1rB, FFLR, PHR, FTF	U93553
	NR5A3	FTZ-F1	M63711
5B	NR5B1	DHR39, FTZ-F1B	L06423
6A	NR6A1	GCNF1, RTR	U14666
0A	NR0A1	KNI, Knirps	X13331
	NR0A2	KNRL, Knirps related	X14153
	NR0A3	EGON, Embryonic gonad, EAGLE	X16631
	NR0A4	ODR7	U16708
	NR0A5	Trithorax	M31617
0B	NR0B1	DAX1, AHCH	S74720
	NR0B2	SHP	L76571

Note: Subfamilies and groups are defined as referred to in the text. The groups contain highly related genes often with a paralogous relationship in vertebrates (e.g. RARA, RARB and RARG). The term isoform is reserved for different gene products originating from the same gene due to alternative promoter usage or splicing, or alternative initiation of translation.
[a] EMBL/GenBank.

classical steroid hormone receptors. Presently, however, there is significant emphasis both from basic scientists and industrial pharmacologists on the systematic screening for agonists and antagonists that bind with high affinity to the ligand-binding domain of several orphan receptors.

Interestingly, some recently identified ligands are metabolic intermediates. It appears therefore that, in certain systems, such as the metabolism of cholesterols or the fatty acid β-oxidation pathways, the control of build-up, breakdown and storage of metabolic active substances is regulated at the level of gene expression, and that in many cases this 'intracrine' signaling is brought about by nuclear receptors (for an example of this type of signaling, see Fig. 4). Furthermore, gene knockout experiments suggest that metabolic intermediates, such as SF1 (NR5A1) or PPARγ (NR1C3) ligands, may have regulatory function in specifying organ development[9,10]. Prominent metabolic ligands are bile and fatty acids, eicosanoids and oxysterols. The group of steroid hormones encompasses estrogens, progestins, mineralocorticoids, glucocorticoids, androgens and ecdysterones. Examples of vitamin–derived nuclear receptor ligands are vitamin D_3 (VDR; NR1I1) or the vitamin A derivative, retinoic acid (RARs and RXRs; NR1B and NR2B, respectively). Thus, nuclear receptors function in endocrine (steroid hormone receptors), autocrine/paracrine (retinoid receptors) and intracrine [metabolic receptors, such as LXRα (NR1H3), SF1 (NR5A1), FXR (NR1H4), PXR (NR1I2), PPARs (NR1C), CARβ (NR1I4)] signaling pathways. Certainly, with the identification of more ligands for orphan receptors, new surprises will be encountered.

An interesting receptor is PXR (NR1I2), which regulates the transcription of cytochrome P450 enzymes (such as CYP3A4) in response to binding of a large variety of steroids, xenobiotics and therapeutic drugs (e.g. glucocorticoids, RU486, rifampicin, taxol, PCN, etc.), thus playing a central role in steroid hormone homeostasis and drug metabolism (see recent reviews[5,6]).

Also of interest is CARβ, which has been reported to correspond to a constitutively active receptor whose activity is repressed by certain androstane metabolites[11]. It remains to be seen whether CARβ is active due to the presence of an unidentified andogenous agonist or whether this is the prototype of a constitutively active orphan receptor. Crystal structure evaluation may provide

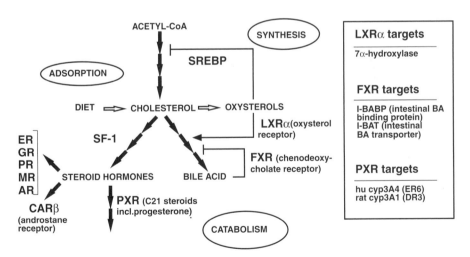

Figure 4 *Cholesterol metabolism: an example of intracrine regulation by nuclear receptors for which ligands were discovered only recently. Members of the nuclear receptor family that regulate cholesterol metabolism comprise LXRα, FXR and SF-1. PXR regulates the catabolism of steroids and certain xenobiotics. Some known transcriptional targets of these receptors are depicted in the box. For further details and original references, see recent reviews in references 5, 6 and 67.*

important clues. Note also that agonists for this receptor exist, which stimulate its transactivation capability far beyond the level seen in absence of a ligand.

GENETICS OF NUCLEAR RECEPTORS

Genetic programs consist typically of several hundred genes that are expressed in a spatially and temporally controlled fashion. Nuclear receptors act as master 'switches' to initiate specific genetic programs that, for example, lead to cell differentiation, proliferation or apoptosis, or regulate homeostasis. In the context of other programs these genetic activities support or initiate complex physiological phenomena, such as reproduction and organ function. Once activated by the cognate ligand, nuclear receptors regulate the primary and secondary target gene expressions that make up the corresponding physiological event. Throughout the life cycle of a multicellular organism, the coordinate interplay between programs defining cell fates in different tissues, organs and finally the entire body is at the foundation of the organism's development and subsistence. This is fully supported by the analysis of mice bearing mutations or deletions of one or several receptors (a searchable mouse knockout and mutation database with PubMed can be found at http://www.biomednet.com/db/mkmd). Several nuclear receptor knockout animals, in particular compound knockout animals, die *in utero* or around birth, displaying severe malformations of organs that render them inviable (for examples, see references 12 and 13, and references; for reviews on earlier work, see references 14 and 15). Others are viable under laboratory conditions but display a reduced life span and are often infertile[9,16]. These knockout animal models have been of great help in deciphering the physiology of nuclear receptor action. Often they provided initial or additional evidence for new, as yet undiscovered, functions exerted by the receptor, and thus initiate further research on previously unknown signaling pathways. One example is the involvement of retinoic acid receptor, RARβ (NR1B3) and retinoid X receptor RXRγ (NR2B3) in long-term memory potentiation and depression[17,17a]. Furthermore, knockouts have also provided insight into the distinct modes of transcriptional regulation by nuclear receptors. An example is the mouse NR3C1 gene encoding the glucocorticoid receptor (GR). GR[−/−] mice die at birth due to respiratory failure[18]. However, replacement of the GR gene by a GR mutant (GRA458T[19]) that impairs binding to consensus GR response elements generates mice (termed GR[dim/dim] with 'dim' indicating DNA binding domain dimerization deficiency) that were fully viable and exhibited only minor phenotypic alteration[20]. These results demonstrated that the indirect transcriptional effects ('signaling cross-talk') of the glucocorticoid receptor are as crucial for GR-mediated signaling as the direct regulation of cognate target genes[20,21]. These indirect effects of glucocorticoid receptor comprise the transrepression of activating protein AP-1, a mechanism discussed further below.

An important result of studies with nuclear receptor gene deletion models has been the discovery of redundancy and adaptivity among family members of the same group. In this respect, the interpretation of vitamin A signaling, which is of remarkable complexity and displays a high degree of apparent retinoic acid receptor redundancy, may serve as an example. Vitamin A derivatives are bound by two different groups of nuclear receptors. The retinoic acid receptors RARα, RARβ and

RARγ (NR1B1, NR1B2, NR1B3) bind both all-*trans* retinoic acid and the isomer 9-*cis* retinoic acid, whereas the retinoic X receptors RXRα, RXRβ and RXRγ (NR2B1, NR2B2, NR2B3) bind exclusively 9-*cis* retinoic acid. Retinoic X receptors are heterodimerization partners for a great number of nuclear receptors and they also heterodimerize with retinoic acid receptors to form active signaling molecules. All six retinoid receptor genes give rise to at least two different isoforms through alternative splicing or differential promoter usage. Considering the necessity of heterodimerization between one member of each group, theoretically 36 different combinations can be formed from the existing pool of genes.

As indicated through single and compound knockouts, the removal of a single gene results in rather restricted phenotypes, while the removal of multiple isotypes or of RAR and RXR group members results in animals displaying the full vitamin A deficiency (VAD) syndrome[12,15,22–25]. This indicates functional redundancy between different retinoid receptor genes *in vivo*. However, molecular and biochemical analyses have demonstrated significant functional differences amongst group members. For example, cell differentiation assays have revealed that one receptor is able to induce a certain differentiation event, while another is not, even though it is equally expressed[26,27]. These seemingly contradictory findings reveal another important feature that contributes to the complexity of nuclear receptor signaling adaptivity *in vivo*. It is assumed, although not yet convincingly demonstrated, that the organism (in this case the developing mouse) overcomes the lack of a given receptor gene by adjusting the activities of the remaining genes/gene products to cope with this deficiency. Thereby, the organism compensates for most of the deleterious effects resulting, at least under laboratory conditions, in absent or weak phenotypes. This phenomenon assures that the organism can survive under standard conditions (e.g. the spontaneous somatic mutation of a given receptor gene will not necessarily lead to the loss of the affected cell), but may fail to deal with more extreme physiological conditions. This phenomenon is therefore distinct from true functional redundancy. It is possible that redundancy and adaptivity can 'buffer' mutations, thus allowing a more rapid evolution of phenotypes based on the accumulation of multiple and/or more complex gene alterations.

MODULAR STRUCTURE AND FUNCTION

The N-terminal region A/B harbors cell-specific activation function(s) AF-1 of unknown structure

As schematically depicted in Fig. 3, nuclear receptors are composed of 5–6 regions (A–F; originally defined by Krust *et al.*[28]) that have modular character. The N-terminal A/B region harbors one (or more) autonomous transcriptional activation function (AF-1), which, when linked to a heterologous DNA binding domain, can activate transcription in a constitutive manner. Note, however, that in the context of full-length steroid receptors AF-1 is silent in the absence of an agonist and certain antagonists. When comparing nuclear receptors from different subfamilies and groups, the A/B region displays the weakest evolutionary conservation, and the distinction between the A and B regions is not always evident. A/B regions differ significantly in their length, ranging from 23 (vitamin D receptor, NR1I1) to 550 (androgen, NR3C4, mineralocorticoid, NR3C2, and glucocorticoid receptors,

NR3C1) amino acids. No three-dimensional (3D) structure of a nuclear receptor A/B region has been solved up to now and the structure prediction is not straightforward. A/B regions are subject to alternative splicing and differential promoter usage and the majority of known nuclear receptor isoforms differ in their N-terminal region. Through alternative splicing and differential promoter usage (PR forms A + B) the absence or presence of different activation functions found in the A/B regions can be regulated (for a review, see reference 1). Moreover, the N-terminus of nuclear receptors has reportedly been found as subject of post-translational events such as phosphorylation[29-34] (reviewed in Shao and Lazar[35]). The role of phosphorylation of the A/B domains for the transactivation potential of the activation function AF-1 as well as for synergy and cooperativity with the second activation function AF-2 located in the E domain of the receptor is currently being investigated in several laboratories. Finally, the activation function(s) AF-1 display cell, DNA-binding domain and promoter specificity[36-38], the origin of which is still elusive but may be related to the cell-specific action and/or expression of AF-1 coactivators.

Table 2 *List of presently reported three-dimensional structures of nuclear receptor DNA-binding domains together with their Brookhaven Protein Data Bank (PDB) assignments*

PDB ID	Receptors	Response element	Reference
Homodimers			
1GLU	GR	GRE	69
1LAT	GR mutant	non-cognate	70
1HCQ	ER	ERE	71
1A6Y	Rev-erb	DR2	72
1BY4	RXR	DR1	73
Heterodimers			
2NLL	5'-RXR-TR-3'	DR4	74
	5'-RAR-RXR-3'	DR1	75
Monomers			
1CIT	NGFI-B	NBRE	76
NMR structures			
2GDA	GR	—	77
1GDC	GR	—	77
1RGD	GR	—	78
	GR	—	79
1HCP	ER	—	80
1RXR	RXR	—	81
1HRA	RARβ	—	82
Molecular dynamics simulations			
	GR	+/− GRE	83

NMR, nuclear magnetic resonance.

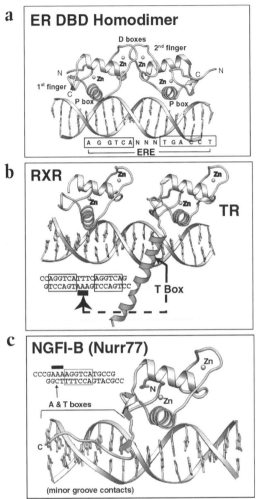

Figure 5 *Three-dimensional structures of the three prototype DNA-binding domains (DBDs) of nuclear receptors obtained from X-ray crystal structure analyses of the DBD–DNA response element cocrystals. (a) Structure of the estrogen receptor a (ER) DBD on a estrogen response element (ERE); one strand of the canonical ERE DNA sequence is given at the bottom. The various structural elements (Zn^{2+} fingers, D- and P-boxes) are indicated. Note that the D-boxes form a DNA-induced DBD dimerization interface, while the P-box α-helices establish the selective base contacts in the major groove. (b) Crystal structure of the 5'-RXR-TR-3' heterodimer on a cognate direct repeat response element spaced by four base pairs (DR4), depicted as a double-stranded DNA sequence at the bottom left. Note that the T-box makes minor groove contacts, thus specifying to some degree the DR4 spacer nucleotide sequence (arrow). (c) Crystal structure of the monomer NGFI-B on its response element (NBRE). The double-stranded NBRE sequence is given at the top left. Note that the A and T boxes define the 5' AAA sequence that contacts the minor groove.*

The DNA-binding domain encompasses region C

The highly conserved domain C harbors the DNA-binding domain (DBD) of nuclear receptors that confers sequence-specific DNA recognition. This domain has been extensively investigated, especially with respect to its selective response element recognition and dimerization properties (for details, see below). Several X-ray and nuclear magnetic resonance (NMR) data sets are available for different nuclear receptor C domains in their DNA complexed and uncomplexed forms (see Table 2 for PDB file names). The DNA-binding domain is mainly composed of two zinc-finger motifs, the N-terminal motif Cys-X2-Cys-X13-Cys-X2-Cys (CI) and the C-terminal motif Cys-X5-Cys-X9-Cys-X2-Cys (CII); in each motif, four cysteine residues chelate one Zn^{2+} ion. Within the C domain several sequence elements (termed P-, D-, T- and A-boxes) have been characterized that define or contribute to: (1) response element specificity; (2) a dimerization interface within the DBD; and (3) contacts with the DNA backbone and residues flanking the DNA core recognition sequence. Figure 5 illustrates the three prototypic DNA-binding modes of nuclear receptors: (1) the estrogen receptor DBD as an example of a homodimer that binds to a palindromic response element (Fig. 5a); (2) the RXR-TR as an example for a anisotropic (5'-RXR-TR-3') heterodimeric complex on a DR1 direct repeat (Fig. 5b); and (3) NGFI-B as an

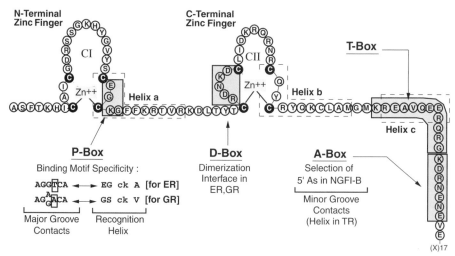

Figure 6 *Schematic illustration of the RXR DNA-binding domain (DBD). Boxes illustrate regions involved in response element selection. The P-box is part of the DNA recognition helix (helix a; framed by broken lines) and swapping of the EGA residues of the ER P-box with the corresponding GSV residues of GR switches ERE/GRE recognition as is outlined below the illustration. The D-box is responsible for PAL3/PAL0 selection by estrogen or thyroid hormone receptors and contributes to the homodimerization interfaces of ER and GR DBDs. The T-box region forms a helix (helix c) and corresponds to a dimerization surface in RXR homodimers. The A-box in solution CI and CII are the two zinc-finger motifs. PRF, pre-finger region. For further details, see Gronemeyer and Laudet (1995)[1] and references therein.*

example for a monomer that binds to an extended hexameric motif, the so-called NBRE (Fig. 5c). Note that the so-called A- and T-boxes of TR and NGFI-B are involved in additional minor groove DNA contacts (see below for details). Figure 6 displays a schematic illustration of the retinoic X receptor DBD with an illustration of the different Zn^{2+} fingers and the various boxes (for details, see the figure legend and below).

Region D, a hinge with compartmentalization functions

The D region of nuclear receptors is less conserved than the surrounding regions C and E. This domain appears to correspond to a 'hinge' between the highly structured C and E domains. It might allow the DNA and ligand-binding domains to adopt several different conformations without creating steric hindrance problems. Note in this respect that the C and the E regions contribute dimerization interfaces allowing some receptors to accommodate different heterodimerization partners and different types of response elements. Region D contains a nuclear localization signal (NLS), or at least some elements of a functional NLS (reviewed in reference 1). The intracellular localization of nuclear receptors is a result of a dynamic equilibrium between nuclear cytoplasmatic and cytoplasmatic nuclear shuttling[39]. At equilibrium the large majority of receptors are nuclear, although there is some controversy in the case of corticoid receptors (glucocorticoid and mineralocorticoid receptors), which have been reported to reside at cytoplasmic locations in the absence of their cognate ligands and translocate to the nucleus in a ligand-induced fashion[40].

Region E encompasses the ligand-binding domain and AF-2

The hallmark of a nuclear receptor is its ligand-binding domain (LBD) in the E region. This domain is highly structured, and encodes a wealth of distinct functions, most of which operate in a ligand-dependent manner. The LBD harbors the ligand-dependent activation function AF-2, a major dimerization interface and often a repression function[1,2,41,42]. Detailed molecular insights into the structure–function relation of signal integration by nuclear receptors have been gained by the elucidation of the crystal structures of the E region alone or in the presence of agonists, antagonists and coregulator peptides. The first 3D structures reported for nuclear receptor LBDs were those of the unliganded RXRα (NR2B1), the all-*trans* retinoic acid-bound RARγ (NR1B3) and the agonist-bound thyroid receptor TRβ (NR1A2)[43–45] (see also Table 3). Unliganded receptors are frequently referred to as apo-receptor forms, while the liganded receptor corresponds to the holo- form. These structures demonstrated that apo- and holo- ligand binding domains have a common fold. Moreover, a structure-based sequence alignment of all known nuclear receptor primary amino-acid sequences strongly supported a common fold for all nuclear receptor LBDs[46]. This hypothesis was fully confirmed when the crystal structures of multiple other NR LBDs were solved.

Table 3 compiles all currently reported LBD structures and the corresponding PDB assignments. The general fold of nuclear receptors consists of a three-layered α-helical sandwich. Further structural features are one β-hairpin and connecting loops of variable lengths. The helices have been designated H1 to H12, starting with the most N-terminal H1, and form a hydrophobic cavity that accommodates

Table 3 *List of all presently reported three-dimensional structures of nuclear receptor ligand-binding domains together with their PDB assignments*

Receptors	Ligands	Remarks	PDB ID	Reference
Monomers				
RARγ	T-RA	Agonist	2lbd	44
	9C-RA, BMS961	Agonists	3lbd, 4lbd	48
	BMS394, BMS395	Agonists	—	84
RXRα	9C-RA	Agonist	—	85
TRα	T3, Dimit	Agonists	—	86
TRβ	T3	NR box complex	1bsx	52
PPARδ	GW2433, EPA	Agonists	1gwx, 2gwx	87
VDR	Vitamin D_3	Agonist	1db1	88
PR	Progesterone	Agonist	1a28	89
PPARγ	Apo	—	3prg	90
Homodimers				
RXRα	Apo	—	1lbd	43
ERα	Estradiol, RAL	Agonist, antagonist	1ere, 1err	91
	DES	NR box complex	3erd	54
	OHT	Antagonist	3ert	54
	Estradiol	Agonist	1a52	92
ERβ	RAL, Genistein	Antagonist, partial agonist	1qkn, 1qkm	93
PPARγ	Apo	—	1prg	53
	Rosiglitazone	NR box complex	2prg	53
	GW0072	Partial agonist	4prg	94
Heterodimers				
RARα/RXRα	BMS614/oleic acid	Antagonist/partial agonist	1dkf	55
PPARγ/RXRα	Rosiglitazone/9C-RA	NR box complex	—	59
	GI262570/9C-RA	NR box complex	—	59

PDB, Brookhaven Protein Data Bank.

the hydrophobic ligands. For an example of a detailed analysis of ligand–LBD interactions in the case of the retinoic acid receptors and the structural basis of isoform specificity, see references 47 and 48. The different crystal structures of apo- and holo- forms of receptor LBDs, as well as extensive mutagenesis, demonstrated that the E domain undergoes a major conformational change upon ligand binding (Fig. 7).

The structural transition upon ligand binding has been described as a 'mousetrap' mechanism[42,44]: pushed by the ligand, the helix H11 is repositioned in the continuity of helix H10, and the concomitant swinging of helix H12 unleashes the ω-loop, which flips over underneath helix H6, carrying along the N-terminal part of helix H3. In its final position, helix H12 seals the 'lid' in the ligand-binding pocket and further stabilizes ligand binding by contributing to the hydrophobic environment, in some cases by making additional contacts with the ligand itself. The repositioning of helix H12 back on to the core of the LBD is apparent from the structures of the apo-RXRα and holo-RARγ LBDs shown in Fig. 7. The critical implication in transcription activation of helix H12, also referred to as the

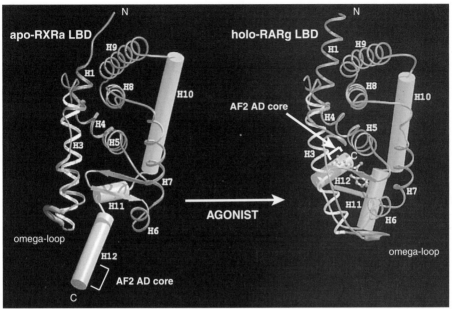

Figure 7 *A comparison of the crystal structures of the RXRα apo-ligand binding domain (apo-LBD) and the RARγ holo-LBD reveals the ligand-induced transconformation generating the transcriptionally active form of a nuclear receptor. In this model, ligand binding induces a structural transition, which triggers a mousetrap-like mechanism: pushed by the ligand, H11 is repositioned in the continuity of H10, and the concomitant swinging of H12 unleashes the Ω-loop, which flips over underneath H6, carrying along the N-terminal part of H3. In its final position, H12 seals the ligand binding cavity as a 'lid' and further stabilizes ligand binding by contributing to the hydrophobic pocket. Note that helix H12, which encompasses the core of the AF-2 activation domain (AF-2 AD core), is now positioned in an entirely different environment of the LBD, where it is able to interact with the LxxLL NR box motif of bona fide coactivators (see recent reviews in references 42 and 68).*

activating domain (AD) of the AF-2 function (the features of this AF-2 AD are conserved in all transcriptionally active receptors), was previously deduced from multiple mutagenesis studies (for reviews and references, see Gronemeyer and Laudet[1] and Chambon[2]). Moreover, helix H12 not only coincided with residues critically required for transcriptional activation but also for coactivator recruitment (coactivators and their interaction with the LBD are discussed below).

The structural data reveal that helix H12, when folded back on to the core of the LBD, forms a hydrophobic cleft together with other surface-exposed amino acids that accommodates the 'NR box' of coactivators[49–51]; (for a more detailed description, see below). Crystal structure data have revealed the details of the interaction of the LXXLL NR box motif with the cognate surface on the LBD (originally described in references 52–54). As an example, the interaction of the TIF2/GRIP1 NR box 2 peptide with the agonist diethylstilbestrol (DES)-bound estrogen receptor LBD is illustrated in Fig. 8a. Note that helix H12 is crucial in

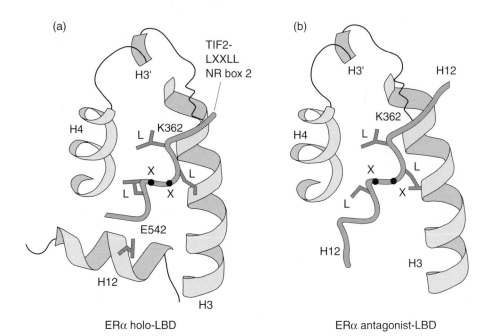

(a)

TIF2-
LXXLL
NR box 2

H3'

H4

K362

L

X L

L X

E542

H12

H3

ERα holo-LBD

(b)

H3'

H12

K362

H4

L

X L

L

X

H12

H3

ERα antagonist-LBD

Figure 8 *Mechanism of antagonist action. Comparison of the crystal structures of the complex between the estrogen receptor (ER) LBD in the presence of the agonist diethylstilbestrol (DES) and the TIF2/GRIP1 NR box peptide (a) with the ER LBD structure in the presence of the antagonist tamoxifen (b). Note that H12 in the antagonist structure occupies the identical position as the NR box peptide in the agonist LBD. The same position of H12 was observed in the crystal structure of the RARα LBD-BMS614 antagonist structure[55], indicating a more general mechanism of antagonism. In the RAR case, steric hindrance problems are responsible for the 'antagonistic' positioning of helix H12. The structures presented here were originally described by Shiau and colleagues[54] and are derived from PDB Ids 3ERD and 3ERT in the Brookhaven Protein DataBank.*

stabilizing this interaction (for details, see Shiau *et al.*[54]). The corresponding crystal structure of the (tamoxifen) antagonist-bound estrogen receptor LBD was particularly illuminating, as it demonstrated that this type of antagonist induces a positioning of H12, which is incompatible with the binding of coactivator NR boxes; indeed, H12 binds to the identical cleft to which the NR box binds (compare the two structures in Fig. 8; see also references 54 and 55). Hence, the information generated by the synthesis or secretion of a nuclear receptor ligand is converted first into a molecular recognition (ligand-receptor) process and subsequently into an allosteric event that allows the secondary recognition of a 'downstream' mediator that is more closely connected to the cellular machineries involved in transcription activation.

Some nuclear receptors can also function as DNA-bound repressors of transcription. This phenomenon occurs in the absence of agonists, may be enhanced by certain antagonists and is attributable to the recruitment of corepressors. Corepressor binding also occurs on the surface of the non-liganded

LBD and is similar to coactivator binding in that it also is mediated by a short signature motif, the so-called 'CoRNR box', which binds to a surface topologically related to that involved in coactivator interaction[56-58]. Upon ligand binding the conformational transconfiguration of the α-helical sandwich displaces the corepressors by shifting the equilibrium of the E domain from the apo- to the holo-form.

Some nuclear receptors form only homodimers (e.g. steroid hormone receptors), while others can form both homodimers (e.g. RXR, TR) and heterodimers with the promiscuous heterodimerization partner RXR (e.g. RAR-RXR, TR-RXR, VDR-RXR, PPAR-RXR, etc.). The LBD is the major domain that contributes to dimerization. The recent crystallization of the RAR-RXR ligand-binding domain heterodimer and the comparison with homodimer interfaces has provided detailed information on the structural elements governing homodimerization and heterodimerization[55,59]. Based on studies *in vitro* it has also been reported that ligand binding can affect the dimerization properties of nuclear receptors. For the thyroid hormone and vitamin D receptors, which can both form homodimers and RXR heterodimers, ligand binding appears to favor heterodimerization[60-62].

The role of the C-terminal region F is unknown

Some receptors possess at the C-terminus of the ligand binding domain a region F, which displays little evolutionary conservation. Note that the LBD is structurally defined as the domain generated by the elements between the beginning of helix H1 and the end of helix H12. This sequence is not necessarily identical to that which is commonly referred to as region E from sequence alignments and also receptors such as the progesterone receptor possess some kind of F region. This region is, however, much longer in the cases of, for example, estrogen (NR3A) and retinoic acid (NR1B) receptors. There are no clues as to the function of the C-terminal sequence. Recent literature suggests that the F region might play a role in coactivator recruitment to the E domain and in determining the specificity of the ligand binding domain coactivator interface[63,64]. It is clear that this domain also inherits little structural features. It is tempting to speculate that it 'fine tunes' the molecular events associated with the transcriptional properties of the E domain, or the entire receptor, as it may affect antagonist action[65,66].

As with the A/B domains, the E and F domains are also targets of post-translational modifications. This adds another level of complexity to nuclear receptor signal integration, since such events might influence the properties of the encoded functions.

References

[1] Gronemeyer, H. and Laudet, V. (1995) Protein Profile 2, 1173–1308.

[2] Chambon, P. (1996) FASEB J. 10, 940–954.

[3] Mangelsdorf, D.J. and Evans, R.M. (1995) Cell 83, 841–850.

[4] Giguere, V. (1999) Endo. Rev. 20, 689–725.

[5] Kliewer, S.A. et al. (1999) Science 284, 757–760.

[6] Kliewer, S.A et al. (1999) Rec. Prog. Hormone Res. 54, 345–367.

[7] Nuclear Receptors Nomenclature Committee (1999) Cell 97, 161–163.

[8] Escriva, H. et al. (1997) Proc. Natl Acad. Sci. USA 94, 6803–6808.

[9] Luo, X., et al. (1994) Cell 77, 481–490.

[10] Barak, Y. et al. (1999) Mol. Cell 4, 585–595.
[11] Forman, B.M. et al. (1998) Nature 395, 612–615.
[12] Kastner, P. et al. (1997) Development 124(23), 313–326.
[13] Mascrez, B. et al. (1998) Development 125(23), 4691–4707.
[14] Beato, M. et al. (1995) Cell 83, 851–857.
[15] Kastner, P. et al. (1995) Cell 83, 859–869.
[16] Parker, K.L. et al. (1996) Steroids 61, 161–165.
[17] Chiang, M.Y. et al. (1998) Neuron 21, 1353–1361.
[17a] Krezel et al. (1998) Science 279, 863–867.
[18] Cole, T.J. et al. (1995) Genes Dev. 9, 1608–1621.
[19] Heck, S. et al. (1994) EMBO J. 13, 4087–4095.
[20] Reichardt, H.M. et al. (1998) Cell 93, 531–541.
[21] Tronche, F. et al. (1998) Curr. Opin. Genet. Dev. 8, 532–538.
[22] Sucov, H.M. et al. (1994) Genes Dev. 8, 1007–1018.
[23] Kastner, P. et al. (1997) Development 124(2), 4749–4758.
[24] Smith, S.M. et al. (1998) J. Nutr. 128, 467–470.
[25] Mark, M. et al. (1999) Proc. Nutr. Soc. 58, 609–613.
[26] Chen, J.Y. et al. (1996) Nature 382, 819–822.
[27] Benoit, G. et al. (1999) EMBO J. 18, 7011–7018.
[28] Krust, A. et al. (1986) EMBO J. 5, 891–897.
[29] Kato, S. et al. (1995) Science 270, 1491–1494.
[30] Rochette-Egly, C. et al. (1997) Cell 90, 97–107.
[31] Taneja, R. et al. (1997) EMBO J. 16, 6452–6465.
[32] Adam-Stitah, S. et al. (1999) J. Biol. Chem. 274, 18932–18941.
[33] Hammer, G.D. et al. (1999) Mol. Cell. 3, 521–526.
[34] Tremblay, A. et al. (1999) Mol. Cell. 3, 513–519.
[35] Shao, D. and Lazar, M.A. (1999) J. Clin. Invest. 103, 1617–1618.
[36] Bocquel, M.T. et al. (1989) Nucl. Acids Res. 17, 2581–2595.
[37] Berry, M. et al. (1990) EMBO J. 9, 2811–2818.
[38] Sartorius, C.A. et al. (1994) Mol. Endocrinol. 8, 1347–1360.
[39] Guiochon-Mantel, A. et al. (1994) Proc. Natl Acad. Sci. USA 91, 7179–7183.
[40] Baumann, C.T. et al. (1999) Cell Biochem. Biophys. 31, 119–127.
[41] Tsai, M.J. and O'Malley, B.W. (1994) Annu. Rev. Biochem. 63, 451–486.
[42] Moras, D. and Gronemeyer, H. (1998) Curr. Opin. Cell Biol. 10, 384–391.
[43] Bourguet, W. et al. (1995) Nature 375, 377–382.
[44] Renaud, J.P. et al. (1995) Nature 378, 681–689.
[45] Wagner, R.L. et al. (1995) Nature 378, 690–697.
[46] Wurtz, J.M. et al. (1996) Nature Struct. Biol. 3, 87–94.
[47] Gehin, M. et al. (1999) Chem. Biol. 6, 519–529.
[48] Klaholz, B.P. et al. (1998) Nature Struct. Biol. 5, 199–202.
[49] Le Douarin, B. et al. (1996) EMBO J. 15, 6701–6715.
[50] Heery, D.M. et al. (1997) Nature 387, 733–736.
[51] Voegel, J.J. et al. (1998) EMBO J. 17, 507–519.
[52] Darimont, B.D. et al. (1998) Genes Dev. 12, 3343–3356.
[53] Nolte, R.T. et al. (1998) Nature 395, 137–143.
[54] Shiau, A.K. et al. (1998) Cell 95, 927–937.
[55] Bourguet, W. et al. (2000) Mol. Cell 5, 289–298.
[56] Hu, X. and Lazar, M.A. (1999) Nature 402, 93–96.
[57] Nagy, L. et al. (1999) Genes Dev. 13, 3209–3216.

58 Perissi, V. (1999) Genes Dev. 13, 3198–3208.

59 Gampe, R.T. et al. (2000) Mol. Cell. 5, 545–555.

60 Cheskis, B. and Freedman, L.P. (1994) Mol. Cell Biol. 14, 3329–3338.

61 Collingwood, T.N. et al. (1997) J. Biol. Chem. 272, 13060–13065.

62 Kakizawa, T. et al. (1997) J. Biol. Chem. 272, 23799–23804.

63 Peters, G.A. and Khan, S.A. (1999) Mol. Endocrinol. 13, 286–296.

64 Sladek, F.M. et al. (1999) Mol. Cell Biol. 19, 6509–6522.

65 Montano, M.M. et al. (1995) Mol. Endocrinol. 9, 814–825.

66 Nichols, M. (1998) EMBO J. 17, 765–773.

67 Repa, J.J. and Mangelsdorf, D.J. (1999) Curr. Opin. Biotechnol. 10, 557–563.

68 Bourguet, W. et al. (2000) Trends Pharmacol. Sci. 21, 381–388.

69 Luisi, B.F. et al. (1991) Nature 352, 497–505.

70 Gewirth, D.T. and Sigler, P.B. (1995) Nature Struct. Biol. 2, 386–394.

71 Schwabe, J.W. et al. (1993) Cell 75, 567–578.

72 Zhao, Q. et al. (1998) Mol. Cell 1, 849–861.

73 Zhao, Q. et al. (2000) J. Mol. Biol. 296, 509–520.

74 Rastinejad, F. et al. (1995) Nature 375, 203–211.

75 Rastinejad, F. et al. (2000) EMBO J. 19, 1045–1054.

76 Meinke, G. and Sigler, P.B. (1999) Nature Struct. Biol. 6, 471–477.

77 Baumann, H. et al. (1993) Biochemistry 32, 13463–13471.

78 Hard, T. et al. (1990) Science 249, 157–160.

79 van Tilborg, M.A. et al. (1995) J. Mol. Biol. 247, 689–700.

80 Schwabe, J.W. et al. (1993) Structure 1, 187–204.

81 Holmbeck, S.M. et al. (1998) J. Mol. Biol. 281, 271–284.

82 Katahira, M. et al. (1992) Biochemistry 31, 6474–6480.

83 Eriksson M.A. et al. (1995) Biophys. J. 68, 402–426.

84 Klaholz, B.P. et al. (2000) Proc. Natl Acad. Sci. USA (in press).

85 Egea, P.F. et al. (2000) EMBO J. 19, 2592–2601.

86 Wagner, R.L. et al. (1995) Nature 378, 690–697.

87 Xu, H.E. et al., 1999 Mol. Cell 3, 397–403.

88 Rochel, N. et al. (2000) Mol. Cell 5, 173–179.

89 Williams, S.P. and Sigler, P.B. (1998) Nature 39, 392–396.

90 Uppenberg, J. et al. (1998) J. Biol. Chem. 273, 31108–31112.

91 Brzozowski, A.M. et al. (1997) Nature 389, 753–758.

92 Tanenbaum, D.M. et al. (1998) Proc. Natl Acad. Sci. USA 95, 5998–6003.

93 Pike, A.C. et al. (1999) EMBO J. 18, 4608–4618.

94 Oberfield, J.L. et al. (1999) Proc. Natl Acad. Sci. USA 96, 6102–6106.

2 DNA recognition by nuclear receptors

RESPONSE ELEMENTS OF NUCLEAR RECEPTORS

The common principle

All nuclear receptors recognize derivatives of the same hexameric DNA core motif, 5'-PuGGTCA (Pu = A or G). However, mutation, extension and duplication, and, moreover, distinct relative orientations of repeats of this motif generate response elements that are selective for a given (class of) receptors (Fig. 9). Apparently, nuclear receptors coevolutionarily devised mechanisms for optimal interaction with these sequences: they either modified residues, which establish contacts to the nucleotides that specify a given response element, or generated homodimerization or heterodimerization interfaces that are adapted to response elements.

Spacer 'rules' derived from synthetic response elements

To describe the preference of the various direct repeat (DR)-recognizing receptors for elements with a certain spacer length, an (over)simplified rule has been proposed[1,2] that has the advantage of being easily memorized. According to an updated version of this rule, DRn elements, with n spacer nucleotides, exhibit the specifications shown in Table 4.

Table 4 *Specifications for DR elements*

Spacer NTs	Systematic name	Acronym	Receptor complex
1	DR1	RXRE, PPARE, ...	RXR–RXR, PPAR–RXR, RAR–RXR, ...
2	DR2	RARE	RAR–RXR
3	DR3	VDRE	RXR–VDR
4	DR4	TRE	RXR–TR
5	DR5	RARE	RXR–RAR

Albeit a reasonable approximation, this rule does not consider a number of important aspects of receptor–DNA interaction. For example: (1) it is unclear whether functional DR1 RXREs do exist in natural genes; (2) DR1 elements have been shown to act as RAREs[3] and PPAREs (for reviews, see references 4 and 5); (3) several orphan receptors bind [as homodimers (HNF4) or heterodimers (COUP-TF/arp-1) with RXR] to certain DR elements, which 'belong' to other receptors according to the above spacer rule; (4) the rule does not distinguish between homodimers and heterodimers, which may bind to distinct DR options (DR3 and DR6 VDREs); and (5) it does not consider the polarity of the receptor–DNA complexes. Note also that, in addition to DR elements, several other types of (more complex) response elements exist (Fig. 9).

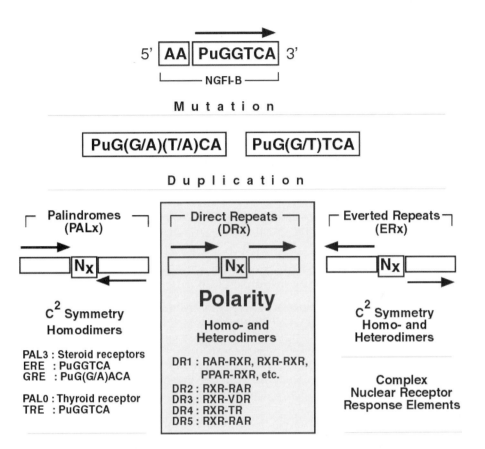

Widely spaced DRs can act as promiscuous response elements

Figure 9 *Response elements of nuclear receptors. The canonical core recognition sequence is 5'-PuGGTCA (arrows indicate the 5' to 3' direction), which, together with two 5' As, is a response element of the orphan receptor NGFIB. Duplication of the core sequence generates symmetrical palindromes (PALx) and everted repeats (ERx), and polar direct repeats (DRx), with x bp separating the two half-sites. PAL3 is an estrogen response element (ERE), while PAL0 corresponds to a thyroid hormone response element (TRE). A single mutation at position 4 of the core sequence from T to A leads to PAL3 response elements (GREs) recognized by the glucocorticoid receptor (and also androgen, progesterone and mineralo-corticoid receptors). Note that most response elements are far from ideal; often one of the half-sites contains one or more mutations. Whereas PALs bind homodimers, DRs can bind homodimers or heterodimers with the specificities given in the shaded box (polarity of the receptors on their cognate DRs: left, 5'; right, 3'). Note that the 3'-positioned receptor makes minor grove DNA contacts in the spacer. Some ERs are response elements for homodimers or heterodimers of the thyroid or retinoid receptors. Response elements are known that are comprised of complex arrangements of the core motif.*

Variability of the binding motif, spacer sequence and flanking nucleotides

It is important to point out that there is considerable degeneration in the sequence of half-site motifs of a given type of natural retinoid response element and that there is a distinct preference of the various receptors for a certain motif. For example, the preference for the half-site motif 5'-PuGGTCA over 5'-PuGTTCA follows the order TR > RXR > RAR[6].

In addition to a distinct preference for certain nucleotides in the half-site motif, there is also a receptor-specific preference for certain nucleotides in the DR spacer. A DNA-binding site selection with RAR-RXR, TR-RXR and VDR-RXR heterodimers to identify the optimal 3' positioned motif and spacer sequence has been reported[7] and is easily rationalized in view of the crystallographic data[8] (see Table 2). See in this respect also the NGFI-B DNA complex, which illustrates the binding of A- and T-box residues to the 5' minor groove of the NBRE[9] (see Table 2).

Steroid hormone receptor response elements (HREs)

Steroid hormone receptors bind to 3 bp-spaced palindromic arrangements (3 bp-spaced inverted repeats; generally termed IR3) of the prototypic recognition motif (for reviews, see references 10–15) (see Figs 5a and 9). The mutation of a single nucleotide at position 4 in each motif from T to A (5'-PuGG_T_CA to 5'-PuGG_A_CA) will convert an estrogen response element (ERE) into a glucocorticoid response element (GRE)[16]. Note, however, that the classical GRE is 5'-PuG_AA_CA[12], which corresponds to the mutation of two nucleotides. Steroid receptors do not recognize palindromes with a spacing other than 3 bp (in contrast to TR, see section on TREs), but ER can bind to, and transactivate from, widely spaced direct repeats (see below).

The following examples illustrate the complexity and sequence variability of steroid hormone response elements. The consensus ERE [5'-(G/A)GGTCA-N$_3$-TGACC(T/C)-3'] is present in the *Xenopus laevis* vitellogenin A2 promoter[17], while the promoter of the estrogen-inducible pS2 gene of MCF-7 breast cancer cells contains a degenerate ERE[18,19]. The first described GRE is that found in the mouse mammary tumor virus (MMTV) long terminal repeat (LTR). This GRE was the first sequence responding to a nuclear receptor ligand and the first demonstration of a receptor–response element interaction was by *in vitro* footprinting of the MMTV LTR GRE[20]. The GRE of the MMTV LTR is a complex response element, comprising of a promoter-distal degenerated palindrome and a promoter-proximal arrangement of three half-site motifs[12]. Like consensus GREs, also the complex MMTV GRE responds not only to glucocorticoids, but also to androgens, progestins and mineralocorticoids[21-25] (for a review, see reference 12). *In vivo* footprinting of the MMTV GRE/PRE[26], the GRE of the tyrosine amino transferase (TAT) promoter[27], and the ERE of the apoVLDLII promoter[28] have confirmed that steroid receptors bind directly to these sequences *in vivo*. In the few cases that have been studied, binding occurred in a ligand-dependent manner.

Although progesterone (PR), androgen (AR) and mineralocorticoid (MR) receptors bind to GREs, differences in DNA-binding specificities have also been observed, in particular in the case of the GR and PR, for which more studies have been performed (for a review and references, see reference 12). A systematic mutational analysis concluded that GR and PR may not distinguish individual target sites but may use the whole of the response element context differentially[29,30]. Notably, chicken PR was shown to replace GR for the activation of the endogenous tyrosine-amino transferase gene, a cognate GR target gene[31]. This could suggest that, in some cases, only the distinct expression profiles of some NRs and their ligands suffice to generate specificity.

Response elements for retinoid, thyroid, vitamin D and peroxisome proliferator-activated receptors

For extensive discussion of this topic, the reader is referred to a number of reviews[4,5,14,15,32,33]; please compare also the published crystal structure data on complexes between various nuclear receptor DNA-binding domains and the cognate DNA-response elements (see Table 2 in Chapter 1 and below). The characteristics of the major retinoid response elements to which retinoid receptors can bind, either as an RAR-RXR heterodimer or through RXR as the heterodimeric partner, are reviewed below.

Retinoid response elements

The classical retinoic acid response element (RARE), which was found in the P2 promoter of the RARβ gene and gives rise to the RARβ2 mRNA, is a 5 bp-spaced direct repeat (generally referred to as DR5) of the motif 5'-PuGTTCA (cf. Fig. 9). In addition, response elements with a DR5 containing the motif 5'-PuGGTCA (also termed DR5G to distinguish it from the DR5T of the RARβ2 promoter) act as perfect RAREs[34-36] as well as direct 5'-PuGGTCA repeats spaced by 1 bp (DR1) or 2 bp (DR2) (see Fig. 9). RAR-RXR heterodimers bind to, and activate transcription from, these three types of RAREs, provided target cells express both RARs and RXRs. DR1 elements also bind RXR homodimers *in vitro*, in addition to RAR-RXR heterodimers, and RXRs can transactivate in response to an RXR ligand target gene containing DR1 elements. That DR1 elements can, in principle, act as functional retinoid X receptor response elements (RXREs) *in vivo* is supported by their activity in yeast cells[37] in which any contribution of endogenous RAR via heterodimerization with RXR can be excluded. However, no natural RXRE has been found up to now. The only reported natural RXRE is a DR1-related element found in the rat CRBPII promoter[38]. However, RXR-specific induction of this CRBPII promoter *in vivo* has not yet been demonstrated and the lack of conservation of the CRBPII RXRE in the mouse homologue[39] casts some doubt on its physiological role as an RXRE.

Thyroid hormone receptor response elements

The following three aspects define a thyroid hormone receptor response element (TRE) as a distinct element among the large variety of other nuclear hormone response elements.

The consensus sequence of the core motif

The TRE consensus sequence found is 5'-AGGTCA as for RARs, RXRs, PPARs or VDRs, which are all able to recognize the same DNA sequence (reviewed in references 14 and 15). However, there is evidence for some differences in the (natural) response element repertoires of these receptors. It has been shown, for example, that TRα is able to bind to both 5'-AGGTCA and 5'-AGGACA motifs[40–42]. Such differences could be further enhanced by cooperative DNA binding with other promoter-bound factors and could contribute to the ability of a given target gene to respond preferentially to a particular signaling pathway.

The orientation of the half-site motifs

Like other receptors, TRs are able to bind to a palindromic element called the TREpal[42–44], but such an element has not been reported in cellular genes. Furthermore, this element confers no hormonal specificity, since it can be recognized by a large number of other receptors. The most commonly found TREs are either direct repeats (DRs) or everted repeats (ERs). Examples of direct repeat TREs are those found in the LTR of the Moloney murine leukemia virus[45], in the rat and human myosin heavy-chain genes[46,47] and in the rat malic enzyme[48]. The sequence of the spacer between the two repeats is not purely random, since TR recognizes nucleotides 5' of the hexameric core motif[7,8,49–51]. In everted repeats, the two core motifs display a tail-to-tail orientation. Such elements are found, for example, in the myelin basic protein[52]. In addition, it should be noted that TRα is also able to bind DNA as a monomer on a 5'-AGGTCA half-site extended 5' to the two neighboring bases, which are in this case T and A[49]. Some monomeric half-sites present in cellular genes have been shown to respond to T3. This is the case in the human prolactin and EGF receptor gene promoters[53,54]. It has to be emphasized here that – at least *in vitro* – the ability to bind to DNA as a monomer is peculiar to TRα.

Spacing between the half-sites

On direct repeats TRs have a strong preference for DR4, i.e. direct repeats spaced by four nucleotides (for the corresponding crystal structure, see Table 2 and references therein). Nevertheless, as with other receptors, TRs are able to bind to direct repeats with a spacing other than four, such as DR5 (human alcohol dehydrogenase 3[55]), DR2 (mouse βTSH) or DR0 (rat αTSH[56]). TRs also exhibit a preferred spacing for inverted palindromes of six nucleotides (myelin basic protein[52]).

Inverse TREs

The existence of a peculiar type of TRE has been reported[57,58] to activate transcription in the absence and repress transcription in the presence of ligand. A similar mechanism has been proposed for the former orphan receptor CARβ[59]. Note, however, that the existence of endogenous or serum-derived agonistic ligands cannot be rigorously excluded in these cases; this notion has to be considered as agonistic ligands for CARβ have also been reported to exist[60].

Vitamin D response elements

Only a few natural vitamin D response elements (VDREs) are known, several of which contain DR3 elements. Studies with 'optimized' synthetic response elements assembled from 5'-PuGGTCA motifs have confirmed that DR3 elements bind VDR-RXR heterodimers and that the corresponding promoters are transactivated by the cognate ligands vitamin D and 9-*cis* retinoic acid (9C-RA). The promoter of the human/rat osteocalcin gene contains a complex VDRE with several possible combinations of the recognition motifs, including that of a DR6. For more extensive discussion on VDREs and their action, see the reviews in references 61–63.

Peroxisome proliferator-activated receptor response elements (PPAREs)

Natural PPAREs, which have been found in enzymes that catalyze the peroxisomal β-oxidation and microsomal ω-hydroxylation in response to peroxisome proliferators, usually contain (degenerate) DR1 elements, but more complex PPAREs have also been reported (for reviews, see references 4 and 64–66).

Ecdysone response elements (EcREs)

Ecdysone, the insect molting hormone, was the first steroid hormone shown to be genetically active by inducing puffing in dipteran polytene chromosomes in a hierarchial and temporal order (for reviews, see references 67 and 68). A 23-bp prototypic EcRE has been identified in the promoter of the Hsp27 gene[69,70] and it has been shown that its critical feature is an imperfect palindrome spaced by one nucleotide (5'-GGGTTCAaTGCACTT)[71]. Related binding sites have been found in the ecdysone-inducible Eip28/29[72] and fat body Fbp1 gene promoters[73,74]. Several mutational studies have been performed to identify the significance of the inverted motif and its spacing[71,73,75,76]. Note that *in vivo* footprinting demonstrated that the Fbp1 EcRE was occupied in a EcR- and ecdysteroid-dependent manner[74], which is yet another example of ligand-dependent DNA-binding *in vivo*. Note that this implies that EcR-USP binds to the Fbp1 EcRE, as ecdysteroid-binding ability is acquired only upon heterodimerization[77]. Note that bacterially produced USP (also called CF-1, chorion factor 1) can bind on its own to specific DNA sequences. *In vitro* selection revealed that the USP target sequences are single half-site motifs[78]. For additional reviews and references on the ecdysone regulatory system, see references 68 and 79–81.

Receptors recognizing single extended half-site motifs

Some orphan receptors recognize response elements composed of single half-sites, such as NGFIB[82,83], Rev-erb[84–86], FTZ-F1/Ad4BP/SF-1/ELP[87] and RORs[32]. These receptors are believed to bind as monomers to their cognate response elements. In all of these cases, additional residues 3' of the hexameric half-site motif are specifically recognized. Thus, NGFIB recognizes the sequence 5'-AAAGGTCA-3'[82,83] and Rev-erbα binds to 5'-(A/T)A(A/T)NTPuGGTCA-3'[84]. In the case of NGFIB and ROR, a regional C-terminal of the zinc fingers, termed the A-box, was shown to be necessary for the recognition of the two adenine–thymidine (A/T) pairs 5' of the PuG-GTCA motif. For a review on this topic, see reference 88. Note that NGFIB can also heterodimerize with RXR and bind to DR5 elements[89,90].

Widely spaced direct repeats are promiscuous response elements

The regulatory regions of some target genes harbor complex response elements containing several widely spaced directly repeated 5'-AGGTCA motifs (see Fig. 7). In the chicken ovalbumin promoter region, such direct repetitions, separated from each other by more than 100 bp, have been shown to act synergistically as a complex ERE[91]. At least two other natural response elements are also composed of atypically spaced DRs. One is the retinoic acid response element of the laminin B1 gene, which contains three degenerate DR motifs spaced by 3 and 4 bp[92], the other is the RARE of the oxytocin gene promoter consisting of four directly repeated 5'-AGGTCA-related motifs spaced by 14, 47 and 0 bp[93]. A systematic study of widely spaced direct repeats indicated that these elements constitute a novel class of promiscuous response elements for retinoic acid, vitamin D and estrogen receptors, but not thyroid hormone receptors[94]. These data are supported by the observation that the oxytocin RARE is not only inducible by retinoic acid, but appears also to constitute a promiscuous response element for ER and TR[95-97].

Silencers, everted repeats, 'reversed ligand' and non-autonomous response elements

An 8 bp-spaced everted repeat (ER8, Fig. 9; also referred to as inverted palindrome or 'lap') functions as an RARE in the promoter of the γF crystallin gene[98]. In the chicken lysozyme promoter, an ER8 motif was detected during the characterization of its silencer, which synergized with a second motif for silencing. Surprisingly, this ER8 was shown to act as TRE and RARE in the presence of the cognate ligand, while the ligand-free TR and RAR were required for silencing[99,100]. Note that the ability of TRs and RARs to silence gene transcription was at the basis of the discovery of nuclear receptor corepressors[101,102].

An unusual TRE was identified in the Rous sarcoma virus LTR. It consists of an ER4 on which TR activates heterologous promoters in the absence of the ligand, while ligand exposure results in silencing of the ligand-independent activation[57,58]. It is unclear whether this type of 'reversed ligand TRE' is used as a regulatory element of natural genes.

RA-induction of the Pit-1 gene is mediated by a response element, which contains only one unambiguous half-site motif[103]. However, this motif does not act autonomously, since RA responsiveness appears to be dependent on the presence of Pit-1.

Synergy between hormone response elements and other enhancers

Multiple HREs have been shown to synergize[104-108]. In addition, they can synergize with other enhancer elements of target gene promoters, as has been shown in a systematic analysis with chimeric promoters[109-111]. It appears that the distance from the core promoter and the stereoalignment of the response element on the DNA influence the magnitude of the synergistic effect[108].

PRINCIPLES OF DNA RECOGNITION BY NUCLEAR RECEPTOR DNA-BINDING DOMAINS

Homodimerization and heterodimerization

Nuclear receptors can bind their cognate response elements as monomers, homodimers, or heterodimers with another family member (see Fig. 5; for a review, see reference 14). Dimerization is a general mechanism to increase binding site affinity, specificity and diversity due to: (1) cooperative DNA binding (an extreme case of cooperative binding is the existence, in solution, of stable dimers); (2) the lower frequency of two hexamer binding motifs separated by a defined spacer compared to that of single hexamers (statistically, a hexameric repeat like the estrogen response element is 4^6 times less frequent than a single half-site motif); and (3) heterodimers may have recognition sites distinct from those of homodimers.

Steroid hormone receptors generally bind as homodimers to their response elements, while RAR, RXR, TR and VDR can homodimerize and heterodimerize. RXRs play a central role in these various signal transduction pathways, since they can both homodimerize and act as promiscuous heterodimerization partners for RAR, TR, VDR and orphan receptors. Heterodimerization has a three-fold effect: it leads to a novel response element repertoire, increases the efficiency of DNA binding relative to the corresponding homodimers, and allows two signaling inputs, that of the ligands of RXR and its partner. Note that a phenomenon called 'RXR subordination' maintains signaling pathway identity for retinoic acid, thyroid and vitamin D signaling (see below). Whether some RXR complexes may allow signaling for RXR ligands in the absence of a ligand for the partner of RXR is not entirely clear. Initially, biochemical studies have shown that two dimerization interfaces can be distinguished in nuclear receptors, a weak one by the DBDs and a strong one by the LBDs. LBDs dimerize in solution; the DBD interface is apparently only seen when bound to DNA. Crystal structures of DBD homodimers and heterodimers have defined the sufaces involved in dimerization (see reference 112 and references therein). It is important to point out that the response element reportoire described above for receptor homodimers and heterodimers (Fig. 9) is dictated by the DBD, while the interface formed by the LBDs stabilizes the dimers but does not play any role in response element selection.

Specificity of DNA recognition (P-box, D-box, T-box, A-box)

The DNA response element specificity (half-site sequence, spacing and orientation) is generated by: (1) the recognition of the actual 'core' or 'half-site' motif; and (2) the dimerization characteristics (no dimerization, homodimerization or heterodimerization; structure of the actual dimerization interface) of the receptor(s).

Identification of the residues involved in distinguishing the hexameric half-site motif of EREs (5'-AGGTCA) and GREs (5'-AGAACA) was done by a series of refined swapping experiments. Initially, DBD swaps showed that specific half-site recognition depends on DBD identity[113], subsequently the N-terminal finger was found to differentiate between ERE and GRE recognition[114]. Finally, three studies identified 2–3 residues at the C-terminal 'knuckle' of the N-terminal finger, commonly referred to as the P-box (proximal box; see Fig. 6), to be responsible for ERE *vs* GRE recognition[115-117].

A second region, the D-box (distal box; N-terminal 'knuckle' of the C-terminal finger; see Figs 5 and 6), was found to be involved in differentiating between the binding to a 3 bp- (characteristic for steroid receptor REs) and a 0 bp-spaced (one type of TRE) palindrome[117]. As was later confirmed by the crystal structures of GR and ER DBDs, this region does indeed contribute to the DBD dimerization interface.

Two other boxes have been described within the DBDs of heterodimerizing receptors. The A-box was originally described for NGFI-B as the sequence responsible for the recognition of two or three additional A nucleotides in the minor groove 5' of the hexameric core motif, thus generating an NGFI-B response element (NBRE; 5'-(A/s)AAAGGTCA)[82,83]. This A-box was later found to play a similar role in heterodimers, such are 5'-RXR-TR on DR4 elements, where it specifies to some extent the spacer 5' of TR[7] and sets a minimal spacing by steric hindrance phenomena[50,51]. Interestingly, in the 3D structure, the A-box presents as a helix contacting the minor groove and modeling is in keeping with its role in setting a minimal distance between the half-sites[8].

The T-box (Fig. 5) was originally defined in RXRβ (then H-2RIIBP) as a sequence required for dimerization on a DR1 element[118]. Its role as an RXR homodimerization and heterodimerization surface has subsequently been confirmed[50,51,119].

THREE-DIMENSIONAL STRUCTURE OF NUCLEAR RECEPTOR DNA-BINDING DOMAINS

A significant amount of structural information has been accumulated during the past 10 years, providing information about the solution structure of the GR[120-124], ER[125], RAR[126-130] and RXR[119] DNA-binding domains (DBDs). Moreover, the 3D crystal structures have been solved for the GR DBD homodimer bound to non-cognate DNAs[131,132], the crystal structure of ER DBD homodimer bound to consensus[133] and non-consensus natural[134] EREs, the crystal structure of the RXR homodimer on a DR1 element[135] and the TR-RXR DBD heterodimer bound to its cognate DR4 element[8], and the structures of the NGFI-B-NBRE[9], Rev-erb-DR2[136] and RAR-RXR-DR1[137] complexes. Notably, the opposite orientation of the RAR-RXR heterodimer on DR5 (5'-RXR-RAR-3') and DR1 (5'-RAR-RXR-3') turned out to be as predicted by Zechel et al.[50] on the basis of DBD dimerization interface mapping and gapped-oligonucleotide cross-linking. The corresponding PDB accession numbers are given in Table 1.

The global structure

The 3D structure of the ER DBD-ERE cocrystal is shown in Fig. 5a[133]. The structure consists of a pair of amphipathic α helices packed at right angles and crossing near their midpoints. A zinc-binding pocket lies near the N-terminus of each of the two helices. Hydrophobic side-chains form an extensive hydrophobic core between the two helices. The residues N-terminal to the first helix are folded to form two loops. Hydrophobic residues at the tips of the two loops pack with hydrophobic residues in the core between the two helices.

Two ER DBD molecules bind to adjacent major grooves from one side of the DNA double helix. The protein makes extensive contacts to the phosphate backbone on one side, orienting the DBD such that the recognition helix enters the major groove, allowing surface side-chains to make sequence-specific contacts to the base pairs. Although ER or GR DBDs are monomers in solution, they bind cooperatively to the cognate response elements, owing to the DNA-induced formation of a dimerization interface, which comprises also D-box residues. For further details, see the original publications referred to in Table 2.

Modeling on the DNA of the family of conformers determined by NMR reveals that the region encompassing the ER D-box is the least well-defined structure in solution[138]. It is thus possible that these residues are flexible in solution, and that only upon DNA binding are the dimerization surfaces configured to allow optimal protein–protein interaction. Note, however, that no such flexibility was noted for the GR DBD from backbone dynamics determinations[139].

Half-site specificity

The ER DBD interacts with the central four base pairs of the hexameric ERE motif using four amino acid side-chains on the surface of the recognition helix (E25, K28, K32 and R33). Each of these four side-chains makes hydrogen bonds with the bases. Ordered water molecules greatly increase the number of intramolecular interactions by forming a network of hydrogen bonds between the side-chains and base pairs, and four additional ordered water molecules extend the network to the phosphate backbone of the DNA.

The amino acid contacts within the DNA-binding interface are as follows[133]: E25 of the ER DBD establishes a hydrogen bond with the N4 of C33. K28 donates a hydrogen bond to the O6 of the G4 and forms a salt bridge with E25. S15 donates a bifurcated hydrogen bond to E25 and H18, which itself establishes a hydrogen bond with the phosphate oxygen of A3. Thus, the orientation of the E25 and K28 side-chains are fixed by a buttressing network of hydrogen bonds. K32 generates hydrogen bonds with four acceptors, the O4 of T6, the N7 of G5, and two water molecules. The side-chain of R33 interacts with both the N7 of G31 and the phosphate of T30, and three ordered water molecules. One of these interacts with the O6 of G31 and the two others with backbone phosphate groups. Thus, R33 is also very precisely positioned, as all five guanidinium protons are involved in hydrogen bonding. Note that the DNA-binding interface involves purely polar interactions.

Which interactions are responsible for the differential specificity of ER and GR DBDs? In the GR DBD three amino acids make base-specific contacts with the DNA: K28, V29 and R33[131]. Thus, within the P-box (ER: E25-G26-A29; GR: G25-S26-V29) V29 is the only non-conserved P-box residue that contacts T32, which is A in an ERE. However, A29 of the ER does not establish base contacts that could generate ERE specificity. Instead, it is the complex network of interactions between conserved residues described above that can be established only in the ER DBD-ERE interface, and thus generates ERE specificity (for further details, see reference 133).

Non-cognate response element recognition

Natural response elements are rarely of the consensus type. Instead, one half-site is frequently 'mutated'. Based on crystal structures mechanisms have been described recently by which receptor DBDs can accommodate binding to a non-consensus/non-cognate response element. One highlights the flexibility and/or promiscuity of the ER DBD, since a G to A mutation at position 2 within the 5′ half of the ERE (5′-AGGTCA), which occurs in the *Xenopus* vitamin B1 ERE can be recognized by the ER DBD, owing to a rearrangement of a lysine side-chain so as to make an alternative base contact[134]. As a result of this flexibility/adaptivity, even the DNA complex between an ER-like DBD and a (non-cognate) GRE could be crystallized[132]. In this case, additional water molecules fill the gap generated by the absence of side-chain contacts, resulting in a substantial entropic burden on the stability of the protein-DNA interface.

Differential polarities of heterodimers on different DR response elements

Biochemical studies predicted not only that the DBDs of RAR, RXR and TR dictated the distinct response element repertoires of the various homodimers and heterodimers, but also that the polarity of heterodimers on DR elements was both DBD- and DR-specific (i.e. 5′-RXR-RAR on DR2 and DR5; 5′-RXR-TR on DR4; 5′-RAR-RXR on DR1[7,50]). From these results, two mechanisms were proposed to generate specificity and polarity: (1) DBD dimerization resulting in the observed cooperative DNA-binding to cognate elements; and (2) steric hindrance precluding the concomitant binding of two monomers. It was possible to define by swapping techniques particular regions within these receptor DBDs, which forced a chimeric DBD to adopt a distinct DNA-binding specificity [50,51].

Indeed, the 3D structures of the RXR-TR-DR4[8] and RAR-RXR-DR1[137] complexes, and the models derived from these data confirmed the biochemical data.

1 *The polarity*: the RXR DBD does indeed occupy the 5′ position.
2 *The asymmetric dimerization interfaces*: the crystallographic data show how a dimerization interface is formed to fit precisely to a given DR spacing. The RXR surface is generated by CII residues, including a D-box arginine, and the TR side contributes a surface comprising residues of the prefinger (PRF), CI and T-box.
3 *Two principles generate DR spacer selectivity*. First, steric hindrance: a helix that encompasses the TR A-box and forms interesting backbone and minor groove interactions is the key element that precludes the formation of RXR-TR on DRs spaced by less than 4 bp. Because of this helix, TR accounts for two-thirds of the DNA contacts within the complex. Second, dimer interfaces can only form on the cognate DR elements.

Role of the C-terminal extension of nuclear receptor DBDs

The crystal structure of the heterodimeric RXR-TR-DR4, homodimeric Rev-erbα-DR2 and monomeric NGFI-B–NBRE complexes (Table 2) demonstrated that, in all cases, the C-terminal extension (CTE), comprising the T- and A-boxes, plays an important role in response element binding affinity and selectivity. In keeping with the previous biochemical studies, which indicated a role of the T- and A-boxes in dimerization and binding to the 5′-(A/T)AA flanking sequence of the NBRE, respectively (see above), the C-terminal DBD extension of either the NGFI-

B monomer or the 3' subunit of the RXR-TR heterodimer and Rev-erbα homodimer interacted with the DNA's minor groove upstream of the 5'-AGGTCA motif[8,9,135,136]. This interaction is apparently sufficient to stabilize NGFI-B binding to the NBRE as a monomer, whereas it binds to consensus DR5 elements as a heterodimer with RXR[89,90]. In addition, the A-box of the 3' subunit of Rev-erbα establishes important intersubunit contacts with the CII finger of the 5' subunit in the Rev-erbα–DR2 homodimer[136].

Thus, the CTEs of these receptors fulfill multiple roles in response element recognition: (1) stabilization of DNA binding, thus allowing monomer–DNA interaction; (2) selection of response elements carrying cognate 5' extensions of the core recognition motif (in the case of RXR heterodimers, this concerns the 3' subunit that specifies, at least to some extent, the spacer sequence in DR elements); and (3) discrimination between DR elements due to steric hindrance phenomena.

Monomeric nuclear receptor–response element interaction

Two subclasses of monomeric DNA-binding nuclear receptors can be distinguished[9], the NGFI-B and SF-1 subfamilies. The two families can be distinguished by the preferences for 5'-AAA or 5'-TCA extensions, respectively, of the 5'-AGGTCA half-site motif.

Meinke and Sigler[9] rationalize that it is the CTE, in particular the A-box, of the two receptor families which accounts for the discrimination of the two types of 5' extension. The authors propose that the RGR motif in the A-box of NGFI-B family members mediates 5'-AAA recognition, while the corresponding A-box of members of the SF-1 subfamily dictates binding to core motifs with 5'-TCA extension.

DNA-induced structure of nuclear receptor DBDs

The NMR structure of the ERR2 orphan DBD shows a disordered CTE[140]. The NMR structures of all NR DBDs are monomers with a defined core but otherwise poorly ordered, including the potential dimer interface[8,131,132,138]. In contrast, the CTE and the dimer interface are highly ordered in DNA complexes, suggesting that the DNA serves as a template on which the CTE and other DBD structures can fold. It is possible that distinct response elements may differentially affect NR functionality in an allosteric fashion (for a detailed discussion of this phenomenon, see reference 141).

References

[1] Umesono, K. et al. (1991) Cell 65, 1255–1266.

[2] Kliewer, S.A. et al., (1992) Nature 358, 771–774.

[3] Durand, B. et al. (1992) Cell 71, 73–85.

[4] Desvergne, B. and Wahli, W. (1999) Endocrinol Rev. 20, 649–688.

[5] Kliewer, S.A. et al. (1999) Recent Prog. Horm. Res. 54, 345–367.

[6] Mader, S. et al. (1993) EMBO J. 12, 5029–5041.

[7] Kurokawa, R. et al. (1993) Genes Dev. 7, 1423–1435.

[8] Rastinejad, F. et al. (1995) Nature 375, 203–211.

[9] Meinke, G. and Sigler, P.B. (1999) Nature Struct. Biol. 6, 471–477.

[10] Evans, R.M. (1988) Science 240, 889–895.

[11] Green, S. and Chambon, P. (1988) Trends Genet. 4, 309–314.

[12] Beato, M. (1989) Cell 56, 335–344.

[13] Truss, M. and Beato, M. (1993) Endocrinol. Rev. 14, 459–479.

[14] Glass, C.K. (1994) Endocrinol. Rev. 15, 391–407.

[15] Gronemeyer, H. and Laudet, V. (1995) Protein Profile 2, 1173–1308.

[16] Klock, G. et al. (1987) Nature 329, 734–736.

[17] Klein-Hitpass, L. et al. (1986) Cell 46, 1053–1061.

[18] Berry, M. et al. (1989) Proc. Natl Acad. Sci. USA 86, 1218–1222.

[19] Nunez, A.M. et al. (1989) EMBO J. 8, 823–829.

[20] Scheidereit, C. et al. (1983) Nature 304, 749–752.

[21] Darbre, P.D. et al. (1985) J. Steroid Biochem. 23, 379–384.

[22] von der Ahe, D. et al. (1985) Nature 313, 706–709.

[23] Cato, A.C. et al. (1986) EMBO J. 5, 2237–2240.

[24] Arriza, J.L. et al. (1987) Science 237, 268–275.

[25] Ham, J. et al. (1988) Nucl. Acids Res. 16, 5263–5276.

[26] Truss, M. et al. (1995) EMBO J. 14, 1737–1751.

[27] Archer, T.K. et al. (1992) Science 255, 1573–1576.

[28] Wijnholds, J. et al. (1988) EMBO J. 7, 2757–2763.

[29] Nordeen, S.K. (1990) Mol. Endocrinol. 4, 1866–1873.

[30] Lieberman, B.A. et al. (1993) Mol. Endocrinol. 7, 515–527.

[31] Strähle, U. et al. (1989) Nature 339, 629–632.

[32] Giguere, V. et al. (1994) Genes Dev. 8, 538–553.

[33] Tsai, M.J. and O'Malley, B.W. (1994) Annu. Rev. Biochem. 63, 451–486.

[34] de The, H. et al. (1990) Nature 343, 177–180.

[35] Hoffmann, B. et al. (1990) Mol. Endocrinol. 4, 1727–1736.

[36] Sucov, H.M. et al. (1990) Proc. Natl Acad. Sci. USA 87, 5392–5396.

[37] Heery, D.M. et al. (1994) Nucl. Acids Res. 22, 726–731.

[38] Mangelsdorf, D.J. et al. (1991) Cell 66, 555–561.

[39] Nakshatri, H. and Chambon, P. (1994) J. Biol. Chem. 269, 890–902.

[40] Chen, H. et al. (1993) Mol. Cell Biol. 13, 2366–2376.

[41] Chen, H.W. and Privalsky, M.L. (1993) Mol. Cell Biol. 13, 5970–5980.

[42] Forman, B.M. et al. (1992) Mol. Endocrinol. 6, 429–442.

[43] Glass, C.K. (1987) Nature 329, 738–741.

[44] Glass, C.K. et al. (1988) Cell 54, 313–323.

[45] Sap, J. et al. (1989) Nature 340, 242–244.

[46] Flink, I.L. and Morkin, E. (1990) J. Biol. Chem. 265, 11233–11237.

[47] Izumo, S. and Mahdavi, V. (1988) Nature 334, 539–542.

[48] Desvergne, B. et al. (1991) J. Biol. Chem. 266, 1008–1013.

[49] Katz, R.W. and Koenig, R.J. (1993) J. Biol. Chem. 268, 19392–19397.

[50] Zechel, C. et al. (1994) EMBO J. 13, 1425–1433.

[51] Zechel, C. et al. (1994) EMBO J. 13, 1414–1424.

[52] Farsetti, A. (1992) J. Biol. Chem. 267, 15784–15788.

[53] Day, R.N. and Maurer, R.A. (1989) Mol. Endocrinol. 3, 931–938.

[54] Hudson, L.G. et al. (1990) Cell 62, 1165–1175.

[55] Harding, P.P. and Duester, G (1992) J. Biol. Chem. 267, 14145–14150.

[56] Burnside, J. et al. (1989) J. Biol. Chem. 264, 6886–6891.

[57] Saatcioglu, F. et al. (1993) Cell 75, 1095–1105.

[58] Saatcioglu, F. (1994) Semin. Cancer Biol. 5, 347–359.

[59] Forman, B.M. et al. (1998) Nature 395, 612–615.

[60] Tzameli, I. et al. (2000) Mol. Cell Biol. 20, 2951–2958.

[61] Haussler, M.R. et al. (1997) J. Endocrinol. 154 (Suppl), S57–73.

[62] Carlberg, C. and Polly, P. (1998) Crit. Rev. Eukaryot. Gene Expr. 8, 19–42.

[63] DeLuca, H.F. and Zierold, C. (1998) Nutr. Rev. 56, S4–10; discussion S 54–75.

[64] Kliewer, S.A. (1999) Recent Prog. Horm. Res. 54, 345–367.

[65] Kersten, S. et al. (2000) Nature 405, 421–424.

[66] Willson, T.M. (2000) J. Med. Chem. 43, 527–550.

[67] Ashburner, M. et al. (1974) Cold Spring. Harb. Symp. Quant. Biol. 38, 655–662.

[68] Thummel, C.S. (1995) Cell 83, 871–877.

[69] Mestril, R. et al. (1986) EMBO J. 5, 1667–1673.

[70] Riddihough, G. and Pelham, H.R.B. (1987) EMBO J. 6, 3729–3734.

[71] Martinez, E. et al. (1991) EMBO J. 10, 263–268.

[72] Cherbas, L. et al. (1991) Genes. Dev. 5, 120–131.

[73] Antoniewski, C. et al. (1993) Insect Biochem. Mol. Biol. 23, 105–114.

[74] Antoniewski, C. et al. (1994) Mol. Cell Biol. 14, 4465–4474.

[75] Dobens, L. et al. (1991) Mol. Cell Biol. 11, 1846–1853.

[76] Ozyhar, A. and Pongs, O. (1993) J. Steroid Biochem. Molec. Biol. 46, 135–145.

[77] Yao, T.P. et al. (1992) Cell 71, 63–72.

[78] Christianson, A.M. and Kafatos, F.C. (1993) Biochem. Biophys. Res. Commun. 193, 1318–1323.

[79] Segraves, W.A. (1994) Semin. Cell Biol. 5, 105–113.

[80] Thummel, C.S. (1997) Bioessays 19, 669–672.

[81] Henrich, V.C. et al. (1999) Vitam. Horm. 55, 73–125.

[82] Wilson, T.E. et al. (1991) Science 252, 1296–1300.

[83] Wilson, T.E. et al. (1993) Mol. Cell Biol. 13, 5794–5804.

[84] Harding, H.P. and Lazar, M.A. (1993) Mol. Cell Biol. 13, 3113–3121.

[85] Forman, B.M. et al. (1994) Mol. Endocrinol. 8, 1253–1260.

[86] Retnakaran, R. et al. (1994) Mol. Endocrinol. 8, 1234–1244.

[87] Ueda, H. et al. (1992) Mol. Cell Biol. 12, 5667–5672.

[88] Laudet, V. and Adelman, G. (1995) Current Biol. 5, 124–127.

[89] Perlmann, T. and Jansson, L. (1995) Genes Dev. 9, 769–782.

[90] Forman, B.M. et al. (1995) Cell 81, 541–550.

[91] Kato, S. et al. (1992) Cell 68, 731–742.

[92] Vasios, G. et al. (1991) EMBO J. 5, 1149–1158.

[93] Richard, S. and Zingg, H.H. (1991) J. Biol. Chem. 266, 21428–21433.

[94] Kato, S. (1995) Mol. Cell Biol. 15, 5858–5867.

[95] Adan, R.A. et al. (1991) Biochem. Biophys. Res. Commun. 175, 117–122.

[96] Adan, R.A. et al. (1992) J. Biol. Chem. 267, 3771–3777.

[97] Adan, R.A. et al. (1993) Mol. Endocrinol. 7, 47–57.

[98] Tini, M. et al. (1993) Genes. Dev. 7, 295–307.

[99] Baniahmad, A. et al. (1992) EMBO J. 11, 1015–1023.

[100] Baniahmad, A. et al. (1990) Cell 61, 505–514.

[101] Hörlein, A.J. et al. (1995) Nature 377, 397–404.

[102] Chen, J.D. and Evans, R.M. (1995) Nature 377, 454–457.

[103] Rhodes, S.J. et al. (1993) Genes Dev. 7, 913–932.

[104] Jantzen, H.M. et al. (1987) Cell 49, 29–38.

[105] Ankenbauer, W. et al. (1988) Proc. Natl Acad. Sci. USA 85, 7526–7530.

[106] Martinez, E. and Wahli, W. (1989) EMBO J. 8, 3781–3791.

[107] Schmid, W. et al. (1989) EMBO J. 8, 2257–2263.
[108] Ponglikitmongkol, M. et al. (1990) EMBO J. 9, 2221–2231.
[109] Schule, R. et al. (1988) Science 242, 1418–1420.
[110] Schule, R. et al. (1988) Nature 332, 87–90.
[111] Strähle, U. et al. (1988) EMBO J. 7, 3389–3395.
[112] Gronemeyer, H. and Moras, D. (1995) Nature 375, 190–191.
[113] Green, S. and Chambon, P. (1987) Nature 325, 75–78.
[114] Green, S. and Chambon, P. (1988) Trends Genet. 4, 309–314.
[115] Danielsen, M. et al. (1989) Cell 57, 1131–1138.
[116] Mader, S. et al. (1989) Nature 338, 271–274.
[117] Umesono, K. and Evans, R.M. (1989) Cell 57, 1139–1146.
[118] Wilson, T.E. et al. (1992) Science 256, 107–110.
[119] Lee, M.S. et al. (1993) Science 260, 1117–1121.
[120] Hard, T. et al. (1990) Biochemistry 29, 9015–9023.
[121] Hard, T. et al. (1990) Science 249, 157–160.
[122] Pan, T. et al. (1990) Biochemistry 29, 9218–9225.
[123] Remerowski, M.L. et al. (1991) Biochemistry 30, 11620–11624.
[124] Baumann, H. et al. (1993) Biochemistry 32, 13463–13471.
[125] Schwabe, J.W. et al. (1990) Nature 348, 458–461.
[126] Katahira, M. et al. (1992a) Nucl. Acids Symp. Ser. 1992, 65–66.
[127] Katahira, M. et al. (1992b) Biochemistry 31, 6474–6480.
[128] Knegtel, R.M. et al. (1993) Ann. N. York Acad. Sci. 684, 49–62.
[129] Knegtel, R.M. et al. (1993) Biochem. Biophys. Res. Commun. 192, 492–498.
[130] Knegtel, R.M. et al. (1993) J. Biomol. NMR 3, 1–17.
[131] Luisi, B.F. et al. (1991) Nature 352, 497–505.
[132] Gewirth, D.T. and Sigler, P.B. (1995) Nature Struct. Biol. 2, 386–394.
[133] Schwabe, J.W. et al. (1993) Cell 75, 567–578.
[134] Schwabe, J.W. et al. (1995) Structure 3, 201–213.
[135] Zhao, Q. et al. (2000) J. Mol. Biol. 296, 509–520.
[136] Zhao, Q. et al. (1998) Mol. Cell 1, 849–861.
[137] Rastinejad, F. et al. (2000) EMBO J. 19, 1045–1054.
[138] Schwabe, J.W. et al. (1993) Structure 1, 187–204.
[139] Berglund, H. et al. (1997) Biochemistry 36, 11188–11197.
[140] Sem, D.S. et al. (1997) J. Biol. Chem. 272, 18038–18043.
[141] Lefstin, J.A. and Yamamoto, K.R. (1998) Nature 392, 885–888.

3 Ligand binding

STRUCTURE OF NUCLEAR RECEPTOR LIGAND-BINDING DOMAINS

The canonical fold of nuclear receptor ligand-binding domains

To date the crystal structures of 14 distinct NR LBDs have been described (see Table 3), among them the first structures reported in 1995 of an apo (non-liganded) and holo (agonist-bound) LBD, the dimeric apo-RXRα (see Fig. 7[1]) and the monomeric holo-RARγ (Fig. 7[2,3]) and holo-TRα[4]. In addition, crystal structures for the holo-RXRα[5], monomeric, dimeric holo- (estradiol, DES)-bound and antagonist (raloxifen, tamoxifen)-bound ERα[6-8] and raloxifen- and genistein-bound ERβ[9], dimeric holo- (progesterone)-bound PR[10], apo- and holo- (thiazolidinedione)-PPARγ and δ[11-13] and holo-VDR[14] have been reported.

Recently, the first structures of heterodimeric LBDs have been solved. One comprises the RARα LBD bound to the α-selective antagonist BMS614 and the constitutively active[15] RXRαF318A mutant, which turned out to harbor an oleic acid-like ligand, has been reported[16]. The other is the LBD of PPARγ bound to rosiglitazone or another synthetic ligand in a heterodimer with 9*cis* RA-bound RXRα[17]. All these NR LBDs display a common fold, as originally predicted[18], with 12 α-helices (H1 to H12) and one β-turn arranged as an antiparallel α-helical 'sandwich' in a three-layer structure (see Fig. 7). Note that some variability exists; for example, no helix H2 was found in RARγ[2], while an additional short helix H2' is present in PPARγ[7].

The mousetrap model

A comparison of the apo- and holo-LBD structures (see Fig. 7) suggested a common mechanism by which the activation function AF-2 becomes transcriptional competent: upon ligand binding H11 is repositioned in the continuity of H10, and the concomitant swinging of H12 unleashes the loop that flips over underneath H6, carrying along the N-terminal part of H3. In its final position, H12 seals the ligand-binding cavity as a 'lid' and further stabilizes ligand binding (in some but not all NRs) by contributing additional ligand–protein interactions. It is a general and essential feature of the ligand 'activation' of nuclear receptors that the transconformation of H12, together with additional structural changes (such as bending of helix H3), creates distinct surface(s) on the apo- and holo-LBD. The novel surfaces generated upon agonist binding allow bona fide coactivators, such as the members of the SRC-1/TIF2 family, to bind and recruit additional transcription factors (see below). Concomitantly, corepressor proteins, which bind to (presently unknown) surface(s) of the apo-LBD, dissociate upon agonist, but not necessarily antagonist, binding (see below). Notably, as is discussed further below, certain antagonists 'force' H12 in a third position, distinct from the holo position whereby it impairs coactivator binding.

For a given receptor, the equilibrium between the apo and holo (or apo and antagonist) conformational states of a NR LBD can be affected through intramolecular interactions of H12, such as a salt bridge (holo-LBD of RARγ[2]) or hydrophobic contacts (as suggested for apo-ER[19]). This implies that the apo conformation is not necessarily the default state, so that some NRs may be constitutive activators or repressors without possessing a cognate ligand. Moreover, an increase in coactivator concentration can generate a transcriptionally competent RAR under certain conditions[20] and the apo-ER conformation may be destabilized by phosphorylation[19,21]. Thus, overexpression of coactivators or receptor modification may generate ligand-independent receptors. Such scenarios could have significant implications for endocrine cancer therapies.

The dimer interface

The recent crystallization of the RAR-RXR[16] and PPAR-RXR[17] ligand-binding domain heterodimers for the first time allowed the comparison of the homodimerization and heterodimerization interfaces of several nuclear receptor ligand-binding domains. In principle, the overall heterodimeric arrangement closely resembled that of homodimers of RXRα, ERα and PPARγ[1,6,11]. Below, the heterodimeric interface is described based on the results obtained in reference 16: the interfaces involves residues from helices H7, H9, H10 and H11, as well as loops L8–9 and L9–10, with the helices H10 being parallel to the two-fold axis relating the subunits. Within the heterodimer, however, the two protomers do not contribute equally to the heterodimerization interface. For example, RAR provides significantly more polar groups than RXR, while the latter provides the majority of the positively charged atomic groups. Moreover, RAR and RXR exhibit different patterns of contact areas, as helix H7 of RXR contributes four times more surface area to the interface than its RAR counterpart. Conversely, the contribution of RARα loop L8–9 to the interface is three times higher than that of the corresponding loop in RXRαF318A[16].

Except for side-chain rearrangements, the part of the RXR ligand-binding domain structure involved in the heterodimeric interface does not differ significantly from that of the RXRα apo-LBD homodimer. In both dimers, helices H9 and H10 contribute to more than 75% of the total surface and constitute the core of the dimer interfaces. However, some differences in the relative involvement of some RXR structural elements to the homodimer or heterodimer interfaces are observed and originate from the nature of the interacting protomer. For example, in the heterodimer, helix H7 contributes twice as much surface to the interface than in the homodimer. This is due to the different structure of RAR loop L8–9, which makes more extensive contacts with H7 of RXR than its RXR counterpart in the homodimer. The RXR residues involved in the homodimerization or heterodimerization are identical. Of the 25 RARα residues implicated in the dimerization, 21 (85%) are homologous to those of RXR. However, owing to the limited identity, the nature of a number of interactions is different within the homodimer and heterodimer, leading to probable differences in stability. This is likely to account for the observation that mixing of RARs and RXRs leads to the formation of almost exclusively RAR-RXR heterodimers.

When compared to RAR and RXR, the ERα ligand-binding domain structural elements generating the dimerization surface are identical. However, helices H8, H9, H10 and the loop L8–9 in ERα are longer and make additional contacts. As a consequence, the buried surfaces are larger for the ERα homodimer (around 1700 Å^2). The smaller interface in homodimers or heterodimers involving RXR (less than 1000 Å^2) suggests a weaker link between the protomers that could be related to the promiscuous character of this NR. The moderate stability of RXR homodimeric or heterodimeric associations may facilitate the switch between various partners.

The observations that NR LBDs exhibit a common fold and that the dimerization interfaces of all crystallized LBD dimers are topologically conserved suggest that all NRs can establish similar dimer interfaces. Therefore, amino-acid variations at the dimerization surface of the various NRs most probably determine their dimerization characteristics.

An extensive analysis of the dimerization interfaces of nuclear receptor ligand-binding domains, based on crystal structure data and sequence alignment has been reported by Bourget et al.[16]. The reader is referred to this publication for details and an interpretation of the structural basis that accounts for the homodimerization and heterodimerization pattern of distinct members of this family.

The ligand-binding pocket

In all crystal structures presently available, the ligand is embedded within the protein with no clear accessible entry or exit site. PPARγ seems the only exception to that rule, since a potential access cleft to the ligand-binding pocket (LBP) was observed between helix 3 and the β-turn, which may be of sufficient size to allow entry of small ligands without major adaptation. For all other receptors of known structure, significant conformational changes are necessary to generate potential entry sites. The mousetrap model provides an easy solution to the problem: the mobility of H12 opens a channel by removing the 'lid' from the ligand pocket.

The ligand binding pockets are lined with mostly hydrophobic residues. Few polar residues at the deep end of the pocket near the β-turn act as anchoring points for the ligand or play an essential role in the correct positioning and enforce the selectivity of the pocket. Most nuclear receptors contain a conserved arginine attached to helix H5, which points into this part of the cavity. These anchoring residues, conserved within a given subfamily, are indicative of the polar group characteristics of each family of ligands (i.e. carboxylate for retinoids and ketone for steroids).

In the case of retinoid receptors, it is the ligand that adapts to a fairly rigid ligand-binding pocket[3]. The ligand-binding pockets of some other NRs are significantly larger and use precise anchoring points for their ligands; in such cases it is possible to generate ligands, which, owing to additional contacts, have higher binding affinities than the natural ligands and may even act as 'superagonists'. In this context it will be interesting to understand how the ligand-binding pocket of PXR (NR1I2) can accommodate such diverse ligands as dexamethasone, RU486, rifampicin, taxol and others. The PXR LBD structure has been solved very recently[29].

Ligand selectivity

As shown in the cases of RARγ and TRβ, the shape of the ligand-binding pocket matches that of the ligand. The accordance of shape and volume maximize the number of mostly hydrophobic contacts, thus contributing to the stability of the complex and the selectivity of the pocket for the cognate ligand.

RAR possesses an interesting LBD, since it can bind two chemically different ligands equally well: all-*trans* retinoic acid and its 9-*cis* isomer. Crystallographic analysis[3] of the two ligands in the RARγ LBD showed that both adapt conformationally to the LBP, which acts as matrix. Moreover, the conformation of a RARγ-selective agonist was also shown to match closely that of the natural ligands in their bound state[3]. The adaptation of ligands to the protein leads to an optimal number of interactions for binding and selectivity, and justifies modeling approaches for ligand design.

For steroid receptors, the LBP volume is significantly larger than that of the corresponding ligands and the rigidity of the ligand does not allow adaptability. Therefore, selectivity cannot be driven by multiple hydrophobic contacts, which could anyway not suffice to discriminate between small structurally similar ligands. In this case, specific key interactions are more important. Note that very large LBP volumes allow for the binding multiple ligands of different stereochemistry, as in the case of PPAR[11], often at the expense of lower binding affinities.

A structure-based sequence alignment revealed that only three residues diverged in the LBPs of RARα, β and γ, leading to the prediction that these divergent residues were critically involved in differentiating between isotype-selective retinoids[2]. Indeed, swapping of these residues confirmed this hypothesis[22]. Moreover, swapping of these residues not only mediated isotype-selective binding but also the agonistic/antagonistic response of a cognate ligand on to any other RAR isotype, thus emphasizing the importance of these three residues in triggering the ligand-induced transcriptional response.

RXR TETRAMERS

A peculiar aspect of the oligomeric state of nuclear receptors is the observation that apo-RXR, in contrast to other apo-NRs, can form homotetramers that dissociate upon DNA and ligand binding[23-25]. The tetramerization region has been identified[26] and awaits crystallographic scrutiny. The functional meaning and the structure of these entities is not yet fully established[25,27,28]. It is likely that tetramer formation by RXRs not only serves to generate an (inactive) intracellular pool of a promiscuous heterodimerization partner but comprises an additional level of complexity of the regulation of gene transcription mediated by these proteins.

References
[1]   Bourguet, W. et al. (1995) Nature 375, 377–382.
[2]   Renaud, J.P. et al. (1995) Nature 378, 681–689.
[3]   Klaholz, B.P. et al. (1998) Nature Struct. Biol. 5, 199–202.
[4]   Wagner, R.L. et al. (1995) Nature 378, 690–697.
[5]   Egea, P.F. et al. (2000) EMBO J. 19, 2592–2601.
[6]   Brzozowski, A.M. et al. (1997) Nature 389, 753–758.
[7]   Shiau, A.K. et al. (1998) Cell 95, 927–937.
[8]   Tanenbaum, D.M. (1998) Proc. Natl Acad. Sci. USA 95, 5998–6003.
[9]   Pike, A.C. et al. (1999) EMBO J. 18, 4608–4618.
[10] Williams, S.P. and Sigler, P.B. (1998) Nature 393, 392–396.
[11] Nolte, R.T. et al. (1998) Nature 395, 137–143.
[12] Uppenberg, J. et al. (1998) J. Biol. Chem. 273, 31108–31112.

13 Xu, H.E. et al. (1999) Mol. Cell 3, 397–403.
14 Rochel, N. et al. (2000) Mol. Cell 5, 173–179.
15 Vivat, V. et al. (1997) Embo J. 16, 5697–5709.
16 Bourguet, W. et al. (2000) Mol. Cell 5, 289–329.
17 Gampe, R.T. Jr et al. (2000) Mol. Cell 5, 545–555.
18 Wurtz, J.M. et al. (1996) Nature Struct. Biol. 3, 87–94.
19 White, R. et al. (1997) EMBO J. 16, 1427–1435.
20 Voegel, J.J. et al. (1998) EMBO J. 17, 507–519.
21 Weis, K.E. et al. (1996) Mol. Endocrinol. 10, 1388–1398.
22 Gehin, M. (1999) Chem. Biol. 6, 519–529.
23 Kersten, S. et al. (1995) Proc. Natl Acad. Sci. USA 92, 8645–8649.
24 Chen, Z. et al. (1998) J. Mol. Biol. 275, 55–65.
25 Kersten, S. et al. (1998) J. Mol. Biol. 284, 21–32.
26 Kersten, S. et al. (1997) J. Biol. Chem. 272, 29759–29768.
27 Dong, D. and Noy, N. (1998) Biochemistry 37, 10691–10700.
28 Lin, B.C. et al. (1997) J. Biol. Chem. 272, 9860–9867.
29 Watkins, R.E. et al. (2001) Science 292, 2329–2333.

4 Molecular mechanisms of transcriptional regulation

The expression of a given gene can be regulated at several different levels (transcription, translation or RNA processing, post-translation) and also transcription itself has multiple levels at which regulation might occur. The transcriptional activity of a gene can be controlled epigenetically via methylation, at the level of its chromatin structure, and at the level of the assembly and activity of the initiating and elongating polymerase complexes. The reported effects of nuclear receptors on transcription are so far restricted to the initiation of transcription by RNA polymerase II. Below we will summarize our current knowledge about the molecular mechanisms/interactions through which nuclear receptors can positively or negatively regulate the expression of cognate genes. For further details and references, see the most recent reviews by Chen[1] and Glass[2].

MOST NUCLEAR RECEPTORS CONTAIN AT LEAST TWO INDEPENDENT ACTIVATION FUNCTIONS, AF1 AND AF2

Nuclear receptors contain at least two distinct regions, termed activation functions AF-1 and AF-2 (Fig. 3), which, when tethered to a (heterologous) DNA-binding domain, will transactivate transcription from response elements recognized by this DBD[3-12]; for a detailed review and further references see Gronemeyer[13] and Mangelsdorf[14]. In the context of the wild-type receptor, both AF-1 and AF-2 become active in response to the ligand, but AF-1 can act constitutively in fusion proteins with heterologous DBDs. AF-2 remains ligand dependent even in such fusion proteins. Within AF-2, at the C-terminal boundary of the LBD, an autonomous constitutively active transactivation function (AF-2 AD) has been identified whose integrity is crucial for AF-2, as mutations in AF-2 AD abolish AF-2 activity[15-19]. Note, however, that the constitutive activity of AF-2 AD is very weak compared with the full ligand-induced activity of AF-2. AF-2 AD had been proposed to adopt an amphipathic α-helical structure[15,16], which was confirmed by all crystal structure data available to date. At the D–E junction, a weak autonomous activation function has been described in the GR[8]; the corresponding region of ER was active in yeast cells[20] but not in mammalian cells[21].

Within the AF-1-containing A/B region shorter regions have been described to display constitutive activation function[12,22-24]. In the case of the human PR isoforms, the additional N-terminal sequence, which is unique to the larger ('form B') isoform, was found to be able to squelch and to display an autonomous transactivation activity on its own together with the homologous, but not with a heterologous, DNA-binding domain[22,25,26].

Several nuclear receptors exist as isoforms. As was originally shown for the PR forms A and B, both isoforms exhibit a different promoter specificity[27,28]. For a more detailed discussion on this topic, see the recent review by Conneely and colleagues[29], and for a discussion of the differential antagonist action of the two isoforms and the relevance to endocrine cancer therapy, see the corresponding reviews from the Horwitz laboratory[30,31]. Note that the additional N-terminal sequence of PR B may have a peculiar structure[32]. A later study showed that isoform-specific transcription activation is not confined to PR but can be observed also with the retinoic acid receptors[33].

The activation functions of nuclear receptors act in both promoter context- and cell-specific fashion, as was best documented for ER and RAR[33,34]. This selectivity may originate from cooperative/synergism with other promoter-bound transcription factors and/or the cell-specific expression of transcription intermediary factors/coactivators (see below).

CHROMATIN-MODIFYING NUCLEAR RECEPTOR COREGULATORS

So-called squelching[35] experiments paved the way to predict the existence of factors that would transmit the signal generated by the holo-receptor to the transcription machineries. Squelching occurs if a receptor inhibits the activity 'off the DNA' of the same ('autosquelching'[36]) or a different ('heterosquelching'[37,38]) receptor in an agonist-dependent and AF-2 integrity-dependent manner. These squelching data were interpreted as the result of sequestering, by either excess of the same, or addition of another ligand-activated receptor, of so-called transcription intermediary factors (TIFs) that mediate the action of the activation/repression functions of nuclear receptors and are limiting constituents of the mechanisms required for transcription initiation. This concept predicted the existence of TIFs that are shared between, and are critically involved in, the action of different receptors. Indeed, the subsequent cloning and characterization of TIFs, also known as coactivators and corepressors, has fully justified this concept.

According to the squelching studies, bona fide coactivators were predicted to fulfill the following criteria: (1) interact directly with nuclear receptor ligand binding domains in an agonist-dependent and activation function-dependent manner; (2) enhance nuclear receptor-dependent reporter gene activity when transiently expressed in mammalian cells; (3) activate transcription autonomously when tethered to DNA via a heterologous DBD; and (4) relieve squelching. The development of yeast 'two hybrid' and direct cDNA expression library screening approaches has led to the identification of a large number of putative and bone fide coactivators, corepressors and other coregulators that are believed to transmit the nuclear receptor signal to its molecular targets. A summary of these coregulators, their characteristics and their putative mode of action is given in Table 5 (see also the reviews 2, 39–44). The general features of different classes of coregulators are discussed below.

The cloning of coregulators ('bona fide' coactivators and corepressors in Table 5) was followed by the definition of the coactivator signature LxxLL NR box (where x is any amino acid) motifs embedded in a short α-helical peptide[45–47]. These nuclear receptor boxes are necessary and sufficient for ligand-dependent direct interaction with a cognate surface in the nuclear receptor ligand-binding domain that constitutes the transcriptional activation function AF-2. This surface corresponds to a hydrophobic cleft with 'charge clamps' (to which helix H12 contributes when repositioned on the surface of the ligand binding domain upon agonist binding), which, together with surrounding amino acids, accommodates the amphiphatic LxxLL NR box helix as revealed by X-ray crystallography[48–51]. The fact that some

Table 5 *Nuclear receptor-interacting proteins*

Human factors	Synonyms	NID motif	Accession number[a]	Comments	Reference
1. Nuclear receptor-binding subunits of coactivator complexes					
Bona fide NR coactivators					
SRC-1a	hERAP160 mNCoA1	Yes	U90661 NM_010881	First identified coactivator; binds NR and CBP/p300; displays acetyltransferase activity and contacts basal transcription factors	100 101
SRC-1e		Yes	U19179	Isoform of SRC-1a; lacks the C-terminal NR box of SRC-1a	
TIF2	hNCoA2 mGRIP1 SRC-3	Yes	X97674 U39060	Member of the same coactivator family as SRC-1	63, 64 102
ACTR	hAIB1 hRAC3 hTRAM1 mpCIP	Yes	U59302 AF000581	Member of the same coactivator family as SRC-1/TIF2	103 104 105 106
Cointegrators					
CBP	mCBP	Yes	U47741 sw:P45481	Transcriptional cointegrator for several signaling pathways in addition to those involving NRs (CREB, STATs, AP1, etc.); interacts with pCAF, SRC-1, TIF2 and pCIP; displays acetyltransferase activity	106 107
p300		Yes	U01877	Functional homologue of CBP; p300 and CBP functions are not entirely redundant	
pCAF			U57317	Similar to hGCN5; interacts with CBP/p300; possesses intrinsic acetyltransferase activity; present in a 2 MDa complex	108
Others					
SRA	mSRA		AF092038 AF092039	An RNA molecule, identified by using the N-terminal domain of PR as a bait; the SRA gene does not encode a protein; SRA is proposed to be part of a complex containing SRC-1	109

2. Nuclear receptor-binding subunits of corepressor complexes

Bona fide NR corepressors

N-CoR	hNcoR1 mN-CoR RIP13 (partial)	Yes	NM_006311 U35312	Interacts with and corepresses TR, RAR, COUPTFI and Rev-erb	110 111
SMRT	hSMRT hNCoR2 TRAC2 (partial)	Yes	U37146 NM_006312	Corepresses TR and RAR, and has sequence similarity with N-CoR	75
Sun-CoR	mSun-CoR		AF031426	Corepresses TR and Rev-erb	112
Others					
TRUP	SURF-3		M36072	Interacts with the hinge region and the N-terminal portion of the LBD of TR in a hormone-independent manner; exerts its inhibiting activity by interfering with DNA binding of the receptor to this element	113
TSG101				Interacts with AF-1 of GR and represses transcriptional activity of GR in mammalian and in yeast cells	97
REA	BAP-37	Unclear		Isolated by interaction with the dominant negative ER receptor (L540Q); interacts also with the antiestrogen-liganded ER; represses transcriptional activity of ER but not of PR, RAR or VP16; contains an LxxLL NR box motif but its function in interaction has not been demonstrated	114
ALIEN	hALIEN dALIEN		AF120268 U57758	Shows the expected properties of a corepressor of nuclear receptor in *Drosophila*; apparently plays the same role for mammalian nuclear receptors	115

3. Nuclear receptor-binding subunits of the SRB/mediator containing complex (SMCC)

TRAP220	hDRIP205 hDRIP230	Yes	AF055994	Included in the SMCC complex that interacts with liganded TR and VDR, enhances *in vitro* transcription by these receptors	86, 87, 94, 96,

Continued

Table 5 *Continued.*

Human factors	Synonyms	NID motif	Accession number[a]	Comments	Reference
	hTRIP2 (partial) mPBP		AF000294	mPBP was originally isolated by interaction with PPAR	97, 116
TRAP100	hDRIP100	Unclear	AF055995	DRIP100 does not interact directly with VDR in a ligand-dependent manner; overexpression leads to coactivation first and then to squelching; NID does not promote direct interaction with NR	117, 86
TRAP170	hDRIP150 hRGR1 hEXLM1		AF135802	DRIP150 interacts with and coactivates AF-1 of GR; may work in association with DRIP205 (interacts with and coactivates GR AF-2)	97
4. Nuclear receptor-binding subunits of the SWI/SNF complex					
hSNF2a hSNF2b	hbrm BRG1		X72889	Coimmunoprecipitates with GR; both hBRM and BRG1 interact with ER in a ligand-dependent manner in yeast	118–120
5. Factors interacting with nuclear receptors					
In a ligand- and/or AF-2-dependent manner					
RIP140	hERAP140 mRIP140	Yes	sw:P48542 AF053062	Interacts with and coactivates ER, but corepresses TR2 orphan receptor and retinoid receptors	121, 122
ARA70	hELE1a hRFG hELE1b	Unclear	L49399	Reported to coactivate AR in DU145 cells; coactivates PPARγ activity in a ligand-dependent manner; possesses one LxxLL motif but this motif is not implicated in the interaction; isoform of hELE1a	123, 124
ARA55			AF116343	Interacts with AR through LIM domain	125
ARA54			AF049330	Another AR putative coactivator that contains a RING finger domain	126
TRIP1	mSUG1	Yes	L38810 Z54219	TRIP1 interacts with RXR, mSUG1 with different members of the NR superfamily; TRIP1 is a component of the 26S proteasome	127, 128

Name		Coactivator	Accession	Description	Ref.
ASC-1			AF168418	Interacts with basal transcription factors (TBP and TFIIA), transcription integrators (SRC-1 and CBP/p300) and nuclear receptors. Although the interaction domain with NR seems to be the hinge region, AF-2 integrity is required	129
ASC-2		Unclear	AF177388	Interacts with RAR, TR, ERα and GR and binds also TFIIA, TBP, SRC-1 and CBP/p300; microinjection of anti-ASC-2 antibody abrogated the ligand-dependent transactivation by retinoic acid receptor; is amplified in human cancers; contains two LxxLL motifs but implication in the interaction has not been demonstrated	130
p120		Unclear	AF016270	Interacts with TR LBD; possesses a LSELL motif in the NR interacting region	131
TIF1α	mTIF1a	Yes	AF009353 S78221	Interacts with several NR; belongs to the RBCC family of proteins; interacts also with KRAB domain-containing proteins, which are involved in determining heterochromatin structures	132, 128
NRIF3	EnL and EnS	Yes	AAD09135	Claimed to be a specific coactivator for TR and interacts through a variant of the NID motif LxxLL	133
Trip230		Yes		Binds to and coactivates TR; binds also to Rb	134
L7	SPA			Interacts with and coactivates RU486-bound PR and enhances its partial agonist activity	80
Tip60		Unclear	U74667	Interaction with AR LBD enhanced by the ligand; able to coactivate PR and ER in transfections; contains a LxxLL motif but its function has not been demonstrated	135
RAP250		Yes	AF128458	Ligand-dependent or ligand-enhanced interaction with nuclear receptors; interaction involves the only LxxLL NR box; has intrinsic glutamine-rich activation domain; widely expressed in reproductive organs	136
FHL2	DRAL	Unclear	NM_001450	Human four and a half LIM domains protein 2 (FHL2); claimed to act as tissue-specific coactivator of the androgen receptor; nuclear protein; expressed in myocardium and prostate epithelium; contains	137

Continued

Table 5 *Continued.*

Human factors	Synonyms	NID motif	Accession number[a]	Comments	Reference
				autonomous transactivation function; coactivates AR but not other NRs	
In a ligand-independent manner					
MBF1	hMBF1a hMBF1b		AB002282 AB002283	Example of a coactivator conserved throughout evolution; inter-action implicates the DBD and increases DNA binding	138
ARA160	TMF		NM_007114	Interacts with the AB domain of AR; however, the ligand enhances interaction, suggesting that ARA160 may work in association with ARA70 to coactivate AR	139
ORCA	hORCA		U46751	Factor that binds COUP-TFII *in vitro* and allows COUP-TFII to function as a transcriptional activator in mammalian cells; identical to a recently described ligand of the tyrosine kinase signaling molecule p56|lck), suggesting that it mediates cross-talk between mitogenic and NR signal transduction pathways	140
PGC-1	mPGC-1	Unclear	AF049330	First report of an inducible coactivator for a nuclear receptor, originally described for PPARγ but interacts also with other NRs; interaction is ligand independent and implicates the DBD and the hinge region of the receptor; NID motif is not required for PPAR binding	141
PGC-2	mPGC-2		AF017433	Isolated using the AF-1 region of PPARγ as bait; contains a partial SCAN domain, binds to and increases the transcriptional activity of PPARγ, but does not interact with other PPARs or most other nuclear receptors	142
SNURF	rSNURF		AF022081	Isolated by two-hybrid interaction using AR DBD as bait; inter-action is ligand dependent but SNURF is able to stimulate both basal and ligand-induced transcription	143
NCoA-62			AF045184	Fished out with VDR as bait; interacts with the LBD even in the	144

RAP46	hBAG-1 hBAG-1L		Z35491	absence of the ligand and coactivates in mammalian cells vitamin D_3,retinoic acid, estrogen- and glucocorticoid-mediated gene expression Originally identified as a protein able to interact with several nuclear receptors including GR, ER and TR; the ligand was not strictly required for interaction; subsequent cloning of a longer form, BAG-1L that coimmunoprecipitates and coactivates AR in PC3 transfections; RAP46 was described as a negative regulator of GR and RAR	145– 148
MIP224	TBP7		NM_006503	Isolated using the two-hybrid system and the orphan receptor MB67 as a bait; in transfection experiments MIP224 inhibits the transcriptional activation by MB67; MIP224 is a component of the 26S proteasome	149
PIAS1		Unclear	AF167160	Binds in yeast the DBD + LBD of AR in a ligand-dependent manner; in vitro it binds the isolated DBD; coactivates in mammalian cells AR and PR ligand-dependent transactivation but represses PR; first cloned as an inhibitor of STAT1; contains an LxxLL motif but its role in interaction has not been demonstrated	150
ARIP3	hPIASxy rARIP3	Unclear	AF044058	Binds the DBD of AR and modulates AR-dependent transcription. Belongs to the PIAS family of proteins; contains an LxxLL motif but its role in interaction has not been demonstrated	151
p68		Unclear	AF015812	Interacts with the A/B domain of hERα; phosphorylation of hERαSer118 potentiates the interaction with p68; enhances the AF-1 activity; contains an LxxLL motif but its role in interaction has not been demonstrated	66
With both liganded and non-liganded receptors					
NSD1	mNSD1	Yes	AF064553	Interacts through NID-L with unliganded RAR and TR and through NID + L with liganded RAR, TR, RXR and ER has properties of both a coactivator and a corepressor	83
zac1b				Variant of zac1, a putative transcriptional activator involved in	152

Continued

49

Table 5 *Continued.*

Human factors	Synonyms	NID motif	Accession number[a]	Comments	Reference
	zac1		NM_009938	apoptosis and cell-cycle regulation; isolated as a protein that binds to the C-terminal region of the coactivator TIF2/GRIP1; interacts with CBP, p300 and nuclear receptors in yeast two-hybrid and *in vitro* experiments; contains autonomous transactivation domain; acts as coactivator but in some cell lines and with certain promoters zac1b acts as repressor of nuclear receptor action	

6. Other factors that interact directly or indirectly with nuclear receptors

Human factors	Synonyms	NID motif	Accession number[a]	Comments	Reference
STAT3	mSTAT3		U30709	Associates with ligand-bound glucocorticoid receptor to form a transactivating/signaling complex, which can function through either an IL-6-responsive element or a glucocorticoid-responsive element	153
STAT5	mSTAT5		U48730 U21110	GR can act as a transcriptional coactivator for Stat5 and enhance Stat5-dependent transcription	154
E6-AP	mE6-AP	Unclear	AF016708 U96636	Interacts with and coactivates the transcriptional activity of PR in a ligand-dependent manner; however, in the Angelman syndrome, the phenotype results from the defect in the ubiquitin–proteosome-mediated degradation of E6-AP	155
cyclinD1		Unclear	M64349	Direct physical binding of cyclin D1 to the LBD increases binding of the receptor to this element and either activates transcription in the absence of estrogen or enhances transcription in its presence	156
HMG-1	mHMG-1		X12597 U00431	Coregulatory proteins that increase the DNA binding and transcriptional activity of steroid NR	157
HMG-2	mHMG-2		X62534 U00431	cf HMG-1	157
Bcl3		Unclear	U05681	Able to interact with RXR both in a ligand-independent manner	158

Factor	Alias	NID	EMBL/GenBank[a]	Function	Ref
				with a fragment comprising the ABC domain and in a ligand-dependent manner with the ligand-binding domain; interacts with the general transcription factors TFIIB, TBP and TFIIA, but not with TFIIEα in the GST pull-down assays and enhances the 9cis-RA-induced transactivation of RXR; possesses an LxxLL motif but mutation in the motif does not abolish interaction	
Vpr		Yes	U71182	Interacts directly with the GR and general transcription factors, acting as a coactivator via a LxxLL signature motif	159
TLS			AF071213	interaction with the DBD of RXR, ER, TR and GR	160
dUTPase			U62891	PPARα-interacting protein. Interacts in vitro with all three isoforms of mouse PPAR, but not with RXR and TR; this interaction seems to inhibit PPAR/RXR, resulting in an inhibition of PPAR activity in a ligand-independent manner	161
Smad3	mSmad3		AB004930 AF016189.1t	Specific coactivator for ligand-induced transactivation of VDR by forming a complex with a member of the steroid receptor coactivator-1 protein family	162
E1A				Interacts directly with TR via the DBD and carboxy-terminal part of the LBD in a ligand-independent manner	163
hnRNP U	GRIP120 mhnRNP U		NM_004501 AF073992	Scaffold attachment region and RNA-binding protein that coimmunoprecipitates with GR	164
TIF1β	KAP1		NM_005762	Enhances the transcription of the agp gene by GR in a ligand- and GRE-dependent manner, even if the interaction is ligand independent	165
TDG			NM_003211	Implicated in the reparation of T:G mismatch; interacts through a region containing α-helix1 of the LBD with RAR and RXR	166
TSC-2			NM_000548	Cloned by interaction with RXR in yeast; coactivates PPARγ and VDR in mammalian cells but no clear demonstration of a direct interaction	167

NID, nuclear receptor-interacting domain.
[a] EMBL/GenBank.

coactivators contain multiple LxxLL motifs (up to nine in RIP140), most of which appear to be functional in terms of nuclear receptor binding *in vitro*, brought up the still unsolved question of whether this multiplicity reflects redundancy or may confer some specificity to the interface. Indeed, different coactivators, even highly related coactivators from the TIF2/SRC1/RAC3 family, display some degree of nuclear receptor selectivity[48,52-54]. Residues flanking the LxxLL motif are believed to make contact with the charged residues (the 'charge clamp') at both extremities of the hydrophobic cleft on the holo-ligand binding domain and generate some receptor selectivity. However, these selectivities have still to be confirmed by studies *in vivo*. In this respect the analysis of different coactivator knockout mice will be particularly instructive. Initial analysis of the TIF2 and SRC1 knockout mice do actually support the concept that bona fide coactivators are not fully redundant *in vivo*.

Recently, a second contact site for coactivators has been identified in nuclear receptors. Proteins from the TIF2 family are able to interact specifically with the A/B domains of estrogen and androgen receptor[55-59]. These interactions result in a stimulation of the transcriptional activity originating from AF-1. Moreover, it appears that simultaneous interaction of coactivators with both the AF-1 and the AF-2 of a nuclear receptor accounts for the synergy between both transcriptional activation functions[60,61]. It is thus possible that the activation functions AF-1 and AF-2 may not in all cases be independent of each other in the context of the entire receptor, but rather constitute two separate regions of a single coactivator recruiting surface. Hence, the coactivator specificity will have to be reanalyzed, particularly for receptors with proven AF-1 activities, taking into account the effect of the AF-1-binding surface(s). In this respect it should also be noted that, for some nuclear receptors, the N- and C-terminal domains interact with each other even in the absence of coactivators[56,58,62]. In keeping with the great sequence divergence between the different A/B domains of nuclear receptors, the coactivators do not invoke defined short interaction motifs similar to the LxxLL NR box for AF-1 recognition; however, the 'Q-rich' region C-terminal of TIF2 AD1[63,64] appears to be critically involved in AF-1 interaction[61]. The structural features of the A/B domains that are recognized by the coactivator have not yet been defined. Notably, it has been reported for ERβ that specific phosphorylation of the A/B domain by MAPK pathways results in enhanced recruitment of the coactivator SRC-1[65]. In addition to the bona fide coactivators, other putative coactivator proteins have been reported to interact with AF-1 and contribute to its activity (see Table 5). An interesting candidate is p68[66], which interacts with the A/B domain of human ERα; phosphorylation of human ERα Ser$_{118}$ potentiates the interaction with p68.

The identification of specific nuclear receptor coactivators has prompted the question of how they function on a molecular level in transcription. To this end, several observations have been made. It is now generally accepted that nuclear receptor coactivators possess or recruit enzymatic activities, and that they form large coactivator complexes. CBP, p300, P/CAF, SRC-1, P/CIP and GCN5 are reported to act as histone acetyltransferases (for recent reviews, see references 1 and 2). They are capable of acetylating specific residues in the N-terminal tails of different histones, a process that is believed to play an important role in the opening of chromatin during transcription activation, and also non-histone targets[67-70]. Note, however, that the histone acetyltransferase (HAT) activity of

A. REPRESSION

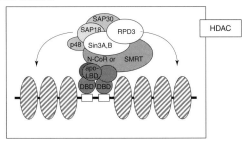

B. DEREPRESSION

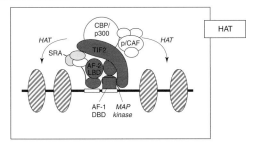

C. TRANSCRIPTION ACTIVATION

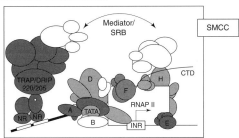

SMCC/SRB/MED-COMPLEX

Figure 10 *Three-step mechanism of nuclear receptor action. Some nuclear receptors act as silencers of target gene transcription in the absence of ligand (or in the presence of certain antagonists). This 'repression' step (A) is due to the recruitment by the apo-NR of a corepressor complex that exerts histone deacetylase (HDAC) activity. Note that for repression to occur by this mechanism the receptor has to be able to interact with its target gene promoter in the absence of ligand. Ligand binding dissociates this complex and recruits the coactivator complex that displays histone acetyltransferase (HAT) activity. The subsequent chromatin decondensation ('derepression'; B) is believed to be necessary, but not sufficient, for target gene activation. In the third step (C), the HAT complex dissociates, possibly due to acetylation of the coactivator, which decreases its ability to interact with the receptor, and the SMCC/DRIP/TRAP complex is assembled through its TRAP220/DRIP205 subunit. The SMCC complex is able to establish contacts with the basal transcription machinery, resulting in transcription initiation. Modified from Freedman (1999)[84].*

SRC-1 and P/CIP, if real, is negligible compared to that of p300 or CBP. Thus, like TIF2 and RIP140, they may rather recruit such activities by physical association with histone acetyl transferases or complexes containing such activities. Specifically, the AD1 of TIF2 has been demonstrated to function via the recruitment of cointegrator CBP[64], which apparently in turn acetylates TIF2[70]. Besides HAT activities, other enzymatic activities have also been attributed to nuclear receptor coactivator complexes. TIF2 proteins are capable of interacting functionally via their activation domain AD2 with a protein methyltransferase[71]. Although unproven, it is believed that the methyltransferase activity changes the activity of the basal transcription mechanism, and it will be of great interest to identify its specific targets.

In conclusion, bona fide coactivators, i.e. members of the TIF2/SRC-1/RAC3 family, together with the CBP/p300 cointegrators (see Table 5) function by rendering the chromatin environment of a nuclear receptor target gene prone to transcription. This opening of the chromatin environment is achieved by intrinsic or recruited HAT activity. The HAT activities of different coactivators/coactivator complexes target: (1) the N-termini of histones, which have reduced DNA-binding activity upon acetylation; (2) certain basal transcription factors; and (3) at least some bona fide coactivators themselves. The chromatin modification step represents the first of two independent steps in transcription activation by nuclear receptors (see Fig. 10 for an illustration of the 'derepression' and further below for the second step).

NUCLEAR RECEPTOR COREPRESSORS, SILENCING AND HDACS

The second class of nuclear receptor coregulators comprises the corepressors (Table 5). Early on, it had been found that some nuclear receptors do actively repress transcription when in the apo-form. This phenomenon had been particularly well established for retinoic acid and thyroid hormone receptors[72-76] (and references therein). Soon after the identification of ligand-recruited coactivators, similar approaches identified proteins that recognize the ligand-free nuclear receptor. To date, several different nuclear receptor corepressors have been identified (Table 5), but by far the most studies have been performed with NCoR and SMRT. For these, it was shown recently that a conserved CoRNR box motif interacts with a surface on the ligand-binding domain that is topologically very similar to that recognized by coactivator LxxLL motifs, but does not involve helix H12[77-79]. Corepressors are believed to reside in, or recruit, high-molecular-weight complexes that display the opposite activity of coactivator complexes. While coactivator complexes acetylate histones, corepressors recruit histone deacetylase activities that reverse this process (illustrated in Fig. 10 as the 'repression' step). Deacetylated histones are associated with silent regions of the genome, and it is generally accepted that histone acetylation and deacetylation shuffle nucleosomal targets between a condensed and relaxed chromatin configuration, the latter being a requisite for transcriptional activation. An unresolved issue is whether all nuclear receptors are capable of active repression. In concert with this observation, recent evidence has been presented that some steroid hormone receptors also bind to corepresssors in the presence of certain antagonists[80–82].

An intriguing example of a nuclear receptor coregulator is NSD-1[83]. This factor has been reported to contain both, transcriptional activation and repression domains. NSD-1 is a SET family protein and contains coactivator-type LxxLL motifs. It will be of interest to further elucidate the molecular mechanisms by which this factor is able to switch from the corepression to the coactivation state upon ligand binding to the associated nuclear receptor.

Given this large number of different coregulators for nuclear receptors, two principal questions emerge: (1) what defines coactivator selectivity, and (2) how is the assembly of different coactivator complexes with different intrinsic transcription activities regulated? Future research will have to address such questions, especially in view of the therapeutic perspectives on disease.

RECRUITMENT OF THE RNA POLYMERASE II HOLOENZYME – THE SECOND STEP OF COACTIVATION

The initial chromatin-modifying step carried out by nuclear receptor coactivators (see above) is believed to be followed by the actual recruitment of the RNA polymerase II holoenzyme and initiation of transcription (illustrated in Fig. 10 as the 'transcription activation' step; for a recent review, see Freedman[84]). Comprehension of the recruitment of the polymerase II holoenzyme by nuclear receptors has only become possible through the identification and cloning of the mammalian mediator complex as a thyroid hormone and vitamin D receptor coactivator[85–87]. The mammalian mediator came in several versions. It was identified as the so-called SMCC (the Srb and Mediator protein-containing complex[88]), the TRAP complex (a thyroid hormone receptor-associated protein complex[85]), or the DRIP complex (a vitamin D receptor interacting protein complex[86]). Furthermore, common subunits are shared with PC2, the so-called positive coactivator 2[89], the ARC[90], CRSP[91] and NAT[92] complexes. Hereafter this complex is referred to as 'SMCC'. It is a large multisubunit protein complex that contains several homologues of the yeast mediator complex (RGR1, MED6, MED7, SRB7, SRB10, SRB11, NUT2, SOH1[87,88] as well as additional proteins of unknown function. As expected for a mediator complex, SMCC associates with the RNA polymerase II to form RNA pol II holoenzymes[87]. However, the SMCC complex is able to interact functionally with different transcription factors such as p53 and VP16[87]. Furthermore, owing to its identification as a thyroid hormone and vitamin D receptor interacting complex, it is believed to function as a nuclear receptor coactivator. This notion is supported by the demonstration that SMCC can enhance thyroid hormone and vitamin D transcription activation in transcription systems *in vitro*[86,93,94]. The subunit of the complex that is responsible for interaction with the agonist-bound ligand-binding domain of nuclear receptors was identified as DRIP205[95], which is identical to TRAP220 and contains a functional LxxLL NR box motif[96]. Interestingly, another subunit of the SMCC complex, DRIP150, interacts with the N-terminal region of the glucocorticoid receptor, which harbors the activation function AF-1[97]. The current working hypothesis is that, once the chromatin environment at target gene promoters has been decondensed by coactivators complexes containing members of the TIF2 and CBP families, the nuclear receptor recruits RNA pol II holoenzymes via its association with the TRAP220/DRIP205 subunit of the SMCC. This switch between coactivators and the SMCC complex might be regulated by the acetylation of coactivators within the HAT complex[70], resulting in their dissociation from the nuclear receptor, thus allowing the recruitment of factors such as SMCC via the LxxLL motif of the TRAP220/DRIP205 subunit.

Table 6 *Factors of the basal machinery reported to interact, directly or indirectly, with nuclear receptors*

Basal factor		References
TBP		
RXR	Interaction requires ligand and intact AF-2	168
ER	Interaction does not require ligand or intact AF-2	169
TR	Interaction could be involved in repression	170
TF$_{II}$B		
COUP-TFI, ER and PR	COUP–TFI interaction is direct; ER interaction seems to involve AF-2 function	171
TR	Interaction seems to promote either AF-1 activation or AF-2 repression	172
VDR	Cell specific; depending on the cell type, in the presence of ligand, promotes either activation of transcription (P19 cells) or repression (3T3 cells); two-hybrid interaction and *in vitro* techniques have been used to show this interaction; however, the requirement for ligand in this interaction is controversial	173–175
TF$_{II}$D		
hTAF$_{II}$30:ER	Interaction requires the DEF region of the receptor; ligand independent; an antibody directed against hTAFII30 inhibits ER transactivation	176
dTAF$_{II}$110: PR	Interaction implicates the DBD of the receptor and is ligand independent	177
dTAF$_{II}$110: RXR	Interact in the two-hybrid system in a AF-2 and ligand-dependent manner but not *in vitro*	168
dTAF$_{II}$110: TR	Interact both *in vitro* and in the two-hybrid system; but in the first case the ligand inhibits *in vitro* but has no effect in the two-hybrid system	178
hTAF$_{II}$28	Stimulates AF-2 function of RXR, ER, VDR and RAR in transfection but no direct interaction has been shown	179
hTAF$_{II}$135	Stimulates AF-2 function of RAR, VDR and TR, but no direct interaction has been shown	180
hTAF$_{II}$55	Interacts with α-helix H3 to H5 of VDR and TR in a ligand-independent manner	181
TF$_{II}$H		
RAR	Binds TF$_{II}$H and free CAK *in vitro*; cdk7 phosphorylation of Ser77, located in the A/B region enhances transcriptional activation	98

The recruitment of the RNA polymerase II holoenzyme might also be enhanced by interactions of nuclear receptors with components of the SWI/SNF complex, which is part of RNA pol II holoenzymes (see Table 5 for a complete list of interactions to date and relevant publications).

INTERACTION OF NUCLEAR RECEPTORS WITH COMPONENTS OF THE BASAL TRANSCRIPTION MACHINERY

The first hypotheses about possible mechanisms of transcription activation by some nuclear receptors were based on *in vitro* studies demonstrating direct interactions with components of the basal machinery. Amongst those are the TATA binding protein (TBP) and the TBP-associated factors (for a detailed list and a short description, see Table 6). These interactions were reported to promote the transcriptional activity of nuclear receptors, at least in transient transfection studies. Whether and how actively the nuclear receptors recruit independently both the holoenzyme and the TAF complex(es) during the second activation step remains to be established. $TF_{II}H$ has been found to phosphorylate certain nuclear receptors in their N-terminal AF-1[98]. This is interesting in view of the fact that nuclear receptor coactivators interact sometimes in a phosphorylation-dependent manner with the receptor AF-1 activation functions[65,99].

Taken together, the allosteric changes induced by nuclear receptor–ligand interaction define protein interaction and recruitment cascades that lead to the assembly of complexes, which can: (1) modify the status of chromatin condensation due to the complex-associated enzymatic activities; and (2) recruit, or allow or enhance the formation of, transcription initiation complexes at target gene promoters.

References

[1] Chen, J.D. (2000) Vitam. Horm. 58, 391–448.
[2] Glass, C.K. and Rosenfeld, M.G. (2000) Genes Dev. 14, 121–141.
[3] Giguere, V. et al. (1986) Cell 46, 645–652.
[4] Godowski, P.J. (1987) Nature 325, 365–368.
[5] Gronemeyer, H. et al. (1987) EMBO J. 6, 3985–3994.
[6] Miesfeld, R. et al. (1987) Science 236, 423–427.
[7] Godowski, P.J. et al. (1988) Science 241, 812–816.
[8] Hollenberg, S.M. and Evans, R.M. (1988) Cell 55, 899–906.
[9] Webster, N.J. et al. (1988) Cell 54, 199–207.
[10] Lees, J.A. et al. (1989) Nucl. Acids Res. 17, 5477–5488.
[11] Lees, J.A. et al. (1989) J. Steroid Biochem. 34, 33–39.
[12] Metzger, D. et al. (1995) J. Biol. Chem. 270, 9535–9542.
[13] Gronemeyer, H. and Laudet, V. (1995) Protein Profile 2, 1173–1308.
[14] Mangelsdorf, D.J. et al. (1995) Cell 83, 835–839.
[15] Zenke, M. et al. (1990) Cell 61, 1035–1049.
[16] Danielian, P.S. et al. (1992) EMBO J. 11, 1025–1033.
[17] Baretino, D. et al. (1994) EMBO J. 13, 3039–3049.
[18] Durand, B. et al. (1994) EMBO J. 13, 5370–5382.
[19] Leng, X. et al. (1995) Mol. Cell Biol. 15, 255–263.

20 Pierrat, B. et al. (1994) Gene 143, 193–200.
21 Webster, N.J. et al. (1989) EMBO J. 8, 1441–1446.
22 Meyer, M.E. et al. (1992) J. Biol. Chem. 267, 10882–10887.
23 Dahlman-Wright, K. et al. (1994) Proc. Natl Acad. Sci. USA 91, 1619–1623.
24 Dahlman-Wright, K. et al. (1995) Proc. Natl Acad. Sci. USA 92, 1699–1703.
25 Shemshedini, L. et al. (1992) J. Biol. Chem. 267, 1834–1839.
26 Sartorius, C.A. et al. (1994) Mol. Endocrinol. 8, 1347–1360.
27 Tora, L. et al. (1988) Nature 333, 185–188.
28 Kastner, P. et al. (1990) J. Biol. Chem. 265, 12163–12167.
29 Conneely, O.M. et al. (2000) J. Soc. Gynecol. Investig. 7, S25–32.
30 Horwitz, K.B. et al. (1995) J. Steroid Biochem. Mol. Biol. 53, 9–17.
31 Lange, C.A. et al. (1999) Mol. Endocrinol. 13, 829–836.
32 Bain, D.L. et al. (2000) J. Biol. Chem. 275, 7313–7320.
33 Nagpal, S. et al. (1992) Cell 70, 1007–1019.
34 Berry, M. et al. (1990) EMBO J. 9, 2811–2818.
35 Gill, G. and Ptashne, M. (1988) Nature 334, 721–724.
36 Bocquel, M.T. (1989) Nucleic Acids Res. 17, 2581–2595.
37 Meyer, M.E et al. (1989) Cell 57, 433–442.
38 Tasset, D. et al. (1990) Cell 62, 1177–1187.
39 Chen J.D. and Li, H. (1998) Crit. Rev. Eukaryot. Gene Expr. 8, 169–190.
40 Moras, D. and Gronemeyer, H. (1998) Curr. Opin. Cell Biol. 10, 384–391.
41 Torchia, J. et al. (1998) Curr. Opin. Cell Biol. 10, 373–383.
42 Collingwood, T.N. et al. (1999) J. Mol. Endocrinol. 23, 255–275.
43 McKenna et al. (1999) Endocrinol. Rev. 20, 321–344.
44 Xu, L. et al. (1999) Curr. Opin. Genet. Dev. 9, 140–147.
45 Le Douarin, B. et al. (1996) EMBO J. 15, 6701–6715.
46 Heery, D.M. et al. (1997) Nature 387, 733–736.
47 Torchia, J. et al. (1997) Nature 387, 677–684.
48 Darimont, B.D. et al. (1998) Genes Dev. 12, 3343–3356.
49 Nolte, R.T. et al. (1998) Nature 395, 137–143.
50 Shiau, A.K. et al. (1998) Cell 95, 927–937.
51 Gampe, R.T. Jr et al. (2000) Mol. Cell 5, 545–555.
52 Ding, X.F. et al. (1998) Mol. Endocrinol. 12, 302–313.
53 McInerney, E.M. et al. (1998) Genes Dev. 12, 3357–3368.
54 Hong, H. et al. (1999) J. Biol. Chem. 274, 3496–3502.
55 Berrevoets, C.A. et al. (1998) Mol. Endocrinol. 12, 1172–1183.
56 Alen, P. et al. (1999) Mol. Cell Biol. 19, 6085–6097.
57 Bevan, C.L. et al. (1999) Mol. Cell Biol. 19, 8383–8392.
58 He, B. et al. (1999) J. Biol. Chem. 274, 37219–37225.
59 Ma, H. et al. (1999) Mol. Cell Biol. 19, 6164–6173.
60 McInerney, E.M. et al. (1996) Proc. Natl Acad. Sci. USA 93, 10069–10073.
61 Benecke, A. et al. (2000) EMBO Rep. 1, 151–157.
62 Tetel, M.J. et al. (1999) Mol. Endocrinol. 13, 910–924.
63 Voegel, J.J. et al. (1996) EMBO J. 15, 3667–3675.
64 Voegel, J.J. et al. (1998) EMBO J. 17, 507–519.
65 Tremblay, A. et al. (1999) Mol. Cell 3, 513–519.
66 Endoh, H. et al. (1999) Mol. Cell Biol. 19, 5363–5372.
67 Bayle, J.H. and Crabtree, G.R. (1997) Chem. Biol. 4, 885–888.
68 Gu, W. and Roeder, R.G. (1997) Cell 90, 595–606.

[69] Imhof, A. et al. (1997) Curr. Biol. 7, 689–692.

[70] Chen, H. et al. (1999) Cell 98, 675–686.

[71] Chen, D. et al. (1999) Science 284, 2174–2177.

[72] Baniahmad, A. et al. (1990) Cell 61, 505–514.

[73] Baniahmad, A. et al. (1992) EMBO J. 11, 1015–1023.

[74] Burcin, M. et al. (1994) Semin Cancer Biol. 5, 337–346.

[75] Chen, J.D. and Evans, R.M. (1995) Nature 377, 454–457.

[76] Hörlein, A.J. et al. (1995) Nature 377, 397–404.

[77] Hu, X. and Lazar, M.A. (1999) Nature 402, 93–96.

[78] Nagy, L. et al. (1999) Genes Dev. 13, 3209–3216.

[79] Perissi, V. et al. (1999) Genes Dev. 13, 3198–3208.

[80] Jackson, T.A. et al. (1997) Mol. Endocrinol. 11, 693–705.

[81] Wagner, B.L. et al. (1998) Mol. Cell Biol. 18, 1369–1378.

[82] Zhang, X. et al. (1998) Mol. Endocrinol. 12, 513–524.

[83] Huang, N. et al. (1998) EMBO J. 17, 3398–3412.

[84] Freedman, L.P. (1999) Cell 97, 5–8.

[85] Fondell, J.D. et al. (1996) Proc. Natl Acad. Sci. USA 93, 8329–8333.

[86] Rachez, C. et al. (1998) Genes Dev. 12, 1787–1800.

[87] Ito, M. et al. (1999) Mol. Cell 3, 361–370.

[88] Gu, W. et al. (1999) Mol. Cell 3, 97–108.

[89] Kretzschmar, M. et al. (1994) Mol. Cell Biol. 14, 3927–3937.

[90] Näär, A.M. et al. (1999) Nature 398, 828–832.

[91] Ryu, S. and Tjian, R. (1999) Proc. Natl Acad. Sci. USA 96, 7137–7142.

[92] Sun, X. (1998) Mol. Cell 2, 213–222.

[93] Fondell, J.D. et al. (1999) Proc. Natl Acad. Sci. USA 96, 1959–1964.

[94] Rachez, C. et al. (1999) Nature 398, 824–828.

[95] Yang, W. and Freedman, L.P. (1999) J. Biol. Chem. 274, 16838–16845.

[96] Yuan, C.X. et al. (1998) Proc. Natl Acad. Sci. USA 95, 7939–7944.

[97] Hittelman, A.B. et al. (1999) EMBO J. 18, 5380–5388.

[98] Rochette-Egly, C. et al. (1997) Cell 90, 97–107.

[99] Hammer, G.D. et al. (1999) Mol. Cell 3, 521–526.

[100] Cavaillès, V. et al. (1994) Proc. Natl Acad. Sci. USA 9, 10009–10013.

[101] Onate, S.A. et al. (1995) Science 270, 1354–1357.

[102] Hong, H. et al. (1997) Mol. Cell Biol. 17, 2735–2744.

[103] Anzick, S.L. et al. (1997) Science 277, 965–968.

[104] Li, H. et al. (1997) Proc. Natl Acad. Sci. USA 94, 8479–8484.

[105] Takeshita, A. et al. (1997) J. Biol. Chem. 272, 27629–27634.

[106] Torchia, J. et al. (1996) Nature 387, 677–684.

[107] Kamei, Y. et al. (1996) Cell 85, 403–414.

[108] Blanco, J.C. et al. (1998) Genes Dev. 12, 1638–1651.

[109] Lanz, R.B. et al. (1999) Cell 97, 17–27.

[110] Hörlein, A.J. et al. (1995) Nature 377, 397–404.

[111] Zamir, I. et al. (1996) Mol. Cell Biol. 16, 5458–5465.

[112] Zamir, I. et al. (1997) Proc. Natl Acad. Sci. USA 94, 14400–14405.

[113] Burris, T.P. et al. (1995) Proc. Natl Acad. Sci. USA 92, 9525–9529.

[114] Montano, M.M. et al. (1999) Proc. Natl Acad. Sci. USA 96, 6947–6952.

[115] Dressel, U. et al. (1999) Mol. Cell Biol. 19, 3383–3394.

[116] Zhu, Y. et al. (1997) J. Biol. Chem. 272, 25500–25506.

[117] Zhang, J. and Fondell, J.D. (1999) Mol. Endocrinol. 13, 1130–1140.

[118] Chiba, H. et al. (1994) Nucl. Acids Res. 22, 1815–1820.

[119] Ichinose, H. et al. (1997) Gene 188, 95–100.

[120] Muchardt, C. and Yaniv, M. (1993) EMBO J. 12, 4279–4290.

[121] Cavailles, V. et al. (1995) EMBO J. 14, 3741–3751.

[122] Halachmi, S. et al. (1994) Science 264, 1455–1458.

[123] Gao, T. et al. (1999) Mol. Endocrinol. 13, 1645–1656.

[124] Yeh, S. and Chang, C. (1996) Proc. Natl Acad. Sci. USA 93, 5517–5521.

[125] Fujimoto, N. et al. (1999) J. Biol. Chem. 274, 8316–8321.

[126] Kang, H.Y. et al. (1999) J. Biol. Chem. 274, 8570–8576.

[127] Lee, J.W. et al. (1995) Nature 374, 91–94.

[128] vom Baur, E. et al. (1996) EMBO J. 15, 110–124.

[129] Kim, H.J. et al. (1999) Mol. Cell Biol. 19, 6323–6332.

[130] Lee, S.K. et al. (1999) J. Biol. Chem. 274, 34283–34293.

[131] Monden, T. et al. (1999) Mol. Endocrinol. 13, 1695–1703.

[132] Le Douarin, B. et al. (1995) EMBO J. 14, 2020–2033.

[133] Li, D. et al. (1999) Mol. Cell Biol. 19, 7191–7202.

[134] Chang, K.H. et al. (1997) Proc. Natl Acad. Sci. USA 94, 9040–9045.

[135] Brady, M.E. et al. (1999) J. Biol. Chem. 274, 17599–17604.

[136] Caira, F. et al. (2000) J. Biol. Chem. 275, 5308–5317.

[137] Muller, J.M. et al. (2000) EMBO J. 19, 359–369.

[138] Kabe, Y. et al. (1999) J. Biol. Chem. 274, 34196–34202.

[139] Hsiao, P.W. and Chang, C. (1999) J. Biol. Chem. 274, 22373–22379.

[140] Marcus, S.L. et al. (1996) J. Biol. Chem. 271, 27197–27200.

[141] Puigserver, P. et al. (1998) Cell 92, 829–839.

[142] Castillo, G. et al. (1999) EMBO J. 18, 3676–3687.

[143] Moilanen, A.M. et al. (1998) Mol. Cell Biol. 18, 5128–5139.

[144] Baudino, T.A. et al. (1998) J. Biol. Chem. 273, 16434–16441.

[145] Froesch, B.A. et al. (1998) J. Biol. Chem. 273, 11660–11666.

[146] Kullmann, M. et al. (1998) J. Biol. Chem. 273, 14620–14625.

[147] Liu, R. et al. (1998) J. Biol. Chem. 273, 16985–16992.

[148] Zeiner, M. et al. (1997) EMBO J. 16, 5483–5490.

[149] Choi, H.S. et al. (1996) J. Steroid Biochem. Mol. Biol. 56, 23–30.

[150] Tan, J. et al. (2000) Mol. Endocrinol. 14, 14–26.

[151] Moilanen, A.M. et al. (1999) J. Biol. Chem. 274, 3700–3704.

[152] Huang, S.M. and Stallcup, M.R. (2000) Mol. Cell Biol. 20, 1855–1867.

[153] Zhang, Z. et al. (1997) J. Biol. Chem. 272, 30607–30610.

[154] Stocklin, E. et al. (1996) Nature 383, 726–728.

[155] Nawaz, Z. et al. (1999) Mol. Cell Biol. 19, 1182–1189.

[156] Zwijsen, R.M.L. et al. (1997) Cell 88, 405–415.

[157] Boonyaratanakornkit, V. et al. (1998) Mol. Cell Biol. 18, 4471–4487.

[158] Na, S.Y. et al. (1998) J. Biol. Chem. 273, 30933–30938.

[159] Kino, T. et al. (1999) J. Exp. Med. 189, 51–62.

[160] Powers, C.A. et al. (1998) Mol. Endocrinol. 12, 4–18.

[161] Chu, R. et al. (1996) J. Biol. Chem. 271, 27670–27676.

[162] Yanagisawa, J. et al. (1999) Science 283, 1317–1321.

[163] Wahlstrom, G.M. et al. (1999) Mol. Endocrinol. 13, 1119–11129.

[164] Eggert, M. et al. (1997) J. Biol. Chem. 272, 28471–28478.

[165] Chang, C.J. et al. (1998) Mol. Cell Biol. 18, 5880–5887.

[166] Um, S. et al. (1998) J. Biol. Chem. 273, 20728–20736.

167 Henry, K.W. et al. (1998) J. Biol. Chem. 273, 20535–20539.
168 Schulman, I.G. et al. (1995) Proc. Natl Acad. Sci. USA 92, 8288–8292.
169 Sadovsky, Y. et al. (1995) Mol. Cell Biol. 15, 1554–1563.
170 Fondell, J.D. et al. (1993) Genes Dev. 7, 1400–1410.
171 Ing, N.H. et al. (1992) J. Biol. Chem. 267, 17617–17623.
172 Baniahmad, A. et al. (1993) Proc. Natl Acad. Sci. USA 90, 8832–8836.
173 Blanco, J.C. et al. (1995) Proc. Natl Acad. Sci. USA 92, 1535–1539.
174 MacDonald, P.N. et al. (1995) J. Biol. Chem. 271, 19774–19780.
175 Masuyama, H. et al. (1997) Mol. Endocrinol. 11, 218–228.
176 Jacq, X. et al. (1994) Cell 79, 107–117.
177 Schwerk, C. et al. (1995) J. Biol. Chem. 270, 21331–21338.
178 Petty, K.J. et al. (1996) Mol. Endocrinol. 10, 1632–1645.
179 May, M. et al. (1996) EMBO J. 15, 3093–3104.
180 Mengus, G. et al. (1997) Genes Dev. 11, 1381–1395.
181 Lavigne, A.C. et al. (1996) J. Biol. Chem. 271, 19774–19780.

5 RXR subordination in heterodimers

While RAR agonists can autonomously activate transcription through RAR–RXR heterodimers, RXR is unable to respond to RXR-selective agonists in the absence of a RAR ligand. This phenomenon is generally referred to a RXR 'subordination' or RXR 'silencing' by apoRAR. In the presence of a RAR agonist or certain RAR antagonists, RXR agonists further stimulate the transcriptional activity of RAR–RXR heterodimers in a synergistic manner (Chen[1]; and references therein). RXR subordination solves a potential problem that could arise from the ability of RXR to act as a promiscuous heterodimerization partner for numerous nuclear receptors. This could create a problem for signaling pathway identity because, in the absence of subordination, RXR ligands could simultaneously activate multiple heterodimeric receptors, such as RAR–RXR, TR–RXR and VDR–RXR and confuse retinoic acid, thyroid hormone and vitamin D_3 signaling, respectively. However, retinoic acid (RA)-deprived animals do not exhibit abnormalities that could be readily related to impaired thyroid hormone or vitamin D_3 signaling. Moreover, RXR-selective ligands on their own were unable to trigger RXR–RAR heterodimer-mediated RA-induced events in various cell systems[1-3]. This is not due to an inability of the RXR partner to bind its cognate ligand in DNA-bound heterodimers, as has been previously suggested[4], because RXR ligand binding has been demonstrated to occur in such complexes in several studies *in vitro*, and synergistic transactivation induced by RAR and RXR-selective ligands has been observed *in vivo*[1-3,5-8]. Moreover, the AF-2 of RXR is important for signaling *in vivo*[12].

All of these observations rule out the initial concept that RXR is a priori a transcriptionally 'silent' partner in RAR–RXR heterodimers[4,9]. Rather, RAR apparently 'controls' the activity of RXR–RAR heterodimers in two ways: (1) it induces transcription in response to its own ligand; and (2) it silences RXR activity in the absence of an RAR ligand. Consequently, the only way for RXR to affect transactivation in response to its ligand in RXR–RAR heterodimers is through synergy with RAR ligands. This concept of RXR silencing may not apply to all NR partners, as the ligand-induced RXR activity was permissive in heterodimers with NGFI-B, even leading to a synergistic response[9,10]. However, neither the existence of an endogenous NGFI-B ligand nor a weak constitutive activity of the NGFI-B AF-2 can be excluded; both of these scenarios would readily explain RXR activity and NGFI-B–RXR synergy, owing to the absence of RXR silencing. Note that the same argument would apply to the important observation of a seemingly RXR ligand-induced activity of RXR–PPAR heterodimers[11,] which lead to insulin sensitization by RXR agonists in an animal model of non insulin-dependent diabetes.

References

1 Chen, J.Y. et al. (1996) Nature 382, 819–822.
2 Clifford, J. et al. (1996) EMBO J. 15, 4142–4155.
3 Taneja, R. et al. (1996) Proc. Natl Acad. Sci. USA 93, 6197–6202.
4 Kurokawa, R. (1994) Nature 371, 528–531.
5 Apfel, C.M. et al. (1995) J. Biol. Chem. 270, 30765–30772.
6 Kersten, S. et al. (1996) Biochemistry 35, 3816–3824.
7 Li, H. et al. (1997) Proc. Natl Acad. Sci. USA 94, 8479–8484.

8 Minucci, S. et al. (1997) Mol. Cell Biol. 17, 644–655.

9 Forman, B.M. et al. (1995) Cell 81, 541–550.

10 Perlmann, T. and Jansson, L. (1995) Genes Dev. 9, 769–782.

11 Mukherjee, R. et al. (1997) Nature 386, 407–410.

12 Mascrez, B. et al. (1998) Development 125, 4694–4707.

6 Antagonist action

In order to define or understand the agonistic/antagonistic features of a ligand, the following aspects need to be considered/elucidated.

1 Ligands may positively or negatively interfere with receptor activities at various levels. In the case of steroid hormone receptors, they may affect: (a) the stability of the so-called hetero-oligomeric 8S complex, which is proposed to exist in hormonally naive cells and comprises, in addition to steroid receptors, the Hsp90 heat-shock protein and additional associated factors (note that certain receptors, such as TR and RAR do not form complexes with Hsp90; (b) the homodimerization or heterodimerization ability of the receptor; and (c) its interaction with the cognate DNA response element.

2 Nuclear receptors harbor two activation functions (AF-1 and AF-2). These activation functions operate in a cell-type and promoter environment-dependent fashion. Thus, a given antagonist may antagonize only one or both AFs and an AF-2 antagonist can act as an AF-1 agonist.

3 While the structural basis of AF-1 activity is still unknown, AF-2s correspond to agonist-induced surfaces that can interact with coactivators. Conversely, non-liganded receptors express a surface(s) that can accommodate corepressors. A given ligand may more or less precisely generate these surfaces, and lead to different coregulator recruitment efficiencies. Thus 'superagonists' may enhance coactivator binding more efficiently than the natural ligand, while 'inverse agonists' may stabilize the receptor–corepressor complex.

4 Ligands may act at various levels in the sequence of events that lead to transcriptional activation or silencing. Theoretically, the same ligand may stimulate the recruitment of SMCC but not of the HAT complex. Indeed, initial evidence for the existence of ligands that differentially affect HAT and SMCC subunit recruitment has been provided[1].

SEVERAL TYPES OF ANTAGONISTS

From the above consideration, there are multiple aspects that have to be considered when analyzing the mechanism of action of an antagonist or when the aim is to design an antagonist with certain characteristics. Below we will discuss first some general principles originating from structural studies and then discuss particular mechanisms and individual antagonists. Note that some analyses should be considered in the context of the experimental setting because some receptor activities, such as DNA-binding, can be ligand independent *in vitro*, but are ligand dependent *in vivo* (e.g. see Dey[2]). Therefore, some of the results/categories described below may have to be reconsidered as more *in vivo* data accumulate.

STRUCTURAL BASIS OF AF-2 ANTAGONISTS: STERIC HINDRANCE BY THE LIGAND PRECLUDES THE HOLO-POSITION OF H12

Helix H12 is a crucial component of the nuclear receptor ligand-binding domains, because its ligand-induced repositioning in the holo-NR contributes in a critical manner to the surface recognized by the LxxLL NR boxes of coactivators and thereby generates the transcriptional activity of the AF-2 domain. The critical importance of the core of the AF-2 activity domain, a C-terminal LBD motif that displays features conserved in all AF-2-containing NRs, had been demonstrated by mutational analyses before the LBD structure was solved and was subsequently shown to correspond to the H12 amphipathic α-helix (for a review and references, see Gronemeyer and Laudet[3]). The first structures of apo- and holo-LBDs (Fig. 7) stressed the importance of the ligand-induced conformational changes, and immediately suggested that the interactions between H12 or residues in its proximity and the ligand were critical for the control of agonist–antagonist properties of nuclear receptors[4–6]. The crystal structures of the ERα LBD complexes with raloxifen and tamoxifen confirmed this hypothesis, and showed an alternative stable antagonist 'position' for H12 (cf. Fig. 8b), nested between H4 and H3[7–9]. Apparently, steric hindrance upon binding of the bulkier ligands (compared with the agonists) prevents the proper positioning of H12 in its 'agonistic' site.

In the antagonist conformation, a most important feature is the lengthening of the loop L11–12, resulting from the unwinding of the C-terminus of helix H11. This enables helix H12 to adopt a second low-energy position by binding to the coactivator LxxLL recognition cleft. In contrast to agonists that stabilize a long H11 helical conformation, different ligand–receptor interactions at the level of H11 and of the surrounding regions (loop L6–7 and H3) most likely explain the antagonist-induced unwinding of the C-terminal part of this helix. Note that these structural features are found in all antagonist-bound LBD complexes crystallized so far. Hence, it appears that the above described type of action of pure AF-2 antagonists originates from at least two structural principles. The major feature is the presence of a large 'antagonistic' ligand extension that sterically prevents the alignment of helix H12 in the holo-position. Without a holo-H12, no LBD–coactivator interface can be formed. The second structural principle is the unwinding of helix H11, which allows H12 to bind to the coactivator NR box LxxLL motif binding groove. Thus, the second feature of antagonism is the competition between H12 and the NR boxes of coactivators for a common LBD surface. For a comprehensive review see Bourguet et al.[32].

Although the necessity of a unique location for all antagonist situations is not obvious, in view of the existence of distinct types of antagonists that differentially modulate coregulator interaction, this 'flip-flop' mechanism between two positions is appealing. It appears that H12 in a dynamic fashion can adopt one of (at least) two stable positions (Fig. 8), dictated by the binding characteristics (agonist and antagonist) of the ligand, and be removed from either one in the absence of ligand to adopt the apo-form. Going from the latter to the agonist form necessitates conformational changes that are most likely to be responsible for disrupting the interactions with the corepressors. Whether transconformation of H12 is directly involved in agonist-induced corepressor dissociation is not proven but suggested by the observation that deletion of H12 in RARαΔ403 results in a constitutive binding of the SMRT corepressor[10].

While the position of H12 in its active (holo) agonist conformation creates the correct surface that allows the binding of coactivators, its positioning in any other place changes the shape and the charge distribution of the cognate recognition surface(s), and no coactivator interaction can occur. Indeed, crystallization of complexes between the RARγ or ERα LBDs and peptides comprising NR boxes of SRC-1 and TIF2, respectively, demonstrated that H12 in the antagonist position occupies the identical surface to which the LxxLL NR boxes of coactivators bind[8,11,12]. Thus, in the presence of antagonists like tamoxifen and raloxifen (ER) or BMS614 (RARα), helix H12 occupies the coactivator binding cleft and, owing to steric hindrance by the ligand, cannot flip back into the holo-position, which would allow coactivator interaction (compare Fig. 8a and b).

STRUCTURAL BASIS ACCOUNTING FOR FULL AND PARTIAL AF-2 AGONISTIC ACTIVITY OF A LIGAND

In addition to complete antagonists of the AF-2 function (e.g. raloxifen or tamoxifen for ER; BMS614 for RARα; see above), AF-2 partial agonists/antagonists (Table 3) have been crystallized with the corresponding receptors. While for the complete AF-2 antagonists there was a clear correlation between H12 positioning in the coactivator cleft (also referred to as the 'antagonist groove'), two crystal structures have been described in which there is an apparent discrepancy between the orientation of the AF-2 helix and the biological activity of the corresponding ligand. The ERβ/genistein and RXRαF318A/oleic acid LBD structures (for references, see Table 3) show that H12 can adopt the antagonist conformation even though the corresponding ligand elicits a weak but clear transcriptional AF-2 activity. A likely explanation for the discrepancy between the antagonist location of H12 and the transcriptional activity of these complexes is that these ligands display some but not all features of pure AF-2 agonists or antagonists. They can thus be classified as partial AF-2 agonists/antagonists. A major difference between pure and partial antagonists, lies in their steric properties. In contrast to full antagonists, genistein and oleic acid do not bear a bulky extension. Thus, they do not sterically preclude the agonist position of H12 and are in this respect similar to agonists. However, they induce unwinding of helix H11 which permits the positioning of helix H12 in the antagonist groove; in this respect, these ligands are similar to antagonists. In the presence of such mixed ligands, the equilibrium between the agonist holo-position of H12 and its antagonist position in the coactivator binding groove is likely to depend on the intracellular concentration of coactivators and corepressors, and these ligands may act as either AF-2 agonists or antagonists depending on the cellular context.

ANTAGONISTS WITH ALTERED DISSOCIATION KINETICS

A distinct mechanism of antagonism was suggested by recent mutational and modeling studies of mineralocorticoid receptor antagonists[13]. These antagonists exhibit a smaller size and faster off-rate kinetics when compared with MR agonists. Based on homology modeling, the authors proposed a mechanism according to which antagonism is due to a destabilization of the AF-2–AD core region by the lack of contacts between the antagonist and the helix H12 region.

ADDITIONAL CHARACTERISTICS OF PROMINENT ANTAGONISTS

4-Hydroxytamoxifen (OHT) and RU486 act as complete AF-2 antagonists and AF-1 agonists

Neither OHT (anti-estrogen[14]) nor RU486 (antiprogestin and antiglucocorticoid) impair dimerization of the corresponding receptors, nor DNA-binding *in vitro* in gel-retardation experiments. Moreover, in transient transfections, both antagonists can induce transcriptional activity of their cognate receptors. However, the ER–OHT complex is transcriptionally active only when AF-1 is present and when AF-1 can be active based on its promoter environment and cell specificity. Thus, in HeLa cells, where AF-1 is very weakly active, ER exhibits little transcriptional activity in the presence of OHT, while it activates efficiently in chicken embryo fibroblasts (CEF) where AF-1 is active. Deletion of region AB, which harbors AF-1, abolishes OHT-dependent stimulation of ER activity[15]. These results indicate that OHT does not support AF-2 activity (see above for the structural basis of the AF-2 antagonism), while AF-1 can be active. This is important, as it indicates the possibility of designing cell-specific ligands via the cell specificity of AF-1. In fact, OHT has already been considered as a partial antagonist, as it could induce ER target genes, such as PR[16,17].

Several groups have shown that RU486 induces DNA-binding of PR *in vitro* and *in vivo*[18–21]. Indeed, in transient transfection experiments and in yeast cells PR–RU486 activated transcription from palindromic response elements with an efficiency similar to that of the AF-2-truncated PR. Note, however, that the complex response element present in the LTR of the mouse mammary tumor virus was not active under these conditions[19]. Thus, RU486 also does not support AF-2 activity, but AF-1 can be active.

In vivo footprinting data appear to be at variance with these findings, as RU486 did not induce the footprints seen with the agonist on the MMTV response element[22]. It remains to be established, whether this is a promoter-related (MMTV–RE versus palindromic RE) difference or an indication for the existence of additional (anti)ligand-dependent activities on chromatin-embedded response elements, which cannot be studied in transient transfection experiments.

Note that effects of RU486 on the stability of the 8S steroid-free PR hetero-oligomer have also been reported[23].

ICI164,384 (ICI) acts as a complete AF-1 and AF-2 antagonist

In contrast to OHT, ICI[24] has always been considered a pure antagonist and no agonistic activities have been reported. Although it has been found that ICI may inhibit ER dimerization[25], overwhelming evidence is accumulating from various types of *in vivo* and *in vitro* studies that the ER–ICI complex can efficiently interact with cognate DNA response elements (for a detailed discussion and references, see Metzger[26]).

Note, however, that *in vivo* footprinting studies, such as those with the agonist[27], are lacking and that additional mechanisms may be superimposed, as it has been shown that ICI increases ER turnover [28].

Antagonists can affect the efficiency of nuclear receptor DNA binding

The existence of antagonists for PR and GR that negatively affect the interaction of these receptors with cognate response elements has been reported[29,30]. Note, however, a contradictory view in the case of the antiprogestin ZK98,299[31].

References
1 Yang, W. and Freedman, L.P. (1999) J. Biol. Chem. 274, 16838–16845.
2 Dey, A. et al. (1994) Mol. Cell Biol. 14, 8191–8201.
3 Gronemeyer, H. and Laudet, V. (1995) Protein Profile 2, 1173–1308.
4 Bourguet, W. et al. (1995) Nature 375, 377–382.
5 Renaud, J.P. et al. (1995) Nature 378, 681–689.
6 Wurtz, J.M. et al. (1996) Nature Struct. Biol. 3, 87–94.
7 Brzozowski, A.M. et al. (1997) Nature 389, 753–758.
8 Shiau, A.K. et al. (1998) Cell 95, 927–937.
9 Bourguet, W. et al. (2000) Mol. Cell 5, 289–298.
10 Chen, J.D. and Evans, R.M. (1995) Nature 377, 454–457.
11 Darimont, B.D. et al. (1998) Genes Dev. 12, 3343–3356.
12 Nolte, R.T. et al. (1998) Nature 395, 137–143.
13 Fagart, J. et al. (1998) EMBO J. 17, 3317–3325.
14 Jordan, V.C. (1984) Pharmacol. Rev. 36, 245–270.
15 Berry, M. et al. (1990) EMBO J. 9, 2811–2818.
16 Janssens, J.P. et al. (1984) Anticancer Res. 4, 157–162.
17 Noguchi, S. et al. (1988) Cancer 61, 1345–1349.
18 el Ashry, D. et al. (1989) Mol. Endocrinol. 3, 1545–1558.
19 Meyer, M.E. et al. (1990) EMBO J. 9, 3923–3932.
20 Turcotte, B. et al. (1990) Mol. Cell Biol. 10, 5002–5006.
21 DeMarzo, A.M. et al. (1992) Biochemistry 31, 10491–10501.
22 Truss, M. et al. (1994) Proc. Natl Acad. Sci. USA 91, 11333–11337.
23 Renoir, J.M. et al. (1989) Steroids 53, 1–20.
24 Wakeling, A.E. et al. (1991) Cancer Res. 51, 3867–3873.
25 Fawell, S.E. et al. (1990) Proc. Natl Acad. Sci. USA 87, 6883–6887.
26 Metzger et al. (1995) Mol Endocrinol 9, 579–591.
27 Wijnholds, J. et al. (1988) EMBO J. 7, 2757–2763.
28 Dauvois, S. et al. (1992) Proc. Natl Acad. Sci. USA 89, 4037–4041.
29 Klein Hitpass, L. et al. (1991) Nucl. Acids Res. 19, 1227–1234.
30 Takimoto, G.S. et al. (1992) Proc. Natl Acad. Sci. USA 89, 3050–3054.
31 Delabre, K. et al. (1993) Proc. Natl Acad. Sci. USA 90, 4421–4425.
32 Bourguet, W. et al. (2000) Trends Pharmacol. Sci. 21, 381–388.

7 Nuclear receptors: platforms for multiple signal integration

It has become increasingly well documented in the past few years that nuclear receptor action is not confined to the positive and negative regulation of the expression of cognate target genes. Indeed, these receptors, and most likely also their 'downstream' mediators, are targets of other signaling pathways and, reciprocally, can modify the activity of such pathways. The best known examples of such a signal transduction 'cross-talk' is the mutual repression of nuclear receptor and AP1 (c-Fos/c-Jun) activities. A distinct type of cross-talk is the modification of nuclear receptor AF activity by phosphorylation, e.g. by the MAP kinase pathway. The existence of signal transduction 'cross-talks' is likely to reflect the integration of nuclear receptor action in the context of the functional state of the cell in which it is expressed. The importance of signal transduction 'cross-talk' in 'real life' was recently demonstrated impressively by the observation that glucocorticoid receptor null mice die at birth, whereas mice harboring a GR mutant[1] (GR[dim/dim]) that can still cross-talk with AP-1 but not activate target genes with consensus GR response elements are viable[2].

SIGNAL TRANSDUCTION CROSS-TALK BETWEEN NUCLEAR RECEPTORS AND AP1

In addition to transactivation and silencing of their own target genes, certain nuclear receptors were shown to cross-talk with other signal transduction pathways. The original observation was made in 1990 when it was observed that GR could inhibit, in a ligand-dependent manner, the ability of AP1 (the heterodimer composed of the proto-oncogene products c-Fos and c-Jun) to transactivate its target gene promoters[3-6]. This transrepression is mutual and requires an unknown state of the receptor, which can be induced by both agonists and certain, but not all, antagonists (for reviews and references to the original work, see references 7–9). It is important to point out that the NR–AP1 cross-talk does not *per se* imply negative regulation of transcription; several reports show that, under certain conditions, this cross-talk can lead to positive transcriptional effects[10-12].

AP1 transrepression has been studied intensively because of its likely implication in important events, such as anti-inflammation and cell proliferation. AP1 transrepression is found to be dependent on the promoter structure and on the dimer composition of the AP1 transcription factor[12]. Furthermore, AP1 nuclear receptor cross-talk is also cell-type dependent[10], which probably reflects the involvement of other cellular factors with varying expression as mediators of this mutual transrepression event.

The mechanism(s) of NR–AP1 cross-talk has remained elusive despite, or because of, several contradictory reports and discrepancies between studies using *in vivo* and *in vitro* approaches. For example, the original proposal that AP1 and GR form DNA-abortive complexes (derived from *in vitro* evidence obtained in gel retardation experiments) appears to represent an *in vitro* artifact, as *in vivo* footprinting did not support this concept[13]. Some studies have proposed that sequestration of the coactivator CBP (or its homologue p300), which functions as a mediator of the transcriptional activities of both AP1 and nuclear receptors, accounts for the cross-talk phenomenon, as transrepression could be relieved when CBP was overexpressed in cells with limiting endogenous levels of this protein[14,15]. Several other studies, however, have shown that this mechanism can at best only partially account for the observed phenomena. In particular, the use of synthetic ligands that dissociate transrepression from coactivator recruitment to nuclear receptors argue against the involvement of coactivators, since it was shown that antagonists can still transrepress AP1 activity[16] (see below). A second mechanism has been proposed that is based on the observation that estrogen receptors are capable of down-regulating the activity of Jun-kinase, leading to reduced AP1 activity[17]. Thirdly, nuclear receptor-mediated effects on the dimerization of the AP1 subunits have been observed[18]. Further analysis is required to understand the contribution of these various mechanisms to the receptor–AP1 cross-talk. A fourth mechanism proposed to be involved in AP1 nuclear receptor cross-talk is direct physical contact between both factors when bound to so-called composite elements on a promoter[12,19]. Composite elements are thought to recruit both AP1 proteins and nuclear receptors, bringing them into close physical contact. Depending on the nature of such response elements and the activity of the participating proteins, steric effects might lead to differential regulation. It is currently not clear whether this mechanism applies only to promoters that carry composite elements or is of more general significance.

In addition to GR, mutual interference has been observed between the transcriptional activities of AP1 and RARs and RXRs[20-23], ER[24], TR/v-ErbA[25-31] PML-RARα[32], while MR appears to be insensitive to AP1[33]. In addition, interference between the transactivation abilities of nuclear receptors and other transcription factors has been reported[34-38]. For reviews and references to the original work, the reader is referred to the reviews in references 8, 9 and 39.

An interesting addition to the AP1 signaling cross-talk theme was the recent discovery that the two estrogen receptors ERα and ERβ, differ in their ability to modulate transcription driven from AP1 sites in response to synthetic estrogens, notably raloxifen[39-41]. It will be of interest to find molecular explanations for such different activities in transcriptional interference, which will provide ideas as to how synthetic ligands might discriminate between, and hence differentially affect, the various functions associated with nuclear receptors.

NFκB AND NUCLEAR RECEPTOR CROSS-TALK

The second best-studied example of transcription factor cross-talk is the mutual interference between glucocorticoid receptors and NFκB proteins. This pathway is again of significant importance, since it may also contribute to the anti-inflammatory, as well as osteoporotic action of glucocorticoids. Although some researchers have suggested that glucocorticoid action can be attributed to the increased production of the NFκB inhibitory molecule IκB, which in turn would remove active NFκB from the cell nucleus[42,43], studies with mutant receptors and 'dissociated' glucocorticoids showing IκB-independent repression of NFκB activity are incompatible with a simple IκB-mediated mechanism[44]. The underlying molecular events are still elusive, but may be related to those discussed above for the AP1 cross-talk. Note that, as in the case of AP1[16], squelching of limiting amounts of CBP, which also coactivates NFκB, is unlikely to be involved, as glucocorticoid receptor ligands that dissociate transactivation from transrepression still induce transcriptional interference with NFκB signaling[44].

Other transcription factors that are cross-regulated in their activities by nuclear receptors are Oct2A, RelA (another NFκB family member), STAT5 and Spi-1/PU.1 (for references to original work, see Göttlicher[8]).

DISSOCIATION OF RAR–RXR HETERODIMER-DEPENDENT TRANSACTIVATION AND AP1 TRANSREPRESSION BY SYNTHETIC RETINOIDS

Both RARs and RXRs (RAR and RXRα, β and γ isotypes) can act as ligand-dependent transrepressors of AP1 (c-Jun/c-Fos) activity and, reciprocally, AP1 can inhibit transactivation by RARs and RXRs. In the case of RAR, mutant analyses have shown that the integrity of both the DNA and ligand-binding domains is required for efficient AP1 repression. Notably, C-terminal truncation mutants lacking helix H12 fail to transrepress, suggesting that H12 transconformation is involved in both transactivation and transrepression.

Retinoids are of particular pharmacological interest, as they show great promise for both prevention and treatment of various types of cancer[45]. Indeed, acute promyelocytic leukemia is generally considered to be the first cancer that can be cured involving a combination of retinoic acid 'differentiation' therapy and chemotherapy[46]. These perspectives have inspired several pharmaceutical and academic groups to generate ligands: (1) that are RAR isotype or RXR-selective agonists and antagonists; and (2) that separate the transactivation and transrepression functions of retinoid receptors. The aim of these ongoing studies is to find ligands that are largely devoid of the side effects of all-*trans* retinoic acid, and function as either apoptosis and/or differentiation inducers or inhibitors of cell proliferation, preferably in a cell-specific manner by exploiting the isotype selectivity of certain synthetic retinoids.

Interestingly, a large panel of synthetic retinoids has been generated that exhibit a great variety of RAR isotype or RXR selectivity and functional specificity. Using these tools it has been possible to provide evidence that both subunits of the RAR–RXR heterodimer synergistically contribute to differentiation and apoptosis in the model APL NB4 and HL60 myeloid leukemia cell line[47,48] and that retinoids that can transrepress AP1 but are largely devoid of the ability to transactivate cognate target genes can inhibit the growth of tumor cells[16]. Notably, using *in vivo* mouse models for tumor promoter-induced papilloma formation, it has been shown that the antitumor effect of synthetic retinoids requires the AP1 repressive but not the 'classical' transactivation ability of the cognate receptor[49]. However, depending on the cell type, anti-AP1 activity may not be required for the antitumor effect of retinoids. HL60 cells differentiated and ceased proliferation in response to RAR–RXR agonists but not to a retinoid with anti-AP1 activity that did not transactivate target genes[48]. Similar AP1 antagonism-independent antitumor effects of retinoids have been reported for neuroblastoma[50] and prostate cancer cells[51]. The above results suggest that it may be possible to target the antitumor effect of retinoids to specific cell types or tissues by designing synthetic retinoids with a defined pattern of RAR isotype and RXR selectivity, and transactivation or transrepression ability.

In addition to the antitumor effect, dissociation of the multiple functions associated with retinoid receptors are also attractive for other pharmacological targets, such as anti-inflammation, scarring, and skin aging[52].

The exciting message emanating from multiple studies in the past few years is that we have been able to develop tools that will allow us to understand the individual aspects of the multifaceted NR activities. Synthetic ligands can differentiate between receptor isotypes and act as agonists or antagonists, revealing the contribution of individual NR isotypes to signaling. 'Dissociated' ligands have been found that allow differentiation between the contribution of transactivation and transrepression to NR action. In addition, genetic analyses with NR mutant mice, pioneered in the case of the GR[2], allow the role of NR-mediated transactivation and transrepression to be elucidated *in vivo*. Most importantly, however, these various isotype-selective and 'dissociated' ligands have promise as novel pharmacological tools in the treatment and prevention of a large variety of diseases in which NRs play a critical role, ranging from inflammatory diseases and osteoporosis to cancer.

POST-TRANSLATIONAL MODIFICATION BY PHOSPHORYLATION

Phosphorylation is increasingly recognized as a signaling cross-talk that affects most if not all nuclear receptors. The major phosphorylation targets are the A/B, E and F regions. Phosphorylation has been reported to occur via the MAPK, PKA and PKC pathways, as well as by the CDK7 subunit of the general transcription factor $TF_{II}H$ and the cyclinA/cdk2 complex[53-61]. Only in some cases the functional consequences of receptor phosphorylation have been defined. Nevertheless, phosphorylation of estrogen receptors in their A/B regions by MAP kinases enhances the transactivation potential of the associated activation function AF-1 (e.g. Kato et al.[52]). Inversely, MAP kinase-dependent phosphorylation of PPAR-γ inhibited its transcriptional activity[55]. It is tempting to speculate that phosphorylation of the A/B region of nuclear receptors can, positively or negatively, modulate coregulator association. Indeed, it has been recently reported that phosphorylation by MAP kinase of ERβ region A/B led to a constitutive recruitment of SRC-1[62]. Notably, enhanced receptor–SRC-1 binding was also induced *in vivo* in cells treated with epidermal growth factor or expressing activated Ras[62]. On the other hand, it cannot be concluded that phosphorylation is always required for AF-1 activity. Retinoic acid receptors can be phosphorylated by $TF_{II}H$ and phosphorylation has some enhancing effects on the receptor-associated transcriptional activity. However, reintroduction of retinoic acid receptors mutated at the corresponding serine residues into retinoic acid receptor knockout F9 cell lines rescues some, but not all, of the observed phenotypes[63]. It is tempting to speculate that the phosphorylation fine-tunes the activity of AF-1. However, the observations that both AF-1 activity[64] and the phosphorylation of serine residues in the A/B domains are cell and promoter-context dependent, warrant further investigations.

Other sites of phosphorylation are found in the E and F domains of nuclear receptors[63,67]. The functional consequences of region E/F phosphorylation is not understood. Preliminary data suggest a role in establishing coregulator specificity[65] and affecting protein degradation[66]. One remarkable example of a phosphorylation event has been described for the estrogen receptor ERα. This receptor is phosphorylated by a protein kinase A-dependent pathway on Ser537. This serine is located in the helix H3 of the LBD. Once targeted, the phosphoserine induces a conformational change that renders the estrogen receptor ERα constitutively active (see Yudt[67] and references therein). Alternatively, in view of the results concerning the enhanced recruitment of coactivators to phosphorylated A/B domains, the phosphoserine might also constitute together with residues in the vicinity a novel coactivator binding surface. This result is of importance, since it might be interpreted as a way to bypass the need of a specific ligand to activate the receptor.

References

1   Heck, S. et al. (1994) EMBO J. 13, 4087–4095.
2   Reichardt, H.M. et al. (1998) Cell 93, 531–541.
3   Diamond, M.I. et al. (1990) Science 249, 1266–1272.
4   Jonat, C. et al. (1990) Cell 62, 1189–1204.
5   Schule, R. et al. (1990) Cell 62, 1217–1226.
6   Yang-Yen, H.F. et al. (1990) Cell 62, 1205–1215.
7   Pfahl, M. (1993) Endocrinol. Rev. 14, 651–658.
8   Göttlicher, M. et al. (1998) J. Mol. Med. 76, 480–489.
9   Resche-Rigon, M. and Gronemeyer, H. (1998) Curr. Opin. Chem. Biol. 2, 501–507.
10   Shemshedini, L. et al. (1991) EMBO J. 10, 3839–3849.
11   Bubulya, A. et al. (1996) J. Biol. Chem. 271, 24583–24589.
12   Pearce, D. et al. (1998) J. Biol. Chem. 273, 30081–30085.
13   Konig, H. et al. (1992) EMBO J. 11, 2241–2246.
14   Kamei, Y. et al. (1996) Cell 85, 403–414.
15   Fronsdal, K., et al. (1998) J. Biol. Chem. 273, 31853–31859.
16   Chen, J.Y. et al. (1995) EMBO J. 14, 1187–1197.
17   Caelles, C. et al. (1997) Genes Dev. 11, 3351–3364.
18   Zhou, X.F. et al. (1999) Mol. Endocrinol.13, 276–285.
19   Pearce, D. and Yamamoto, K.R. (1993) Science 259, 1161–1165.
20   Nicholson, R.C. et al. (1990) EMBO J. 9, 4443–4454.
21   Schule, R. et al. (1991) Proc. Natl Acad. Sci. USA 88, 6092–6096.
22   Yang-Yen, H.F. et al. (1991) New. Biol. 3, 1206–1219.
23   Salbert, G. et al. (1993) Mol. Endocrinol. 7, 1347–1356.
24   Doucas, V. et al. (1991) EMBO J. 10, 2237–2245.
25   Desbois, C. et al. (1991) Cell 67, 731–740.
26   Desbois, C. et al. (1991) Oncogene 6, 2129–2135.
27   Zhang, X.K. et al.. (1991) Mol. Cell Biol. 11, 6016–6025.
28   Sharif, M. and Privalsky, M.L. (1992) Oncogene. 7, 953–960.
29   Lopez, G. et al. (1993) Mol. Cell Biol. 13, 3042–3049.
30   Schmidt, E.D. et al. (1993) Biochem. Biophys. Res. Commun. 192, 151–160.
31   Wondisford, F.E. et al. (1993) J. Biol. Chem. 268, 2749–2754.
32   Doucas, V. et al. (1993) Proc. Natl Acad. Sci. USA 90, 9345–9349.
33   Perez, P. et al. (1991) Biochem. Biophys. Res. Commun. 181, 9–15.
34   Stein, B. et al. (1993) EMBO J. 12, 3879–3891.
35   Ray et al. (1994) Proc. Natl Acad. Sci. USA 91, 752–756

36 Caldenhoven, E. et al. (1995) Mol. Endocrinol. 9, 401–412.
37 Scheinmann, R.I. et al. (1995) Mol. Cell Biol. 15, 943–953.
38 Stein, B. and Yang, M.X. (1995) Mol. Cell. Biol. 15, 4971–4979.
39 Webb, P. et al. (1995) Mol. Endocrinol. 9, 443–456.
40 Webb, P. et al. (1999) Mol. Endocrinol. 13, 1672–1685.
41 Paech, K. et al. (1997) Science 277, 1508–1510.
42 Auphan, N. et al. (1995) Science. 270, 286–290.
43 Scheinmann, R.I. et al. (1995) Mol. Cell Biol. 15, 943–953.
44 Heck, S. et al. (1997) EMBO J. 16, 4698–4707.
45 Hong, W.K. and Sporn, M.B. (1997) Science 278, 1073–1077.
46 Warrell, R.P. et al. (1993) N. Engl. J. Med. 329, 177–189.
47 Chen, J.Y. et al. (1996) Nature 382, 819–822.
48 Kizaki, M. et al. (1996) Blood 87, 1977–1984.
49 Huang, C. et al. (1997) Proc. Natl Acad. Sci. USA 94, 5826–5830.
50 Giannini, G. et al. (1997) J. Biol. Chem. 272, 26693–26701.
51 de Vos, S. et al. (1997) Prostate 32, 115–121.
52 Fisher, G.J. et al. (1996) Nature 379, 335–339.
53 Joel P.B. et al. (1995) Mol. Endocrinol. 9, 1041–1052.
54 Kato, S. et al. (1995) Science 270, 1491–1494.
55 Hu, E. et al. (1996) Science. 274, 2100–2103.
56 Adams, M. et al. (1997) J. Biol. Chem. 272, 5128–5132.
57 Rochette-Egly, C. et al. (1997) Cell 90, 97–107.
58 Shao, D. et al. (1998) Nature 396, 377–380.
59 Camp, H.S. et al. (1999) Endocrinology 140, 392–397.
60 Chen, D. et al. (1999) Mol. Cell Biol. 19, 1002–1015.
61 Rogatsky, I. et al. (1999) J. Biol. Chem. 274, 22296–22302.
62 Tremblay, A. et al. (1999) Mol. Cell 3, 513–519.
63 Taneja, R. et al. (1997) EMBO J. 16, 6452–6465.
64 Berry, M. et al. (1990) EMBO J. 9, 2811–2818.
65 Hong, S.H. et al. (1998) Mol. Endocrinol. 12, 1161–1171.
66 Zhang, J. et al. (1998) Genes Dev. 12, 1775–1780.
67 Yudt, M.R. et al. (1999) Biochemistry 38, 14146–14156.

8 Deregulation in disease and novel therapeutic targets

Given the major impact of nuclear receptor signaling on animal physiology, it is no surprise to find aberrant nuclear receptor function as the basis of multiple pathologies. Indeed, synthetic agonists and antagonists have been developed and are in clinical use for endocrine therapies of cancer as well as hormone replacement therapies in osteoporosis. Thiazolidinediones, which were known as insulin 'sensitizers' in the treatment of non-insulin-dependent diabetes, have been recognized as PPARγ agonists some time ago, and the recent establishment of a link between human type 2 diabetes and PPARγ mutation has proved that PPARγ malfunction can lead to severe insulin resistance, diabetes mellitus and hypertension[1]. Moreover, the impact of the contraceptive pill on social life is evident, as is that for RU486, for pregnancy termination. However, this is likely to represent only the tip of the iceberg, and novel types of nuclear receptor-based drugs are expected to be developed based on our increasing knowledge of the structural and molecular details of nuclear receptor and ligand function, and the elucidation of the signaling pathways involved in (patho)physiological events.

NOVEL PERSPECTIVES FOR THERAPY

In addition to the well-established endocrine therapies of breast and prostate cancers with estrogen and androgen receptor antagonists, respectively, and to the more recent differentiation therapy of acute promyelocytic leukemia by retinoids, novel synthetic nuclear receptor ligands are of considerable interest for the therapy and prevention of different types of cancers[2,3]. The development of novel types of nuclear receptor ligands is facilitated by recent pharmacological and chemical developments, for example: (1) combinatorial chemistry, computer-assisted ligand docking based on LBD crystal structure and ultra-high throughput screening with nuclear receptor-based reporter systems; (2) the possibility of dissociating nuclear receptor-associated functions, such as transactivation and cross-talk with other signaling pathways; and (3) the possibility of generating receptor and receptor isotype-selective ligands.

It is tempting to speculate that coactivators are not entirely promiscuous in their choice of nuclear receptors. For example, only AIB1/RAC3 is found to be amplified in breast cancer cells, while the expression level of the other two family members remains constant, reflecting estrogen receptor specificity[4]. Furthermore, in some types of acute myeloid leukemia (AML), a chromosomal translocation specifically fuses a monocytic zinc finger protein of unknown function (MOZ) to the C-terminus of TIF2 and not to any of the other two TIF2 family members, again reflecting a bias towards one specific coactivator[5,6]. Interestingly, the observation has been made that overexpression of coactivators of the TIF2/AIB1 family can lead in some systems to ligand-independent activity under certain conditions[7]. This suggests that transcriptional mediators may possibly be involved in the origin and/or progression of proliferative diseases, and may become novel pharmacological targets. Indeed, mutations of the CBP gene have been implicated in the cause of Rubinstein–Tabi syndrome, and alterations of the p300 gene were found associated with gastric and colorectal carcinomas[8].

Also of interest are studies in which the oncogenic capacity of E1A was compromised by coexpression of p300[9]. Based on these various data, it is tempting to speculate that alterations in the cellular abundance of coregulators or altered substrate specificity of the associated enzymatic functions may lead to pathological states.

Synthetic ligands of nuclear receptors are classified as agonist and antagonist with respect to a particular receptor-associated function[10]. This discrimination is not always obvious, since a particular ligand might antagonize some activities while functioning as an agonist for other activities. Examples are the ERα antagonists hydroxytamoxifen and ICI164,384. Hydroxytamoxifen antagonizes the activation function AF-2 but it acts as an agonist for AF-1, whereas ICI164,384 in turn antagonizes both AF-1 and AF-2[11]. Similarly, certain retinoid receptor antagonists are agonists for AP1 repression[12]. It is therefore important to consider, where possible, the molecular basis of the anticipated action of a nuclear receptor-based drug to increase drug efficacy and limit side-effects. If the molecular mechanism is unknown, it may be wise to use screening paradigms that consider the multiple dimensions of receptor activities.

Another twist to the classification of synthetic nuclear receptor ligands results from the availability of isotype-specific ligands. These compounds affect one isotype of a nuclear receptor but not another. The interspecies conservation of retinoid receptor isoforms, together with results obtained with isotype-selective retinoids and gene ablation studies, have established that each of the three retinoic acid receptor genes has a cognate spectrum of functions[13]. Given the pharmacological potential of retinoids, the development of isotype-specific ligands has attracted much attention. Today a wealth of synthetic retinoids exist that display either isotype specificity or act as mixed agonists/antagonists for the three retinoic acid receptors[14]. Some of these retinoids were found to display cell specificity and their pharmacological potential is currently investigated. Retinoid X receptor specific ligands are also being developed, which is of particular interest in view of the role of the retinoid X receptor as the promiscuous heterodimerization partner in a number of signaling pathways. A recent report suggests that retinoid X receptor ligands may stimulate insulin action in non-insulin-dependent diabetes[15] through a PPARγ–RXR heterodimer that is responsive to thiazolidinediones. It is thus conceivable that pathway-specific RXR ligands can be generated.

The recent gain in understanding of NR coregulator function at the molecular level provides the basis for new strategies of pharmacological interference within NR signaling pathways. The fact that the NR–coactivator and NR–corepressor interfaces are composed of precisely defined features – an amphipatic α-helical chain containing either the LxxLL NR box motif of the coactivator or the CoRNR box motif of the corepressor, and a hydrophobic cleft plus 'charge clamp' on the surface of the LBD of the NR[16–20] – raise the possibility of disrupting such interactions with small synthetic molecules. On the basis of the structural information available now, it is feasible to screen combinatorial peptides containing the core LxxLL motif for high-affinity binding to the hydrophobic groove on the LBD. A phage display approach has provided a library of nuclear receptor interaction peptides[21]. These peptides have been found to be active (in transfected cells) and could be used in probing surfaces of NRs differentially generated in the presence of agonists and antagonists[22,23].

References

1   Barroso, I. et al. (1999) Nature 402, 880–883.
2   Lotan, R. (1996) FASEB J. 10, 1031–1039.
3   Hong, W.K. and Sporn, M.B. (1997) Science 278, 1073–1077.
4   Anzick, S.L. et al. (1997) Science 277, 965–968.
5   Carapeti, M. et al. (1998) Blood 91, 3127–3133.
6   Liang, J. et al. (1998) Blood 92, 2118–2122.
7   Voegel, J.J. et al. (1998) EMBO J. 17, 507–519.
8   Giles, R.H. et al. (1998) Trends Genet. 14, 178–183.
9   Chakravarti, D. et al. (1999) Cell 96, 393–403.
10 Resche-Rigon, M. and Gronemeyer, H. (1998) Curr. Opin. Chem. Biol. 2, 501–507.
11 Berry, M. et al. (1990) EMBO J. 9, 2811–2818.
12 Chen, J.Y. et al. (1995) EMBO J. 14, 1187–1197.
13 Kastner, P. et al. (1995) Cell 83, 859–869.
14 Gehin, M. et al. (1999) Chem. Biol. 6, 519–529.
15 Mukherjee, R. et al. (1997) Nature 386, 407–410.
16 Darimont, B.D. et al. (1998) Genes Dev. 12, 3343–3356.
17 Feng, W.F. et al. (1998) Science 280, 1747–1749.
18 Nolte, R.T. et al. (1998) Nature 395, 137–143.
19 Shiau, A.K. et al. (1998) Cell 95, 927–937.
20 Hu, X. and Lazar, M.A. (1999) Nature 402, 93–96.
21 Chang, C. (1999) Mol. Cell Biol. 19, 8226–8239.
22 Norris, J.D. et al. (1999) Science 285, 744–746.
23 Paige, L.A. et al. (1999) Proc. Natl Acad. Sci. USA 96, 3999–4004.

9 Genes and evolution

Owing to their great diversity and to the extent of sequence conservation shown between distantly related members, nuclear hormone receptors are interesting tools for evolutionary studies. One of the most interesting questions regarding nuclear receptor evolution is the existence of orphan receptors and their relationships with classic liganded receptors. The wide diversity of chemically and structurally unrelated ligands that are known for nuclear receptors is striking given the high conservation of the structure and mode of action of the receptors. This has prompted many researchers to speculate on the origins of such a signaling pathway. No clear sequence signature reminiscent of nuclear receptors has been found in other protein families. Recently, however, low scores of sequences identities of the DBD and LBD of nuclear receptors with a LIM/GATA zinc-finger domain and peroxisomal membrane protein Pex11p, respectively, have prompted some groups to suggest that the first nuclear receptor has been constructed by the fusion of two genes encoding these various proteins[1,2]. In the absence of structural data that can confirm the significance of these low similarity scores, it is nevertheless hard to draw firm conclusions on this matter. The question of the origin of the first nuclear receptor is thus still open.

MOLECULAR PHYLOGENY

The question of orphan receptors can be addressed with the help of molecular phylogeny. The first studies conducted almost 10 years ago have mainly considered either the C domain alone or the C and E domains separately[3-5]. From these data it has been suggested that the nuclear receptors evolved by duplication of a unique common ancestor that contains the C and E domains. Swapping between C and E domains during evolution has been proposed[3,6] but it is in fact extremely difficult to see such an event given the very poor resolution of phylogenetic trees based on the C domain alone (discussed in Escriva[7]).

As a large number of receptors have been cloned since the first attempts to understand the phylogeny of the family, and because phylogenetic methods have greatly improved, more recent studies have started to resolve the evolutionary history of the superfamily. Phylogenetic analysis was carried out using a larger set of sequences on the evolutionary history of the NR superfamily, which resulted in classification of the family into six subfamilies of unequal size[8]:

1. a large subfamily containing TR, RAR, VDR, EcR and PPAR receptors, as well as numerous orphan receptors (ROR, Rev-erb, CAR, etc.);
2. a subfamily containing RXR, COUP-TF and HNF-4;
3. the steroid subfamily with ER, GR, MR, PR and AR, as well as the ERRs orphan receptors;
4. a subfamily containing the NGFIB group of orphan receptors;
5. the SF1 and *Drosophila* FTZ-F1 subfamily;
6. a small subfamily containing only the GCNF1 receptor.

The phylogenetical relationships between subfamilies have been tested and confirmed by high bootstrap values in distance and parsimony analysis. These data have been independently confirmed by several groups[6,9].

EVOLUTION OF LIGAND BINDING

Several hypotheses have been proposed to explain the origins of classical liganded NRs and of orphan receptors. The longest held view suggests that orphan receptors evolved as liganded molecules, which diversified through gene duplication during evolution in order to reach their present-day diversity. According to this view, orphan receptors would have lost, subsequently, their ability to bind ligands. An alternative hypothesis suggests that ligand binding was acquired during the evolution of NRs. Accordingly, the ancestral receptor was an orphan receptor and the various liganded receptors would have gained ligand recognition independently during evolution. The arguments for these two scenarios have been discussed recently[7].

The phylogenetic analysis has given an indication in favor of the gain of ligand-binding ability during the evolution of the superfamily[7,10]. When the DNA-binding and dimerization abilities of each given NR are compared to its phylogenetic position, it is clear that there is a correlation. For example, it is known that Rev-erb and ROR orphan receptors, which are related, bind as monomers and dimers to the same response elements. Steroid receptors such as ER or GR are all able to bind as homodimers to palindromic elements. Most, if not all, RXR interacting receptors belong to subfamilies I (TR, RAR, PPAR, VDR, etc.) and IV (NGFIB). Strikingly, such a correlation is not found when the ligand-binding abilities of the receptors are compared to the phylogeny. In fact, as discussed below, it clearly appears that the phylogenetic position of a receptor is not directly correlated with the chemical nature of its ligand[8,11]. For instance, the closely evolutionary related receptors of subfamily I (TRs, RARs, PPARs and VDRs) bind ligands originating from totally different biosynthetic pathways. Conversely, RAR and RXR (subfamily II), which are not evolutionarily related, bind similar ligands (all-*trans* and 9-*cis* retinoic acid, respectively). This situation could be explained as an independent gain of ligand-binding capacity during NR evolution.

DIVERSIFICATION OF THE NUCLEAR RECEPTOR SUPERFAMILY

The phylogeny of the superfamily suggests that nuclear receptors are specific to metazoans and that the superfamily undergoes an explosive expansion during early metazoan evolution[8]. Indeed, an extensive PCR survey established that NRs appeared very early on during metazoan evolution and are present in all the metazoans but not in non-metazoans since no NR signatures were found in fungi, plants or unicellular eukaryotes[11]. Moreover, the diversification of the superfamily followed two waves of a gene duplication model: a first wave during the emergence of the various metazoans, leading to the acquisition of the present subfamilies and groups of receptors, and later, after the arthropod–vertebrate split, a second wave, specifically in vertebrates, producing the paralogous groups within each subfamily e.g. TRα and β, RARα, β and γ, etc).

Nuclear receptors were reported from a large number of metazoans: diploblasts (hydra and *Tripedalia*[11,12]), parasitic helminth worms[13], mollusc, segmented worms or echinoderms[11,14]. Interestingly, the split of the protostomian metazoans into two classes[15], Ecdysozoa (molting animals, including arthropods and nematodes among others) and Lophotrochozoa (molluscs and annelids among others), suggests that the ecdysone receptor, which controls molting in arthropods, has played an important role in metazoan evolution (see De Mendonça[14] for a discussion). Nevertheless, we are still far from the proof of a direct regulatory role of nuclear receptors in metazoan evolution. To provide this proof is a challenge for future research in the next years.

GENOMICS OF NUCLEAR RECEPTORS

The availability of complete genome sequences provides a unique opportunity for understanding the various steps that have led to the present-day diversity of nuclear receptors. From these data, it is clear that fungi (with *Saccharomyces cerevisiae*) and plants (with *Arabidopsis thaliana*) do not contain nuclear receptors, even if some short matches with high scores were reported with FASTA homology searches. This confirms that nuclear receptors are specific to metazoans. If the *Drosophila* genome apparently contains a reasonable number (21) of nuclear receptors[16], it is not the same as the genome of the nematode worm *Caenorhabditis elegans*. Indeed, more than 200 nuclear receptor sequences have been found in this genome[17,18]. If clear homologues with mammalian or *Drosophila* sequences were found for a minority of these receptors, most of them harbor no peculiar similarity with nuclear receptors from other organisms. This and the clustering of large fractions of these receptors in one chromosome in the nematode genome suggest that these numerous receptors evolved through an extensive wave of gene duplication that arose specifically in the nematode lineage. The reason for this striking proliferation is still a matter of speculation. In contrast to the nematode, and in accordance with the data available in *Drosophila*, the human genome appears to contain 49 ± 1 nuclear receptors[20].

NOMENCLATURE

Because of the high sequence conservation between all types of nuclear receptors and their important physiological role, the number of known superfamily members strongly increased after the discovery of the first nuclear receptor in 1986. Over the last decade, workers in the field have described more than 1000 sequences of nuclear receptors in many species using an increasingly complex and baroque nomenclature. The existence of several names for the same gene is an acute problem in the case of the orphan receptors, which often could not be described by their function, particularly at the time of their discovery. For researchers in this field it became obvious that this plethora of names has become increasingly confusing and constitutes a barrier for understanding to researchers outside as well as in the field. For this reason, a committee for the nomenclature of nuclear receptors has been set up and has recommended names for the subfamilies and groups of receptors based on a phylogenetic tree connecting all known nuclear receptor sequences[8]. This system, based on the evolution of the two well-conserved domains of nuclear receptors (the DNA-binding C domain and the ligand-binding E domain), offers a practical and significant framework to which subsequent genes can easily be added. The principle that was used was to propose an official name that should be indicated in each relevant publication and which provides information on the position of the given receptor in the phylogeny (see Figure 2). For example, RXRα is called NR2B1, since it is a member of subfamily 2, group B, gene 1. Each author can then use the name of his or her choice in any relevant paper given that the official name is indicated once in the paper. This nomenclature has been described in the literature[19] and is implemented regularly on the following web site: http://www.ens-lyon.fr/lbmc/laudet/nomenc.html). In the present book, the nomenclature for the nuclear receptors is indicated in each specific entry.

References

1   Clarke, N.D. and Berg, J.M. (1998) Science 282, 2018–2022.
2   Barnett, P. et al. (2000) TIBS 25, 227–228.
3   Laudet, V. et al. (1992) EMBO J. 11, 1003–1013.
4   Amero, S.A. et al. (1992) Mol. Endocrinol. 6, 3–7.
5   Detera-Wadleigh, S.D. and Fanning, T.G. (1994) Mol. Phyl. Evol. 3, 192–205.
6   Thornton, J. and DeSalle, R. (2000) Syst. Biol. 49, 183–201.
7   Escriva, H. et al. (2000) BioEssays 22, 717–727.
8   Laudet, V. (1997) J. Mol. Endocrinol. 19, 207–226.
9   Garcia-Vallvé, S. and Palau, J. (1998) Mol. Biol. Evol. 15, 665–682.
10 Escriva, H. et al. (1999) In Nuclear Receptors: A Practical Approach (Picard, D., ed.). Oxford University Press, Oxford, 1999.
11 Escriva, H. et al. (1997) Proc. Natl. Acad. Sci. USA 94, 6803–6808.
12 Kostrouch, Z. et al. (1998) Proc. Natl Acad. Sci. USA 95, 13442–13447.
13 De Mendonça, R.L. et al. (2000) Parasitol. Today 16, 233–240.
14 De Mendonça, R.L. et al. (1999) Am. Zool. 39, 704–713.
15 Aguinaldo, A.M. et al. (1997) Nature 387, 489–493.
16 Adams, M.D. et al. (2000) Science 287, 2185–2224.
17 Sluder, A.E. et al. (1999) Genome Res. 9, 103–120.
18 Enmark, E. and Gustafsson, J.A. (2000) Trends Pharmacol. Sci. 21, 85–87.
19 Nuclear Receptor Nomenclature Committee (1999) Cell, 97, 161–163.
20 Robinson-Rechaui, M. (2001) Trends Genet. in press.

10 Conclusions

Nuclear receptors are ligand-regulated transcription factors that have evolved from an ancestral orphan receptor into a highly diverse family present throughout the entire animal kingdom and encompassing receptors for steroid and non-steroid hormones, vitamins, corticoids and metabolic intermediates. These receptors signal through endocrine, paracrine, autocrine and intracrine modes of action to regulate multiple aspects of animal physiology, such as homeostasis, development and reproduction. They regulate target genes that they either bind directly as mono-, homo- or heterodimers at cognate response elements, and have the ability to modulate other gene expression programs indirectly ('signal transduction cross-talks'). Through the coordinated expression of genetic programs, nuclear receptors contribute to cell fate-determining processes, thereby shaping and sustaining the organism. The inducing signal – binding of the ligand – induces a major allosteric change in the ligand-binding domain, which is transformed into cascades of protein–protein recognition paradigms including coregulator and cointegrator proteins.

Direct transcriptional repression in the absence of ligand or the presence of certain antagonists by some nuclear receptors is mediated by corepressor complexes that are associated with the unliganded receptor, and condense the chromatin environment at the promoter region through histone deacetylation. Corepressors interact by virtue of their CoRNR boxes with non-liganded nuclear receptors. Upon ligand binding, the allosteric change in the ligand-binding domain induces corepressor dissociation, and coactivator complexes are recruited. Bona fide coactivators recognize the active nuclear receptor ligand-binding domain (AF-2) via conserved LxxLL NR boxes and often the N-terminal activation function AF-1. The NR boxes of coactivators and CoRNR boxes of corepressor bind to topologically similar sites in the ligand-binding domain but the surfaces are entirely distinct owing to the agonist-induced conformational changes. In particular the holo-H12 is required for coactivator but incompatible with corepressor binding. Coactivator complexes reverse the repressive effects of chromatin by specific histone acetylation, and allow access of the basal transcription machinery. In a subsequent step, the mammalian SMCC mediator is recruited to the nuclear receptor and possibly stabilizes the formation of the preinitiation complex at target gene promoters. SMCC recruitment might be regulated by the acetylation and subsequent dissociation of TIF2 family members thus allowing SMCC–receptor association.

Despite their direct actions on the chromatin environment and the transcription machinery, nuclear receptors also regulate transcription by positive and negative interference with other signaling pathways. Different mechanisms for such transcription factor cross-talk have been described, but none is fully accepted and can explain all aspects of the particular cross-talk. The activity of nuclear receptors is regulated by phosphorylation that may serve to fine tune the signaling or/and to established a link to other signaling pathways. Finally, the promoter context, and the temporal order of incoming signals on a particular promoter have the likelihood of adjusting the transcriptional potential of nuclear receptors to particular situations. Taken together, nuclear receptors serve as platforms to coordinate cognate signals with those emanating from other signaling pathways, thereby integrating the nuclear receptor signal into the functional context of cellular state and activity.

Nuclear receptors and their coregulators have been implicated in several diseases. Their role as key regulatory molecules in a vide variety of signaling pathways qualifies them as novel pharmacological targets. The ongoing improvement of synthetic nuclear receptor ligands with altered specificity is likely to improve therapy and reduce side-effects.

Future research on nuclear receptors still has important questions to answer. What are the constituents of the genetic programs that are governed by a given nuclear receptor? How are the nuclear receptor signals matched and complemented with other signaling cascades? What are the precise molecular events leading to the variety of transcriptional effects exerted by nuclear receptors? Once these questions have been addressed adequately, specific interference into these immensely complex systems might lead to the successful control and reprogramming of an organism's (patho)physiological state. Understanding of nuclear receptor-controlled transcription will shed light on the general and signaling pathway-selective control of gene expression. In this respect, the use of gene arrays together with the information derived from the genome sequencing will certainly have an enormous impact.

THE
NUCLEAR
RECEPTORS

Notes on the amino acid sequences

1. All the accession numbers provided are from Genbank/EMBL.
2. The sequences underlined represent the C domain implicated in DNA binding. The bold Cs are cysteine residues that form the two zinc fingers coordinating zinc ions, and the short sequences in italic represent the P-box, i.e. the α-helical region that directly contacts specific bases of the DNA response element.

Names

The thyroid hormone receptors are encoded by two genes, TRα and TRβ, which are also known as c-erbA-1 and c-erbA-2 (or c-erbAα and c-erb-Aβ). The c-erbA-1 gene is the cellular homologue of the v-erbA oncogene of the avian erythroblastosis virus (AEV; erb stands for erythroblastosis). These names are now almost totally replaced by the name of the products that give clear information on their function. Their human versions are called THRA and THRB.

Species	Latin name	Gene product	Accession number	References
Official name: NR1A1				
Human	*Homo sapiens*	TRα	M24748	1
Mouse	*Mus musculus*	TRα	X51983	4
Rat	*Rattus* sp.	TRα	M18028	38
Sheep	*Ovis aries*	TRα	Z68308	2
Pig	*Sus scrofa*	TRα	AJ005797	3
Chicken	*Gallus gallus*	TRα	Y00987	36
Virus AEV	*Oncovirus*	v-erbA	K02006	32
Muscovy duck	*Cairina moschata*	TRα	Z50080	5
King penguin	*Aptenodytes patagonicus*	TRα	AJ002363	Unpub.
Adelie penguin	*Pygocelis adeliae*	TRα	AJ002300	Unpub.
Xenopus	*Xenopus laevis*	TRαA	M35343	6
Xenopus	*Xenopus laevis*	TRαB	M35344	6
Frog	*Rana castebeiana*	TRα	L06064	7
Mudpuppy	*Necturus maculosus*	TRα	Y16623	13
Zebrafish	*Danio rerio*	TRα	U54796	10
Fugu	*Takifugu rubripes*	TRα	AF302243	9
Rainbow trout	*Oncorhynchus mykiss*	TRα	AF302245	9
Atlantic salmon	*Salmo salar*	TRαA	AF302250	9
Atlantic salmon	*Salmo salar*	TRαB	AF146775	Unpub.
Tilapia	*Oreochromis niloticus*	TRα	AF302248	9
Japanese flounder	*Paralichthys olivaceus*	TRαA	D16461	8
Japanese flounder	*Paralichthys olivaceus*	TRαB	D16462	8
Atlantic halibut	*Hippoglossus hippoglossus*	TRα	AF143296	Unpub.
Turbot	*Scophthalmus maximus*	TRα	AF302253	9
Eel	*Anguilla anguilla*	TRα	AF302240	9
Official name: NR1A2				
Human	*Homo sapiens*	TRβ	X04707	37
Mouse	*Mus musculus*	TRβ	S62756	154
Rat	*Rattus rattus*	TRβ	J03933	149
Sheep	*Ovis aries*	TRβ	Z68307	2
Pig	*Sus scrofa*	TRβ	AJ238614	3
Chicken	*Gallus gallus*	TRβ	X17504	166
Muscovy duck	*Cairina moschata*	TRβ	Z49151	11
Xenopus	*Xenopus laevis*	TRαA	M35360	6
Xenopus	*Xenopus laevis*	TRαB	M35361	6
Frog	*Rana catesbeiana*	TRβ	L27344	12
Mudpuppy	*Necturus maculosus*	TRβXX	Y16623	13
Zebrafish	*Danio rerio*	TRβ	AF109732	15
Fugu	*Takifugu rubripes*	TRβ	AF302244	9
Rainbow trout	*Oncorhynchus mykiss*	TRβ	AF302246	9
Atlantic salmon	*Salmo salar*	TRβ	AF302251	9
Tilapia	*Oreochromis niloticus*	TRβ	AF302247	9
Japanese flounder	*Paralichthys olivaceus*	TRβ	D45245	14
Turbot	*Scophthalmus maximus*	TRβ	AF302254	9

| Eel | *Anguilla anguilla* | TRβ | AF302241 | 9 |
| Tunicate | *Ciona intestinalis* | TR, CiNR1 | AF077403 | 16 |

Official name: NR1A3

Isolation

The story of the cloning of the thyroid hormone receptor cDNAs shows how two different fields, namely molecular endocrinology and oncology, can fuse and generate an immediate breakthrough in our knowledge. This story starts with the identification of the v-erbA oncogene from the AEV, which causes erythroblastosis in infected chickens[30]. This virus was shown to contain two oncogenes, v-erbA and v-erbB, coming from two distinct cellular progenitors[31,35]. The sequence of v-erbA gene by Dominique Stéhelin's laboratory was in fact the first sequence of a member of the nuclear receptor superfamily to be published[32]. This was made clear when GR and ER cDNAs were cloned by Ron Evans' and Pierre Chambon's laboratories, respectively, and shown to exhibit a striking level of sequence identity with v-erbA[33,34]. The characterization of the cellular counterpart of v-erbA by the laboratories of Björn Vennström in chicken and Ron Evans in human reveals that it was encoding a thyroid hormone receptor[36,37]. In fact, two different genes were found, called c-erbA-1 and c-erbA-2, which encode the two different thyroid hormone receptors, TRα and TRβ. The v-erbA oncogene is derived from the c-erbA-1 gene and encodes a fusion protein with the retroviral gag gene, P75gag-erbA, which is unable to bind T3[36]. Mammalian homologues of chicken TRα were rapidly found but were shown to differ at their C-terminal end[38,39]. It later became clear that c-erbA-1 gene encodes two isoforms, TRα1 and TRα2, which correspond to a non-T3-binding variant that behaves as a dominant negative inhibitor of the bona fide gene product TRα1[40,41]. The structure of the c-erbA-1 loci is even more complex, since an overlap was found between the last exon of this gene, encoding the C-terminal part of TRα2 and the Rev-erbα orphan receptor (see below)[42,43]. The partial purification of the thyroid hormone receptor and photoaffinity labeling experiments with modified thyroid hormones have provided tools to confirm the difference between c-erbA gene products and TRs (see references 44 and 45, and references therein). TRs were found in a number of other species, including *Xenopus*, in which it controls metamorphosis, by low-stringency screening with v-erbA or c-erbA probes.

Recently, a TR-related sequence was found outside vertebrates, in the tunicate *Ciona intestinalis*[16]. In a phylogenetic tree, this sequence is clearly related to the TRs despite the fact that the LBD is only 31% identical to the LBD of human TRs. The DBDs exhibit 75% sequence identity with the DBDs of human receptors. The amino acids involved in T3 binding in the vertebrates receptors are not well conserved and 200 extra amino acids are present in the C-terminus of the protein. Strikingly, this protein did not bind T3 and did not transactivate. It is unable to bind to HREs but it nevertheless acts in transient transfection assays as a dominant negative inhibitor of human receptors. It is only expressed in embryos and larvae, which may suggest a role during development and metamorphosis.

The two human TRs exhibit 88% and 82% sequence identity in their

DBD and LBD, respectively. The nuclear receptors most strongly related to the TRs are the RARs, which display 59% and 34% sequence identity with the DBD and LBD, respectively, of the TRs.

DNA binding

The first DNA target described for the TRs was a TRE located in rat growth hormone gene promoter[46], but it rapidly became clear that TRs can bind as homodimers to a palindromic element, called TREpal based on two AGGTCA motifs[47] (for reviews, see references 72 and 73). This element exhibits the same structure and sequence as the estrogen receptor element (ERE) except that the distance between the two half-sites is 0 nucleotides for the TREpal and 3 nucleotides for the ERE. It was shown that TR can bind to a ERE but that it is then in a transcriptionally inactive form[47]. RARs are also able to bind and transactivate through the TREpal, and heterodimers between RAR and TR were detected on the TREpal[48,51]. The determinants for the discrimination between TREpal and GRE (or between ERE and GRE) were shown to be localized in the C-terminal part of the first zinc finger, a region called the P-box, which forms a recognition helix that interacts with the major groove of the DNA double helix[49,50]. It was proposed that the N-terminal part of the second zinc finger as well as the region of the LBD, called the ninth heptad repeat but now known as helix 10, were important for TR–RAR interaction in the heterodimer[49,71]. The preferred binding site for TRα1 as a monomer was shown to be the TA AGGTCA octamer motif, in which the classical half-site sequence is preceded by the two bases T and A[69].

It was immediately clear that TRs can bind to a wide variety of enhancer elements (see reference 73 for a review). Beside the palindromic arrangement of the AGGTCA core motif, direct repeats were shown to be efficient DNA targets[53,54]. In addition it was shown that the spacing between the two AGGTCA motifs dictates the specificity of either TR (DR4; i.e. a direct repeat in which the half-sites are spaced by four nucleotides), VDR (DR3) or RAR (DR5). This rule should be considered as a simplification since numerous exceptions (TR binding to a DR5, for example[147]) exist[74,241]. In addition, TR was shown to bind to inverted palindromes spaced by three or six nucleotides (IR3 or IR6 elements)[52,60]. Half-site motifs were also described (see reference 73 for references). Three points need to be considered to define a TRE: the consensus hexanucleotide, the orientation of the two half-sites and the spacing between them[73]. Depending on these three points a given TRE can sustain a classical positive response to T3 (i.e. repression with the apo-receptor and activation with the holo-receptor) or a negative activity (repression when hormone is present such as in the TSH genes). This flexibility of TR binding to DNA is well illustrated by the case of TRα1, which is able to bind to DNA as monomer, homodimer and heterodimer (see below)[65]. In contrast, TRβ binds only as homodimer or heterodimer[65,72]. Binding of TR monomer and homodimer is relatively weak, owing to rapid dissociation of the TR–TRE complex when compared to the RXR–TR heterodimer, which is much more stable (see reference 73 for a review). The TR–RAR heterodimer is also much more weakly bound to DNA than the RXR–TR heterodimer. In case of the RXR–TR heterodimer, the preferred binding site was determined by the polymerase chain reaction (PCR) and an electrophoretic mobility shift assay (EMSA) selection of binding sites[83]. The receptor showed marked preference for a DR4 element with a non-random spacer that exhibits a pyrimidine in position three. The best selected spacer sequence was ATCA[83].

Several researchers observed that TR produced *in vitro* was not able to bind DNA efficiently and that a nuclear protein was necessary to enhance the DNA-binding activity of both TRα or TRβ[64,65]. The identification of RXR as a common partner not only of TRs but also of RARs or VDR to help their recognition of correctly spaced direct repeats was a breakthrough in the field[55-59], (reviewed in reference 63). Footprint experiments revealed that TR monomer, homodimer and heterodimer contact different nucleotides in the TREs[62]. It was shown that in direct repeats, RXR occupies the 5' element, whereas TR, RAR or VDR occupies the 3'. In addition a region of the first zinc finger of TR was shown to interact with the second zinc finger in the RXR DBD to promote selective binding to the direct repeat[61,66,81]. These data have also shown that no homodimerization domain exists in the DBD of TRs, an observation that supports the idea that the RXR–TR heterodimer is the physiological unit responsible for the T3 response[66]. In addition it was shown that the LBD further stabilizes but does not change the repertoire of possible binding sites dictated by the DBD[66,82]. The precise heterodimerization surfaces inside the DBDs of RXR and TR were determined by an extensive structure–function analysis[67,68]. These data were fully confirmed by the determination of the crystallographic structure of the RXR–TR DBDs bound to DNA and will be discussed under 'Three-dimensional structure'. The DNA-binding abilities of the RXR–TR heterodimer in the context of chromatin were also determined[70]. It was shown that nucleosome assembly plays an active role in the positive or negative activity of the RXR–TR heterodimer (see below and reference 75 for a review).

TR induces a bending of the DNA molecule when bound to its binding site. Both the RXR–TR heterodimer and the TR–TR homodimer induced bends of *ca.* 10° in the DR4 element. TR homodimers and heterodimers bend DR4 in opposite directions. Thus RXR binding may have an influence on the DNA molecule itself when associated to TR (see reference 76 and references therein).

What is the influence of thyroid hormone in the TR DNA-binding ability? It is clear that T3 induces a conformational change in its receptor bound to DNA that is visible on gel-shift assays[77-79]. Although results obtained by different teams were somehow conflicting, the actual consensus suggests that T3 can attenuate TR homodimer binding to some but not all TREs without affecting the binding of TR monomer and heterodimers[77-79] (see also reference 73 for a review).

Localization of TRs within the cells using specific antibodies reveals that they are located in the nucleus even in the apo-form[30]. Nevertheless, recent studies performed with green fluorescent protein (GFP)–TRβ1 fusion proteins reveal that in the absence of hormone a significant portion of the molecule is in the cytoplasm[84]. The addition of T3 induces a shift of the receptor inside the nucleus, suggesting that nuclear entry may regulate TR activity[84]. A nuclear localization signal has been localized in v-erbA in a region fully conserved in its cellular counterpart[80]. This signal exhibits a strong similarity with the NLS of the SV40 T antigen and is located in the T and A boxes in the C-terminal part of the DBD. It was shown to be active in living cells using a mutated GFP–TRβ1 chimera that remains in the cytoplasm even in the presence of T3[84]. A N-terminally truncated version of TRα, called p43, has been found inside mitochondria and was shown to regulate mitochondrial gene expression[160,161].

Partners

The major partners of TRs are the RXRs, which strongly enhance their ability to bind to specific DNA sequences and contribute to the specificity of action of TR, since the RXR–TR binds specifically to a DR4 element (see reference 73 for a review and references). The region of TRs implicated in the dimerization interface with RXR is located in the DBD and the LBD. The region of helix 10 only stabilizes the dimer, whereas the interaction between DBDs allows specific interactions between monomers and with the correct binding site. For RXR, it has been shown that the N-terminal part of the second zinc finger containing the D-box forms the dimerization interface. For TR, the pre-finger region (i.e. just before the first zinc finger) forms the dimerization interface. In addition the A-box, located just downstream of the T-box, plays an important part, since it impairs binding on other response elements owing to steric hindrance with RXR[67,68] (see section on three-dimensional structure).

Other nuclear receptors have been shown to interact with TRs. The first were the RARs, which were shown to bind in association with TR on the TREpal element[48,51,85]. The dimerization interface existing between TR and RAR was studied and used as a model to describe the ninth heptad repeat that governs the interaction between LBDs of nuclear receptors[71,86]. It was shown that unliganded TR interferes with RAR transactivation of a TREpal-containing promoter[85]. In fact, most of the action of apo-TR on RAR transactivation has been explained by the titration of their common cofactor, RXR[87]. The physiological relevance of the RAR–TR heterodimer is unclear given the stability of the TR–RXR and RAR–RXR heterodimers that render improbable the existence of a TR–RAR heterodimer when RXR is present[87,88].

COUP-TFs (NR2F1 and 2) were shown to interact with TRs and to disrupt their functions[90,91]. Nevertheless, results from the two-hybrid system suggest that the efficiency of this heterodimer formation is much weaker than that of homodimer formation[89]. It was recently shown in a two-hybrid screen done with TRβ1 as a bait that EAR2 (NR2F6), a COUP-TF-related orphan receptor, interacts efficiently with the C-terminal region of TRβ1[92]. This interaction was confirmed by glutathione S-transferase (GST)-pulldown experiments. It was also shown that EAR2 decreased the binding of TRβ1 to its DNA response elements and inhibited TRβ1-mediated transactivation. An *in vitro* interaction between PPAR and TR has also been described but its *in vivo* significance remains elusive[95].

SHP (NR0B2) an unusual orphan receptor containing only the LBD was found as an heterodimeric partner and is able to bind to a variety of nuclear receptors including TR in the presence but not in the absence of ligand[93,94]. This interaction results in an inhibition of the transcriptional activity of TR. Once again the *in vivo* relevance of this interaction has still to be studied.

TRs also interact with a number of transcription factors, such as AP1, p53 and Stat5. TRα is able to repress in a ligand-dependent fashion the activity of AP1[96,97]. Using the interstitial collagenase and c-fos gene promoters it has been shown that this repression is mutual and that each factor inhibits the DNA-binding activity of its partner. Of note, v-erbA is unable to repress AP1 activity and impair the repression mediated by TRα or RARα[96]. It was suggested that this may play an important role in the oncogenic effect of v-erbA. This mutual antagonism has also been observed on the human prolactine gene promoter[99]. A direct interaction between human TRβ and Jun that may explain this interference was observed in solution[98]. This mutual transrepression may occur at least in part by the sequestration of a common partner, the cointegrator CBP/P300[144,145]. It has also been proposed that hormone-activated nuclear receptors prevent c-Jun phosphorylation on its N-terminal Ser 63 and 73 by Jun N-terminal kinase (JNK) and consequently AP1 activation[146]. This inhibition of the JNK pathway by liganded TRs is in accordance with the observation that nuclear receptors also antagonize other JNK-activated transcription factors, such as Elk1 and activating transcription factor 2 (ATF-2)[146].

TRβ also interacts with p53 via its DNA-binding domain[100,101]. As a result of this physical interaction, binding of TRβ to the TRE as a homodimer or heterodimer with RXR was inhibited, resulting in impaired transcriptional activation[100]. Further studies revealed that, in pituitary cell lines such as GH4C1, p53 blocks TRβ-mediated constitutive activation but has no effect on T3-dependent stimulation[101]. T3 apparently reduced p53–TRβ interaction[101]. A cross-talk between TRβ1 and Stat5, a transcription factor mediating the effects of hormones such as prolactine, has been described recently. It has been shown that liganded TRβ1 inhibited Stat5-dependent gene expression and that the two factors interact[102].

TRs not only interact with transcription factors but also with proteins that are part of the general transcription machinery. One of the most interesting examples is the interaction that exists between TRβ and TFIIB, which is part of the pre-initiation complex[106]. Both unliganded and liganded TRβ bind to TFIIB, but the repressor (apo) and activator (holo) forms of TRβ target distinct TFIIB regions. As an activator, the transcriptional activation domain of TRβ interacts with the basic region of TFIIB, which is also a contact site for other transcriptional activators. As a repressor the transcriptional repression domain of TRβ interacts with the N-terminal part of TFIIB that is required for the entry of general transcription factors in the pre-initiation complex[106]. The model is that TRβ interacts with the TFIIB amino terminus to block further assembly of the pre-initiation complex. TRs also interact with the TFIID component of the pre-initiation complex. TFIID contains the TATA-binding protein (TBP) as well as TBP-associated factors (TAFs). A ligand-reversible interaction was observed between TRα and $TAF_{II}28$ in the region of the LBD encompassing helices 3–5, the region known to contain a conserved signature of the LBD[107]. The role of this interaction, which is weaker in presence of ligand, remains unclear. Of note, another TAF, $TAF_{II}55$, was shown to interact with the same region of TRα in a ligand-independent manner[108]. TAFs may also interact with the AF-2 region, which, in the holo-receptor, forms a unique interaction surface with helix 4. This is the case for TAF$_{II}$135, which potentiates the activation of many NRs, including TRs[109].

TRs have been used in addition to other classical receptors, such as RAR and ER, as baits in two-hybrid screens to search for coactivators and corepressors. This has resulted in a number of proteins that were found to interact with either apo- or holo-TRs. Only the classic cases can be mentioned below. It should be kept in mind that the precise definition of a coactivator or a corepressor is far from trivial. For more information, the interested reader is directed to the introductory chapters and reviews in references 103–105.

Three main types of corepressors, NCoR, SMRT and SUN-CoR were shown to interact with unliganded TRs[110–112]. SUN-CoR is a 16 kDa nuclear protein that has been shown to bind NCoR, SMRT and nuclear receptors such as apo-TR. Regions of TRs that are important for corepressor interactions have been studied by mutagenesis as well as by the study of receptors mutated in thyroid hormone resistance syndrome[110,111,113]. Two regions of TR are important for corepressor binding: one is the CoR box, which overlaps with the first helix of the LBD and the other is the ninth heptad repeat at the C-terminal end of the receptor and which is also implicated in dimerization (helix 10)[110,111,115]. The requirement of this region could be simply due to the fact that the corepressors bind to TR as dimers and not as monomers[114,115]. Additionally the AF-2 domain appears to be important for triggering the release of corepressors when hormone is present[116]. Analysis of the nuclear receptor interaction motif in NCoR and SMRT defined a specific signature motif containing an hydrophobic core (ΦXXΦΦ, where Φ is an hydrophobic amino acid)[120,121]. This domain is reminiscent of the LXXLL motif in coactivators and it was suggested that both motifs interact with a common hydrophobic groove in the surface of the receptor. These corepressors, including SUN-CoR, interact with several proteins implicated in chromatin remodeling that were shown to have histone deacetylase activities (see references 103–105 for reviews). Of note, TRβ was shown to interact directly with the histone deacetylase HDAC2 on the rat TSHβ promoter that is negatively regulated by T3[207]. The repressive activity of apo-TR in the context of chromatin-assembled promoters has been studied by Alan Wolffe's group, emphasizing the role of chromatin in the control of transcription by RXR–TR heterodimers[70,75]. Other corepressors for which an interaction with apo-TR was found are Alien, which contains an autonomous repressor domain[117], TRUP, which is identical to the SURF-3/L7a ribosomal protein[118], and NURD, which couples ATPase activity to histone deacetylation[119].

In contrast, the diversity of coactivators is astonishing. TRs were shown to interact with two main coactivator complexes: the p160 complex, which also contains the cointegrators CBP and P300, and the DRIP/TRAP complex. The relationship between these two complexes are far from being clear but it is hypothesized that they bind successively to liganded receptors[103–105]. The p160 family of coactivators comprises three genes called SRC-1, TIF2 and pCIP. A variety of other names exist, illustrating that these molecules were cloned using different receptors and different approaches by a number of laboratories (see references 103–105). These three molecules share a common organization and a good level of identity. They interact with nuclear receptors via LXXLL motifs with a hydrophobic groove of the receptor surface that is created by the AF2AD region and the H3 region of the liganded receptors[122]. The p160 family activates transcription by recruiting the CBP and P300 proteins that also bind directly to the receptors. Interestingly, SRC-1, CBP and PCAF, a CBP-associated factor, all possess intrinsic histone acetyltransferase activity. The acetylation of histones then results in a more open chromatin configuration that allows the basal transcription machinery to access the promoter of the target gene. Interestingly, the SRC-1 knockout mice are resistant to thyroid hormones, suggesting that a defect in SRC-1 or in another related coactivator may be the cause of thyroid hormone resistance syndrome without mutations in TR genes[127] (see below). The DRIP/TRAP complex was coimmunopurified with epitope-tagged TR from T3-treated cells by Bob Roeder's laboratory[123]. It was also found independently by Leonard Freedman's laboratory using liganded VDR and the two complexes were subsequently shown to be identical[124,125]. This multi-subunit complex shares components with the mediator complex that associates directly with RNA polymerase II. It has been proposed that it targets RNA polymerase II to the promoters that should be activated (for a review, see reference 105). This complex was shown to play a direct role in TR signaling since the knockout of TRAP220, which interacts directly with liganded TR, produces mice that have an impaired thyroid hormone receptor function[126]. Numerous other coactivators were shown to bind to TR, such as PGC-1, an adipose tissue specific coactivator[130]; TRIP-1, also known as SUG-1[128,129]; RIP140[131]; NRIF3[132]; RAP250[133]; PSU1[135]; and ASC-2[134].

Ligands

The ligands for TRs are the thyroid hormones, which exist in two forms: T4 (3,5,3′,5′-tetraiodo-L-thyronine) and T3 (3,5,3′-tri-iodo-L-thyronine). These compounds are high-affinity ligands for the receptors[36,37]. The K_d determined for chicken TRα are 0.21–0.33 nM for T3 and 1.4 nM for T4[36]. For TRβ, these values are 2.3 nM for T3 and c. 50 nM for T4[37,38,136]. These values are in accordance with the K_d of the partially purified TR (6 nM for T3)[45]. A number of synthetic analogues of thyroid hormones were tested for their ability to bind the receptors. Among these compounds, TRIAC (3′-tri-iodothyroacetic acid), D-T3 and reverse T3 (3,3′,5′-tri-iodo-L-thyronine) were tested by competition with the labeled T3 for binding to the receptor. In all cases, the respective order of magnitude of the binding was TRIAC, L-T3, L-T4, D-T3 and reverse-T3[36,37]. TRIAC is apparently more efficient on TRβ than on TRα[38]. Strikingly, there is no known antagonist that is able to bind to the receptor but not to induce a transactivation. This may be due to structural constraints in the three-dimensional structure of the receptor. Ligands specific for either TRα or TRβ are under development but details have not yet been published.

The hormone-binding domain was found to be the E domain[137]. It is nevertheless clear that the D domain plays an important role in hormone binding[138]. The C-terminal part of the D domain contains the first two helices of the LBD. It was immediately obvious that, because of eight mutations and a specific nine-amino-acid deletion at the C-terminal part of the protein, the P75gag–v-erbA protein was unable to bind a ligand[36,137]. The study of specific mutations allowed several groups to separate hormone binding and transcriptional regulation (for example, see references 139 and 140 among others). Overall, these predictions have been confirmed by the analysis of the three-dimensional structure of the LBD of the rat TRα protein.

Three-dimensional structure

The three-dimensional (3D) structure of the human TRβ DNA-binding domain forming an heterodimer with the human RXRα DBD on a DR4 element has been determined by X-ray crystallography[141]. The two DBDs have the classical structural fold of nuclear receptor DBDs with one helix containing the P-box interacting directly with the major groove of the DNA recognition sequence AGGTCA and another helix encompassing the second zinc finger thus stabilizing the structure. An interesting aspect of this study is the dimer interface formed between the upstream RXR and the downstream TR molecules. This asymmetric dimerization interface is formed by residues of the second zinc finger of RXR, including part of the D-box and the pre-finger region of TR[141]. There are also contributions from the tip of the first zinc finger and the T-box of TR[141]. These data confirmed the mutational studies that were done by Hinrich Gronemeyer's laboratory[67,68]. Another important aspect of the TR DBD structure is the existence in the A-box of a long α-helix, which contacts the DNA molecule in the spacer region of the DR4 element. The presence of this helix precludes the formation of RXR–TR heterodimers on direct repeats spaced by less than four base pairs.

The LBD of a TR (rat TRα in the presence of DIMIT, 3,5-dimethyl-3'-isopropylthyronine) has also been crystallized and the 3D structure determined by Robert Fletterick's laboratory[142]. This structure reveals the same canonical fold with 12 α-helices as the LBD of apo-RXR or holo-RAR (see reference 143 for a review). Helices 1 and 2 are in fact part of the poorly conserved D domain, whereas helices 3–12 are part of the bona fide E domain. The region encompassing the C-terminal part of H3 and H4 is the strongly conserved Ti domain that forms a hydrophobic groove in association with H12 that is recognized by the coactivators and corepressors. Of note, the hormone is buried inside the structure and plays an important structural role as part of the central hydrophobic core. The hormone-binding cavity is enclosed from above by H5–H6, from below by H7–H8 and along the sides by H2, by the turn between the two β-sheet S3–S4, and by H3, H11 and H12, which form a lid that completely encloses the ligand. The volume of the hydrophobic pocket (600 Å³) is essentially the volume of the ligand (530 Å³). Several amino acids are directly in contact with the ligand among which are Arg 228 (H3) and Ser 277 (loop between S3 and S4), which form hydrogen bonds with the carboxylate group of the ligand, and His 381 (H11), which forms a hydrogen bond with the phenolic hydroxyl of DIMIT. In addition, numerous van der Waals interactions occurred between the ligand and specific residues in the regions mentioned above. Since the TRα LBD forms only a monomer in the crystal, the dimerization interface of the LBD was not studied in detail. Nevertheless, by comparison with the structure of the RXR LBD dimer, the dimerization interface is formed mainly by helix H10 and to a lesser extent by H9 and the loop between H7 and H8. This region of TRα forms a hydrophobic surface that could form a dimer interface. Finally, the AF-2–AD domain forms an amphipathic helix, H12, which nestles loosely against the receptor to form part of the hormone-binding cavity. Together with the structures of apo-RXR and holo-RAR, this data served to establish a model for the conformational changes induced by agonists (for a review, see reference 143).

Expression

The TRα gene is expressed in mammals as two mRNAs of 5.0 and 2.7 kb that are distributed in a number of tissues, such as brain, kidney, heart, etc.[38,136]. Each of these transcripts corresponds to an isoform of TRα: the 5.0 kb transcript to α1, the bona fide receptor and the 2.7 transcript to α2[148]. The ratio between the two transcripts is variable from one tissue to another but TRα2 is prominent in most cases, especially in the brain, when compared to TRα1[40,171]. The TRβ gene is expressed as a single transcript of 6.5 kb that is observed in most tissues[37,149]. The TRβ1 and TRβ2 isoforms generate transcripts at the same size but the TRβ2 is expressed only in the pituitary[153].

Expression of mRNA transcripts for TRα and TRβ has been studied in detail during chicken development[166,170]. TRα was shown to be nearly ubiquitously expressed with some variation of levels from early embryogenesis to 3 weeks post-hatching[166]. This has been confirmed by the demonstration of its expression during neurulation in the chick embryo[169]. In contrast expression of TRβ is more restricted to the brain, eye, lung, yolk sac and kidney and is developmentally regulated in brain where levels of TRβ increased 30-fold upon hatching[166,170]. Variation of TRβ transcript levels was also noticed in lung, eye and yolk sac[166]. No expression was detected in early embryos[169]. TRα expression was also observed in chicken erythrocytic cells at late differentiation stages[168]. Study of TR expression in mammals gave essentially the same picture with TRα being nearly ubiquitous but with particular abundance in the skeletal muscle, brown fat and brain. Interestingly, the non-T3-binding isoform encoding transcript, TRα2, is strongly expressed in brain[40,171,240]. Knockout experiments in which the β-galactosidase cDNA expression cassette was included in the constructs and expressed from the endogenous TRα promoter allowed to confirm the ubiquitous expression of TRα (see below and reference 181 for examples). Expression was observed at 9 days post-coitum (dpc) in the heart, at 10.5 dpc in the central nervous system and at 11.5 dpc in all organs[181]. A developmentally regulated expression of TRα1 was observed in testis and correlated with that of the ERR1 (NR3B1) orphan receptor, a regulator of the TRα promoter[167,174]. TRβ was found highly expressed in brain, liver and kidney[151,240] (see reference 72 for additional references). These data have been complemented by the study of the distribution of TR proteins in various tissues, which has confirmed the expression of both TRs in many tissues such as brain, liver, kidney and testis[172,173]. Several N-terminally truncated isoforms of TRα, the function of which remains unclear, are expressed in chicken erythrocytes[175].

The expression of the two TRs has been carefully studied in the central nervous system[170,176–179,233]. In chicken, TRα is widely expressed at early embryonic stages, whereas TRβ was sharply induced at embryonic day 19, which coincides with the known hormone-sensitive period[170]. A striking differential expression of the two genes was noticed in the cerebellum suggesting that TRα may be useful in both immature and mature neural cells, whereas TRβ may function in late hormone-dependent glial and neuronal maturation[170]. In mammals, TRα2 is always more frequently expressed than TRα1[176,182]. In rats, in situ hybridization experiments show that TRα1 was found to be developmentally regulated with a peak of expression during the first 3 post-natal weeks in the cerebral cortex, amygdala, hippocampus and cerebellum[177]. The TRβ1 transcript was low or undetectable before birth. A striking expression of TRβ was found during rat inner ear development, consistent with the deafness observed in TRβ knockout mice as well as in some patients with congenital thyroid disorders[178]. Functional experiments have also supported the idea that TRβ may play a role in neuronal differentiation (for examples, see reference 179 and below).

The expression of the two TR genes is differentially regulated during amphibian metamorphosis[242–244]. The level of TRβ message is about 1/30th that of TRα mRNA at stage 44–48 before metamorphosis, but accumulates more rapidly than TRα during later stages[243,244]. The maximal level of TRα is found during pre-metamorphosis (around stage 56), whereas the peak of TRβ is contemporaneous to the climax of the metamorphosis as well as to the rise of T3 concentration. TRβ gene expression can be induced by exogenous T3 and a TRE element was found in the promoter of the gene that is a direct target of TRα and TRβ (see reference 245 for a review).

The question of the regulation of TR gene expression by thyroid hormones has been studied extensively (see reference 72 for a review). In amphibians in which TRs play a major role in the control of metamorphosis, it is clear that TRβ and not TRα is directly regulated by T3 through a TRE element found in one of the TRβ promoters[162–165]. In mammals the situation is less clear. Biochemical evidence of the pre-cloning era suggested that T3 may down-regulate the amount of receptor present in pituitary cells[150]. More recent studies focusing essentially on the levels of TR gene transcripts produced conflicting data[151,154,155]. It appears that thyroid hormone treatment does not affect the quantity of TRβ1 mRNA in rat brain, heart, kidney or liver, but in the pituitary gland the level of mRNA is increased three-fold after T3 treatment[151]. TRβ2 expression is down-regulated by T3 in pituitary cells[151–153]. The promoter for TRα does not contain a TRE sequence and is unresponsive to thyroid hormones[156,157]. In contrast, two TREs have been found in the human TRβ1 promoter and a regulation by T3 of the activity of this promoter has been observed in a pituitary cell line[158,159]. Nevertheless, the expression of TRβ1 does not appear to change in the pituitary of TRβ$^{-/-}$ mice, casting some doubts on the physiological relevance of these findings (see below). Another

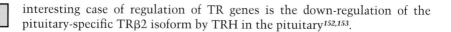

interesting case of regulation of TR genes is the down-regulation of the pituitary-specific TRβ2 isoform by TRH in the pituitary[152,153].

Activity and target genes

Both TRs are repressors in the apo form and transcription activators in the holo form except for some target genes in which the liganded receptor may also behave as a repressor (see references 72 and 73 for reviews). TRs were shown to activate transcription in a ligand-dependent manner both *in vitro*[185,186] and in yeast cells[187,188]. Of note, TRα1 was shown to repress transcription *in vitro* by inhibiting the formation of a functional preinitiation complex[185].

The first structure–function analysis reveals that TRα and TRβ both contain a domain responsible for their transcriptional activity in the C-terminal part of the molecule called AF-2[183]. Further studies demonstrated that the N-terminal region of rat TRα1 and TRβ1 contains an activating function called AF-1[184]. Interestingly, the pituitary-specific TRβ2 isoform, which differs from TRβ1 in the A/B domain, is devoid of an AF-1 function[184]. The AF-2 domain was mapped in the extreme C-terminal part of the receptor[189–191], although other regions, such as the end of the helix 3 that was called Ti domain is also required for transactivation (see e.g. reference 140). Precise mutational analysis revealed that this domain, termed AF-2 AD, is an autonomous activation function like that found in other nuclear receptors such as RAR[189,190]. In the 3D structure this region was shown to form helix 12, which in the holoreceptor forms the lid of the ligand-binding pocket and a surface close to the Ti domain for coactivator binding (see above)[142]. This domain is missing in v-erbA and explains why this oncogenic version does not bind T3 and does not activate transcription[191]. The functional analysis of the v-erbA oncogene has been instrumental in the deciphering of the mode of action of thyroid hormone receptors (see below and reference 197 for reviews).

The discovery that both apo-TR and v-erbA actively repress transcription[192–194] has prompted very strong interest in transcriptional repression and has finally led to the discovery of corepressors[110,111]. It has been shown that a transferable silencing domain is present in TR as in v-erbA and that such a domain can also be found in RAR[195,196]. Both extremities of the LBD play an important role in the repressive activity of apo-TR, in line with the mapping of the NCoR and SMRT interaction domains[110,111,116].

Both v-erbA and TRα are phosphoproteins[24]. The chicken TRα has been shown to be phosphorylated in its N-terminal A/B region by casein kinase II on Ser 12[25]. Chicken TRα is also phosphorylated on Ser 28/29 by protein kinase A and these phosphorylation sites are conserved in v-erbA[24]. It has been shown that this phosphorylation is required for the oncogenic function of v-erbA since mutation of these two serines in non-phosphorylable residues give rise to inactive v-erbA proteins that are unable to block erythroid differentiation[26]. Functional studies have shown that these two residues are implicated in DNA binding of the TRα protein as a monomer but not as a dimer[29]. The TRβ protein is also phosphorylated and its phosphorylation augments its transcriptional activity, by increasing DNA binding efficiency[27,28]. The sites of phosphorylation in the TRβ protein were not mapped.

Target genes of TRs are too numerous and diverse to be discussed here. We will concentrate on major examples such as GH, TRH or TSH regulation, and refer to other examples. The reviews in references 19–23 and 73 are suggested to the interested reader.

The rat growth hormone has been the first TR target gene for which a TRE was found[46,201]. This element was located 180 bp upstream of the transcriptional start site. The regulation of this gene has been scrutinized in great detail and the relationship between TR and other transcription factors such as Pit1 has been studied (see reference 198 for an example). A second TRE located in the third intron of the gene also exists[199]. The two elements exhibit a complex structure with several mutated repeats of the AGGTCA core sequence[199,201]. In contrast to the rat gene, the human GH gene is down-regulated by T3 by a mechanism that is still poorly understood[200].

The hypothalamic–pituitary–thyroid axis and its feedback regulation by thyroid hormones is textbook knowledge, but how do thyroid hormones downregulate the production of TRH in the hypothalamus and TSH in the pituitary? A large number of studies have used these two genes as models for negative regulation by thyroid hormones. TSH is composed of two subunits, α and β, encoded by different genes that are both downregulated by T3. An element that is responsible for this negative regulation and that contains two overlapping mutated AGGTCA motifs as direct repeat was localized adjacent to the TATA box in the human TSHα gene[22]. In addition, TRβ was shown to bind to this TRE[202]. Nevertheless, mutations of this element that impair TR binding do not impair repression by T3, suggesting that the ligand-dependent repression mechanism does not involve direct binding of TR to this TRE[203]. One solution to explain this negative regulation came when Larry Jameson's laboratory observed that corepressors such as NCoR and SMRT activate rather than suppress basal transcription of TSHα and β genes[205,206]. In addition, the same study demonstrated that the DNA-binding activity of TR is not required for the negative regulation of TSHα or β promoters. It has thus been proposed that these promoters are especially sensitive to acetylation and that unliganded TR that binds to corepressors sequesters these molecules from the promoter, resulting in increased acetylation and gene activity. In the presence of T3, TR dissociates corepressors, allowing histone deacetylases to access the promoters. TR also binds CBP that is competed away from the promoter, again reducing its activity. It is believed that factors such as CREB are directly responsible for the activity of the promoter[206].

The situation is also unclear concerning the TSHβ gene, which may eventually follow the same type of complex regulation. An element called NRE (negative regulatory element), located downstream of the transcriptional start site, is apparently responsible for the negative regulation[204]. This element can be placed at several locations inside the rat TSHβ promoter and is recognized by TRs. Keiko Ozato's laboratory recently demonstrated that the TSHβ promoter undergoes a dynamic alteration by T3 consistent with a closing of the promoter by this stimuli[207]. The most striking observation is that the NRE directly recruits histone deacetylases such as HDAC1, an event that is associated with TRβ recruitment to the NBRE[207]. This model of direct ligand-mediated recruitment of histone deacetylases to negatively regulated promoters is clearly different from the model of sequestration of corepressors proposed by Larry Jameson's laboratory[206]. The implications of these two interesting proposals should be further studied before any firm conclusion can be drawn.

The TRH promoter is also down-regulated by T3 and this has been verified *in vivo* by gene transfer experiments in mouse and chick hypothalamic neurons[208,209]. Interestingly, in such studies the rat TRH promoter was efficiently down-regulated by TRβ but not by TRα. This has also been observed in more classical transient transfection assays that mapped two distinct classes of negative TRs in the human TRH promoter[210]. As for TSH genes, it is speculated that liganded TR may bridge an unknown corepressor protein to the promoter. In the absence of ligand this interaction will be disrupted and factors such as CREB may induce gene expression (see reference 23 for a review). This model remains to be experimentally verified.

A number of other target genes have been studied (see references 19–22 and 73). Among this plethora, some are interesting, such as: spot 14[212] and apolipoprotein A1[214] in the liver; EGF receptor, which is down-regulated by T3[213]; the erythroid-specific transcription factor GATA-1[220]; the malic enzyme for which different effects of TRα and TRβ have been studied[217]; and the stromelysin-1 and -2 genes[218]. The mdm2 oncogene, which is positively regulated by p53 and inhibits p53 action, is stimulated by T3[219]. This is striking since p53 regulates the TRα promoter[221] and p53 impairs TRβ transcriptional activity by a direct physical interaction[100,101], suggesting a complex regulatory network between T3 and p53 signaling pathways. Viral genes can also be regulated by T3. This is the case, for example, in the HSV tk promoter which is widely used as a minimal promoter but which contains a TRE[215], and in the HIV LTR, which is regulated by a number of nuclear receptors such as TRs and COUP-TFs[216]. Finally, the case of the apolipoprotein B gene is interesting to mention since editing of mRNA which regulates the formation of two isoforms of ApoB, namely $ApoB_{100}$ and $ApoB_{48}$, is modulated by T3[222]. A systematic survey of T3-regulated genes in the liver has been carried out in mouse using microarray technology[211]. A total of 55 genes were found to be regulated by T3 among the 2225 that were screened. Among these genes, 45 were previously not known to be T3 target genes. Surprisingly, the number of down-regulated genes exceeded the number of up-regulated genes (41 *vs* 14)[211].

Knockout

The knockout of TRα and TRβ genes has been performed. Some isoform-specific knockout mice and mice completely devoid of functional TRα and TRβ genes have also been engineered. The phenotypes of these mice have been described extensively in the literature[229-237]. Reviews of these findings can be found in references 226–228.

Two different kinds of TRα[-/-] mice have been created. In a first study by Jacques Samarut's laboratory, mice with an incomplete knockout of TRα were generated[181]. These mice do not express TRα1 and TRα2 but are still able to generate the small isoforms TRΔα1 and TRΔα2 that arise by an alternative promoter located upstream of exon 8 (see below and reference 239). After birth, the homozygous mutant mice showed a strong reduction of both T4 and T3 associated with a growth arrest and a delayed development of bones and small intestine. These animals generally die within 5 weeks after birth because of their severe hypothyroidy. In line with this notion, they can be rescued by injection of T3 within 1 week. This study demonstrates that TRα plays a crucial role in early post-embryonic development, especially during the weaning period. This is reminiscent of T3-controlled metamorphosis in amphibians and has led to the suggestion that the weaning in mammals may be a T3-controlled switch between a newborn to an adult phenotype[181]. Interestingly, the phenotype of the TRα1[-/-] mice developed by Björn Vennström's laboratory is strikingly different[232]. These animals, which still express TRα2 and TRΔα2 but not TRα1 and TRΔα1, exhibit only a mild hypothyroidism with normal behavior and reproductive ability. The only noticeable features of these animals are a body temperature lowered by 0.5°C and an average heart rate 20% lower than that of control animals. The strong difference between these two types of mice may thus be due to the important role of either TRα2 and/or the two short isoforms TRΔα1 and TRΔα2, two possibilities that are both very surprising[181,232] (for a review, see reference 228). The recent description of a complete knock-out of all the known isoforms encoded by the TRα locus (TRα1, TRα2, TRΔα1, TRΔα2), called TRα[0/0] highlighted the importance of the TRΔα isoforms.

The TRβ[-/-] mice that were generated by Douglas Forrest's laboratory express neither TRβ1 nor TRβ2 isoforms[235]. The homozygous animals exhibit a phenotype different from that of the various TRα[-/-] mice, indicating a differential function of these genes. A goiter and elevated levels of thyroid hormones and TSH suggest a unique role for TRβ that cannot be substituted by TRα in the feedback regulation of the pituitary–thyroid axis. No clear behavioral or neurological abnormalities were detected in the brain. These features were reproduced in TRβ[-/-] mice generated independently by Jacques Samarut's laboratory[231]. In addition, these animals exhibit a strong deficit in auditory function with a primary defect in the cochlea and not in the central nervous system[234]. This phenotype was found in one patient with TRβ deletion and resistance to thyroid hormone and is associated with congenital thyroid disorders. This is consistent with the expression of the TRβ gene found during inner ear development[178]. Interestingly, some aspects of the resulting phenotype, especially the increased TSH production, is suggestive of the pleiotropic abnormalities that are observed in thyroid hormone resistance syndromes. This suggests that the loss of the TRβ gene, a recessive event, results in a similar disorder as that associated with dominant resistance to thyroid hormones[235]. One possible explanation is that in thyroid hormone resistance the dominant negative version of TRβ[-/-] inactivates the second allele, giving rise to a functional TRβ[-/-] knockout. This would imply that, in the resistance syndrome, the mutated allele affects only TRβ function and not TRα, RARs or AP1[235]. Recently, mice were generated with a specific knockout of the TRβ2 isoform, whose expression is restricted to the pituitary, T3-responsive TRH neurons of the hypothalamus, the developing inner ear and the retina[229]. These animals develop similar elevated levels of thyroid hormones and TSH as the TRβ knockout mice, stressing the importance of TRβ2 for the regulation of the pituitary–thyroid axis. In contrast, they have no visible auditory defect, suggesting that this function is governed by TRβ1[229].

Finally, the various TRα and TRβ mice were crossed in order to generate animals with no thyroid hormone receptor function (for reviews, see references 226 and 227). Two genotypes, TRα$^{0/0}$β$^{-/-}$ and TRα1$^{-/-}$β$^{-/-}$, were generated and gave rise to the same striking observation that mice completely devoid of functional thyroid hormone receptors are viable[230,231]. This observation strengthens the notion that thyroid hormones play an important role in post-embryonic development rather than in earlier stages. These animals exhibit an extremely high level of T3 and TSH, suggesting that both TRβ and TRα play a role in the feedback regulation of the pituitary–thyroid axis, although the most important role in this axis is played by TRβ. They also exhibit a bone phenotype that is reminiscent of that observed in TRα single knockout mice, suggesting that most of the action of thyroid hormone in bone maturation is fulfilled by TRα[230,231]. In contrast, the analysis of the various phenotypes observed in the intestine suggests that both receptors play an important role in the development of this organ[231]. These data also reveal that the lack of receptors gives rise to a milder phenotype than that observed in the case of lack of hormone and also suggests that each of the TR genes fulfills a specific function of T3. All these analyses of TR knockout mice provide compelling evidence of the strength of *in vivo* genetic analysis for the definition of the function of genes that play pleiotropic roles in development. Further knockouts, such as those for TRΔ or TRα2 isoforms, will certainly provide more knowledge and surprises about thyroid hormone function.

Transgenic mice expressing the v-erbA oncogene ubiquitously from the human β-actin promoter have been generated[238]. The mice exhibit hypothyroidism, reduced fertility, decreased body mass and behavioral abnormalities. Furthermore, males develop seminal vesicle abnormalities and hepatocellular carcinoma. This work suggests that v-erbA may act as an oncogene in a mammalian organism, consistent with its role in leukemogenesis in chicken. Although it is very different from the phenotype of patients with thyroid hormone resistance, this phenotype is consistent with the notion that v-erbA acts as a dominant negative receptor by transcriptional interference with the bona fide thyroid hormone receptors.

A number of transgenic mice overexpressing mutated versions of TRβ found in thyroid-hormone resistance syndrome were generated[246–249]. Several types of promoters, such as the ubiquitous human β-actin promoter[246,247], the cardiac-specific myosin heavy chain and the cardiac sarcoplasmic reticulum Ca^{2+} adenosine triphosphatase (SERCA2) genes[248], were used to construct these lines. Ubiquitous expression of a dominant negative TRβ product results in a strong perturbation of the pituitary–thyroid axis as well as to weight reduction and hyperactivity that is associated with defective learning[246,247]. These features, as well as the altered cardiac phenotype found when the mutant TRβ is specifically expressed in the heart[248], are reminiscent of the various disorders observed in human patients with T3 resistance syndrome[249].

Associated disease

Mutations of TRβ have been associated with the syndrome of resistance to thyroid hormone[250]. This syndrome, which exhibits a pleiotropic clinical manifestation, is characterized by reduced thyroid hormone action relative to the circulating hormone levels. Most patients are identified by elevation of T4 and T3 levels in association with high TSH levels, suggesting that thyroid hormones are unable to repress the TSH genes. Considerable variation in the severity of the syndrome has led to a classification in generalized or pituitary resistance that are both uniquely associated with mutations in the TRβ gene. Affected patients often present a variable degree of delayed bone maturation, heart abnormalities, hearing defects and mental retardation[250]. This syndrome is autosomal dominant and has been associated with a number of different mutations in the TRβ gene as well as in a unique case in a homozygous deletion of the TRβ gene[250-252]. No mutations in the TRα gene have ever been found[250,255]. The functional study of the TRβ mutations found in patients with resistance syndromes clearly reveal that these mutant receptors are valuable tools to dissect the mode of action of the TRs. Most commonly, mutations are located in the ligand-binding domain of the TRβ, which reduce its affinity for thyroid hormones and interfere with the function of the wild-type TR to produce the dominantly inherited disease[253]. The study of these mutants in the context of transgenic mice have revealed the link between TRβ function and clinical symptoms[246-249]. Recently, some mutant TRβs were found to have impaired interaction with one of the cofactors involved in the regulation of thyroid hormone action[113,254]. For example, a mutant with reduced ligand-dependent transactivation that appears to be due in part to a weaker interaction with the coactivator SRC-1 was described[254]. The identification of patients with clear resistance to thyroid hormone in which no mutations were found in either TRβ or TRα further support the hypothesis that defective cofactors could by themselves cause resistance to thyroid hormones[255]. This has been further supported by the finding that mice in which the SRC-1 gene was inactivated are resistant to thyroid hormones[127].

Another important tool for the study of TR function was the v-erbA oncogene, which is one of the two oncogenes present in the AEV virus[30,32,197]. It has been shown that v-erbA cooperates with v-erbB and other oncogenes products with tyrosine kinase activity to induce full erythroblast transformation[257,258]. In the virus, v-erbA is necessary to block the cellular differentiation that is observed in AEV-transformed erythroblasts. v-erbA is also able alone to transform erythroblasts and fibroblastic cells[256,259]. Further analysis of the action of v-erbA in erythroid transformation revealed that v-erbA substituted for the biological function of the activated estrogen and glucocorticoid receptors, which are essential in normal erythroid progenitors[263].

Functional studies has revealed that v-erbA is unable to bind T3 and acts as transcriptional repressor that exerts a dominant negative activity on the bona fide thyroid hormone receptors[36,193,194,258,260,261]. Examples of genes down-regulated by v-erbA are the erythroid-specific anion transporter Band3[258] and the carbonic anhydrase II gene[261,262]. In addition, v-erbA is able to abrogate the inactivation of AP1 function that is normally performed by TRs and RARs, suggesting that it may alter the balance between proliferation and differentiation in erythroblasts precursor cells[96]. The functional analysis of the v-erbA protein allowed the definition of the AF-2 domain, a region necessary for transactivation in TRα, that was deleted in v-erbA[189,191]. Further studies led to the discovery of a transferable silencing domain present in v-erbA as well as in TRα[195,196]. This silencing domain mediates interactions with the corepressors NCoR and SMRT[110,111,265]. Since v-erbA cannot bind thyroid hormones, the corepressors remain bound to this protein and this complex freezes the TRα protein into its non-liganded repressive conformation[264].

Gene structure, promoter and isoforms

The gene structure of the TRα locus is complex and unique[42,43,148]. The last exon of the TRα gene overlaps for 267 bp with the last exon of the gene encoding the orphan receptor Rev-erbα (NR1D1). Interestingly, the last exon of TRα is specific to an isoform of TRα, called TRα2, that acts as a dominant negative regulator of the normal receptor TRα1 (see below). It appears immediately that the c-erbA-1/Rev-erbα overlap offers a way to regulate the ratio between the α1 and α2 isoforms of TRα[42,43]. The functional consequences of this overlap are discussed in more detail in the Rev-erb section.

Besides the overlap, the TRα gene is split into 10 exons distributed along 27 kb of genomic DNA. The positions of the exons are very well conserved in the chicken and xenopus genes[6,266–268]. The DNA-binding domain is encoded by two exons, the intron between the two fingers being located just downstream from the last cysteine residue of the first zinc finger[148]. The ligand-binding domain is assembled from three exons[148]. The genomic structure of the tunicate TR gene has also been determined and is similar to the one of vertebrates TR, especially in the case of the DNA binding domain[16].

The promoter of TRα has been characterized in human[156,157] and rat[269]. This promoter is TATA-less and rich in GC bases. Its activity is unaffected by thyroid hormones and was shown to be enhanced by glucocorticoids[156]. It has also been shown to be regulated by p53[221]. A role for an intronic sequence located just downstream of exon 1 was deciphered in the rat TRα gene[269]. More recently, a promoter was shown to be present upstream of exon 7 in both mice and human. This promoter controls the expression of short isoform products of the TRα gene, called TRΔα1 and TRΔα2 (see above)[239]. The TRβ promoter is also known in human[158,270] and xenopus[266]. In both species the promoter was shown to be regulated by T3 and to contain bona fide TREs[159,164,165]. In mammals, the *in vivo* significance of this regulation has been challenged by the knockout of the TRβ gene product that does not alter its expression, which was revealed by a β-gal gene inserted in the recombination vector (see above).

Both TRα and TRβ are able to generate several isoforms. The TRα gene is the most complex with at least 4 different forms (see references 72 and 227 for reviews). The most common isoforms are called TRα1 which is the bonafide α-type receptor for thyroid hormones and TRα2 which results from an alternative splicing event in exon 9 of the human gene[41]. The TRα2 isoform exists in all studied mammals[4,39–41,136,148,271] but apparently not in other species such as chicken[166]. A variant of this TRα2, called TRα3 or TRα2vII, contains an internal deletion of 39 amino acids at the junction of the region common to TRα1 and unique to TRα2[40]. These isoforms predominate in the fetal, neonatal and adult brain, with region specific differences in expression[40,171,176,182]. In many other tissues the amount of TRα2 transcript is higher than the amount of TRα1 ones. The protein corresponding to TRα2 has been observed to be expressed in granular cells of the cerebellum as well as in testis[275]. Both TRα2 and TRα3 does not bind T3 but can still bind to DNA albeit with a lower affinity than TRα1[273,274]. This is consistent with the observation that TRα2 does not heterodimerize efficiently with RXR[272,276]. The affinity of TRα2 for T3-responsive elements is decreased by phosphorylation at a site in the unique C-terminus of the protein[272]. It has been clearly shown that TRα2 behaves as a dominant negative inhibitor of TRα1 and TRβ action, most likely through a competition for the occupancy of the TREs[273,274]. The regulation of the DNA binding ability of TRα2 by phosphorylation suggests that the function of this protein in down-regulating TR action is important allowing for rapid, T3-independent modulation of T3-responsive genes[272].

Other mechanisms such as the use of internal ATG codons as translational initiation are able to generate N-terminally truncated TRα isoforms. At least three of them, called p43, p30 and p27 were found in chicken[237]. Of these only p43 contains a complete DNA binding domain. Interestingly this isoform has been found to be able to enter in mitochondria and to regulate mitochondrial gene expression[160,161]. An internal promoter upstream of human exon 8 of TRα controls the formation of two short isoforms, TRΔα1 and TRΔα2[239]. These products can be detected by immunohistochemistry in specific tissues such as intestine (cited in 227). They do not bind T3 or DNA but they strongly inhibit transcription mediated by ligand-activated TR and RAR[239]. The study of the various knock-out alleles that were generated inside the TRα locus suggest that these isoforms have an important role (see 227 for a review).

By contrast the TRβ gene is much more simple since only three isoforms generated from this gene are known up to now[72,227]. In rat a pituitary-specific TRβ2 isoform differs from the regular TRβ1 by an alternative 159 amino acid N-terminal A/B domain[153]. This isoform was also detected in TRH-secreting neurons in the hypothalamus. An identical isoform with a shorter A/B region was also found in the mouse[154]. In contrast to TRβ1, this divergent A/B region did not contain an AF-1 activating function[184]. Its specific knock-out in the mouse highlights the importance of TRβ2 for the regulation of the pituitary-thyroid axis[229]. Two types of N-terminally different isoforms of TRβ, TRβ0 and TRβ2, were found in chicken. TRβ0 contains a very short 14 amino acid long A/B region[166] whereas the TRβ2 isoform contains a 120 amino acid A/B domain that is different from the one of the regular TRβ1 isoform[277]. All these isoforms generate DNA binding and T3-binding molecules that can transactivate gene expression in response to the ligand. Their only difference could thus rely on different AF-1 activity as observed for TRβ2 in mammals[184].

Finally it has to be mentioned that in Xenopus the two TRβ genes are able to produce numerous 5' isoforms, some of which results in different N-terminal ends in the resulting proteins[6]. The auto-regulation of TRβ gene expression by T3 affect only some of these isoforms. Their functional roles remain unknown[266]. Specific deletions of exons in the LBD of either TRα or TRβ genes were also found in various fish resulting in proteins that lack a large part of the ligand binding domain and which may thus be hypothesized to behave as dominant negative[9].

Chromosomal location

Curiously, the chromosomal location of the TRα and TRβ genes has been difficult to establish. It is now clear that human TRα is on chromosome 17 at position q11.2 to q12[17,18,223]. This region also contains the Rev-erbα orphan receptor gene, which overlaps with the TRα gene (see above), as well as the RARα gene. The RARα gene is implicated in a chromosomal translocation with the PML gene in acute promyelocytic leukemia (the t(15;17) translocation). This translocation has been used to locate the TRα gene which is not directly implicated in this event[223].

Human TRβ is on 3p25[37,224]. Some authors have considered c-erbA-2 and c-erbA-β as two different genes, but this has never been confirmed. This region of chromosome 3 is frequently deleted in small cell lung carcinoma and the TRβ gene is sometimes deleted in this malignancy, but there are no data favoring a clear causal implication of TRβ in this disease. Interestingly, the RARβ gene is located also in 3p25 as the orphan receptor Rev-erbβ[225]. Since TRα and Rev-erbα are also linked to RARα on chromosome 17, this argues for two ensembles of old synteny that was duplicated once early on during vertebrate evolution and which give rise to TRα, Rev-erbα and RARα on the one hand, and TRβ, Rev-erbβ and RARβ on the other.

Amino acid sequence for human TRα (NR1A1)

Accession number: M24748

```
  1  MEQKPSKVECGSDPEENSARSPDGKRKRKNGQCSLKTSMSGYIPSYLDKDEQCVVCGDKATGYHYRCITC
 71  EGCKGFFRRTIQKNLHPTYSCKYDSCCVIDKITRNQCQLCRFKKCIAVGMAMDLVLDDSKRVAKRKLIEQ
141  NRERRRKEEMIRSLQQRPEPTPEEWDLIHIATEAHRSTNAQGSHWKQRRKFLPDDIGQSPIVSMPDGDKV
211  DLEAFSEFTKIITPAITRVVDFAKKLPMFSELPCEDQIILLKGCCMEIMSLRAAVRYDPESDTLTLSGEM
281  AVKREQLKNGGLGVVSDAIFELGKSLSAFNLDDTEVALLQAVLLMSTDRSGLLCVDKIEKSQEAYLLAFE
351  HYVNHRKHNIPHFWPKLLMKVTDLRMIGACHASRFLHMKVECPTELFPPLFLEVFEDQEV   410
```

Amino acid sequence for human TRβ (NR1A2)

Accession number: X04707

```
  1  MTENGLTAWDKPKHCPDREHDWKLVGMSEACLHRKSHSERRSTLKNEQSSPHLIQTTWTSSIFHLDHDDV
 71  NDQSVSSAQTFQTEEKKCKGYIPSYLDKDELCVVCGDKATGYHYRCITCEGCKGFFRRTIQKNLHPSYSC
141  KYEGKCVIDKVTRNQCQECRFKKCIYVGMATDLVLDDSKRLAKRKLIEENREKRRREELQKSIGHKPEPT
211  DEEWELIKTVTEAHVATNAQGSHWKQKPKFLPEDIGQAPIVNAPEGGKVDLEAFSHFTKIITPAITRVVD
281  FAKKLPMFCELPCEDQIILLKGCCMEIMSLRAAVRYDPESETLTLNGEMAVIRGQLKNGGLGVVSDAIFD
351  LGMSLSSFNLDDTEVALLQAVLLMSSDRPGLACVERIEKYQDSFLLAFEHYINYRKHHVTHFWPKLLMKV
421  TDLRMIGACHASRFLHMKVECPTELLPPLFLEVFED   456
```

References

1 Nakai, A. et al. (1988) Mol. Endocrinol. 2, 1087–1092.
2 Tucker, M. and Polk, D. (1996) Thyroid 6, 237–243.
3 White, P. and Dauncey, M.J. (1999) J. Mol. Endocrinol. 23, 241–254.
4 Masuda, M. et al. (1990) Nucl. Acids Res. 18, 3055.
5 Lachuer, J. et al. (1996) Poult Sci. 75, 1531–1535.
6 Yaoita, Y. et al. (1990) Proc. Natl Acad. Sci. USA 87, 7090–7094.
7 Schneider, M.J. and Galton,V.A. (1991) Mol. Endocrinol. 5, 201–208.
8 Yamano, K. et al. (1994) Dev. Genet. 15, 378–382.
9 Marchand et al. (2000) J. Mol. Endocrinol. 26, 51–65.
10 Essner, J.J. et al. (1997) Differentiation 62,107–117.
11 Lachuer, J. et al. (1996) Biochim. Biophys. Acta 1310, 127–130.
12 Davey, J.C. et al. (1994) Dev. Genet. 15, 339–346.
13 Safi, R. et al. (1997) J. Mol. Evol. 44, 595–604.
14 Yamano, K. and Inui, Y. (1995) Gen. Comp. Endocrinol. 99, 197–203.
15 Liu, Y.W. et al. (2000) Mol. Cell. Endocrinol. 159, 187–195.
16 Carosa, E. et al. (1998) Proc. Natl Acad. Sci. 95, 11152–11157.
17 Spurr, N.K. et al. (1984) EMBO J. 3, 159–163.
18 Dayton, A.I. et al. (1984) Proc. Natl Acad. Sci. USA 81, 4495–4499.
19 Koibuchi, N. and Chin, W.W. (2000) Trends Endocrinol. Metab. 11, 123–128.
20 Abu, E. (1998) Curr. Opin. Endocrinol. Diabetes 5, 282–287.
21 Muscat, G.E.O. (1995) BioEssays 17, 211–218.
22 Bernal, J. (1998) Curr. Opin. Endocrinol. Diabetes 5, 296–302.
23 Wilber, J.F. et al. (1996) Trends Endocrinol. Metab. 7, 93–100.
24 Goldberg, Y. et al. (1988) EMBO J. 7, 2425–2433.
25 Glineur, C. et al. (1989) Oncogene 4, 1247–1254.
26 Glineur, C. et al. (1990) Genes Dev. 4, 1663–1676.
27 Lin, K.H. et al. (1992) Proc. Natl Acad. Sci. USA 89, 7737–7741.
28 Bhat, M.K. et al. (1994) Proc. Natl Acad. Sci. USA 91, 7927–7931.
29 Tzagarakis-Foster, C. and Privalsky, M.L. (1998) J. Biol. Chem. 273, 10926–10932.
30 Roussel, M. et al. (1979) Nature 281, 452–455.
31 Frykberg, L. et al. (1983) Cell 32, 227–238.
32 Debuire, B. et al. (1984) Science 224, 1456–1459.
33 Weinberger, C. et al. (1985) Nature 318, 670–672.
34 Krust, A. et al. (1986) EMBO J. 5, 891–897.
35 Jansson, M. et al. (1983) EMBO J. 2, 561–565.
36 Sap, J. et al. (1986) Nature 324, 635–640.
37 Weinberger, C. et al. (1986) Nature 324, 641–646.
38 Thompson, C.C. et al. (1987) Science 237, 1610–1614.
39 Benbrook, D. and Pfahl, M. (1987) Science 238, 788–791.
40 Mitsuhashi, T. et al. (1988) Proc. Natl Acad. Sci. USA 85, 5804–5808.
41 Izumo, S. and Mahdavi, V. (1988) Nature 334, 539–542; erratum in Nature 335, 744.
42 Miyajima, N. et al. (1989) Cell 57, 31–39.
43 Lazar, M.A. et al. (1989) Mol. Cell. Biol. 9, 1128–1136.
44 Casanova, J. et al. (1984) J. Biol. Chem. 259, 12084–12091.
45 Ichikawa, K. and DeGroot, L.J. (1987) Proc. Natl Acad. Sci. USA 84,

3420–3424.

46 Glass, C.K. et al. (1987) Nature 329, 738–741.

47 Glass, C.K. et al. (1988) Cell 54, 313–323.

48 Umesono, K. et al. (1988) Nature 336, 262–265.

49 Umesono, K. and Evans, R.M. (1989) Cell 57, 1139–1146.

50 Mader, S. et al. (1989) Nature 338, 271–274.

51 Glass, C.K. et al. (1989) Cell 59, 697–708.

52 Baniahmad, A. et al. (1990) Cell 61, 505–514.

53 Umesono, K. et al. (1991) Cell 65, 1255–1266.

54 Näär, A.M. et al. (1991) Cell 65, 1267–1279.

55 Yu, V.C. et al. (1991) Cell 67, 1251–1266.

56 Kliewer, S.A. et al. (1992) Nature 355, 446–449.

57 Marks, M. et al. (1992) EMBO J. 11, 1419–1435.

58 Zhang, X.K. et al. (1992) Nature 355, 441–446.

59 Leid, M. et al. (1992) Cell 68, 377–395.

60 Saatcioglu, F. et al. (1993) Cell 75, 1095–1105.

61 Perlmann, T. et al. (1993) Genes Dev. 7, 1411–1422.

62 Ikeda, M. et al. (1994) Endocrinology 135, 1628–1638.

63 Laudet, V. and Stéhelin, D. (1992) Curr. Biol. 2, 293–295.

64 Burnside, J. et al. (1990) J. Biol. Chem. 265, 2500–2504.

65 Lazar, M.A. et al. (1991) Mol. Cell. Biol. 11, 5005–5015.

66 Mader, S. et al. (1993) EMBO J. 12, 5029–5041.

67 Zechel, C. et al. (1994) EMBO J. 13, 1425–1433.

68 Zechel, C. et al. (1994) EMBO J. 13, 1414–1424.

69 Katz, R.W. and Koenig, R.J. (1993) J. Biol. Chem. 268, 19392–19397.

70 Wong, J. et al. (1995) Genes Dev. 9, 2696–2711.

71 Forman, B.M. and Samuels, H.H. (1990) Mol. Endocrinol. 4, 1293–1301.

72 Lazar, M.A. (1993) Endocrine Rev. 14, 184–193.

73 Desvergnes, B. (1994) Mol. Cell. Endocrinol. 100, 125–131.

74 Durand, B. et al. (1992) Cell 71, 73–85.

75 Li, Q. et al. (1999) Trends Endocrinol. Metab. 10, 157–164.

76 Shulemovich, K. et al. (1995) Nucl. Acids Res. 23, 811–818.

77 Yen, P.M. et al. (1992) J. Biol. Chem. 267, 23248–23252.

78 Ribeiro, R.C.J. et al. (1992) Mol. Endocrinol. 6, 1142–1152.

79 Andersson, M.L. et al. (1992) Nucl. Acids Res. 20, 4803–4810.

80 Boucher, P. et al. (1988) J. Virol. 82, 534–544.

81 Kurokawa, R. et al. (1993) Genes Dev. 7, 1423–1435.

82 Rosen, E.D. et al. (1993) J. Biol. Chem. 268, 11534–11541.

83 Harbers, M. et al. (1996) Nucl. Acids Res. 24, 2252–2259.

84 Zhu, X.G. et al. (1998) J. Biol. Chem. 273, 27058–27063.

85 Graupner, G. et al. (1988) Nature 340, 653–656.

86 Forman, B.M. et al. (1989) Mol. Endocrinol. 3, 1610–1626.

87 Barettino, D. et al. (1993) EMBO J. 12, 1343–1354.

88 Tini, M. et al. (1994) Mol. Endocrinol. 8, 1494–1506.

89 Butler, A.J. and Parker, M.G. (1995) Nucl. Acids Res. 23, 4143–4150.

90 Berrodin, T.J. et al. (1992) Mol. Endocrinol. 6, 1468–1478.

91 Casanova, J. et al. (1994) Mol. Cell. Biol. 14, 5756–5765.

92 Zhu, X.G. et al. (2000) Mol. Cell. Biol. 20, 2604–2618.

[93] Seol, W. et al. (1996) Science 272, 1336–1339.
[94] Seol, W. et al. (1998) Mol. Endocrinol. 12, 1551–1557.
[95] Bogazzi, F. et al. (1994) J. Biol. Chem. 269, 11683–11686.
[96] Desbois, C. et al. (1991) Cell 67, 731–740.
[97] Zhang, X.K. et al. (1991) Mol. Cell. Biol. 11, 6016–6025.
[98] Lopez, G. et al. (1993) Mol. Cell. Biol. 13, 3042–3049.
[99] Pernasetti, F. et al. (1997) Mol. Endocrinol. 11, 986–996.
[100] Yap, N. et al. (1996) Proc. Natl Acad. Sci. USA 93, 4273–4277.
[101] Qi, J.S. et al. (1997) Mol. Cell. Biol. 17, 7195–7207.
[102] Favre-Young, H. et al. (2000) Mol. Endocrinol. 14, 1411–1424.
[103] Hollenberg, A.N. (1998) Curr. Opin. Endocrinol. Diabetes 5, 314–320.
[104] McKenna, N.J. et al. (1999) Endocrine Rev. 20, 321–344.
[105] Glass, C.K. and Rosenfeld, M.G. (2000) Genes Dev. 14, 121–141.
[106] Baniahmad, A. et al. (1993) Proc. Natl Acad. Sci. USA 90, 8832–8836.
[107] Mengus, G. et al. (2000) J. Biol. Chem. 275, 10064–10071.
[108] Lavigne, A.C. et al. (1999) Mol. Cell. Biol. 19, 5486–5494.
[109] Mengus, G. et al. (1997) Genes Dev. 11, 1381–1395.
[110] Hörlein, A.J. et al. (1995) Nature 377, 397–404.
[111] Chen, J.D. and Evans, R.M. (1995) Nature 377, 454–457.
[112] Zamir, I. et al. (1997) Proc. Natl Acad. Sci. USA 94, 14400–14405.
[113] Safer, J.D. et al. (1998) J. Biol. Chem. 273, 30175–30182.
[114] Zhang, J.S. et al. (1997) Mol. Cell. Biol. 17, 6887–6897.
[115] Zamir, I. et al. (1997) Genes Dev. 11, 835–846.
[116] Baniahmad, A. et al. (1995) Mol. Cell. Biol. 15, 76–86.
[117] Dressel, U. et al. (1999) Mol. Cell. Biol. 19, 3383–3394.
[118] Burris, T.P. et al. (1995) Proc. Natl Acad. Sci. USA 92, 9525–9529.
[119] Xue, Y. et al. (1998) Mol. Cell 2, 851–861.
[120] Perissi, V. et al. (1999) Genes Dev. 13, 3198–3208.
[121] Nagy, L. et al. (1999) Genes Dev. 13, 3209–3216.
[122] Darimont, B.D. et al. (1998) Genes Dev. 12, 3343–3356.
[123] Yuan, C.X. et al. (1998) Proc. Natl Acad. Sci. USA 95, 7939–7944.
[124] Rachez, C. et al. (1999) Nature 398, 824–828.
[125] Ito, M. et al. (1999) Mol. Cell 3, 361–370.
[126] Ito, M. et al. (2000) Mol. Cell 5, 683–693.
[127] Weiss, R.E. et al. (1999) EMBO J. 18, 1900–1904.
[128] Lee, J.W. et al. (1995) Nature 374, 91–94.
[129] Vom Baur, E. et al. (1996) EMBO J. 15, 110–124.
[130] Puigserver, P. et al. (1998) Cell 92, 829–839.
[131] Treuter, E. et al. (1998) Mol. Endocrinol. 12, 864–881
[132] Li, D. et al. (1999) Mol. Cell. Biol. 19, 7191–7202.
[133] Caira, F. et al. (2000) J. Biol. Chem. 275, 5308–5317.
[134] Lee, S.K. et al. (1999) J. Biol. Chem. 274, 34283–34293.
[135] Gaudon, C. et al. (1999) EMBO J. 18, 2229–2240.
[136] Nakai, A. et al. (1988) Proc. Natl Acad. Sci. USA 85, 2781–2785.
[137] Munoz, A. et al. (1988) EMBO J. 7, 155–159.
[138] Lin, K.H. et al. (1991) Mol. Endocrinol. 5, 485–492.
[139] Uppaluri, R. and Towle, H.C. (1995) Mol. Cell. Biol. 15, 1499–1512.
[140] O'Donnell, A. and Koenig, R.J. (1990) Mol. Endocrinol. 4, 715–720.
[141] Rastinejad, F. et al. (1995) Nature 375, 203–211.

142 Wagner, R.L. et al. (1995) Nature 378, 690–697.
143 Weatherman, R.V. et al. (1999) Annu. Rev. Biochem. 68, 559–581.
144 Kamei, Y. et al. (1996) Cell 85, 403–414.
145 Chakravarti, D. et al. (1996) Nature 383, 99–103.
146 Caelles, C. et al. (1997) Genes Dev. 11, 3351–3364.
147 Harding, P.P. and Duester, G. (1992) J. Biol. Chem. 267, 14145–14150.
148 Laudet, V. et al. (1991) Nucl. Acids Res. 19, 1105–1112.
149 Murray, M.B. et al. (1988) J. Biol. Chem. 263, 12770–12777.
150 Samuels, H.H. et al. (1977) J. Biol. Chem. 252, 6052–6060.
151 Hodin, R.A. et al. (1990) J. Clin. Invest. 85, 101–105.
152 Jones, K.E. and Chin, W.W. (1991) Endocrinology 128, 1284–1763.
153 Hodin, R.A. et al. (1989) Science 244, 76–79.
154 Wood, W.W. et al. (1991) Mol. Endocrinol. 5, 1049–1061.
155 Sarapura, V.D. et al. (1993) Mol. Cell. Endocrinol. 91, 75–81.
156 Laudet, V. et al. (1993) Oncogene 8, 975–982.
157 Ishida, T. et al. (1993) Biochem. Biophys. Res. Commun. 191, 831–839.
158 Sakurai, A. et al. (1992) Biochem. Biophys. Res. Commun. 185, 78–95.
159 Suzuki, S. et al. (1994) Mol. Endocrinol. 8, 305–314.
160 Wrutniak, C. et al. (1995) J. Biol. Chem. 270, 16347–16354.
161 Casas, F. et al. (1999) Mol. Cell. Biol. 19, 7913–7924.
162 Yaoita, Y. and Brown, D.D. (1990) Genes Dev. 4, 1917–1924.
163 Kanamori, A. and Brown, D.D. (1992) J. Biol. Chem. 267, 739–745.
164 Machuca, I. et al. (1995) Mol. Endocrinol. 9, 96–107.
165 Ranjan, M. et al. (1994) J. Biol. Chem. 269, 24699–24705.
166 Forrest, D. et al. (1990) EMBO J. 9, 1519–1528.
167 Jannini, E.A. et al. (1994) Mol. Endocrinol. 8, 89–96.
168 Hentzen, D. et al. (1987) Mol. Cell. Biol. 7, 2416–2424.
169 Flamant, F. and Samarut, J. (1998) Dev. Biol. 197, 1–11.
170 Forrest, D. et al. (1991) EMBO J. 10, 269–275.
171 Mitsuhashi, T. and Nikodem, V.M. (1989) J. Biol. Chem. 264, 8900–8904.
172 Tagami, T. et al. (1993) Endocrinology 132, 275–279.
173 Lechan, R.M. et al. (1993) Endocrinology 132, 2461–2469.
174 Vanacker, J.-M. et al. (1998) Oncogene 17, 2429–2435.
175 Bigler, J. and Eisenman, R.N. (1988) Mol. Cell. Biol. 8, 4155–4161.
176 Wills, K.N. et al. (1991) Mol. Endocrinol. 5, 1109–1119.
177 Mellström, B. et al. (1991) Mol. Endocrinol. 5, 1339–1350.
178 Bradley, D.J. et al. (1994) Proc. Natl Acad. Sci. USA 91, 439–443.
179 Lebel, J.-M. et al. (1994) Proc. Natl Acad. Sci. USA 91, 2644–2648.
180 Barres, B.A. et al. (1994) Development 120, 1097–1108.
181 Fraichard, A. et al. (1997) EMBO J. 16, 4412–4420.
182 Bradley, D.J. et al. (1989) Proc. Natl Acad. Sci. USA 86, 7250–7254.
183 Thompson, C.C. and Evans, R.M. (1989) Proc. Natl Acad. Sci. USA 86, 3494–3498.
184 Tomura, H. et al. (1995) Proc. Natl Acad. Sci. USA 92, 5600–5604.
185 Fondell, J.D. et al. (1993) Genes Dev. 7, 1400–1410.
186 Lee, I.J. et al. (1994) Proc. Natl Acad. Sci. USA 91, 1647–1651.
187 Lee, J.W. et al. (1994) Mol. Endocrinol. 8, 1245–1252.
188 Sande, S. and Privalsky, M.L. (1994) Mol. Endocrinol. 8, 1455–1464.

[189] Saatcioglu, F. et al. (1993) Mol. Cell. Biol. 13, 3675–3685.
[190] Barettino, D. et al. (1994) EMBO J. 13, 3039–3049.
[191] Zenke, M. et al. (1990) Cell 61, 1035–1049.
[192] Brent, G.A. et al. (1989) New Biol. 1, 329–336.
[193] Damm, K. et al. (1989) Nature 339, 593–597.
[194] Sap, J. et al. (1989) Nature 340, 242–244.
[195] Baniahmad, A. et al. (1992) EMBO J. 11, 1015–1023.
[196] Damm, K. and Evans, R.M. (1993) Proc. Natl Acad. Sci. USA 90, 10668–10672.
[197] Gandrillon, O. et al. (1995) Int. J. Oncol. 6, 215–231.
[198] Schaufele, F. et al. (1992) Mol. Endocrinol. 6, 656–665.
[199] Sap, J. et al. (1990) EMBO J. 9, 887–896.
[200] Zhang, W. et al. (1992) J. Biol. Chem. 267, 15056–15063.
[201] Brent, G.A. et al. (1991) Mol. Endocrinol. 5, 542–548.
[202] Chatterjee, V.K.K. et al. (1989) Proc. Natl Acad. Sci. USA 86, 9114–9118.
[203] Madison, L.D. et al. (1993) Mol. Cell. Endocrinol. 94, 129–136.
[204] Carr, F.E. et al. (1992) J. Biol. Chem. 267, 18689–18694.
[205] Tagami, T. et al. (1999) J. Biol. Chem. 274, 22345–22353.
[206] Tagami, T. et al. (1997) Mol. Cell. Biol. 17, 2642–2648.
[207] Sasaki, S. et al. (1999) EMBO J. 18, 5389–5398.
[208] Lezoual'ch, F. et al. (1992) Mol. Endocrinol. 6, 1797–1804.
[209] Guissouma, H. et al. (1998) FASEB J. 12, 1755–1764.
[210] Hollenberg, A.N. et al. (1995) Mol. Endocrinol. 9, 540–550.
[211] Feng, X. et al. (2000) Mol. Endocrinol. 14, 947–955.
[212] Liu, H.C. and Towle, H.C. (1994) Mol. Endocrinol. 8, 1021–1037.
[213] Hudson, L.G. et al. (1990) Cell 62, 1165–1175.
[214] Taylor, A.H. et al. (1996) Biochemistry 35, 8281–8288.
[215] Park, H.-Y. et al. (1993) Mol. Endocrinol. 7, 319–330.
[216] Desai-Yajnik, V. and Samuels, H.H. (1993) 13, 5057–5069.
[217] Jeannin, E. et al. (1998) J. Biol. Chem. 273, 24239–24248.
[218] Lopez-Barahona, M. et al. (1995) EMBO J. 14, 1145–1155.
[219] Qi, J.S. et al. (1999) Mol. Cell. Biol. 19, 864–872.
[220] Trainor, C.D. et al. (1995) Mol. Endocrinol. 9, 1135–1146.
[221] Shiio, Y. et al. (1993) Oncogene 8, 2059–2065.
[222] Davidson, N.O. et al. (1990) Mol. Endocrinol. 4, 779–785.
[223] Mitelman, F. et al. (1986) Cancer Genet. Cytogenet. 22, 95–98.
[224] Drabkin, H. et al. (1988) Proc. Natl. Acad. Sci. USA 85, 9258–9262.
[225] Koh, Y.S. et al. (1999) Genomics 57, 289–292.
[226] Forrest, D. and Vennström, B. (2000) Thyroid 10, 41–52.
[227] Chassande, O. et al. (1999) Curr. Opin. Endocrinol. Diabetes 6, 293–300.
[228] Hsu, J.H. and Brent, G.A. (1998) Trends Endocrinol. Metab. 9, 103–112.
[229] Dale–Abel, E. et al. (1999) J. Clin. Invest. 104, 291–300.
[230] Göthe, S. et al. (1999) Genes Dev. 13, 1329–1341.
[231] Gauthier, K. et al. (1999) EMBO J. 18, 623–631.
[232] Wilkstrom, L. et al. (1998) EMBO J. 17, 455–461.
[233] Bradley, D.J. et al. (1992) J. Neurosci. 12, 2288–2302.

234 Forrest, D. et al. (1996) Nature Genet. 13, 354–357.
235 Forrest, D. et al. (1996) EMBO J. 15, 3006–3015.
236 Plateroti, M. et al. (1999) Gastroenterology 116, 1367–1378.
237 Bigler, J. et al. (1992) Mol. Cell. Biol. 12, 2406–2417.
238 Barlow, C. et al. (1994) EMBO J. 13, 4241–4250.
239 Chassande, O. et al. (1997) Mol. Endocrinol. 11, 1278–1290.
240 Strait, K.A. et al. (1990) J. Biol. Chem. 265, 10514–10521.
241 Mader, S. et al. (1993) J. Biol. Chem. 268, 591–600.
242 Baker, B.S. and Tata, J.R. (1990) EMBO J. 9, 879–885.
243 Kawahara, A. et al. (1991) Development 112, 933–943.
244 Yaoita, Y. and Brown, D.D. (1990) Genes Dev. 4, 1917–1924.
245 Shi, Y.B. et al. (1998) Int. J. Dev. Biol. 42, 107–116.
246 Wong, R. et al. (1997) Mol. Med. 3, 303–314.
247 McDonald, M. et al. (1998) Learning Memory, 5, 289–301.
248 Gloss, B. et al. (1999) Endocrinology 140, 897–902.
249 Abel, E.D. et al. (1999) J. Clin. Invest. 103, 271–279.
250 Refetoff, S. et al. (1993) Endocrine Rev. 14, 348–399.
251 Usala, S.J. et al. (1991) Mol. Endocrinol. 5, 327–335.
252 Usala, S.J. et al. (1990) J. Clin. Invest. 85, 93–100.
253 Collingwood, T.N. et al. (1994) Mol. Endocrinol. 8, 1262–1277.
254 Collingwood, T.N. et al. (1998) EMBO J. 17, 4760–4770.
255 Weiss, R.E. et al. (1996) J. Clin. Endocrinol. Metab. 81, 4196–4203.
256 Gandrillon, O. et al. (1987) Cell 49, 687–697.
257 Kahn, P. et al. (1986) Cell 45, 349–356.
258 Zenke, M. et al. (1988) Cell 52, 107–119.
259 Gandrillon, O. et al. (1989) Cell 58, 115–121.
260 Disela, C. et al. (1991) Genes Dev. 5, 2033–2047.
261 Pain, B. et al. (1990) New Biol. 2, 1–11.
262 Rascle, A. et al. (1994) Oncogene 9, 2853–2867.
263 Wessely, O. et al. (1997) EMBO J. 16, 267–280.
264 Bauer, A. et al. (1998) EMBO J. 17, 4291–4303.
265 Busch, K. et al. (2000) Mol. Endocrinol. 14, 201–211.
266 Shi, Y.B. et al. (1992) J. Biol. Chem. 267, 733–738.
267 Zahraoui, A. and Cuny, G. (1987) Eur. J. Biochem. 166, 63–69.
268 Sakurai, A. et al. (1990) Mol. Cell. Endocrinol. 71, 83–91.
269 Lazar, J. et al. (1994) J. Biol. Chem. 269, 20352–20359.
270 Suzuki, S. et al. (1995) Mol. Endocrinol. 9, 1288–1296.
271 Lazar, M.A. et al. (1988) Mol. Endocrinol. 2, 893–901.
272 Katz, D. et al. (1995) Mol. Cell. Biol. 15, 2341–2348.
273 Koenig, R.J. et al. (1989) Nature 337 659–661.
274 Lazar, M.A. et al. (1989) Proc. Natl Acad. Sci. USA 86, 7771–7774.
275 Strait, K.A. et al. (1991) Proc. Natl Acad. Sci. USA 88, 3887–3891.
276 Katz, D. and Lazar, M.A. (1993) J. Biol. Chem. 268, 20904–20910.
277 Sjöberg, M. et al. (1992) Development 114, 39–47.
278 Gauthier, K. et al. (2001) Mol. Cell. Biol. 21, 4748–4750.
279 Plateroti, M. et al. (2001) Mol. Cell. Biol. 21, 4761–4772.

Names

The retinoic acid receptors (RAR) bind to two retinoids, all-*trans* retinoic acid and 9-*cis* retinoic acid. They form the 1B group in the nomenclature of nuclear receptors and are closely related to the thyroid hormone receptors (TR; NR1A). There are three genes, called RARα, β and γ, but in several early papers from the cloning era, other names, such as hap, RARδ or RARε were used. The correspondence with the three known genes is clearly indicated in the table below. They are only distantly related to another group of retinoid receptors, the RXRs (NR2B) that bind 9-*cis* retinoic acid. They are known from a wide variety of vertebrates and were recently cloned from two chordates, Amphioxus and a tunicate. They are not known outside the chordates and there is no clear orthologue in the complete genomes of *Drosophila* or nematode.

Species	Other names	Accession number	References
Official name: NR1B1			
Human	RARα	X06538	1
Mouse	RARα	X57528	4
Mus spretus	RARα	Y10094	Unpub.
Rat	RARα	U15211	7
Chicken	RARα	X73972	3
Xenopus	RARα	X87365	10
Notophtalamus	RARα	X17585	6
Zebrafish	RARαA	S74155	23
Zebrafish	RARαB	L03399	Unpub.
Fugu	RARα	AJ012379	2
Salmon	RARα	AF290610	Unpub.
Official name: NR1B2			
Human	RARβ, hap, RARε	Y00291	11
Mouse	RARβ	S56660	4
Rat	RARβ	AJ002942	Unpub.
Chicken	RARβ	X57340	251
Quail	RARβ	AF110729	252
Official name: NR1B3			
Human	RARγ	M57707	17
Mouse	RARγ	M15848	4
Rat	RARγ	S77804	21
Chicken	RARγ	X73973	253
Xenopus	RARγ	X59396	24
Notophtalamus	RARγ, RARδ	X69944	6
Axolotl	RARγ	AF206315	18
Zebrafish	RARγ	S74156	23
Official name: NR1B4			
Polyandrocarpa	RAR	D86615	294

Isolation

The first retinoic acid receptor, RARα, was cloned in 1987 independently by Pierre Chambon's and Ron Evans's groups[1,5]. It was identified by the screening of cDNA libraries from the breast cancer cell lines MCF-7 and T47D with an oligonucleotide corresponding to a highly conserved sequence in the C domain of steroid receptors and TRs[1]. The resulting clone was shown to bind all-*trans* retinoic acid with an affinity that was estimated to be lower than 10 nM. Using a finger swap experiment with the estrogen receptor, it was further shown that 1 nM retinoic acid was able to activate a reporter construct containing an ERE sequence. Retinol was shown to be poorly active in the same assay. Ron Evans's group performed similar experiments with a clone that was identified by studying an integration site of the hepatitis B virus (HBV) from a human hepatocellular carcinoma. Since the gene in which HBV was integrated revealed a strong level of sequence similarity with other members of the nuclear receptor superfamilly, a cDNA corresponding to this gene was cloned from a human testis cDNA library[5]. Later, the mouse sequence was cloned from an embryonic cDNA library using the human RARα cDNA as a probe[4]. Since retinoic acid was known to have a clear effect on regeneration in amphibians, a RARα cDNA was cloned from the newt *Notophtalamus viridescens*[6]. It was later cloned from a variety of other species, including zebrafish[23], fugu[2] or salmon.

The second retinoic acid receptor, RARβ, has an interesting history. The first sequence of this gene came from Anne Dejean's laboratory, which in 1986 characterized an integration site for the HBV in a case of human hepatocellular carcinoma, before the cloning of RARα or TRs[8]. The cloning of the corresponding cDNA, which was called hap for hepatoma, clearly shows that it was a new member of the nuclear receptor superfamily[11]. Given its high level of sequence identity with RARα, it was later shown by Anne Dejean's and Pierre Chambon's laboratories that the relevant product was a retinoic acid receptor that was thus called RARβ[9]. This receptor exhibits a similar high ability to be activated by all-*trans* retinoic acid, whereas retinol activates it only at high concentrations[9,12]. The mouse RARβ cDNA was cloned by the screening of an embryonic cDNA library using the human RARβ cDNA as a probe[4]. Of note, this gene is known only from mammals and birds, but not in other vertebrates such as amphibians or fishes. The reason for this surprising situation is not known.

The RARγ gene was first identified in the mouse by Pierre Chambon's laboratory during the characterization of the mouse orthologues of the RARα and β genes. By screening a mouse embryonic (11.5 days post-coitum) cDNA library, a new RAR clone was identified and shown to be activated by nanomolar concentrations of retinoic acid in a transient transfection assay using a palindromic TRE-tk-CAT construct as a reporter vector[4]. The human orthologue was soon characterized and shown to produce several N-terminal isoforms[17]. Orthologues in *Xenopus*[24], axolotl[18], newt[6,13] and zebrafish[14,23] were also characterized, showing that, as for RARα, RARγ is present in all vertebrates.

The three RARs exhibit 97–100% sequence identity in their DBDs and 85–90% in their LBDs. The receptors that are most closely related to them are the two TRs, which exhibit 56–59% amino-acid identity in

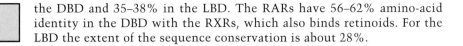

the DBD and 35–38% in the LBD. The RARs have 56–62% amino-acid identity in the DBD with the RXRs, which also binds retinoids. For the LBD the extent of the sequence conservation is about 28%.

DNA binding

Initial studies of the transcriptional properties of RARs were limited by the lack of characterized retinoic acid response elements (RARE). However, RARα was found to activate transcription efficiently from promoters containing the classical palindromic TRE, which is a variant of the TRE found in the rat growth hormone gene[15]. This is not surprising since TR and RAR are closely related and have the same P-box sequence. Further analysis revealed that RAR was able to form heterodimers with TR on such elements[16]. Subsequent analysis of retinoic acid-inducible genes, including the laminin B1[19] and the RARβ[20,22,26] genes, has allowed the characterization of naturally occurring RARE. These elements contain two or more degenerate copies of the AGGTCA motif that are organized in direct repeats or palindromes. Further experiments revealed that a common rule, called the 3-4-5 rule, dictates the specificity of the response elements for nuclear receptors. According to this rule, the spacing between the two AGGTCA motifs dictates the specificity of either RAR (DR5, i.e. direct repeat in which the half-sites are spaced by five nucleotides), VDR (DR3) or TR (DR4)[30,31]. As pointed out in the preceding chapter this rule should nevertheless be used with caution since numerous exceptions exist[28,29] and because these elements are recognized by numerous different complexes of liganded and orphan receptors[41]. Specific proteins present in nuclear extracts were shown to increase the binding efficiency of RARs to these response elements[25,27]. The identification of RXR as a common partner that is required for efficient DNA binding for RAR, TR VDR and other nuclear receptors was an important step in the comprehension of the selection of DNA targets by nuclear receptors[32–36] (reviewed in reference 37). In direct repeats, RXR occupies the 5′ element, whereas RAR occupies the 3′one[38–40,42,46]. RXR–RAR heterodimers can also recognize DR1 and DR2 elements, such as those found in cellular retinoic acid binding protein (CRABP) and cellular retinol binding protein (CRBP)[44,45]. In DR1 elements, the RXR–RAR heterodimer is unresponsive to retinoids and behaves as a repressor of RXR homodimers[47]. In these elements, the polarity of the heterodimer is inverted (5′-RAR–RXR-3′)[47]. Probably linked to this inverted polarity, specific RAR ligands do not induce the dissociation of corepressors from RAR–RXR heterodimers bound to DR1 elements, which explains their repressive action[48]. Elements with very large spacing (DR10–DR200) can also be recognized by RAR–RXR as well as by other heterodimers[47].

The architecture of the RAR–RXR heterodimer was studied in order to locate the heterodimerization interfaces in the DBD and LBD, and to understand the rules governing the interactions between these receptors. A region of the second zinc finger of RAR was shown to interact with the second zinc finger in the RXR DBD to promote selective binding to the direct repeat[38,40,41]. This has revealed that no homodimerization domains exist in the DBD of RARs, an observation that supports the idea that the

RXR–RAR heterodimer is the physiological unit responsible for retinoic acid response[38]. The LBD further stabilizes but does not change the repertoire of possible binding sites dictated by the DBD[38]. The precise heterodimerization surfaces inside the DBDs of RXR and RAR were determined by an extensive structure–function analysis[42,43]. Within RXR the D-box in the second zinc finger forms a surface that interacts with the tip of the second zinc finger of RAR bound to a DR5 element[43]. A second type of dimerization interface, which specifically implicates the RAR T-box and the RXR second zinc finger to the exclusion of the D-box, determines the selective binding to the DR2 element[42].

All the above mentioned studies were conducted on RAR bound to naked DNA but it is clear that *in vivo* RAR should recognize DNA assembled into a complex chromatin structure. In addition, it is known that ligand binding by nuclear receptors induces a reorganization of the chromatin that plays an important role in transcriptional activation (reviewed in references 49 and 53). In this context, it is interesting to note that an RXR–RAR heterodimer was not found bound to DNA by *in vivo* footprint analysis performed on the RARβ2 promoter in the absence of retinoic acid[50]. This contrasts with the occupancy of the TRβ promoter by apo-RXR-TR heterodimer (reviewed in 53). When the ligand was added in the P19 cell culture, the first detectable event was the appearance of a specific footprint in the RARE element of this promoter (called the βRARE)[50]. Rapid chromatin structure alterations were detected by the appearance of DNase 1 hypersensitive sites in this region after RA treatment[51]. Using the *Xenopus* oocyte as an assay system, it was shown that binding of the RXR–RAR heterodimer to the βRARE element induces an alteration of the chromatin assembled on this promoter[52] (see 'Partners' below).

A nuclear localization signal has not been mapped in the RARs, but an Arg- and Lys-rich sequence in the vicinity of the DBD has been outlined in a study of the ER and PR NLS[295]. The nuclear localization of RARs appears to be regulated by several physiological signals that modulate the phosphorylation status of the receptor (see reference 296 and references therein).

Partners

Essential partners for RARs are the *9-cis* retinoic acid receptors RXRs, which are essential for their recognition of response elements and efficient DNA binding, and which play an important role in the integration of the hormonal signal. As discussed above, two dimerization interfaces mediate the interaction between RXR and RAR: one in the DBD, which plays an important role in the selection of the correct response element[42,43], and one in the LBD, which is independent of the DNA-binding activity of the complex. The dimerization interface in the LBD is essentially identical to that for the RXR–TR heterodimer (see the relevant chapter) and has been mapped by detailed mutagenesis studies. Most of our knowledge in this field came from the study of the 3D structure of the heterodimer formed by the isolated LBD of RXR and

RAR (see 'Three-dimensional structure' below). The TR–RAR heterodimers that were detected before the discovery of the important role of RXR are now believed to be physiologically irrelevant[55].

RARs have been shown to interact functionally and/or physically with a number of transcription factors and it is often not easy to understand the molecular basis of these interactions. One particularly interesting case is the mutual antagonistic relationship existing between RAR and the AP-1 transcription factor (see reference 60 for a review). Using a number of genes, such as osteocalcin[56], stromelysine[57], collagenase[58,64] or TGFβ[59], it has been shown that RAR inhibits AP-1 driven transactivation and that AP-1 represses RAR-mediated transcription. Interestingly, selective retinoids that dissociate the inhibition of AP-1 activity from the classical RARE-dependent activation of transcription have been identified[61,62]. These compounds are promising therapeutic agents in a number of pathologies and provide valuable tools to address the mechanism of an AP-1/RAR cross-talk. In some cases, such as the osteocalcin promoter, this mutual repression occurs as a result of competition between the two transcription factors for the same DNA-binding site[56]. This situation is nevertheless relatively rare. The discovery of the important role of CBP as a cointegrator have suggested that RAR and AP-1 may compete for this limiting factor, leading to mutual inhibition[65,66]. Other authors have suggested that RAR (as well as TR or GR) can block AP-1 activity by inhibiting the Jun amino-terminal kinase (JNK), which enhances c-Jun transcriptional activity by specific phosphorylation events[67]. These mechanisms are certainly active in many cases but probably cannot explain some observations[63]. In other cases it has been proposed that RAR inhibits the binding of Jun to target DNA[58,59]. Recently, a specific physical interaction between RAR, but not GR, and c-Jun has been reported[63]. This interaction impairs Jun dimerization activity and thus abolishes the Jun/Jun or Jun/Fos dimerization and DNA-binding activity.

Other transcription factors for which a mutual interference with RAR has been described include the Fli-1 oncogene, a member of the Ets family. The DBD of RARα and the N-terminal region of Fli-1 are necessary for this mutual inhibition that was proposed to occur via a direct or indirect physical interaction[68]. Recently, it was observed that retinoic acid decreases the activity of the β-catenin-LEF/TCF signaling pathway, which is implicated in colon cancer[69]. RARs, but not RXRs interact with β-catenin in a retinoid-dependent manner and RARs compete with TCF for β-catenin binding. The activity of retinoic acid on RAR-responsive promoters is potentiated by β-catenin[69]. Interactions between RARs and other transcription factors, such as Myb[297], Oct factors[298] or Sp1[299], were also demonstrated.

As steroid receptors, RARs can also interact with calreticulin via the conserved GFFKR motif found at the end of the first zinc finger[76]. This interaction abrogates RAR–DNA binding and thus calreticulin inhibits retinoic acid signaling. RARα1 was shown to be phosphorylated by TFIIH and cdk7 in its AF-1 domain and to interact with these two proteins[154] (see 'Activity and target genes' below). RARα was shown to physically interact with the cellular retinoic acid binding protein II (CRABPII), which can enhance the transactivation by RARα–RXRα on DR5 element. Interestingly, CRABP is a target gene of RAR–RXR, suggesting that a positive regulatory loop takes place between these proteins[165].

RARs can regulate transcription through a direct physical interaction with TBP-associated factors, such as TAF$_{II}$28[70] or TAF$_{II}$135[300]. RARβ was shown to interact physically with the adenovirus E1A protein[71]. This protein was previously shown to behave as a coactivator for RARβ in the context of the RA induction of the RARβ2 promoter[72]. A synergistic effect of RAR and of TBP was observed in this promoter and E1A was shown to be required for this synergy[73]. Since E1A directly interacts with TBP[74], the observation of a direct interaction between E1A and RARβ via the AF-2 function of RARβ2 clearly supports the notion that this complex plays an important role in the initiation of transcription pre-initiation[71]. Physical interaction between RAR and other virally encoded proteins involved the BZLF1 protein of the Epstein–Barr virus[75].

RARs have been central in the identification of nuclear receptor coactivators and corepressors since numerous pull-down assays or two-hybrid screens have been performed using RARs as bait. Since RARs bind a very similar set of coactivators and corepressors to those of TRs, which were extensively discussed in the preceding entry, we will not discuss them in detail. The reader can find adequate references in the TR entry as well as in the following reviews: references 77–80. We will just briefly mention the information that is specifically relevant for the retinoic acid field. RARs bind in a AF-2 and ligand-dependent manner to the three members of the p160 family, SRC-1, TIF2 and p/CIP[77], as well as to the common cointegrators P300 and CBP[80]. The role of the p160 coactivators and of the chromatin remodeling mediated by histone acetyl-transferase containing complexes in RAR–RXR activation has been studied in vitro[85]. As discussed above, the interaction with CBP could be part of the cross-inhibition between RAR and AP-1 signaling[65,66]. The inhibition of RAR transactivation by the EIA protein is caused by its ability to prevent the assembly of a CBP–nuclear receptor coactivator complex[81]. CBP and P300 have distinct roles in retinoic acid-induced F9 cells differentiation: P300 is required for this differentiation, whereas CBP is dispensable[82]. In contrast, both P300 and CBP are required for retinoic acid-mediated apoptosis[82]. As TR, RAR interacts with the TRAP220 protein, a member of the DRIP/TRAP/SMCC complex[79,83]. Nevertheless, in contrast to TR–RXR activation, RAR–RXR activation was not impaired in TRAP220 null mice, suggesting that TRAP220 does not play an important role in retinoid signalling in vivo[84]. RARs are also able to interact with the following coactivators (see reference 77 for a review): TIF1, SUG1, RIP140, NSD1, PSU1 and ASC-2.

As apo-TR, apo-RAR binds to the corepressors N-CoR and SMRT (reviewed in references 77 and 78) and this interaction, which persists in the holo-RAR bound to DR1 elements, explains the repressive effect of RAR–RXR heterodimers bound to these elements[48]. The study of the transcriptional repression mediated by retinoic acid receptors has

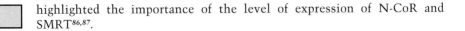

highlighted the importance of the level of expression of N-CoR and SMRT[86,87].

Ligands

The retinoic acid receptors bind all-*trans* retinoic acid and 9-*cis* retinoic acid with a high affinity, exhibiting K_d values in the 0.2–0.7 nM range[88] (see reference 106 for a review). The various subtypes of RAR exhibit similar affinity for these ligands but the kinetics of the binding could be different. For example, 9-*cis*RA is six-fold more rapidly displaced from RARα than from RARβ[90]. Both ligands compete for binding to RAR, suggesting that they recognize the same hydrophobic pocket in the LBD. Nevertheless, using various truncation mutants of RARα, it has been suggested that all-*trans* RA and 9-*cis*RA interact with distinct but overlapping regions of the LBD[89]. Met 406 and Leu 410 in RARα were proposed to be important for the binding of 9-*cis*RA to RARα[92]. Detailed mutagenesis studies and biophysical studies, such as fluorescence quenching, delineate the role of specific amino acids that are located in the N-terminal part of the LBD in the binding of the ligands[91,93,95]. Among these, in RARα, are Ser 232 located in helix 3, Ile 270 in helix 5 and Val 395 in helix 11[91,95,291]. These amino acids are also associated with the recognition of the isoform specific retinoids by the various types of receptors (see below and references 95 and 291). It has been shown by several methods that the LBD of RARs exhibits a conformational change upon ligand binding[93,94]. This has been fully confirmed by the resolution of the 3D structure of the RARγ LBD complexed to all-*trans* RA (see 'three-dimensional structures' below).

Members of the RAR family are activated by a number of physiologically occurring retinoids, among which are all-*trans*-4-oxo retinoic acid, all-*trans*-4-oxo retinal, all-*trans*-4-oxo retinol, and all-*trans*-3,4-didehydro retinoic acid (see reference 106 for a review). This last compound was found in chick wing bud in which its ectopic application causes mirror-image duplications of the digits, as all-*trans* RA, as well as in mammalian tissues[96]. All-*trans* retinol was shown to bind to all three RARs with an affinity *c.* 4–7-fold lower than all-*trans* RA[98]. The compound all-*trans*-4-oxo-retinoic acid was shown to activate RARβ with an IC_{50} value of 4.2 nM, and to be available *in vivo* in which it is a modulator of positional information during *Xenopus* development[99]. 4-oxo-retinol was also shown to bind and activates to the RARs at the nanomolar range and to induce differentiation of F9 cells as all-*trans* RA[100]. Another compound, 14-hydroxy-4,14-retro retinol, was found active *in vivo*[97].

Given the importance of retinoids in a number of pathologies and the very broad expression pattern of RARs and RXRs, it is not surprising that a strong research effort has been directed towards the identification of isoform-specific agonists and antagonists. One of the first antagonists, Ro 41-5253, was shown to counteract RA effects on HL60 cell differentiation and was only active on RARα[101]. This compound does not influence RARα–RXR heterodimer formation and DNA binding, and induces a conformational change of the LBD, which is different from that generated by all-*trans* RA or specific agonists[102]. Other researchers have developed a series of synthetic agonists that are specific to either RARα, β or γ, which allowed the role of each receptor in several processes such as embryonal carcinoma cells (F9 or P19) or differentiation of acute promyelocytic leukemia cells (NB4) to be deciphered[103,104]. These compounds, which are associated with RXR-specific ligands, became valuable tools for the study of the dynamic function of the RAR–RXR heterodimer. Mutagenesis, 3D structure determination and modeling studies have pointed to the role of three specific residues of the LBD (Ser 232, Ile 270 and Val 395 in RARα) that are different in the three subtypes and responsible for their differential recognition[95,291]. Of note, some compounds are able to dissociate the role of RARs as AP-1 inhibitors and their role as classical ligand-induced transcription factors acting through RAREs[61,62].

Three-dimensional structure

The 3D structure of the human RARβ DBD in solution has been determined by the NMR method[105,107]. This structure is very similar to that of other nuclear receptors with two perpendicular α-helices. The only difference with the GR DBD structure is that the second helix is a little shorter in the RARβ DBD and is followed by a region with extended backbone structure[107].

The crystal structure of the RARγ LBD bound to all-*trans* retinoic acid was one of the first 3D structures of a nuclear receptor LBD to be determined[108] (reviewed in reference 109). The comparison of this structure with that of apo-RXR has led to the first model of conformational change of nuclear receptors LBDs induced by the binding of the ligand[109]. The overall fold of the holo-RARγ LBD is similar to that of the apo-RXRα except for the C-terminal part, including the helix H12, which corresponds to the AF-2–AD region, which folds back towards the LBD core, forming a lid that seals the entry point of the ligand[108]. The ligand enters the hydrophobic pocket via electrostatic guidance and, by a 'mousetrap' mechanism, reposition the helix H12. The ligand-binding pocket is delineated by helices H1, H3, H5, the β-turn between H5 and H6, the loop between H6 and H7, H11, the loop between H11 and H12, and H12 itself. Comparison of the RARγ sequence with the two other subtypes immediately shows that the three positions (Ser 232 in H3, Ile 270 in H5 and Val 395 in H11 in RARα) defined as critical for the recognition of isoform-specific ligands were part of the hydrophobic surface that directly contacts the ligand[112]. These three amino acids are the only ones that vary between the various RARs in the ligand-binding pocket. The structural basis of the specific recognition of a given RAR subtype by its specific ligand has been studied by solving the structure of RARγ LBD bound by its specific agonist BMS189961. Interestingly, this compound is a racemic mixture of two enantiomeres, the S form, called BMS270395, and the R form, called BMS270394. The R form is active and binds RARγ strongly, whereas the S-enantiomere shows no measurable binding. The crystal structure of the RARγ LBD bound to these two compounds has allowed to understand how the receptor discriminates the two enantiomeres as well as to better understand the ligand conformation and its effect on ligand activity[110].

A new breakthrough in the structural analysis of the nuclear receptors was provided by the crystal structure of the heterodimeric complex between RARα and RXRα LBDs[111]. In this structure the RXRα LBD was a mutant (F318A) that displays a constitutive activity and that was found to be complexed to a fatty acid likely to account for this constitutive activity (see the entry for RXR). The RARα LBD was bound to a selective antagonist, BMS614. This structure provides crucial information on the dimerization interface formed in heterodimers and on the basis of the antagonists action on RAR activity. In the RARα LBD, the interaction surface with RXRα is very similar to the RXRα homodimerization interface. It involves residues from H7, H9, H10 and H11 as well as the loops H8–H9 and H9–H10, with a major role played by the helices H9 and H10 of each monomer. The analysis of the sequence conservation in residues homologous to those involved in this interface in other receptors provides a structural basis to the specificity of heterodimer formation. The BMS614 ligand recognizes RARα specifically, mainly due to the three divergent amino acids of the pocket Ser 232 in H3, Ile 270 in H5 and Val 395 in H11. Interestingly, in the BMS614-bound RARα LBD, the surface for binding of the coactivator, which involves residues of H12 and of the groove formed by the C-terminal part of H3, the L3–L4 loop and H4, is not generated. This is mainly because the quinolyl extension of BMS614 prevents the positioning of helix 12 in the active conformation due to a clash between Ile 412 of H12 and BMS614. This structure is reminiscent of the ER LBD bound to the antagonists raloxifene or tamoxifene, suggesting that the structural basis for antagonism are similar in different receptors.

Expression

The expression of the three RARs during mouse embryonic development was studied extensively by Pierre Chambon's laboratory by *in situ* hybridization[115-118,130]. The expression patterns of RARs are particularly interesting for study in the light of the highly disputed role of all-*trans* RA in limb-bud development[134] and on its role in hindbrain patterning[135]. The role of retinoids in limb-bud development has been studied using different model organisms, such as mouse, chicken and amphibians in the context of limb regeneration[140]. Numerous other roles of retinoids during embryogenesis are the subject of intense scrutiny (e.g. heart morphogenesis, retina development, identity of motor neurons, etc.) but their description is beyond the scope of this book. For the sake of clarity, we will describe these data separately for each gene, starting first with northern blot experiments, then describing the *in situ* hybridization results in mouse and other species and, finally, discussing the regulation of the expression by retinoic acid and other compounds.

The human RARα gene was found expressed as two transcripts of 3.8 and 2.8 kbp in embryonic but not adult skin, lung and several cell lines, such as the breast cancer cells T47D or the hepatoma cell line HepG2[17]. The expression in various cell lines was precisely quantified using specific antibodies that were developed towards each RAR subtype[114,144]. In the mouse, two transcripts of 3.8 and 2.5 kb were observed in most tissues with high levels in skin, lung, testis and moderate levels in brain, liver, kidney, spleen and intestine[4]. The rat RARα gene was shown to be expressed as a 3.2 kb mRNA in most tissues[5]. The highest levels were found in the hippocampus, adrenals, cerebellum, hypothalamus and testis.

During the period of early morphogenesis of the mouse embryo, RARα expression starts at E8.5 in the neural epithelium of the forebrain, and in the neural crest cells of the frontonasal and first pharyngeal arch regions in which it is associated with migration[117]. Later, the gene is ubiquitously expressed with stronger levels in hindbrain and spinal chord. During organogenesis, RARα transcripts are uniformly distributed in the limb bud at E10 and at later stages were found in all tissues of the growing limb bud[115]. In other organs, RARα transcripts were almost ubiquitous at all stages of organogenesis. The only places in which a distinct low abundance of RARα transcript were detected are the liver and some places of the central nervous system[116]. In the nervous system, RARα was also ubiquitously expressed in a territory caudal to its anterior limit of expression, which corresponds to its expression domain in the early hindbrain neural tube[118]. In the zebrafish, the expression of RARα is found at high levels in the anterior neural tissue as well as in tail bud. Transcripts are also found at low levels in the entire embryo[23]. This pattern is similar to that observed in Xenopus[10]. This is interesting since, in the mouse, no higher expression levels have been detected in the posterior region of the embryo.

The regulation of RARα expression by retinoic acid has been studied by several teams[119-122]. It was shown that, in human hepatoma cells, its expression is not modified by retinoic acid treatment[119]. In a model of retinol-deficient rats, a change in RARα transcript distribution was observed after retinoic acid treatment[120]. In fact, the RARα gene can generate several N-terminal isoforms generated from different promoters (reviewed in reference 113). One of these isoforms, RARα2, is induced by retinoic acid in embryonal carcinoma F9 cells[121]. The promoter governing the expression of this isoform contains a functional RARE[122]. Retinoic acid also has a strong effect on the RARα protein itself. In fact, it was shown that RA triggers degradation of the RARα protein (and of its oncogenic variant PML-RARα) by the proteasome pathway[123]. In the zebrafish, RARα expression is induced by retinoic acid in the eye[23].

Human RARβ gene is expressed as two mRNA species of 3 and 2.5 kb, the size of this smaller transcript being variable from one tissue to another. This gene is expressed at low abundance in ovary, uterus, mammary cell lines, adult and fetal spleen and hematopoietic cells HL60 and K562[11]. Expression was much higher in prostatic adenoma and kidney[11]. In fetal and adult liver only the 3 kb transcript was present at low levels, but the 2.5 kb species was observed in hepatocellular carcinomas and in hepatoma cell lines. RARβ was not found expressed in the HepG2 hepatoma cell line[17]. The two transcripts are different in their 5' region, consistent with the notion that several 5' isoforms are transcribed from the RARβ gene. In the mouse a 3.4 kb RARβ mRNA was found abundantly expressed in the brain, liver, kidney and heart. In muscle a supplementary transcript at 1.9 kb was observed. No transcript was found in the adult skin in both mouse and human[4,17]. In the rat, several transcripts from 2.4 to 9.5 kb were observed with strong expression in brain, pituitary, kidney, colon, uterus, ovary, testis, prostate, adrenal and eye[12].

Expression of RARβ during early mouse embryogenesis was observed in pre-somitic stage embryos with high levels in the lateral region[117]. RARβ was also found expressed in the hindbrain and spinal cord. A shift from RARβ to RARγ expression occurs caudally during neural tube closure. In the limb bud, RARβ transcripts were found at E14.5 restricted to the interdigital mesenchyme, suggesting a link with apoptosis[115]. The expression pattern of RARβ during organogenesis is complex and will not be described extensively. Of note, it is expressed in various epithelia, such as tracheobranchial, intestinal and genital tract epithelia, suggesting a role in mediating retinoid effects on these epithelia[116]. Expression is also noticeable at E10.5 in the anterior facial mesenchyme (around the eye and in the frontonasal region) but not in the maxillary process and rostral part of the mandibular arch. RARβ is also found later at the base of the skull as well as from E10.5 in the eye in which its expression is important in the pigmented retina[116]. The expression in facial primordia was also examined during chicken development and expression was found in a subset of migrating neural crest cells[124,125]. In contrast to RARα, RARβ is not found expressed in mouse heart[116]. A dynamic, complex and specific expression pattern is also observed during nervous system development[118].

An important feature of RARβ expression is its clear up-regulation by retinoic acid in a number of organs, tissues and species[119,120,127-29,144]. This was observed in hepatoma cell lines[119] as well as in a model of retinoid-deficient rats[120]. The importance of this regulation has prompted several teams to characterize the RARβ regulatory regions and it was found that this gene encodes several N-terminal isoforms of which only RARβ2 is regulated by retinoic acid[136]. The promoter governing the expression of this isoform was shown to contain a RARE that rapidly became a paradigm of retinoic acid regulation[20,22,26]. Several groups have described transgenic mice bearing the RARβ2 promoter linked to a lacZ reporter gene to study retinoic acid regulation in vivo[127,128]. The RARE element of the RARβ2 promoter was also used fused to an heterologous minimal promoter in such transgenic studies (see 'Knockout' below)[137-139]. Studies done in transgenic mice have also shown that the 5'UTR of the RARβ2 isoform is very important in the regulation of the gene in vivo, highlighting the importance of a correct regulation of this gene[126].

The human RARγ gene is expressed as a 3.3 kb mRNA and is the predominant RAR subtype expressed in both adult and fetal skin[17]. High expression is also observed in the mammary cell line T47D as well as in human teratocarcinoma cells. A low level was also observed in lung. In mouse, a transcript of the same size was also found in skin as well as at a lower level in lung and in spleen[4].

During mouse early embryogenesis, RARγ was present in the primitive streak region throughout the period of neurulation. Interestingly, caudally, RARγ replaces RARβ expression in the neural tube after closure[117]. The expression pattern of RARγ during organogenesis is clearly distinct from those of RARα and RARγ[116,130]. In the limb bud, RARγ was found as RARα uniformly distributed at E10.5 but later the RARγ transcripts became specific to the cartilage cell lineage and to the differentiated skin, in contrast to RARα[115]. RARγ is expressed throughout the maxillary and mandibulary components of the first arch, in the mesenchymal part of the trachea, in the genital tubercules, and in some restricted parts of the heart in the endocardial cushion and in the vicinity of large blood vessels[116]. RARγ is found at E12.5 in all pre-cartilaginous mesenchymal condensations, and from E13.5 in all cartilages and differentiating squamous keratinizing epithelia[130]. In accordance with this observation, RARγ was found during late gestation in both the dermis and epidermis. It is also expressed in the developing teeth and whisker follicles[130]. RARγ is not expressed in developing neural structures[118]. In amphibian RARγ expression was observed at both the mRNA and protein levels in normal and regenerating limbs of the newt, mostly in mesenchymal tissues as well as in the epidermis[13,131]. In the early development of *Xenopus laevis*, the anterior and posterior ends of the neurula, as well as, later, the head mesenchyme and tail bud express RARγ. This specific localization is consistent with the temporal and spatial pattern of retinoic acid sensitivity of *Xenopus*[24]. These findings were confirmed by the study of the expression of several N-terminal isoforms of RARγ during early development of another amphibian, the axolotl[18].

Several isoforms of RARγ exist in mammals, deriving from alternative splicing and alternative promoter usage[132,141]. A retinoic acid regulation of RARγ expression was observed *in vivo* in a model of retinoid-deficient rats[120]. Indeed, the promoter governing RARγ2 expression was shown to contain a RARE that is responsible for its specific induction by retinoic acid treatment[133].

Activity and target genes

The retinoic acid receptors are transcriptional activators and, as numerous nuclear receptors, including their close relatives the TRs, they contain two autonomous transactivation functions, AF-1 and AF-2, located respectively in the N-terminal A/B region and in the E domain[54,142]. The AF-2 region, which is ligand dependent, seems to be more potent than the AF-1, which is ligand independent, since receptors containing deletions in their C-terminal regions behave as dominant negative regulators of the wild-type receptors[145,146]. This has allowed the demonstration that, as for the RXR–TR heterodimer, the RAR–RXR heterodimer behaves as a repressor in the apo-form. A motif essential for the AF-2 region has been characterized as an amphipatic α-helix, which is located in the C-terminal end of the E domain and corresponds to the helix 12 in the 3D structure (see above)[143]. The role of RXR in the transcriptional activation by RAR has been controversial and is discussed in more detail in the RXR entry.

RARs have been shown to be active in yeast *S. cerevisiae* and *S. pombe*[147,148]. Yeast has been used as a model system to circumvent the problems linked to the wide distribution of RARs and RXRs in mammalian cells, which can lead to misleading conclusions. This has, for example, allowed study of the regulation of the ApoA1 gene by RXR–RAR and demonstration that RARβ but not RARα transactivates the RARE present in region A of the ApoA1 promoter[147]. This has also allowed the respective contributions of RXR and RAR in the activity of the heterodimer activated by all-*trans* RA and 9-*cis* RA to be tested[149]. The RARs also behave as transcriptional activators in *in vitro* transcription assays[150,151]. Using such cell-free systems. it has been shown that ligands induce a conformational change of the LBD. In such a system, however, the RXR–RAR heterodimer is partially active in the absence of ligand[150]. The addition of ligand induces a transcriptional activity, which is dependent on the AF-2 of RAR but not on that of RXR. Nevertheless, in other systems, the RXR AF-2 was shown to be important (see the RXR entry)[150]. Cell-free transcription assays were also done in the context of reconstituted chromatin templates and, in such systems, the activation by a RXR-specific ligand was subordinated to the binding of an agonist ligand to RAR[151]. In these systems the coactivator p300 and other coactivators were shown to enhance the transcriptional activity in a synergistic fashion[85,151].

The characterization of RAR proteins with specific antibodies has shown they are phosphoproteins (see reference 106 for a review). This has, for example, been shown for the RARβ2 that is phosphorylated on tyrosine residues. It was further shown that treatment by retinoic acid does not alter RARβ2 phosphorylation, whereas the phosphorylation of RARβ1 and β3 was greatly enhanced by retinoic acid[144]. In addition, several groups have shown that known protein kinases, such as PKC and PKA, can regulate the transcriptional activity of RARs working on RARE-dependent promoters[152,153,157]. Further characterization of RAR phosphorylation sites have shown that RARs are phosphorylated in their AF-1 domain that contains sites for proline-directed kinases, such as cdk7 or TFIIH[154,155]. In RARα1, the phosphorylated residue in the AF-1 domain has been identified as Ser 77[154]. Furthermore, RARα1 can be phosphorylated by PKA at Ser 369 that is located in the AF-2 domain[159]. RARα1 is phosphorylated in these two residues in both COS cells and F9 cells, and similar phosphorylation sites are present in other RARs, such as RARγ2, which is the predominant form in F9 cells[155,159,160]. The role of these phosphorylation events has been studied in the context of the induction of differentiation of F9 cells in primitive and parietal endoderm by retinoic acid[155,156]. It was shown that AF-2 phosphorylation of RARα was required to induce parietal endodermal differentiation, whereas phosphorylation of AF-1 was apparently dispensable[156]. The opposite situation is observed for RARγ, since AF-1 but not AF-2 phosphorylation was required for primitive endodermal differentiation[155].

A very large number of target genes implicated in a wide variety of functions from embryonic development to liver metabolism were described for retinoic acid receptors. We will thus describe some interesting cases, since it is not feasible to discuss in detail the role and regulation of these genes in this book. One of the first target genes to be discovered for RARs were the laminin B1[19], CRBP[44], CRABP[45] and growth hormone (GH)[164] genes as well as the RARβ gene itself (see 'Gene structure promoter and isoforms' below). The laminin B1 gene is regulated during the differentiation of F9 cells induced by retinoic acid[19]. The RARE found in its promoter contains a DR3 element and a half-site motif located in the same orientation 14 bp upstream[161]. The CRBP and CRABP are regulated by retinoic acid in a number of physiological situations, including P19 cell differentiation[162]. A RARE with a DR2 structure was shown to be present in the mouse cellular retinol binding protein I (CRBPI) promoter and to confer RA inducibility[44]. The mouse CRABPII is regulated by RA via DR1 and DR2 elements located on the proximal promoter (up to c. 1000 bp) on which RAR–RXR heterodimers bind[45]. Interestingly, the human gene is also regulated by RA in skin fibroblasts but via a DR5 element located 5.6 kbp upstream of the transcriptional start site[163]. Of note, RARα physically interact with CRABP that was proposed to behave as a transcriptional regulator of RAR–RXR heterodimers[165]. Other proteins implicated in RA metabolism are regulated by RA. This is the case, for example, of the cytochrome P450RAI (CYP26), which is implicated in the catabolism of RA[166].

Among the most interesting target genes of RARs are the Hox genes, which are key players in the specification of the anteroposterior axis during development (reviewed in references 169 and 170). Several Hox genes have been found as RA-regulated genes in cell culture studies and among them ERA-1 latter identified as HoxA1[167,168]. *In vivo*, systemic or local administration of excess RA to developing embryos has profound effects on axial patterning and specification of regional identities in a number of structures including the central nervous system, axial skeleton and limbs (see e.g. references 173 and 174). This is invariably accompanied by repatterning of Hox expression domains, suggesting a specific role for Hox gene products as molecular transducers of the morphogenetic signals provided by retinoids[171-175]. Of note, a sequential activation of Hox genes by retinoic acid has been observed with the anterior genes responding faster and more strongly than posterior genes, reflecting the colinearity of embryonic expression of Hox genes[171,172]. Molecular analysis both *in vivo*, using mainly transgenic mices and *in vitro* have started to decipher the molecular mechanisms implicated in RA regulation of the Hox gene. RAREs have been identified in Hox D4[176,181], HoxB1[177-180,185], HoxA1[183], HoxA4[184], as well as in enhancers located inside the HoxA and HoxB clusters[182]. Hox gene regulation have been found not only in mouse but also in chicken, *Xenopus* and zebrafish, and even in chordates such as amphioxus[187] or tunicates[188] (reviewed in reference 186).

Many other transcription factors are RA-regulated genes. It is the case of the Oct 4 gene, the expression of which is inhibited by RA during differentiation of ES cells. This down-regulation is governed by an upstream enhancer that is inhibited by RA by an indirect mechanism. However, it has also been shown that a promoter proximal sequence harboring three repeats of the AGGTCA motif, including one DR1 element juxtaposed with a Sp1 site is responsible for the RA-mediated down-regulation of the promoter in ES cells[189-191]. Interestingly, this element is also recognized by orphan receptors, such as members of the COUP-TF group (NR2F), which repress the Oct4 promoter through this sequence[190,192]. RXR–RAR heterodimers activate this promoter in a RA-dependent fashion.

Nevertheless, since the affinity of COUP-TF for the RARE element is high and since COUP-TF factors are up-regulated during ES cell differentiation, the net result is an RA-mediated down-regulation of Oct4 expression during this process[192]. The situation is even more complex since the upstream element also plays an important part in this regulation. It has been shown using *in vivo* footprinting that RA-mediated down-regulation coincides with a loss of promoter occupancy *in vivo* in both the distal and proximal elements[193]. Among the other transcription factors regulated by RARs it is interesting to cite c-*fos*[194,195] and c-*jun*[196,197], which are repressed and activated, respectively, by RA. AP-2[198], Pit1[199] or Cdx1[200] are among other targets of RARs.

RA has been considered as a morphogen, being responsible for the morphogenetic effects of the zone of polarizing activity (ZPA) in the limb bud (reviewed in reference 207). Nevertheless, further experiments have shown that the effects of RA in this system are indirect and that RA action appears to be mediated by the morphogen encoded by the sonic hedgehog gene. It has been shown in a number of systems including chicken and zebrafish, that RA controls sonic hedgehog expression[201,202,205,206]. The zebrafish sonic hedgehog promoter contains a DR5 element, which is regulated by RXR–RAR[202]. Nevertheless, in *Xenopus* primary neurogenesis, there is a down-regulation of SHH by retinoic acid[203]. Interestingly, both Shh and RA controls the expression of Lefty-1, a gene that is implicated in left-right asymmetry[204].

The regulation of matrix metalloproteinase genes, such as collagenase and stromelysine, have been widely used as a model system for understanding the role of RARs in gene regulation. RA down-regulation of collagenase and stromelysine has been shown to be mediated through AP-1 sites located in these promoters, and it is clear that RAR and AP-1 exhibit mutual inhibitory effects (see 'Partners' above)[57,58,64,208]. In contrast, stromelysine 3 gene expression is directly activated by retinoic acid[209]. Other cases of antagonistic effects of RAR and AP-1 include the human octeocalcin gene in which Fos-Jun and RAR–RXR compete for the occupancy of the same target sequence[56]. RARs also regulate a number of factors implicated in metabolism. This is the case for the uncoupling proteins UCP1 and UCP2 in brown adipose tissue[210], the 17β-hydroxysteroid deshydrogenase, an enzyme implicated in steroidogenesis[211], medium-chain acyl coenzyme A deshydrogenase (MCAD) involved in fatty acid metabolism[212], the PEPCK gene in the liver[213], the alcohol dehydrogenase ADH3[214], the liver bile acid transporter gene NTCP[215] or the ApoA1 gene[147].

Several viral genes are the targets of RARs. Human papillomavirus 18 (HPV18) transcription is repressed by RA[222], whereas the cytomegalovirus (CMV) enhancer is induced by RARs[223,226]. Interesting links between TGFβ, RA regulation and HIV-1 expression in monocytic cells have been underscored[224,225]. Lastly, several genes implicated in very diverse physiological or regulatory processes are worth mentioning, since their analysis reveals interesting observations on RAR gene regulation. It is the case of the γF crystallin gene[216], the MHC class I genes[217], the IL2 gene[218], the oxytocin gene[219], several members of the TGFβ superfamily[59,204,220,224], platelet-activating factor receptor gene[221]. Of note, the interspersed repetitive Alu element present in the primates genomes at 500 000 to 1 million copies contains a functional RA element[227]. RAREs have also been searched for in whole genomes using the PCR approach in order to isolate new RA-responsive genes (see reference 228 as an example).

Knockout

The knockout of all the members of the RAR has been performed in the mouse (see references. 158 and 254 reviews). Given that each of the RAR genes may encode several isoforms that have been selectively inactivated and that each strain bearing an inactivation of a given RAR isoform has been crossed with another RAR- or RXR-knockout strain, mainly by Pierre Chambon's laboratory, the complete description of these results is not possible here. We will just briefly summarize the main results obtained by these in-depth genetic analyses. We will first describe the phenotype of the RAR single knockout mice and then some of the features of the compound mutant mice.

RARα mice (i.e. mice in which all known isoforms generated by the RARα locus have been inactivated) display some of the defects of the post-natal vitamin A deficiency syndrome with a decreased viability, growth deficiency and male sterility, caused by a degeneration of the seminiferous epithelium[230]. Most of these defects can be reversed by the administration of RA. In addition, these mice exhibit various congenital malformations with low penetrance, such as homeotic transformation and malformation of the cervical vertebrae or webbed digits[230]. In contrast, mouse lacking only the RARα1 isoform are viable and showed no apparent phenotype, with no evidence for a compensatory increase of RARα2[229,230]. This suggests that the RARα2 isoform carries most of the functions that are altered in the complete RARα knockout.

The RARβ or RARβ2 knockout mice were described as phenotypically normal[255,256]. Nevertheless, it was later shown that RARβ2$^{-/-}$ mice display abnormalities in the vitreous body in the eye, suggesting that this isoform plays an important role in the formation of the vitreous body[238]. This defect was more pronounced in RARβ2/γ2 double knockout mice. The RARβ mice were also shown to exhibit locomotor defects that were strongly increased in RARβ$^{-/-}$RXRβ$^{-/-}$ or RARβ$^{-/-}$RXRγ$^{-/-}$ mice[237]. This defect was correlated to impaired dopamine signaling, suggesting that retinoids are directly implicated in the regulation of brain function. In addition, the RARβ and RARγ knockout mice display an abrogation of hippocampal long-term potentiation and depression, two widely studied forms of synaptic plasticity, suggesting an important role for retinoids in higher cognitive functions[301].

The situation for RARγ is reminiscent to that existing for RARα: the RARγ null mice exhibit a phenotype with some of the defects known in the vitamin A deficiency syndrome, whereas the mice carrying an inactivated RARγ2 isoform are phenotypically normal[231]. Among the defects observed in the RARγ mouse that are interesting to note are growth deficiency, early lethality and male sterility, owing to abnormalities in the prostate and seminal vesicles. These defects can be prevented or reversed by the administration of RA. In addition, several congenital abnormalities are observed: homeotic transformation and malformations of cervical vertebrae, fusion of the trachea rings, webbed digits and agenesis of the Harderian gland[231]. These mice were also resistant to the teratogenic effect of RA described in various studies[173,174,231].

In contrast to the single knockout mice, the mice bearing mutations in two RAR genes exhibit a number of defects leading to a dramatically reduced viability[232,233]. The animals present all the known manifestations of the fetal vitamin A deficiency syndrome, with respiratory tract and heart defects, genital tract abnormalities in the male and the female, as well as dramatic ocular abnormalities (reviewed in reference 158). In addition, these animals display congenital malformations that were not seen in the fetal vitamin A syndrome, probably because it is difficult to achieve a complete dietary deprivation of vitamin A[232,233]. Among these defects it is interesting to note various skeletal abnormalities (agenesis or malformation of cranio-facial skeletal elements, of limb bones, as well as homeotic transformations of cervical vertebrae). The appearance of atavistic skeletal structures, such as the pterygoquadrate cartilage, normally present in reptilians was also noticed, although the interpretation of such finding is far from being obvious (see reference 254 for a review). Among this collection of phenotypes and developmental defects, some were studied in more detail. RARβ2/γ2 was shown to exhibit retinal dysplasia and degeneration, and the analysis of this phenotype has revealed the important and specific role of RARβ2 in eye formation[238]. The RARα and RARβ2 compound mutants display renal malformations, suggesting that, within a given renal cell type, their concerted function is required for renal morphogenesis[236]. Another compound mutant, RARα[-/-]RARβ[-/-], exhibits malformations in structures derived from neural crest cells that allows the study of the role of RARs in the patterning of pharyngeal arches and caudal hindbrain[235].

Another important conclusion obtained by the analysis of the RAR and RXR knockout mice is the clear genetic demonstration that the retinoid signal is transduced *in vivo* by heterodimers of RAR and RXR[234]. Indeed, a marked synergy was observed between the effects of mutations in RXRα and RARs. No synergy was observed between the effects of RXRβ and RXRγ mutations and those of any RAR mutations, suggesting that RXRα is the main RXR implicated in the developmental functions of RARs[234]. Nevertheless, in the brain, a synergy between RARβ and RXRβ or γ for locomotor activity and dopamine signaling was observed[237].

All the analysis of the *in vivo* function of each RAR in the mouse clearly reveals that each RAR gene performs a specific function, even if some redundancy exists, suggesting an overlap between these functions (discussed in reference 158). The specific role of each RAR has also been observed by studying their role in retinoic acid-induced differentiation of F9 embryonic teratocarcinoma cells[239,240]. This has been confirmed by the use of RAR isoform specific ligands[241]. The use of these ligands in the context of wild-type RARγ[-/-] F9 cells also reveals that knockouts can induce artefictual redundancies between individual RAR genes[241].

Normal or dominant negative versions of RARs have been used in a number of studies to generate transgenic mice. Overexpression of a constitutively active version of RARα in the ocular lens of mouse induces cataracts and microphtalmia that are also found in RA-induced teratogenesis[242]. The differentiation of neutrophils was blocked at the promyelocyte stage by a dominant negative RARα mutant deleted of its AF-2 function at amino acid 403 (RARα403)[243]. This finding has interesting implications for the study of acute promyelocytic leukemia that harbors a translocation in the RARα locus (see below). The dominant negative mutant was also used to study hematopoietic cells differentiation[245,250] or the role of RARs in the formation and maintenance of a functional epidermis[246,247]. It has been shown that this role in epidermis formation is linked to an altered lipid processing as well as to a disruption of epidermal barrier function[248]. The overexpression of a RARβ4 isoform (which exhibits a truncated A/B domain), under the control of an MMTV promoter, induces hyperplasia and tumors in lung, breast and other tissues, such as salivary gland[244]. This finding is interesting since RARβ4 overexpression was noticed in human lung tumor cell lines when compared to normal lung tissue, suggesting a direct role of this isoform in lung cancer.

Lastly, transgenic mice were also used to study RA distribution and RARE activation *in vivo*. In many cases this has been achieved by the construction of RARE minimal promoter LacZ transgenic mice that were used to define specific domains of transcriptional activity of RARs during mouse embryogenesis[137,139]. By using such a construct, RA signaling was shown to be regulated during lung development, further stressing the importance of mechanisms that control regional RA availability[138]. The RARβ2 promoter, which contains a potent and well-known RARE, has often been used in such studies[127,128]. By an elegant approach using *in vivo* a UAS-GAL4-RAR-LBD and a UAS-LacZ reporter, Thomas Perlman's group have studied the *in vivo* activation pattern of RARs[249]. This system appears powerful for locating the regions in which RARs are transcriptionaly active in the developing transgenic mice.

Associated disease

The RARs have been associated with several diseases, among which cancer is one of the most important (see below). It has been shown that two RARβ isoforms, RARβ2 and RARβ4, may have antagonistic roles in lung cancer formation (see reference 244 for references and discussion). Several lines of evidence also link RAR function to skin diseases[280]. It has been suggested that ultraviolet irradiation of human skin causes reduced levels of RARγ and RXRα, and a loss of RA induction of RAR target genes[281]. It is thus clear from these results as well as from the study of knockout and transgenic mice phenotypes as well as from RAR expression pattern that retinoids play an important role in many skin diseases (see references 280 and 281 for references).

Nevertheless, the most direct implication of RAR in human disease is given by acute promyelocytic leukemia (APL), which is caused by several translocations that all implicate the human RARα gene (see reference 274 for a review). The specific chromosomal translocation found in APL, t(15q17)(q22;q11–q21), fuses the RARα gene (located in 17q21) to a gene called PML (for promyelocytes)[257,258]. This event results in a fusion protein called PML–RARα that contains the DBD and LBD and RARα fused in their N-terminal part to the PML gene product. The fusion protein is still RA responsive but displays several functional differences with the normal RARα gene product[259,260]. The discovery of the PML–RARα fusion has received much attention, since it has been shown that RA treatment induces partial remissions in APL patients by inducing differentiation of the APL cells (see reference 274 for a review and below). When overexpressed in U937 myeloid precursor cells, the PML–RARα protein induces a differentiation block reminiscent of that observed in APL patients, showing that it plays an important role in the development of the disease[264]. This is consistent with the differentiation block also observed in transgenic mice overexpressing a dominant negative version of RARα[243]. The direct role of the PML–RARα in APL has also been shown by the analysis of transgenic mice bearing a PML–RARα transgene under the control of a myeloid-specific promoter. After several months, these animals develop leukemias that were characterized as APL, recapitulating the features of human disease[270,271]. Nevertheless, several lines of evidence suggest that, in these animals, other events have to occur in order to induce the complete APL phenotype[278,279].

The RARα gene can translocate with other genes, such as the PLZF gene product, located on chromosome 11, which exhibits a Krüppel-like zinc-finger domain[262]. This t(11;17)(q23–q21) chromosomal translocation is reciprocal and also gives rise to a RARα–PLZF product that has recently been shown to be important for the development of the disease[278]. Other genes implicated in RARα translocation in APL are NPM on chromosome 5[282] and NuMA on chromosome 11[283].

The functional characterization of the PML–RARα fusion protein has revealed that it interferes with normal retinoid-mediated transactivation[259-261]. The PML–RARα protein can heterodimerize with RXR but, in the absence of RXR, it can also form homodimers as well as heterodimers with PML protein through its coiled-coil domain[261]. The PML–RARα homodimers have been shown to play an important role in the association with corepressors as well as in the transformation efficiency of the protein[277]. In contrast to RARα, the PML–RARα protein can bind strongly to DNA as an homodimer to direct repeat sequences with 1–5 spacers[263]. In addition to its altered transcriptional activity on RAREs, the PML–RARα protein also behaves abnormally on AP-1, since, in contrast to the inhibitory effect of RARα on AP-1, PML–RARα activates AP-1-mediated transcription[265]. PML–RARα and PLZF–RARα proteins behave as transcriptional repressors and are able to interact with corepressors, such as SMRT or NCoR[271-273]. Interestingly, treatment with high doses of RA disrupt the PML–RARα/SMRT complex, leading to a transcriptionally active protein. This may explain the potent differentiating effect of pharmacological doses of RA on APL cells in culture. In contrast, the PLZF–RARα protein is insensitive to RA treatment because the PLZF moiety of the fusion protein constitutively interacts with corepressors, independently of RA[271-273]. In accordance with these findings, transgenic mice that harbor the PML–RARα transgene are sensitive to RA treatment, whereas mice containing the PLZF–RARα protein are resistant[271] (see reference 274 for a review). It has also been shown that other partners, such as the AML1 oncogene, found in a translocation associated with a leukemia, also play an important role in PML–RARα transforming effects[276].

The subcellular location of PML-RARα protein in normal or RA-treated cells has been under intense scrutiny and has led to the molecular characterization of nuclear structures called nuclear bodies or PODs (for PML oncogenic domains), which appear to play an important role in the nucleus (see references 274 and 275 for reviews). In normal cells, the PML protein is a part of these nuclear bodies, whereas in APL cells the nuclear bodies are disrupted into an aberrant microparticulate pattern[266–268]. Treatment of APL cells with RA triggers a reorganization that generates normal nuclear bodies. This suggests that the PML–RARα fusion protein could exert a dominant negative effect by diverting a subset of proteins from their natural sites of action. The nuclear bodies are also the sites of replication of several viruses, such as the adenovirus, which trigger a reorganization of these nuclear structures[269]. Interferon treatment enhances the expression of PML and reduces nuclear bodies reorganization promoted by viral infection. Thus, the study of the PML–RARα fusion protein has provided a fascinating set of findings that allow better understanding of the structural organization of the nucleus[275].

Gene structure, promoter and isoforms

The genomic organization, promoter structure and isoform expression of the three RAR genes have been characterized in detail in a number of species from mouse to amphibians (reviewed in reference 113).

The genomic organization of the human RARα gene has been determined[285]. The cDNA generating the well-known RARβ2 isoform is split into eight exons among which exon 3 codes for the N-terminal A region and exon 4–10 code for the common part of the receptor. As for TR, the DBD is encoded by two exons with the intron lying at the same position as in TR. The structure of the human RARα and mouse RARγ gene is identical[9,132].

The promoter controlling the RARα1 isoform has been characterized as a GC-rich promoter containing no TATA or CCAAT boxes, and no RARE. This promoter contains a site for the Krox20 transcription factor, which plays an important role in the early development of the mouse central nervous system[284]. A total of seven isoforms (α1–α7), differing in their 5′UTR as well as in the A region, are generated from the RARα gene and have been characterized[121]. These isoforms occur from alternative splicing and promoter usage of at least eight different exons that are located upstream of the exon encoding the domain B, i.e. the C-terminal part of the A/B domain. The two major isoforms generating N-terminally different proteins exhibit distinct expression patterns as well as distinct regulation. In contrast to the α1 isoform, α2 is transcribed from a promoter that contains a RARE and which is inducible by retinoic acid[122]. The differences in the A domain may generate different AF-1 functions that are able to regulate target genes differentially.

The RARβ gene also codes for several isoforms, among which three contain different A regions linked to identical B regions[136]. These isoforms are controlled by different promoters and exhibit distinct expression patterns and regulation. The RARβ2 isoform is inducible by RA and is controlled by a specific promoter that contains the well-known βRARE element[20,22,26,286]. Another isoform, called RARβ4, is generated from β2 promoter and differs from the β2 protein by its N-terminal part that initiates in a non-AUG codon, CUG[289]. The RARβ2 promoter is also regulated by the TLX orphan receptor which modulate its retinoic acid inducibility in the retina[287]. A cAMP response element is also involved in RARβ2 activation by RA in P19 embryonal carcinoma cells[288]. RARβ2 inducibility is also regulated by other RARs, such as RARγ1, which neutralize the effects of RA on RARβ2 induction[129]. These findings suggest that several factors play important roles in the autologous regulation of the RARβ gene in different organs. In addition, the expression of this isoform is regulated post-transcriptionally by the structure of its 5′UTR that contains short open reading frames that control its translation[126].

As for the two others, the mouse RARγ gene generates several (at least seven) N-terminal isoforms that harbor specific expression patterns[141,290]. For example, skin contains almost exclusively RARγ1 transcripts. The RARγ2 isoform is regulated through a RARE element that is embedded in Sp1 sites, suggesting, as in the case of RARβ, a complex regulation of RA inducibility[133].

Chromosomal location

The locations of the three RAR genes in the human, mouse and rat genomes have been mapped[292]. As discussed above, the human RARα gene is located at 17q21.1, the position implicated in the t(15:17) chromosomal translocation responsible of APL[257,258]. RARα is close to the TRα/Rev-erbα overlapping locus. In mouse, RARα is on chromosome 11 band D.

The human RARβ gene is located on chromosome 3 on p24, close to the TRβ and Rev-erbβ gene (see the relevant entries)[292]. This corresponds to chromosome 14 band A in the mouse.

RARγ gene is on chromosome 12q13 in human[292,293] and chromosome 15 band F in the mouse.

Amino acid sequence for human RARα (NR1B1)

Accession number: X06538

```
  1  MLGGLSPPGALTTLQHQLPVSGYSTPSPATIETQSSSSEEIVPSPPSPPPLPRIYKPCFVCQDKSSGYHY
 71  GVSACEGCKGFFRRSIQKNMVYTCHRDKNCIINKVTRNRCQYCRLQKCFEVGMSKESVRNDRNKKKKEVP
141  KPECSESYTLTPEVGELIEKVRKAHQETFPALCQLGKYTTNNSSEQRVSLDIDLWDKFSELSTKCIIKTV
211  EFAKQLPGFTTLTIADQITLLKAACLDILILRICTRYTPEQDTMTFSDGLTLNRTQMHNAGFGPLTDLVF
281  AFANQLLPLEMDDAETGLLSAICLICGDRQDLEQPDRVDMLQEPLLEALKVYVRKRRPSRPHMFPKMLMK
351  ITDLRSISAKGAERVITLKMEIPGSMPPLIQEMLENSEGLDTLSGQPGGGGRDGGGLAPPPGSCSPSLSP
421  SSNRSSPATHSP  432
```

Amino acid sequence for human RARβ (NR1B2)

Accession number : Y00291

```
  1 MFDCMDVLSVSPGQILDFYTASPSSCMLQEKALKACFSGLTQTEWQHRHTAQSIETQSTSSEELVPSPPS
 71 PLPPPRVYKPCFVCQDKSSGYHYGVSACEGCKGFFRRSIQKNMIYTCHRDKNCVINKVTRNRCQYCRLQK
141 CFEVGMSKESVRNDRNKKKKETSKQECTESYEMTAELDDLTEKIRKAHQETFPSLCQLAKYTTNSSADHR
211 VRLDLGLWDKFSELATKCIIKIVEFAKRLPGFTGLTIADQITLLKAACLDILILRICTRYTPEQDTMTFS
281 DGLTLNRTQMHNAGFGPLTDLVFTFANQLLPLEMDDTETGLLSAICLICGDRQDLEEPTKVDKLQEPLLE
351 ALKIYIRKRRPSKPHMFPKILMKITDLRSISAKGAERVITLKMEIPGSMPPLIQEMMENSEGHEPLTPSS
421 SGNTAEHSPSISPSSVENSGVSQSPLVQ 448
```

Amino acid sequence for human RARγ (NR1B3)

Accession number : M57707

```
  1 MATNKERLFAAGALGPGSGYPGAGFPFAFPGALRGSPPFEMLSPSFRGLGQPDLPKEMASLSVETQSTSS
 71 EEMVPSSPSPPPPPRVYKPCFVCNDKSSGYHYGVSSCEGCKGFFRRSIQKNMVYTCHRDKNCIINKVTRN
141 RCQYCRLQKCFEVGMSKEAVRNDRNKKKKEVKEEGSPDSYELSPQLEELITKVSKAHQETFPSLCQLGKY
211 TTNSSADHRVQLDLGLWDKFSELATKCIIKIVEFAKRLPGFTGLSIADQITLLKAACLDILMLRICTRYT
281 PEQDTMTFSDGLTLNRTQMHNAGFGPLTDLVFAFAGQLLPLEMDDTETGLLSAICLICGDRMDLEEPEKV
351 DKLQEPLLEALRLYARRRRPSQPYMFPRMLMKITDLRGISTKGAERAITLKMEIPGPMPPLIREMLENPE
421 MFEDDSSQPGPHPNASSEDEVPGGQGKGGLKSPA 454
```

References

1 Petkovich, M. et al. (1987) Nature 330 , 444–450.
2 Wentworth, J.M. et al. (1999) Gene 236, 315–323.
3 Michaille, J.J., et al. (1995) Int. J. Dev. Biol. 39, 587–596.
4 Zelent, A. et al. (1988) Nature 339, 714–717.
5 Giguère, V. et al. (1987) Nature 330, 624–629.
6 Ragsdale, C.W. et al (1989) Nature 341, 654–657.
7 Akmal, K.M. et al. (1996) Biol. Reprod. 54, 1111–1119.
8 Dejean, A. et al. (1986) Nature 322, 70–72.
9 Brand, N. et al. (1988) Nature 332, 850–853.
10 Sharpe, C.R. (1992) Mech. Dev. 39, 81–93.
11 de Thé, H. et al. (1987) Nature 330, 667–670.
12 Benbrook, D. et al. (1988) Nature 333, 669–672.
13 Giguère, V. et al. (1989) Nature 337, 566–569.
14 White, J.A. et al. (1994) Development 120, 1861–1872.
15 Umesono, K. et al. (1988) Nature 336, 262–265.
16 Glass, C.K. et al. (1989) Cell 59, 697–708.
17 Krust, A. et al. (1989) Proc. Natl Acad. Sci. USA 86, 5310–5314.
18 Wirtanen, L. and Seguin, C. (2000) Biochim. Biophys. Acta 1492, 81–93.
19 Vasios, G.W. et al. (1989) Proc. Natl Acad. Sci. USA 86, 9099–9103.
20 de Thé, H. et al. (1990) Nature 343, 177–180.
21 Lopes da Silva, S. et al. (1995) Endocrinology 136, 2276–2283.
22 Sucov, H.M. et al. (1990) Proc. Natl Acad. Sci. USA 87, 5392–5396.
23 Joore, J. et al. (1994) Mech. Dev. 46, 137–150.
24 Ellinger-Ziegelbauer, H. and Dreyer, C. (1991) Genes Dev. 5, 94–104.
25 Glass, C.K. et al. (1990) Cell 63, 729–738.
26 Hoffmann, B. et al. (1990) Mol. Endocrinol. 4, 1727–1736.

[27] Yang, N. et al. (1991) Proc. Natl Acad. Sci. USA 88, 3559–3563.
[28] Durand, B. et al. (1992) Cell 71, 73–85.
[29] Mader, S. et al. (1993) J. Biol. Chem. 268, 591–600
[30] Umesono, K. et al. (1991) Cell 65, 1255–1266.
[31] Näär, A.M. et al. (1991) Cell 65, 1267–1279.
[32] Yu, V.C. et al. (1991) Cell 67, 1251–1266.
[33] Kliewer, S.A. et al. (1992) Nature 355, 446–449.
[34] Marks, M. et al. (1992) EMBO J. 11, 1419–1435.
[35] Zhang, X.K. et al. (1992) Nature 355, 441–446.
[36] Leid, M. et al. (1992) Cell 68, 377–395.
[37] Laudet, V. and Stéhelin, D. (1992) Curr. Biol. 2, 293–295.
[38] Mader, S. et al. (1993) EMBO J. 12, 5029–5041.
[39] Perlmann, T. et al. (1993) Genes Dev. 7, 1411–1422.
[40] Kurokawa, R. et al. (1993) Genes Dev. 7, 1423–1435.
[41] Nakshatri, H. and Chambon, P. (1994) J. Biol. Chem. 269, 890–902.
[42] Zechel, C. et al. (1994) EMBO J. 13, 1425–1433.
[43] Zechel, C. et al. (1994) EMBO J. 13, 1414–1424.
[44] Smith, W.C. et al. (1991) EMBO J. 10, 2223–2230.
[45] Durand, B. et al. (1992) Cell 71, 73–85.
[46] Predki, P. et al. (1994) Mol. Endocrinol. 8, 31–39.
[47] Kurokawa, R. et al. (1994) Nature 371, 528–531.
[48] Kurokawa, R. et al. (1995) Nature 377, 451–454.
[49] Kingston, R.E. and Narlikar, G.J. (1999) Genes Dev. 13, 2339–2352.
[50] Dey, A. et al. (1994) Mol. Cell. Biol. 14, 8191–8201.
[51] Bhattacharyya, N. et al. (1997) Mol. Cell. Biol. 17, 6481–6490.
[52] Minucci, S. et al. (1998) Mol. Endocrinol. 12, 315–324.
[53] Li, Q. et al. (1999) Trends Endocrinol. Metab. 10, 157–164.
[54] Nagpal, S. et al. (1993) EMBO J. 12, 2349–2360.
[55] Tini, M. et al. (1994) Mol. Endocrinol. 8, 1494–1506.
[56] Schüle, R. et al. (1990) Cell 61, 497–504.
[57] Nicholson, R.C. et al. (1990) EMBO J. 9, 4443–4454.
[58] Schüle, R. et al. (1991) Proc. Natl Acad. Sci. USA 88, 6092–6096.
[59] Salbert, G. et al. (1993) Mol. Endocrinol. 7, 1347–1356.
[60] Pfahl, M. (1993) Endocrine Rev. 14, 651–658.
[61] Fanjul, A. et al. (1994) Nature 372, 107–111.
[62] Chen, J.Y. et al. (1995) EMBO J. 14, 1187–1197.
[63] Zhou, X.F. et al. (1999) Mol. Endocrinol. 13, 276–285
[64] Lafyatis, R. et al. (1990) Mol. Endocrinol. 4, 973–980.
[65] Kamei, Y. et al. (1996) Cell 85, 403–414.
[66] Chakravarti, D. et al. (1996) Nature 383, 99–103.
[67] Caelles, C. et al. (1997) Genes Dev. 11, 3351–3364.
[68] Darby, T.G. et al. (1997) Oncogene 15, 3067–3082.
[69] Easwaran, V. et al. (1999) Curr. Biol. 9, 1415–1418.
[70] May, M. et al. (1996) EMBO J. 15, 3093–3104.
[71] Folkers, G.E. and van der Saag, P.T. (1995) Mol. Cell. Biol. 15, 5868–5878.
[72] Kruyt, F.A.E. et al. (1993) Mol. Endocrinol. 7, 604–615.
[73] Berkenstam, A. et al. (1992) Cell 69, 401–412.
[74] Keaveney, M. et al. (1993) Nature 365, 562–566.
[75] Sista, N.D. et al. (1995) Nucl. Acids Res. 10, 1729–1736.

[76] Desai, D. et al. (1996) J. Biol. Chem. 271, 15153–15159.

[77] McKenna, N.J. et al. (1999) Endocrine Rev. 20, 321–344.

[78] Glass, C.K. and Rosefeld, M.G. (2000) Genes Dev. 14, 121–141.

[79] Lemon, B.D. and Freedman, L.P. (1999) Curr. Opin. Genet. Dev. 9, 499–504.

[80] Goodman, R.H. and Smolik, S. (2000) Genes Dev. 14, 1553–1577.

[81] Kurokawa, R. et al. (1998) Science 279, 700–703.

[82] Kawasaki, H. et al. (1998) Nature 393, 284–289.

[83] Yuan, C.X. et al. (1998) Proc. Natl Acad. Sci. USA 95, 7939–7944.

[84] Ito, M. et al. (2000) Mol. Cell 5, 683–693.

[85] Dilworth, F.J. et al. (2000) Mol. Cell 6, 1049–1058.

[86] Söderström, M. et al. (1997) Mol. Endocrinol. 11, 682–692.

[87] Baniahmad, A. et al. (1998) Mol. Endocrinol. 12, 504–512.

[88] Allenby, G. et al. (1993) Proc. Natl Acad. Sci. USA 90, 30–34.

[89] Tate, B.F. et al. (1994) Mol. Cell. Biol. 14, 2323–2330.

[90] Allenby, G. et al. (1994) J. Biol. Chem. 269, 16689–16695.

[91] Ostrowski, J. et al. (1995) Proc. Natl Acad. Sci. USA 92, 1812–1816.

[92] Tate, B.F. and Grippo, J.F. (1995) J. Biol. Chem. 270, 20258–20263.

[93] Lupisello, J.A. et al. (1995) J. Biol. Chem. 270, 24884–24890.

[94] Driscoll, J.E. et al. (1996) J. Biol. Chem. 271, 22969–22975.

[95] Ostrowski, J. et al. (1998) J. Biol. Chem. 273, 3490–3495.

[96] Thaller, C. and Eichele, G. (1990) Nature 345, 815–819.

[97] Buck, J. et al. (1997) Science 254, 1654–1656.

[98] Repa, J.J. et al. (1993) Proc. Natl Acad. Sci. USA 90, 7293–7297.

[99] Pijnappel, W.W.M. et al. (1993) Nature 366, 340–344.

[100] Achkar, C.C. et al. (1996) Proc. Natl Acad. Sci. USA 93, 4879–4884.

[101] Apfel, C. et al. (1992) Proc. Natl Acad. Sci. USA 89, 7129–7133.

[102] Keidel, S. et al. (1994) Mol. Cell. Biol. 14, 287–298.

[103] Roy, B. et al. (1995) Mol. Cell. Biol. 15, 6481–6487.

[104] Chen, J.Y. et al. (1996) Nature 382, 819–822.

[105] Katahira, M. et al. (1992) Biochemistry 31, 6474–6480.

[106] Chambon, P. (1996) FASEB J. 10, 940–954.

[107] Knegtel, R.M.A. et al. (1993) J. Biomolec. NMR 3, 1–17.

[108] Renaud, J.P. et al. (1995) Nature 378, 681–689.

[109] Moras, D. and Gronemeyer, H. (1998) Curr. Opin. Cell Biol. 10, 384–391.

[110] Klaholz, B.P. et al. (2000) Proc. Natl Acad. Sci. USA 97, 6322–6327.

[111] Bourguet, W. et al. (2000) Mol. Cell 5, 289–298.

[112] Klaholz, B.P. et al. (1998) Nature Struct. Biol. 5, 199–202.

[113] Leid, M. et al. (1992) TIBS 17, 427–433.

[114] Titcomb, M.W. et al. (1994) Mol. Endocrinol. 8, 870–877.

[115] Dollé, P. et al. (1989) Nature 342, 702–705.

[116] Dollé, P. et al. (1990) Development 110, 1133–1151.

[117] Ruberte, E. et al. (1991) Development 111, 45–60.

[118] Ruberte, E. et al. (1993) Development 118, 267–282.

[119] de Thé, H. et al. (1989) EMBO J. 8, 429–433.

[120] Haq, R.U. et al. (1991) Proc. Natl Acad. Sci. USA 88, 8272–8276.

[121] Leroy, P. et al. (1991) EMBO J. 10, 59–69.

[122] Leroy, P. et al. (1991) Proc. Natl Acad. Sci. USA 88, 10138–10142.

123 Zhu, J. et al. (1999) Proc. Natl Acad. Sci. USA 96, 14807–14812.
124 Rowe, A. et al. (1992) Development 114, 805–813.
125 Smith, S.M. and Eichele, G. (1991) Development 111, 245–252.
126 Zimmer, A. et al. (1994) J. Cell. Biol. 127, 1111–1119.
127 Mendelsohn, C. et al. (1991) Development 113, 723–734.
128 Zimmer, A. and Zimmer A. (1992) Development 116, 977–983.
129 Ferrari, N. et al. (1998) Mol. Cell. Biol. 18, 6482–6492.
130 Ruberte, E. et al. (1990) Development 108, 213–222.
131 Hill, D.S. et al. (1993) Development 117, 937–945.
132 Lehman, J.M. et al. (1991) Nucl. Acids Res. 19, 573–578.
133 Lehman, J.M. et al. (1992) Mol. Cell. Biol. 12, 2976–2985.
134 Thaller, G. and Eichele, G. (1996) Ann. N.Y. Acad. Sci. 785, 1–11.
135 Gavalas, A. and Krumlauf, R. (2000) Curr. Opin. Genet. Dev. 10, 380–386.
136 Zelent, A. et al. (1991) EMBO J. 10, 71–81.
137 Rossant, J. et al. (1991) Genes Dev. 5, 1333–1344.
138 Malpel, S. et al. (2000) Development 127, 3057–3067.
139 Wagner, M. et al. (1992) Development 116, 55–66.
140 Brockes, J.P. (1997) Science 276, 81–87.
141 Kastner, P. et al. (1990) Proc. Natl Acad. Sci. USA 87, 2700–2704.
142 Folkers, G.E. et al. (1993) Mol. Endocrinol. 7, 616–627.
143 Durand, B. et al. (1994) EMBO J. 13, 5370–5382.
144 Rochette-Egly, C. et al. (1992) Mol. Endocrinol. 6, 2197–2209.
145 Pratt, M.A.C. et al. (1990) Mol. Cell. Biol. 10, 6445–6453.
146 Damm, K. et al. (1993) Proc. Natl Acad. Sci. USA 90, 2989–2993.
147 Salerno, A.J. et al. (1996) Nucl. Acids Res. 24, 566–572.
148 Sande, S. and Privalski, M.L. (1994) Mol. Endocrinol. 8, 1455–1464.
149 Allegretto, E.A. et al. (1993) J. Biol. Chem. 268, 26625–26633.
150 Valcarcel, R. et al. (1994) Genes Dev. 8, 3068–3079.
151 Dilworth, F.J. et al. (1999) Proc. Natl Acad. Sci. USA 96, 1995–2000.
152 Tahayato, A. et al. (1993) Mol. Endocrinol. 7, 1642–1653.
153 Huggenvik, J.I. et al. (1993) Mol. Endocrinol. 7, 543–550.
154 Rochette-Egly, C. et al. (1997) Cell 90, 97–107.
155 Taneja, R. et al. (1997) EMBO J. 16, 6452–6465.
156 Rochette-Egly, C. et al. (2000) Mol. Endocrinol. 14, 1398–1410.
157 Yang, Y. et al. (1994) Mol. Endocrinol. 8, 1370–1376.
158 Kastner, P. et al. (1995) Cell 83, 859–869.
159 Rochette-Egly, C. et al. (1995) Mol. Endocrinol. 9, 860–871.
160 Bastien, J. et al. (2000) J. Biol. Chem. 275, 21896–21904.
161 Vasios, G. et al. (1991) EMBO J. 10, 1149–1158.
162 Wei, L.N. et al. (1989) Mol. Endocrinol. 3, 454–463.
163 Aström, A. et al. (1994) J. Biol. Chem. 269, 22334–22339.
164 Bedo, G. et al. (1989) Nature 339, 231–234.
165 Delva, L. et al. (1999) Mol. Cell. Biol. 19, 7158–7167.
166 Loudig, O. et al. (2000) Mol. Endocrinol. 14, 1483–1497.
167 La Rosa, G.J. et al. (1988) Proc. Natl Acad. Sci. USA 85, 329–333.
168 La Rosa, G.J. and Gudas, L.J. (1988) Mol. Cell. Biol. 8, 3906–3917.
169 Boncinelli, E. et al. (1991) Trends Genet. 7, 329–334.
170 Tabin, C. (1995) Cell 80, 671–674.

[171] Simeone, A. et al. (1990) Nature 346, 763–766.
[172] Conlon, R.A. and Rossant, J. (1992) Development 116, 357–368.
[173] Kessel, M. and Gruss, P. (1991) Cell 67, 89–104.
[174] Kessel, M. (1992) Development 115, 487–501.
[175] Marshall, H. et al. (1992) Nature 360, 737–741.
[176] Pöpperl, H. and Featherstone, M.S. (1993) Mol. Cell. Biol. 13, 257–265.
[177] Studer, M. et al. (1994) Science 265, 1728–1732.
[178] Marshall, H. et al. (1994) Nature 370, 567–571.
[179] Ogura, T. and Evans, R.M. (1995) Proc. Natl Acad. Sci. USA 92, 387–391.
[180] Ogura, T. and Evans, R.M. (1995) Proc. Natl Acad. Sci. USA 92, 392–396.
[181] Morrison, A. et al. (1996) Development 122, 1895–1907.
[182] Langston, A.W. et al. (1997) J. Biol. Chem. 272, 2167–2175.
[183] Dupé, V. et al. (1997) Development 124, 399–410.
[184] Packer, A.I. et al. (1998) Development 125, 1991–1998.
[185] Huang, D. et al. (1998) Development 125, 3235–3246.
[186] Shimmeld, S.M. (1996) BioEssays, 18, 613–616.
[187] Holland, L.Z. and Holland, N.D. (1996) Development 122, 1829–1838.
[188] Katsuyama, Y. et al. (1995) Development 121, 3197–3205.
[189] Schoorlemmer, J. et al. (1994) Mol. Cell. Biol. 14, 1122–1136.
[190] Pikarsky, E. et al. (1994) Mol. Cell. Biol. 14, 1026–1038.
[191] Sylvester, I. and Schöler, H.R. (1994) Nucl. Acids Res. 22, 901–911.
[192] Ben-Shushan, E. et al. (1995) Mol. Cell. Biol. 15, 1034–1048.
[193] Minucci, S. et al. (1996) EMBO J. 15, 888–899.
[194] Busam, K.J. et al. (1992) J. Biol. Chem. 267, 19971–19977.
[195] Talmage, D.A. and Lackey, R.S. (1992) Oncogene 7, 1837–1845.
[196] Kitabayashi, I. et al. (1992) EMBO J. 11, 167–175.
[197] De Groot, R.P. et al. (1991) Nucl. Acids Res. 19, 1585–1591.
[198] Lüscher, B. et al. (1989) Genes Dev. 3, 1507–1517.
[199] Cohen, L.E. et al. (1999) Mol. Endocrinol. 13, 476–484.
[200] Houle, M. et al. (2000) Mol. Cell. Biol. 20, 6579–6586.
[201] Stratford, T. et al. (1996) Curr. Biol. 6, 1124–1133.
[202] Chang, B.E. et al. (1997) EMBO J. 16, 3955–3964.
[203] Franco, P.G. et al. (1999) Development 126, 4257–4265.
[204] Tsukui, T. et al. (1999) Proc. Natl Acad. Sci. USA 96, 11376–11381.
[205] Helms, J. et al. (1994) Development 120, 3267–3274.
[206] Riddle, R. et al. (1993) Cell 75, 1401–1416.
[207] Tabin, C.J. (1991) Cell 66, 199–217.
[208] Pan, L. et al. (1992) Nucl. Acids Res. 20, 3105–3111.
[209] Guérin, E. et al. (1997) Proc. Natl Acad. Sci. USA 272, 11088–11095.
[210] Rial, E. et al. (1999) EMBO J. 18, 5827–5833.
[211] Piao, Y.S. et al. (1995) Mol. Endocrinol. 9, 1633–1644.
[212] Carter, M.E. et al. (1994) Mol. Cell. Biol. 14, 4360–4372.
[213] Hall, R.K. et al. (1992) Mol. Cell. Biol. 12, 5527–5535.
[214] Duester, G. et al. (1991) Mol. Cell. Biol. 11, 1638–1646.
[215] Denson, L.A. et al. (2000) J. Biol. Chem. 275, 8835–8843.
[216] Tini, M. et al. (1993) Genes Dev. 7, 295–307.
[217] Segars, J.H. et al. (1993) Mol. Cell. Biol. 13, 6157–6169.
[218] Felli, M.P. et al. (1991) Mol. Cell. Biol. 11, 4771–4778.
[219] Adan, R.A.H. et al. (1993) Mol. Endocrinol. 7, 47–57.

[220] Mahmood, R. et al. (1992) Development 115, 67–74.
[221] Mutoh, H. et al. (1996) Proc. Natl Acad. Sci. USA 93, 774–779.
[222] Bartsch, D. et al. (1992) EMBO J. 11, 2283–2291.
[223] Ghazal, P. et al. (1992) Proc. Natl Acad. Sci. USA 89, 7630–7634.
[224] Poli, G. et al. (1992) Proc. Natl Acad. Sci. USA 89, 2689–2693.
[225] Maciaszek, J. et al. (1994) Mol. Cell. Biol. 68, 6598–6604.
[226] Angulo, A. et al. (1996) Mol. Endocrinol. 10, 781–793.
[227] Vansant, G. and Reynolds, W.F. (1995) Proc. Natl Acad. Sci. USA 92, 8229–8233.
[228] Costa-Giomi, M.P. et al. (1992) Nucl. Acids Res. 20, 3223–3232.
[229] Li, E. et al. (1993) Proc. Natl Acad. Sci. USA 90, 1590–1594.
[230] Lufkin, T. et al. (1993) Proc. Natl Acad. Sci. USA 90, 7225–7229.
[231] Lohnes, D. et al. (1993) Cell 73, 643–658.
[232] Lohnes, D. et al. (1994) Development 120, 2723–2748.
[233] Mendelsohn, C. et al. (1994) Development 120, 2749–2771.
[234] Kastner, P. et al. (1997) Development 124, 313–326.
[235] Dupé, V. et al. (1999) Development 126, 5051–5059.
[236] Mendelsohn, C. et al. (1999) Development 126, 1139–1148.
[237] Krezel, W. et al. (1998) Science 279, 863–867.
[238] Grondona, J.M. et al. (1996) Development 122, 2173–2188.
[239] Boylan, J.F. et al. (1993) Proc. Natl Acad. Sci. USA 90, 9601–9605.
[240] Boylan, J.F. et al. (1995) Mol. Cell. Biol. 15, 843–851.
[241] Taneja, R. et al. (1996) Proc. Natl Acad. Sci. USA 93, 6197–6202.
[242] Balkan, W. et al. (1992) Dev. Biol. 151, 622–625.
[243] Tsai, S. et al. (1993) Proc. Natl Acad. Sci. USA 90, 7153–7157.
[244] Bérard, J. et al. (1994) EMBO J. 13, 5570–5580.
[245] Tsai, S. et al. (1994) Genes Dev. 8, 2831–2841.
[246] Imakado, S. et al. (1995) Genes Dev. 9, 317–329.
[247] Saitou, M. et al. (1995) Nature, 374, 159–162.
[248] Attar, P.S. et al. (1997) Mol. Endocrinol. 11, 792–800.
[249] Mata de Urquiza, A. et al. (1999) Proc. Natl Acad. Sci. USA 96, 13270–13275.
[250] Tsai, S. et al. (1992) Genes Dev. 6, 2258–2269.
[251] Noji, S. et al. (1991) Nature 350, 83–86.
[252] Fu, Z. et al. (2001) Biol. Reprod. 64, 231–241.
[253] Michaille, J.J. et al. (1994) Dev. Dyn. 201, 334–343.
[254] Mark, M. et al. (1995) Int. J. Dev. Biol. 39, 111–121.
[255] Luo, J. et al. (1995) Mech. Dev. 53, 61–71.
[256] Mendelsohn, C. et al. (1994) Dev. Biol. 166, 246–258.
[257] Borrow, J. et al. (1990) Science 249, 1577–1580.
[258] de Thé, H. et al. (1990) Nature 347, 558–561.
[259] Kakizuka, A. et al. (1991) Cell 66, 663–674.
[260] de Thé, H. et al. (1991) Cell 66, 675–684.
[261] Kastner, P. et al. (1992) EMBO J. 11, 629–642.
[262] Chen, Z. et al. (1993) EMBO J. 12, 1161–1167.
[263] Perez, A. et al. (1993) EMBO J. 12, 3171–3182.
[264] Grignani, F. et al. (1993) Cell 74, 423–431.
[265] Doucas, V. et al. (1993) Proc. Natl Acad. Sci. USA 90, 9345–9349.
[266] Dyck, J.A. et al. (1994) Cell 76, 333–343.

267 Weis, K. et al. (1994) Cell 76, 345–356.

268 Koken, M.H.M. et al. (1994) EMBO J. 13, 1073–1083.

269 Doucas, V. et al. (1996) Genes Dev. 10, 196–207.

270 Brown, D. et al. (1997) Proc. Natl Acad. Sci. USA 94, 2551–2556.

271 He, L.Z. et al. (1998) Nature Genet., 18, 126–135.

272 Grignani, F. et al. (1998) Nature 391, 815–818.

273 Lin, R.J. et al. (1998) Nature 391, 811–814.

274 Lin, R.J. et al. (1999) Trends Genet. 15, 179–184.

275 Seeler, J.S. and Dejean, A. (1999) Curr. Opin. Genet. Dev. 9, 362–367.

276 Minucci, S. et al. (2000) Mol. Cell. 5, 811–820.

277 Lin, R.J. and Evans, R.M. (2000) Mol. Cell. 5, 821–830.

278 He, L.Z. et al. (2000) Mol. Cell. 6, 1131–1141.

279 Zimonjic, D.B. et al. (2000) Proc. Natl Acad. Sci. USA 97, 13306–13311.

280 Fisher, G.J. et al. (1996) Nature, 379, 335–339.

281 Wang, Z. et al. (1999) Nature Med. 5, 418–422.

282 Redner, R.L. et al. (1996) Blood 87, 882–896.

283 Wells, R.A. et al. (1997) Nature Genet. 17, 109–113.

284 Brand, N.J. et al. (1990) Nucl. Acids Res. 18, 6799–6806.

285 Van der Leede, B.J.M. et al. (1992) Biochem. Biophys. Res. Commun. 188, 695–702.

286 Vivanco Ruiz, M.D.M. et al. (1991) EMBO J. 10, 3829–3838.

287 Kobayashi, M. et al. (2000) Mol. Cell. Biol. 20, 8731–8739.

288 Kruyt, F.A.E. et al. (1992) Nucl. Acids Res. 20, 6393–6399.

289 Nagpal, S. et al. (1992) Proc. Natl Acad. Sci. USA 89, 2718–2722.

290 Giguère, V. et al. (1990) Mol. Cell. Biol. 10, 2335–2340.

291 Gehin, M. et al. (1999) Chem. Biol. 6, 519–529.

292 Mattei, M.G. et al. (1991) Genomics 10, 1061–1069.

293 Ishikawa, T. et al. (1990) Mol. Endocrinol. 4, 837–844.

294 Hisata, K. et al. (1998) Dev. Genes Evol. 208, 537–546.

295 Ylikomi, T. et al. (1992) EMBO J. 11, 3681–3694.

296 Braun, K.W. et al. (2000) J. Biol. Chem. 275, 4145–4151.

297 Pfitzner, E. et al. (1998) Proc. Natl Acad. Sci. USA 95, 5539–5544.

298 Préfontaine, G.G. et al. (1999) J. Biol. Chem. 274, 26713–26716.

299 Suzuki, Y. et al. (1999) Blood 93, 4264–4276.

300 Mengus, G. et al. (1997) Genes Dev. 11, 1381–1395.

301 Chiang, M.Y. et al. (1998) Neuron 21, 1353–1361.

Names

The peroxisome proliferator-activated receptors (PPARs) were named because the first member of this group to be discovered, mouse PPARα, was shown to be activated by a diverse class of rodent hepatocarcinogens that causes proliferation of peroxisomes[1]. Three members were found in *Xenopus* and called PPARα, β and γ[2]. Since the relationships existing between the *Xenopus* PPARβ and its putative orthologue in mammals were unclear, the latter received several names, such as PPARδ, NUC1 or FAAR (fatty acid-activated receptor)[3-5], etc. PPARγ was shown to be a component with RXRα of ARF6, an adipocyte-specific transcription factor that was known to regulate the adipocyte P2 enhancer[11,14]. These receptors form the NR1C group in the nomenclature of nuclear receptors[15]. For recent reviews, see references 16 and 128.

Species	Other names	Accession number	References
Official name: NR1C1			
Human	PPARα	L02932	9
Mouse	PPARα	X57638	1
Rat	PPARα	M88592	6
Guinea pig	PPARα	AJ006218	139
Cow	PPARα	AF229356	Unpub.
Pig	PPARα	AF228696	Unpub.
Chicken	PPARα	AF163809	147
Xenopus	PPARα	M84161	2
Nile crocodile	PPARα	AJ011515	140
Official name: NR1C2			
Human	PPARβ, NUC1, PPARδ	L07592	3
Mouse	PPARβ, FAAR, PPARδ	U10375	4, 5, 8
Rat	PPARβ, PPARδ	U40064	12
Rabbit	PPARβ, PPARβ/δ, PPARδ	AB033614	91
Cow	PPARβ, PPARδ	AF229357	Unpub.
Chicken	PPARβ, PPARβ/δ, PPARδ	AF163810	147
Xenopus	PPARβ	M84162	2
Nile crocodile	PPARβ	AJ243131	140
Official name: NR1C3			
Human	PPARγ	L40904	13
Macaca	PPARγ	AF033103	141
Mouse	PPARγ, ARF6	U01664	7, 8, 11
Rat	PPARγ	AF156666	137
Hamster	PPARγ	Z30972	10
Rabbit	PPARγ	U84893	144
Cow	PPARγ	Y12419	143
Pig	PPARγ	AJ006756	142
Chicken	PPARγ	AF163811	147
Xenopus	PPARγ	M84163	2
Nile crocodile	PPARγ	AJ243132	140
Turtle	PPARγ	AJ243133	140
Plaice	PPARγ	AJ243956	Unpub.

Isolation

Mouse PPARα was the first member of the PPAR group to be discovered[1]. The cDNA was isolated by Stephen Green's laboratory from a mouse liver cDNA library by screening with oligonucleotides based on the consensus sequence of the DNA-binding domain of nuclear receptors. Later, the three PPARs were found by Walter Wahli's laboratory using a similar strategy in a *Xenopus* ovary cDNA library[2]. The PPARα cDNAs cloned in other mammals were identified by screening cDNA libraries with degenerated oligonucleotides[6,9]. The three PPAR genes were recently cloned in chicken[147].

The first mammalian PPARβ homologue, called NUC1, was found using a PCR strategy that was set up to identify new nuclear receptors from a human osteosarcoma cell line[3]. The mouse gene, PPARδ, was found by Ron Evans' laboratory by screening of the adult mouse liver cDNA library with an oligonucleotide corresponding to a PPAR-conserved sequence[4]. The same gene was identified as a FAAR expressed during adipocyte differentiation[5].

The first mammalian PPARγ homologue was identified by searching mammalian PPARs using a RT-PCR strategy with degenerated oligonucleotides as well as by low stringency screening with a COUP-TF DNA-binding domain[7,8]. Interestingly, mouse PPARγ was identified as a part of the ARF6 transcription factor that appears to be implicated in adipocyte differentiation[11,14]. ARF6 was shown by competition experiments on gel shift assay to be able to bind a DR1 element. It was shown that ARF6 contains both RXRα and an adipocyte-specific N-terminal isoform of PPARγ, called PPARγ2[11]. The components of ARF6 were also identified by biochemical purification and peptide sequencing[14]. Human PPARγ was isolated from a bone marrow cDNA library screened with degenerated oligonucleotides[13].

When the three PPARs are considered, it is clear that they exhibit a very rapid evolutionary rate[53,54]. In fact, in the DNA-binding domain, PPARα, β and γ exhibit 78–86% amino acid identity[53]. Strikingly, these levels are much smaller in the ligand-binding domain (63–71%)[53,54]. In the LBD, within each gene the evolution is much more rapid than for TR and RAR. For example, in the LBD, the difference between the mammal and *Xenopus* sequences are 14%, 29% and 22% for PPARα, β and γ, respectively, whereas for TR and RAR the values are around 4–8%[53,54]. The reason for this high-evolutionary speed is unclear but it is tempting to link it to the wide diversity of ligands that are able to bind PPARs as well as to the species specificity in ligand recognition that is observable in some cases (reviewed in reference 16).

DNA binding

PPARs bind to DR1 elements as obligate heterodimers with RXR[25]. PPARs are unable to bind DNA as homodimers or monomers[16]. The first natural PPRE to be determined was the one found in the acyl-CoA oxidase[2,26] and most natural PPREs subsequently found were DR1 elements. Nevertheless a closer examination of the PPRE sequences allows additional PPRE determinants[16,27–29] to be defined: (1) an extended 5′ half-site rich in A and T bases; (2) an imperfect core DR1; and (3) adenine as the spacing nucleotide between the two AGGTCA core motifs. Thus, a consensus

PPRE should harbor the following sequence: AACT AGGNCA A AGGTCA[16]. On such elements, PPAR interact with the 5' motif whereas RXR binds to the downstream one[28,30]. It has been shown that the C-terminal extension of the DNA-binding domain of PPAR interacts specifically with the 5'-extension of the PPRE which is rich in A and T bases, as the 5'-extension of the monomeric receptors HREs[29].

Interestingly, it has recently been shown that PPAR–RXR heterodimers can also bind to a subset of DR2 elements that harbor the features of the Rev-DR2 sequence recognized by the Rev-erb orphan receptors, i.e. an A/T-rich 5'-extension and a C and a T between the two AGGTCA repeats[31,32]. This suggests that a cross-talk may exist between the PPAR and Rev-erb orphan receptors for the regulation of some target genes, such as Rev-erbα itself[32].

PPARs are also able to bind to ERE elements and to repress estrogen-activated transcription of a classical ER target such as the vitellogenin gene A2 promoter[33]. No target gene positively regulated by PPAR via an ERE has been described yet[16].

Through their binding to DR1 elements a number of cross-talks exist between PPARs and orphan receptors, such as HNF4, TR2/4 and COUP-TFs (see reference 16 for an exhaustive reference list). Since PPAR binds efficiently only to certain DR1 elements, the cross-talk existing for a given target gene will depend on the sequence of the DR1 elements as well as the relative concentrations of PPAR, RXR and other receptors, such as HNF4 or COUP-TF. Cross-talk with liganded receptors such as TR has also been described[34,35].

Partners

RXR is an obligate partner for PPAR[25]. The interaction surface between both molecules was recently shown by structural analysis to be asymmetric with the AF-2 region of the PPAR moiety interacting with helices 7 and 10 of RXR[131]. An *in vitro* interaction between PPAR and TR has been described but its *in vivo* significance remains elusive[65].

PPARs are able to bind in solution with corepressors such as N-CoR and SMRT but not when bound to DNA complexed with RXR. This is in accordance with a lack of the signature motif for corepressor binding that was noticed for PPAR and with the lack of repressive activity of unliganded PPARs[30].

The first coactivator that was found for PPARs was SRC-1[56,138]. The interaction surface between PPARγ and SRC-1, which contains two LXXLL motifs able to recognize PPAR, has been studied by crystallographic analysis[38,56]. Interestingly, it was shown that distinct amino acids C-terminal to the LXXLL motif are required for PPARγ activation by different ligands, suggesting that these ligands induce a somewhat different conformation of PPARγ LBD[66]. The functional importance of SRC-1 in mediating PPAR transcriptional activity has been proved by the effect of overexpression of the receptor-interacting domain of SRC-1 that led to an inhibition of PPAR-dependent transactivation[30]. Nevertheless, other coactivators are likely to be used to transduce PPAR transcriptional activity, since PPARγ still is transcriptionally active in SRC-1 knockout mice[67].

It has been proposed that PPAR preferentially interacts with the CBP and p300 cointegrators and that the interaction may explain some of the synergisms observed between PPARs and other transcription factors[68,69].

Other coactivators for PPARs were described (reviewed in reference 16) among which PGC-1 (for PPARγ coactivator-1) is of particular interest[70]. This molecule binds to PPARγ in a ligand-independent manner, but its expression is strongly induced in muscle and brown fat upon cold exposure. It has been shown that PGC-1 plays a major role in activating the brown fat-specific uncoupling protein-1 and -2 (UCP-1 and UCP-2) as well as mitochondrial biogenesis and respiration in muscle cells through the regulation of nuclear respiratory factors[70,71,136]. It has recently been shown that PGC-1 couple transcription and RNA processing on genes implicated on thermogenesis and mitochondrial biogenesis[148]. In addition PGC-1 is able to recruit other coactivators such as SRC-1 and CBP/P300[150]. The docking of PGC-1 to PPARγ induces a conformational change in PGC-1 that allows the recruitment of SRC-1 and CBP. Another coactivator for PPARγ, PGC-2, has been recently described[72]. Interestingly it is able to bind to the AF-1 domain of PPARγ and to promote adipocyte differentiation when transfected into PPARγ-expressing pre-adipocytes. PGC-2 does not interact with PPARα or β[72]. Finally, PPARs interact with other known coactivators, such as PPAR-binding protein (PBP), also known as TRAP220 or DRIP230[73], but also with RIP140 or ARA70 (reviewed in reference 16).

An interaction between PPARα and the pituitary transcription factor GHF-1 has been described[36]. PPARα interacts with GHF-1 in a ligand-dependent manner and this gives rise to stimulation of the prolactin promoter in GH4C1 pituitary cells. Interestingly, RXR may down-regulate this gene by titrating out PPAR from the complex with GHF-1[36]. PPARα has also been shown to interact and synergize with the transcription factor Sp1 in the context of the acyl-CoA oxidase gene promoter[74].

Ligands

The first indication that PPARs were not orphan receptors but may be specifically recognized by a ligand came from activation studies that show that fibrates, hypolipidemic compounds as well as fatty acids were able to regulate PPAR transcriptional activity[1,2,6,37]. Nevertheless, such activation may be indirect and it was difficult to find the direct natural ligands for PPARs[16].

Given their implication in several important diseases, an impressive amount of work has been devoted to the search for PPAR ligands. Several different methods were used to identify such ligands, such as competition with radioligands, fluorescence or scintillation assays, limited proteolysis assays or interaction with coactivators[16,38,51]. Overall, three general observations can be made on the PPAR ligands[16,39] :

1   In contrast to other liganded receptors, such as TR, RAR or ER, PPARs accommodate several types of ligand and most PPAR activators directly interact with the receptor.

2   Most known PPAR ligands bind to the receptor with a relatively low affinity (K_d from 10 nM for some synthetic compounds, such as GW2433 on PPARβ, to 1 mM for 8(S)-hydroxyeicosatetraenoic acid (HETE) binding on PPARα).

3   There is an overlap in ligand-binding specificity between the three PPARs, some ligands binding to more than one PPAR type, although with different affinities. It remains unclear whether natural ligands able to bind with a strong affinity to each of the three PPARs exist or not.

Fatty acids were the first natural activators to be proposed for PPARs[6,37]. They are active on PPARs in a range of concentration that is consistent with their physiological levels (2–20 μM) and are able to directly bind to the three PPAR types[38,40,41]. Only the unsaturated fatty acids are efficient ligands for PPARs, saturated ones interacting only marginally. Of the three types of PPARs, PPARα exhibit the highest affinity with unsaturated fatty acids, linolenic and linoleic acids being among the most efficient fatty acids for PPARα binding. Of note, oxidized metabolites of linoleic acid, 9-hydroxyoctadecadienoic acid (9-HODE) and 13-HODE are activators and ligands of PPARγ[42,43].

Eicosanoids are derivatives of fatty acids obtained either via the lipoxygenase (leukotrienes, HETEs) or cycloxygenase (prostaglandins) pathways[16]. Several of these compounds are PPAR ligands, among which 15-deoxy-Δ12,14-prostaglandinJ2 for PPARγ[44,45], 8(S)-HETE for PPARα[38,40,41], leukotriene B4 (LTB4) for *Xenopus* PPARα[38,44,45]. All these compounds bind the PPARs with binding affinities in the micromolar range, although higher binding affinities have been reported, depending on the method used to measure the affinity (reviewed in reference 16).

Many synthetic molecules have been described as PPAR ligands. Among those are the fibrates the more efficient of which are clofibrate and the Wy-14643 compounds that preferentially bind PPARα[38]. Interestingly, antidiabetic drugs, such as the thiazolidinediones (TZDs), including troglitazone, pioglitazone, BRL 49653 (rosiglitazone) which are all derived from clofibric acid, selectively bind to PPARγ[46–48]. The fact that various PPARs may be able to transduce the effects of hypolipidemic and antidiabetic drugs has important implications for the pharmacological use of these compounds (reviewed in references 16, 39 and 48). Other synthetic compounds that bind to PPARs include ETYA, an arachidonic acid analogue[39], various agonists and antagonists of the membrane leukotriene receptor[45] and non-TZD fibrate derivatives[49].

In this multitude of ligands, the case of PPARβ remains obscure since no specific natural ligand has been described for this receptor. Among synthetic compounds, bezafibrate is a specific PPARβ ligand (in *Xenopus* but not in mammals) as well as GW2433 or the substituted fatty acid α-bromopalmitate[38,41,50]. Novel series of non-TZD fibrate derivatives have been described as human PPARβ agonists that may be useful for the better understanding of PPARβ ligand-binding specificity[49].

Although numerous agonists of PPARs have been described, antagonists are, to date, very rare. Recently, a synthetic compound, GW0072, was identified as a high-affinity PPARγ ligand[52]. In transient transfection, it alone has a very weak agonist activity and it antagonizes the effect of strong agonists such as rosiglitazone. In cell culture, it behaves as a potent antagonist of adipocyte differentiation. These results suggest that GW0072 may be the first antagonist (although partial) described to date for a PPAR[52].

Given the high evolutionary rates exhibited by the various PPARs both between and within the three types, it is not surprising to find a species specificity in ligand binding. This has been observed for PPARα in *Xenopus* and human versus mouse, which respond differentially to Wy-14643 and ETYA[55]. This specificity is caused by specific mutations arising in helix 3. Whether this reflects an adaptation of the various PPAR orthologues to a different spectrum of ligands because of different nutritional patterns of the species is still open to question[16,54]. The recent characterization of the three PPAR genes in chicken will allow the deciphering of the origin of ligand selectivity of each subtype[147].

Three-dimensional structure

To date, there is no 3D structure available for the DNA-binding domain of any PPAR. In contrast, several structures were determined for the ligand-binding domain for PPARβ and γ, either alone in the apo or holo form or even complexed with a peptide encompassing the receptor interaction domain of coactivators[52,56–58]. No structure of the PPARα LBD has yet been described.

The basic structure of the PPAR LBD has the common fold found in other nuclear receptors (reviewed in reference 59). Nevertheless, the PPAR LBD exhibits several distinct features:

1 In the apo form, the C-terminal helix containing the AF-2 core activation domain is folded back against the ligand-binding pocket in a conformation similar to that observed for holo-RAR or holo-TR[56,58].
2 The entry of the ligand is facilitated by the existence of an extra helix, called helix 2' as well as a different placement of helix 2[56–58].
3 The ligand-binding pocket is much larger than for any other known nuclear receptor. The ligand binding pocket of PPARγ has a volume of 1300 Å^3 compared with 600 Å^3 for TR[56,59]. The ligand (rosiglitazone[56], eicosapentaenoic acid[57], GW2433[57] or the partial antagonist GW0072[52]) occupies only 40% of the pocket in contrast to that for T3, which occupies *c.* 90% of the volume of the pocket[16,59].
4 The ligand-binding pocket has either a T (PPARγ[56]) or Y (PPARβ[57]) shape.

The precise mode of interaction of the carboxylic part of the ligand together with the relatively free non-specific interactions of the hydrophobic part explain the promiscuous behavior of PPARs with respect to ligand binding.

A ternary complex between the PPARγ LBD, the antidiabetic ligand rosiglitazone (BRL 49653) and 88 amino acids of the human coactivator SRC-1, containing two LXXLL motifs has been described[56]. In this structure, a homodimer of PPARγ was observed, with each monomeric unit being complexed with an LXXLL region of SRC-1. A charge clamp containing a glutamate residue of the AF-2 core domain (E471 of human PPARγ) and a lysine located in helix 3 (K301) allows the orientation and placement of LXXLL motifs into the coactivator binding cleft, which forms a hydrophobic pocket delineated by H3, H4, H5 and AF-2[56].

Recently, the structure of a cocrystal between PPARγ LBD liganded with rosiglitazone or the synthetic GI262570 compound, and RXRα LBD liganded with 9-*cis* retinoic acid together with SRC-1 peptides complexed with each monomer was determined[131]. The structure of this heterodimer reveals that the heterodimer is asymmetric with each LBD deviating by 10° from the symmetry axis. Consequently, the interaction surface between each monomer is not equivalent. It was shown that the PPAR AF-2 helix interacts with helices 7 and 10 of RXR[131].

Expression

A complete description of PPAR expression and regulation can be found in reference 16.

In *Xenopus*, PPARα is expressed as a *c.* 4-kb transcript present during oogenesis as well as during early embryogenesis as a maternal mRNA up to the gastrula stage[2,53]. Zygotic transcripts are observed at the tail bud stage. In adult frogs, PPARα is expressed in all tissues examined. In contrast, expression of PPARα in mammals appears relatively late (E13.5 for rat) as transcripts of 1.8 and 2.0 kb in mouse[1,75]. An expression is observed during embryogenesis in the central nervous system and the skin[75]. In adult rodents, PPARα expression is observed in brown fat, liver, kidney, heart, mucosa of the stomach and duodenum, retina, adrenal gland, skeletal muscle, pancreatic islets and smooth muscle cells[1,4,7,75,76,83]. An expression has been observed in human liver[79] as well as in rat cartilage[135]. Overall, the expression of PPARα correlates with tissues in which a high mitochondrial and peroxisomal β-oxidation activity is observed.

The expression of PPARα is regulated according to the circadian rhythm and this is due at least in part to a positive regulation of PPARα by glucocorticoids[77,78]. In addition, a negative regulation by insulin in rat liver has been observed[80]. In accordance with these data, stress and fasting up-regulate PPARα mRNA levels in the liver[77,78]. Regulation of PPARα expression by other factors, including fibrates or GH has also been demonstrated. Apparently, the expression of PPARα and γ in brown adipose tissue decreases during cold exposure in rat, whereas that for PPARβ increases[137].

As PPARα, PPARβ is a maternal mRNA in *Xenopus*, expressed in even higher levels than PPARα as a 1.5 kb transcript. Its expression persists throughout embryogenesis and is ubiquitous in the adult[2,53]. In mammals the major transcript is at 4.0 kb. The expression starts very early (E8.5 in rat), peaks in the neural tube between E13.5 and E18.5, and then remains ubiquitous at a lower level[75]. In the adult the expression is nearly ubiquitous with an expression level higher than PPARα, especially in the neural tube, where it is the most prominent PPAR gene expressed[3,4,75,76]. Of note, the expression is low in liver when compared to other tissues such as skeletal muscles. A high expression level has been noticed in Sertoli cells as well as in osteoclasts[16,91]. Interestingly, PPARβ transcripts have been observed during embryonic implantation in the uterus and a persistent expression was found in human placenta[81].

Given its ubiquitous expression pattern, the regulation of PPARβ expression has been somewhat neglected. It has been proposed that, as for PPARγ, PPARβ expression is increased during adipocyte differentiation in several models such as Ob17 or 3T3C2[5,82]. It has even been claimed that the induction of PPARβ expression occurred before that of PPARγ. It has also been proposed that PPARβ regulates PPARγ expression and that this event induces terminal differentiation into adipocytes[82]. Nevertheless, of the three PPARs, PPARβ exhibits the lowest capacity to trigger adipocyte differentiation[118]. Recently, an increased PPARβ expression was observed in colon cancer. Interestingly, APC reduce the level of PPARβ, suggesting a direct link between PPARβ and colon cancer[129].

PPARγ is probably the most intensively studied PPAR at the expression level. In contrast to α and β, PPARγ is not expressed during early embryogenesis. In the adult, the 2.2 kb transcript is more prominent in fat body and kidney, and less abundant in liver[2,53]. In rodents the major site of expression of PPARγ is the brown adipose tissue as a 2.0 kb mRNA[11,75,76] (see reference 82 for a review). An expression has also been noticed in white adipose tissue, intestinal mucosa, especially in colon[84–86]. PPARγ levels are also high in lymphoid tissues such as spleen in rodent (but not in human), monocytes, macrophages, in retina, in cartilage, in osteoclast and in skeletal muscle[42,43,75,76,88–90,134,135].

Regulation of PPARγ expression has been submitted to detailed studies because of its implication in several diseases (see below). An induction of PPARγ levels during adipocyte differentiation has been observed and it was demonstrated that PPARγ is a key gene in promoting terminal differentiation of adipocytes[11,96] (see reference 132 for a recent review). PPARγ2 expression is regulated at the promoter level during adipocyte differentiation by transcription factors of the C/EBP family, such as C/EBPβ[98] or C/EBPα[102]. PPARγ1 and γ3 promoters are regulated by SREBP1, a transcription factor that responds to sterols[101,103]. PPARγ expression is regulated by interleukin-4 (IL4), which also induces the production of a PPARγ ligand through the regulation of 12/15 lipooxygenase[90]. Insulin and glucocorticoids increase PPARγ levels, whereas tumor necrosis factor-α (TNFα), which triggers dedifferentiation of mature adipocytes, down-regulates PPARγ[100]. High-fat regimen and fasting, respectively, increase and decrease PPARγ levels in rodent adipose tissue. These data suggest that PPARγ may play a role in obesity in humans, a hypothesis that has been verified by the observation of an increased PPARγ2 expression in obese humans versus controls[92,93].

Activity and target genes

PPARs are potent transcriptional activators and thus have a strong transcriptional activating function, AF-2, in the extreme C-terminal part of the receptor, which is conserved in other nuclear receptors[16]. In addition, the A/B region contains an activation function, AF-1, which has been defined for both PPARα and γ[64]. This AF-1 function may be regulated by phosphorylation events (see below and reference 16). A specific coactivator molecule, PGC-2, is able to interact specifically with the AF-1 region of PPARγ, but not with either PPARα or β[72]. As discussed above, PPARs are unable to repress transcription in the absence of ligand, since they do not interact with corepressors when bound to DNA as an heterodimer with RXR[30].

PPAR are not only activated by their bona fide ligands but also by RXR ligands, since, in contrast to RAR, the PPAR–RXR heterodimer may be regulated both by PPAR-specific and RXR-specific agonists, the presence of two ligands resulting in an additive effect[25,37,69]. This has led to the demonstration that RXR-specific agonists, called rexinoids, may display an antidiabetic activity comparable to that of TZDs[63].

In addition to being regulated by ligand binding, PPAR activity may be regulated by phosphorylation[16]. Conflicting results for phosphorylation in regulating PPAR transcriptional activity have been described. PPARα is phosphorylated and it has been shown that insulin increases both the phosphorylation and transcriptional activity of PPARα and γ (reviewed in reference 16). The phosphorylation sites were mapped in the A/B region at position 12 and 21 of the human PPARα two sites that are the consensus for MAP kinase[60]. A synergy between insulin and PPARγ specific ligands for the expression of a target gene, aP2, have been demonstrated. In contrast, activation of MAP kinase by growth factors such as EGF, led to a phosphorylation of PPARγ2 on serine 112 and to a decrease of PPARγ transcriptional activity, which results in an inhibition of adipogenesis[61]. It has been suggested that communication exists between this phosphorylation site and the LBD, since the mutation of serine 112 to aspartic acid led to a decreased ligand-binding affinity and coactivator recruitment[62]. This suggests that the A/B domain plays a role in the conformation of the LBD and thus regulates the affinity of PPARγ for its ligand. Interestingly, this phosphorylation event on serine 112 also regulates the AF-1 activating function of PPARγ2.

There is now a group of target genes described for PPARs. A list of target genes with known PPREs can be found in reference 16. Only some examples will be presented here, since a description of the physiological functions of PPARs is far outside the scope of this chapter. The first described target genes, which are both regulated by PPARα in the liver, were the acyl-CoA oxidase[2,26] and CYP4A6, which encodes a cytochrome P450 fatty acid ω-hydroxylase[27]. Numerous other PPARα target genes are implicated in lipid and cholesterol homeostasis in liver. This is the case for the apolipoprotein AI, AII and CIII[94,95], of the enzymes of the mitochondrial and peroxisomal β-oxydation pathways[16] and of the malic enzyme[28]. Among the numerous target genes for PPARα in liver, the mitochondrial ketogenic enzyme, hydroxymethylglutaryl-CoA synthase (mHMG-CoAS), which is responsible for the formation of ketone bodies from acetoacetate, is particularly intriguing, since this protein is able to interact with PPARα, to translocate to the nucleus and to potentiate PPARα-dependent transcription of its own gene via a PPRE[117]. Thus, the product of a PPARα target gene specifically autoregulates its own transcription by a fascinating autoregulatory loop! During adipogenesis, PPARγ regulates the aP2 gene, which encodes a lipid-binding protein specific to the adipocyte[11,97], the uncoupling proteins UCP-1 and -2[70–72], the insulin-dependent glucose transporter GLUT4[119] or the PEPCK gene[99] (see references 87 and 132 for reviews). PPARγ also regulates several target genes such as the scavenger receptor CD36, during monocyte/macrophage differentiation[42,43,88,89].

Several target genes were shown to be repressed by PPARs. This is the case of the leptin gene, which is down-regulated by PPARγ2 through a functional antagonism with C/EBPα[133]. This is also the case for the nitric oxide synthase gene, which is also repressed by PPARγ[130]. Other examples are known (see reference 16).

Knockout

The knockout of the mouse PPARα gene has been performed[104]. The resulting mice are viable and fertile, and do not exhibit gross phenotypical defects. However, these animals do not display the peroxisome proliferator response when treated by PPARα agonists such as clofibrates or Wy-14643[104,111]. These results have suggested that the effect of PPARα is much stronger in mitochondrial versus peroxisomal β-oxidation of fatty acids[104,111]. It has been shown that the inflammation mediated by arachidonic acid or leukotriene B4, natural ligands of PPARα, is prolonged in PPARα knockout mice, suggesting that this gene controls the duration of the inflammatory response[44]. The study of these mice also reveals that PPARα plays a critical role in the cellular fasting response and that they may be used as a model for human fatty-acid oxidation disorders[105,110]. During fasting, PPARα knockout mice exhibit a severe hypoglycemia and hypothermia (in accordance with the strong expression of PPARα in brown fat), which are accompanied by an enhanced lipid accumulation in the liver and no increase of ketone body production, suggesting a dramatic impairment of fatty-acid oxidation[105–110].

The knockout of the PPARβ gene has been recently described[149]. PPARβ−/− fetuses are smaller than those of wild-type animals and they display myelination defects in the brain. In addition, these mice harbor an increased hyperplastic response in the epidermis after TPA application, suggesting a role for PPARβ in epidermal cell proliferation.

The PPARγ gene was also inactivated in the mouse[106–109]. This knockout results in the following two independent lethal phases.

1 An early one around E10. Indeed, it was shown that PPARγ1 is expressed in the trophoblast lineage and that its deficiency leads to placental dysfunction and death. This placental deficiency indirectly causes a severe myocardial thinning, presumably because of an impaired supply of fuel and/or oxygen[106–108].

2 After birth, rescued animals exhibit a complete absence of white and brown adipose tissue (complete lipodystrophy), a fatty liver secondary to this lipodystrophy, abnormal sebaceous glands in the skin, and several hemorrhages in the brain and intestine[106–108].

This phenotype thus confirms the key role played by PPARγ in adipocyte differentiation. This was further substantiated by the observation that PPARγ[-/-] cells do not contribute to the formation of adipose tissue in chimeric mice[109]. Interestingly, the study of cells derived from the PPARγ[-/-] animals supports the view that PPARγ is downstream of C/EBPβ and δ, but upstream of C/EBPα and terminal differentiation[109]. This further highlights that PPARγ plays a role in the differentiation of adipocytes but also in their correct function once differentiated (reviewed in references 87 and 106). In line with this notion, the study of the heterozygous animals also reveals that PPARγ plays a part in insulin resistance, since these animals were protected from the development of insulin resistance under a high fat diet[108]. These mice showed overexpression of the leptin gene, suggesting that PPARγ may play a direct role in the regulation of food intake[108]. The relationship between insulin resistance, leptin secretion and impaired adipogenesis is still far from understood, but these PPARγ knockout mice clearly provide an invaluable model to study the regulation of energy balance in mammals and its possible pathological disorders in human (see also reference 133 for a description of the negative effect of PPARγ on the leptin promoter).

Associated disease

The PPARs, especially PPARα and γ have been implicated in several important metabolic diseases, such as dyslipidemia, diabetes, obesity, atherosclerosis and inflammation control. More recently, PPARs have also been implicated in cancers (see reference 128 for a recent review).

One of the first indications that PPARs may be implicated in a wide range of metabolic diseases came from the initial demonstration that fibrates, well-known hypolipidemic drugs, are PPARα ligands[1]. Fibrate treatment both enhances the catabolism of triglyceride-rich particles and reduces the production of very-low-density lipoprotein (VLDL), which is implicated in their transport in the blood. It has been shown that fibrates act by regulating key players of the cholesterol transport from lipoprotein lipase (LPL), which regulates the delivery of fatty acids to target cells to apolipoproteins such as ApoA1, AII and CIII[32,121]. It has been shown that fibrates down-regulate the ApoA1 in rat via the Rev-erbα (NR1D1) orphan receptor, which is a transcriptional repressor. Rev-erbα is a target for PPARα in liver and may be instrumental for many down-regulation events of gene expression mediated by fibrates[32,120]. Nevertheless, the regulation of apolipoproteins is a complex event that also involved other nuclear receptors, such as HNF4, COUP-TF or EAR2. The molecular effects of fibrates are far from completely understood, especially because fibrates are not strictly specific for a type of PPAR. In addition, a PPAR-independent pathway for fibrates effects cannot be completely ruled out (reviewed in references 16 and 121).

In line with these data, PPARs has recently been implicated in atherosclerosis. This pathology results from three pathological processes: (1) differentiation of foam cells from macrophages that infiltrated the intima of the arterial wall, (2) inflammatory reaction and (3) cell proliferation,which altogether induce the formation of atheromatous plaque in arterial wall. The production of inflammatory cytokines by foam cells that accumulate high concentrations of intracellular lipids promotes the proliferation of smooth muscle cells that also participate in the pathological formation of the plaque. Given their role in the regulation of cholesterol levels in blood, in the inflammatory response, in macrophage differentiation and in cell proliferation it is likely that PPARα and γ should play an important role in this disease (reviewed in references 16 and 128). Indeed, PPARγ is expressed in atherosclerotic lesions as well as in foam cells, whereas PPARα is expressed in smooth muscle cells[43,83,122]. Interestingly, TZDs have been shown to have a positive effect in preventing the progression of the atherosclerotic plaque, which is possible via their inhibition of inflammatory cytokine production by monocytes/macrophages[42,43,88,89]. The respective role of each PPAR type now need to be studied carefully in order to develop better therapeutic tools (see reference 123 for a review).

The fact that TZDs are specific PPARγ ligands has prompted much effort in finding the role that PPARγ may have in insulin resistance and non-insulin-dependent diabetes mellitus[46,47]. TZDs are widely used as agents that improve insulin sensitivity in patients with insulin resistance syndrome (reviewed in reference 48). Obesity and hyperlipidemia are associated with this syndrome (see references 16 and 87 for reviews). A mutation of human PPARγ in the MAP kinase target sequence of the A/B region (P115Q) has been described in obese patients but other contradictory results were reported on the association between polymorphisms in the human PPARγ gene and obesity[127,145]. Two mutations that destabilize helix 12 have recently been found in the LBD of PPARγ in patients with diabetes mellitus, insulin resistance and hypertension[145,146]. A common PPARγ polymorphism, Pro 12 changed to Ala, has been associated with a decreased risk of type 2 diabetes[151]. The regulation of several genes, such as the glucose transporter GLUT4[119] by PPARγ, may provide a molecular framework to explain the effects of TZDs. Recent results have focused on the intricate and complex relationship existing between TNFα, leptin and PPARγ signaling. An antagonism between TNFα and PPARγ has been demonstrated and mice heterozygous for the PPARγ knockout exhibit an increased level of leptin[100,106,108,133]. PPARγ and TZDs provide a unique opportunity for better understanding of this network of hormonal controls interregulating lipid and glucose homeostasis from feeding behavior to basal metabolism[16].

The first indication of a role for PPARα in controlling inflammation was the demonstration that leukotriene B4, a potent chemotactic inflammatory eicosanoid whose activity is mediated by a membrane receptor, also binds PPARα and induces its target genes, such as the ones of the β- and ω-oxidation pathways that can degrade LTB4 itself[44,45]. Therapeutic control of inflammation can be achieved by blocking the receptors of inflammatory signaling molecules or by negatively regulating their synthesis and positively regulating their catabolism. In this regard, PPARα provides a large therapeutic avenue for the therapeutic control of inflammation and its role in this process has been recently scrutinized.

The relationships between PPARα and pro-inflammatory cytokines, such as IL6, have been studied, and it has been shown that PPARα decreases the production of IL6 via a negative cross-talk with the NFκB transcription factor[83]. In addition it has been shown that fibrates, in addition to their hypolipidemic effect, lead to a decrease in the blood of acute phase proteins, levels of which are an indication of inflammation[83]. Recent studies have also implicated PPARγ in the control of inflammation, since it has been shown that PPARγ-specific agonists inhibit macrophage activation and the production of inflammatory cytokines by monocytes[88–90]. In addition, it has been proposed that IL4 induces the production of PPARγ ligands in macrophages via the regulation of an enzyme, 12/15 lipooxygenase[90]. Taken together, these observations suggest that the relationships between PPARs and inflammatory cytokines can be of clinical use.

Recently, PPAR and more specifically PPARγ, have been implicated in human cancers. The first realization that PPAR may be linked to this disease came from the observation that peroxisomes proliferators induce hepatocarcinomas in rodents[1]. Nevertheless, this appears to be a rodent-specific effect and no link between PPARα and human hepato-carcinogenesis has been yet found. In another set of studies, PPARγ has been implicated in cell growth arrest, an observation that is consistent with its role in promoting adipocyte differentiation[112,124]. Although the precise way in which PPARγ inhibits cell-cycle progression is unknown several reports have implicated Rb and its partners, the E2F factors, in this arrest[112,124]. The ability of PPARγ to promote terminal differentiation into adipocytes of several cell types is now exploited in order to cure cancer. This has been done in human liposarcoma[116,125], breast cancer[113,114] and colon carcinomas[86] in which PPARγ is expressed at high levels. In some of these studies it has also been shown that PPARγ may induce apoptosis[114]. These effects suggest that PPARγ ligands may be developed as anticancer drugs. However, in sharp contrast, an effect of PPARγ in favor of tumor formation has been described in a mouse model that is susceptible to colon cancer because of a mutation in the APC gene[84,85]. Treatment of these mice with PPARγ agonists significantly increases the frequency and size of colon tumors. The reason for these conflicting results are still unclear (see reference 126 for a review). The recent observation that the human PPARγ gene was inactivated by point mutations or frameshifts in cases of sporadic colon cancer leads to the proposal that PPARγ is a tumor supressor gene and is in accordance with its antiproliferative effect[115]. The point mutations were located in the ligand-binding domain, resulting in proteins with reduced ligand-binding ability, whereas the frameshift mutation generates a truncated protein that contains only the A/B region and the DNA-binding domain. All these mutations were sporadic events likely to result from genetic instability during tumor progression, since they are not found in other tissues of these patients[115]. Furthermore, these mutations affect only one allele, suggesting either haploinsufficiency or a dominant negative activity of the mutated protein[115]. These results further stress the potential of PPARγ ligands as anticancer reagents. It has recently been shown that the expression of PPARβ increased in colon cancer and that this gene is down-regulated by APC[129]. Furthermore, non-steroidal anti-inflammatory drugs (NSAIDs) that are known for their action inhibiting tumorigenesis, inhibit the DNA-binding activity of the PPARβ–RXRα heterodimer by an unknown mechanism. These results suggest a direct link between PPARβ expression and colon cancer, but they need to be integrated with the data concerning PPARγ in order to have a complete understanding of the role of PPARs in colon cancer.

Gene structure, promoter and isoforms

The PPAR genes all contain a similar genomic organization with one exon for the A/B region, two exons for the DNA-binding domain, one exon for the hinge region and, finally, two exons for the ligand-binding domain[19-22]. The location of the introns is very well conserved in the various gene structures known in human, mouse or *Xenopus*[19-22]. The mouse PPARα gene appears smaller (30 kbp) than the PPARγ gene, which in both human and mouse spans more than 100 kbp[20-22].

An isoform has been identified recently from the human PPARα gene[23]. This isoform, which corresponds to the alternative usage of exon 6 in the ligand-binding domain encoding region, apparently does not exist in the rodent PPARα gene. This alternative splicing event results in the production of a truncated product, called hPPARα$_{tr}$, lacking part of the hinge region and all the ligand-binding domain, because of the presence of a premature stop codon. Interestingly, this isoform is able to interfere with the wild-type PPARα receptor through a dominant negative activity[23].

Three N-terminal isoforms, called γ1, γ2 and γ3, are known to arise by alternative splicing and promoter usage from the PPARγ gene. The γ1 and γ3 isoforms, differing in their promoter usage and 5′ UTR, generate the same protein product[24]. The γ2 isoform contains an alternative coding exon, called exon B, coding a distinct N-terminal domain that is fused to the same region as that contained in γ1 and γ2, and is controlled by a distinct promoter[11,22,31]. The γ2 isoform, which possesses an extra 30 N-terminal amino acids, contains a more potent AF-1 activating function than the γ1 isoform[64].

Chromosomal location

The chromosomal location of the PPAR genes has been identified both in human and mouse. The human PPARα gene is located in chromosome 22, slightly telomeric to 22q12–q13.1[13]. Its mouse homologue is located on chromosome 15[17]. Human PPARβ is located on chromosome 6p21.1–p21.2[18], close to RXRβ, which is located at 6p21.3 and its mouse homologue to chromosome 17[17]. Human PPARγ is located on chromosome 3p25[13] close to RARβ, TRβ and Rev-erbβ, which are on 3p24, 3p21 and 3p21, respectively. Mouse PPARγ is located on chromosome 6[17]. These chromosomal locations were not known to be implicated in diabetes or obesity, diseases in which the PPARs are implicated[16].

Amino acid sequence for human PPARα (NR1C1)

Accession number: L02932

```
  1 MVDTESPLCPLSPLEAGDLESPLSEEFLQEMGNIQEISQSIGEDSSGSFGFTEYQYLGSCPGSDGSVITD
 71 TLSPASSPSSVTYPVVPGSVDESPSGALNIECRICGDKASGYHYGVHACEGCKGFFRRTIRLKLVYDKCD
141 RSCKIQKKNRNKCQYCRFHKCLSVGMSHNAIRFGRMPRSEKAKLKAEILTCEHDIEDSETADLKSLAKRI
211 YEAYLKNFNMNKVKARVILSGKASNNPPFVIHDMETLCMAEKTLVAKLVANGIQNKEVEVRIFHCCQCTS
281 VETVTELTEFAKAIPAFANLDLNDQVTLLKYGVYEAIFAMLSSVMNKDGMLVAYGNGFITREFLKSLRKP
351 FCDIMEPKFDFAMKFNALELDDSDISLFVAAIICCGDRPGLLNVGHIEKMQEGIVHVLRLHLQSNHPDDI
421 FLFPKLLQKMADLRQLVTEHAQLVQIIKKTESDAALHPLLQEIYRDMY  468
```

Amino acid sequence for human PPARβ (NR1C2)

Accession number: L07592

```
  1 MEQPQEEAPEVREEEEKEEVAEAEGAPELNGGPQHALPSSSYTDLSRSSSPPSLLDQLQMGCDGASCGSL
 71 NMECRVCGDKASGFHYGVHACEGCKGFFRRTIRMKLEYEKCERSCKIQKKNRNKCQYCRFQKCLALGMSH
141 NAIRFGRMPEAEKRKLVAGLTANEGSQYNPQVADLKAFSKHIYNAYLKNFNMTKKKARSILTGKASHTAP
211 FVIHDIETLWQAEKGLVWKQLVNGLPPYKEISVHVFYRCQCTTVETVRELTEFAKSIPSFSSLFLNDQVT
281 LLKYGVHEAIFAMLASIVNKDGLLVANGSGFVTREFLRSLRKPFSDIIEPKFEFAVKFNALELDDSDLAL
351 FIAAIILCGDRPGLMNVPRVEAIQDTILRALEFHLQANHPDAQYLFPKLLQKMADLRQLVTEHAQMMQRI
421 KKTETETSLHPLLQEIYKDMY  441
```

Amino acid sequence for human PPARγ (NR1C3)

Accession number: L40904

```
  1 MTMVDTEIAFWPTNFGISSVDLSVMEDHSHSFDIKPFTTVDFSSISTPHYEDIPFTRTDPVVADYKYDLK
 71 LQEYQSAIKVEPASPPYYSEKTQLYNKPHEEPSNSLMAIECRVCGDKASGFHYGVHACEGCKGFFRRTIR
141 LKLIYDRCDLNCRIHKKSRNKCQYCRFQKCLAVGMSHNAIRFGRIAQAEKEKLLAEISSDIDQLNPESAD
211 LRQALAKHLYDSYIKSFPLTKAKARAILTGKTTDKSPFVIYDMNSLMMGEDKIKFKHITPLQEQSKEVAI
281 RIFQGCQFRSVEAVQEITEYAKSIPGFVNLDLNDQVTLLKYGVHEIIYTMLASLMNKDGVLISEGQGFMT
351 REFLKSLRKPFGDFMEPKFEFAVKFNALELDDSDLAIFIAVIILSGDRPGLLNVKPIEDIQDNLLQALEL
421 QLKLNHPESSQLFAKLLQKMTDLRQIVTEHVQLLQVIKKTETDMSLHPLLQEIYKDLY  478
```

References

[1] Issemann, I. and Green, S. (1990) Nature 347, 645–650.
[2] Dreyer, C. et al. (1992) Cell 68, 879–887.
[3] Schmidt, A. et al. (1992) Mol. Endocrinol. 6, 1634–1641.
[4] Kliewer, S.A. et al. (1994) Proc. Natl Acad. Sci. USA 91, 7355–7359.
[5] Amri, E.Z. et al. (1995) J. Biol. Chem. 270, 2367–2371.
[6] Göttlicher, M. et al. (1992) Proc. Natl Acad. Sci. USA 89, 4653–4657.
[7] Zhu, Y. et al. (1993) J. Biol. Chem. 268, 26817–26820.
[8] Chen, F. et al. (1993) Biochem. Biophys. Res. Commun. 196, 671–677.
[9] Sher, T. et al. (1993) Biochemistry 32, 5598–5604.
[10] Aperlo, C. et al. (1995) Gene 162, 297–302.
[11] Tontonoz, P. et al. (1994) Genes Dev. 8, 1224–1234.
[12] Xing, G. et al. (1995) Biochem. Biophys. Res. Commun. 217, 1015–1025.
[13] Greene, M.E. et al. (1995) Gene Expression 4, 281–299.
[14] Tontonoz, P. et al. (1994) Nucl. Acids Res. 22, 5628–5634.
[15] Nuclear Receptor Nomenclature Committee (1999) Cell, 97, 161–163.

16 Desvergnes, B. and Wahli, W. (1999) Endocrine Rev. 20, 649–688.

17 Jones, P.S. et al. (1995) Eur. J. Biochem. 233, 219–226.

18 Yoshikawa, T. et al. (1996) Genomics 35, 637–638.

19 Krey, G. et al. (1993) J. Steroid Biochem. Mol. Biol. 47, 65–73.

20 Gearing, K.L. et al. (1994) Biochem. Biophys. Res. Commun. 199, 255–263.

21 Zhu, Y. et al. (1995) Proc. Natl Acad. Sci. USA 92, 7921–7925.

22 Beamer, B.A. et al. (1997) Biochem. Biophys. Res. Commun. 233, 756–759.

23 Gervois, P. et al. (1999) Mol. Endocrinol., 13, 1535–1549.

24 Fajas, L. et al. (1998) FEBS Lett. 438, 55–60.

25 Kliewer, S.A. et al. (1992) Nature 358, 771–774.

26 Tugwood, J.D. et al. (1992) EMBO J. 11, 433–439.

27 Muerhoff, A.S. et al. (1992) J. Biol. Chem. 267, 19051–19053.

28 IJpenberg, A. et al. (1997) J. Biol. Chem. 272, 20108–20117.

29 Juge–Aubry, C. et al. (1997) J. Biol. Chem. 272, 25252–25259.

30 DiRenzo, et al. (1997) Mol. Cell. Biol. 17, 2166–2176.

31 Hsu, M.H. et al. (1998) J. Biol. Chem. 273, 27988–27997.

32 Gervois, P. et al. (1999) Mol. Endocrinol. 13, 400–409.

33 Keller, H. et al. (1995) Mol. Endocrinol. 9, 794–804.

34 Chu, R. et al. (1995) Proc. Natl Acad. Sci. USA 92, 11593–11597.

35 Juge-Aubry, C.E. et al. (1995) J. Biol. Chem. 270, 18117–18122.

36 Tolon, R. et al. (1998) J. Biol. Chem. 273, 26652–26661.

37 Keller, H. et al. (1993) Proc. Natl Acad. Sci. USA 90, 2160–2164.

38 Krey, G. et al. (1997) Mol. Endocrinol. 11, 779–791.

39 Wilson, T.M. and Wahli, W. (1997) Curr. Opin. Chem. Biol. 1, 235–241.

40 Forman, B.A. et al. (1997) Proc. Natl Acad. Sci. USA 94, 4312–4317.

41 Kliewer, S.A. et al. (1997) Proc. Natl Acad. Sci. USA 94, 4318–4323.

42 Nagy, L. et al. (1998) Cell 93, 229–240.

43 Tontonoz, P. et al. (1998) Cell 93, 241–252.

44 Devchand, P.R. et al. (1996) Nature 384, 39–43.

45 Devchand, P.R. et al. (1999) J. Biol. Chem. 274, 23341–23348.

46 Lehmann, J.M. et al. (1995) J. Biol. Chem. 270, 12953–12956.

47 Berger, J. et al. (1996) Endocrinology 137, 4189–4195.

48 Reginato, M.J. and Lazar, M.A. (1999) Trends Endocrinol. Metab. 10, 9–13.

49 Berger, J. et al. (1999) J. Biol. Chem. 274, 6718–6725.

50 Brown, P.J. et al. (1997) Chem. Biol. 4, 909–918.

51 Lin, Q. et al. (1999) Biochemistry 38, 185–190.

52 Oberfield, J.L. et al. (1999) Proc. Natl Acad. Sci. USA. 96, 6102–6106.

53 Dreyer, C. et al. (1993) Biol. Cell 77, 67–76.

54 Laudet, V. (1997) J. Mol. Endocrinol. 19, 207–226.

55 Keller, H.J. et al. (1997) Biol. Chem. 378, 651–655.

56 Nolte, R.T. et al. (1998) Nature 395, 137–143.

57 Xu, H.E. et al. (1999) Mol. Cell 3, 397–403.

58 Uppenberg, J. et al. (1998) J. Biol. Chem. 273, 31108–31112.

59 Weatherman, R.V. et al. (1999) Ann. Rev. Biochem. 68, 559–581.

60 Juge-Aubry, C.E. et al. (1999) J. Biol. Chem. 274, 10505–10510.

61 Hu, E. et al. (1996) Science 274, 2100–2103.

62 Shao, D. et al. (1998) Nature 396, 377–380.
63 Mukherjee, R. et al. (1997) Nature 386, 407–410.
64 Werman, A. et al. (1997) J. Biol. Chem. 272, 2023–2025.
65 Bogazzi, F. et al. (1994) J. Biol. Chem. 269, 11683–11686.
66 McInerney E.M. et al. (1998) Genes Dev. 12, 3357–3368.
67 Qi, C. et al. (1999) Proc. Natl Acad. Sci. USA 96, 1585–1590.
68 Dowell, P. et al. (1997) J. Biol. Chem. 272, 33435–33443.
69 Schulman, I.G. et al. (1998) Mol. Cell. Biol. 18, 3483–3494.
70 Puigserver, P. et al. (1998) Cell 92, 829–839.
71 Wu, Z. et al. (1999) Cell 98, 115–124.
72 Castillo, G. et al. (1999) EMBO J. 18, 3676–3687.
73 Zhu, Y. et al. (1997) J. Biol. Chem. 272, 25500–25506.
74 Krey, G. et al. (1995) Mol. Endocrinol. 9, 219–231.
75 Braissant, O. and Wahli, W. (1998) Endocrinology 139, 2748–2754.
76 Braissant, O. et al. (1996) Endocrinology 137, 354–366.
77 Lemberger, T. et al. (1994) J. Biol. Chem. 269, 24527–24530.
78 Lemberger, T. et al. (1996) J. Biol. Chem. 271, 1764–1769.
79 Auboeuf, D. et al. (1997) Diabetes 48, 1319–1327.
80 Steineger, H.H. et al. (1994) Eur. J. Biochem. 225, 967–974.
81 Lim, H. et al. (1999) Genes Dev. 13, 1561–1574.
82 Bastie, C. et al. (1999) J. Biol. Chem. 274, 21920–21925.
83 Staels, B. et al. (1998) Nature 393, 790–793.
84 Saez, E. et al. (1998) Nature Med. 4, 1058–1061.
85 Lefebvre, A.M. et al. (1998) Nature Med. 4, 1053–1057.
86 Sarraf, P. et al. (1998) Nature Med. 4, 1046–1052.
87 Wu, Z. et al. (1999) Curr. Opin. Cell Biol. 11, 689–694.
88 Ricote, M. et al. (1998) Nature 391, 79–82.
89 Jiang, C. et al. (1998) Nature 391, 82–86.
90 Huang, J.T. et al. (1999) Nature 400, 378–382.
91 Mano, H. et al. (2000) J. Biol. Chem. 275, 8126–8132.
92 Vidal-Puig, A. et al. (1996) J. Clin. Invest. 97, 2553–2561.
93 Vidal-Puig, A. et al. (1997) J. Clin. Invest. 99, 2416–2422.
94 Vu-Dac, N. et al. (1994) J. Biol. Chem., 269, 31012–31018.
95 Vu-Dac, N. et al. (1995) J. Clin. Invest. 96, 741–750.
96 Chawla, A. and Lazar, M.A. (1994) Proc. Natl Acad. Sci. USA 91, 1786–1790.
97 Tontonoz, P. et al. (1994) Cell 79,147–156.
98 Wu, Z. et al. (1995) Genes Dev. 9, 2350–2363.
99 Tontonoz, P. et al. (1995) Mol. Cell. Biol. 15, 351–357.
100 Zhang, B. et al. (1996) Mol. Endocrinol. 10, 1457–1466.
101 Fajas, L. et al. (1999) Mol. Cell. Biol. 19, 5495–5503.
102 Wu, Z. et al. (1999) Mol. Cell 3, 151–158.
103 Kim, J.B. et al. (1998) Proc. Natl Acad. Sci. USA 95, 4333–4337.
104 Lee, S.S.T. et al. (1995) Mol. Cell. Biol. 15, 3012–3022.
105 Leone, T.C. et al. (1999) Proc. Natl Acad. Sci. USA 96, 7473–7478.
106 Lowell, B.B. (1999) Cell 99, 239–242.
107 Barak, Y. et al. (1999), Mol. Cell 4, 585–595.
108 Kubota, N. et al. (1999) Mol. Cell 4, 597–609.
109 Rosen, E.D. et al. (1999) Mol. Cell 4, 611–617.

[110] Kersten, S. et al. (1999) J. Clin. Invest. 103, 1489–1498.
[111] Aoyama, T. et al. (1998) J. Biol. Chem. 273, 5678–5684.
[112] Altiok, S. et al. (1997) Genes Dev. 11, 1987–1998.
[113] Mueller, E. et al. (1998) Mol. Cell 1, 465–470.
[114] Elstner, E. et al. (1998) Proc. Natl Acad. Sci. USA 95, 8806–8811.
[115] Sarraf, P. et al. (1999) Mol. Cell 3, 799–804.
[116] Demetri, G.D. et al. (1999) Proc. Natl Acad. Sci. USA 96, 3951–3956.
[117] Meertens, L.M. et al. (1998) EMBO J. 17, 6972–6978.
[118] Brun, R.P. et al. (1996) Genes Dev. 10, 974–984.
[119] Wu, Z. et al. (1998) J. Clin. Invest. 101, 22–32.
[120] Vu-Dac, N. et al. (1998) J. Biol. Chem. 273, 25713–25720.
[121] Schoonjans, K. et al. (1996) J. Lipid Res. 37, 907–925.
[122] Ricote, M. et al. (1998) Proc. Natl Acad. Sci. USA 95, 7614–7619.
[123] Spigelman, B.M. (1998) Cell 93, 153–155.
[124] Shao, D. and Lazar, M.A. (1997) J. Biol. Chem. 272, 21473–21478.
[125] Tontonoz, P. et al. (1997), Proc. Natl Acad. Sci. USA, 94, 237–241.
[126] Seed, B. (1998) Nature Med. 4, 1004–1005.
[127] Ristow, M. et al. (1998) N. Engl. J. Med. 339, 953–959.
[128] Kersten, S. et al. (2000) Nature 405, 421–424.
[129] He, T.C. et al. (1999) Cell 99, 335–345.
[130] Li, M. et al. (2000) Mol. Cell. Biol. 20, 4699–4707.
[131] Gampe, R.T. et al. (2000) Mol. Cell 5, 545–555.
[132] Rosen, E.D. et al. (2000) Genes Dev. 14, 1293–1307.
[133] Hollenberg, A.N. et al. (1997) J. Biol. Chem. 272, 5283–5290.
[134] Mbalaviele, G. et al. (2000) J. Biol. Chem. 275, 14388–14393.
[135] Bordji, K. et al. (2000) J. Biol. Chem. 275, 12243–12250.
[136] Vega, R.B. et al. (2000) Mol. Cell. Biol. 20, 1868–1876.
[137] Guardiola-Diaz, H.M. et al. (1999) J. Biol. Chem. 274, 23368–23377.
[138] Zhu, Y. et al. (1996) Gene Expression 6, 185–195.
[139] Bell, A.R. et al. (1998) Biochem. J. 332, 689–693.
[140] Hughes, S. et al. (1999) Mol. Biol. Evol. 16, 1521–1527.
[141] Hotta, K. et al. (1998) Int. J. Obesity Related Metabol. Disord. 22, 1000–1010.
[142] Grindflek, E. et al. (1998) Biochem. Biophys. Res. Commun. 249, 713–718.
[143] Sundvold, H. et al. (1997) Biochem. Biophys. Res. Commun. 239, 857–861.
[144] Michael, L.F. et al. (1997) Endocrinology 138, 3695–3703.
[145] Schwartz, M.W. and Kahn, S.E. (1999) Nature 402, 860–861.
[146] Barroso, I. et al. (1999) Nature 402, 880–883.
[147] Takada, I. et al. (2000) Mol. Endocrinol. 14, 733–740.
[148] Monsalve, M. et al. (2000) Mol. Cell 6, 307–316.
[149] Peters, J.M. et al. (2000) Mol. Cell. Biol. 20, 5119–5128.
[150] Puigserver, P. et al. (1999) Science 286, 1368–1371.
[151] Altshuler, D. et al. (2000) Nature Genet. 26, 76–80.

Rev-erb

Names

The Rev-erbα orphan receptor was first described as a gene related to the c-erbA-1 proto-oncogene, which encodes the thyroid hormone receptor α (NR1A1) and was thus called earl (erbA-related gene)[1,2]. Strikingly, the Rev-erbα gene was independently shown to overlap in a reverse orientation the c-erbA-1 gene and this has been the rational for its current name Rev-erbα or Rev-erbAα[2,3]. A second gene has been found called Rev-erbβ. The two *Drosophila* homologues E75 and E78 were named according to their cytological localization[4-6].

Species	Other names	Accession number	References
Official name: NR1D1			
Human	Rev-erbα, earl, Rev-erbAα	M24898	1, 2, 7
Rat	Rev-erbα, Rev-erbAα	M25804	3
Zebrafish	Rev-erbα	Not available	8
Official name: NR1D2			
Human	Rev-erbβ, BD73	L31785	9
Mouse	Rev-erbβ, RVR	U12142	10, 11
Rat	Rev-erbβ, Rev-erbAβ, HZF-2	X82777	12, 13
Chicken	Rev-erbβ	X80258	14
Official name: NR1D3			
Drosophila	E75, DmE75, Eip75	X51548	4, 5
Manduca	E75, MsE75	S60732	15
Galleria	E75, GmE75	U02620	16
Choristoneura	E75, CfE75	U63930	17
Metapenaeus	E75, MeE75	AF092946	18
Official name: NR1E1			
Drosophila	E78, Eip78	U01087	6
Official name: NRIGI			
Caenorhabditis	CNR14, SEX-1	U13074	19

Isolation

Rev-erbα (NR1D1) was discovered by two groups at the same time using two completely different approaches. It was first found in a classic low-stringency screening strategy conducted with a probe encompassing the v-erbA oncogene[1,2]. The resulting cDNA was called ear-1 and exhibited *c.* 30% amino acid identities with the ligand-binding domain of TRα[1]. Later, it was observed that ear-1 overlapped with the last exon of the c-erbA-1/TRα gene[2]. Another group, by studying the isoforms generated by the TRα gene found a cDNA encoding Rev-erbα and discovered the overlapping structure[3]. The phylogenetic analysis of the Rev-erbα sequence revealed that, in fact, it is not particularly closely related to TRα, but rather to the ROR (NR1F) and PPAR (NR1C) groups[20].

Rev-erbβ (NR1D2) was also found by a number of research groups by different strategies involving either RT-PCR with degenerated primers or low-stringency screening[9-14]. The analysis of the Rev-erbβ sequence reveals a strong level of identity with Rev-erbα: 97% in the DNA-binding domain and 68% in the ligand-binding domain. In addition, the two Rev-erb genes have structural characteristics in common, such as a long D domain and a truncated ligand-binding domain. This last observation is of great functional importance, since this domain ends in the middle of the AF-2 AD region, which is thus unable to recruit coactivators, explaining the repressive activity of Rev-erb on transcription (see below). In phylogenetical trees, Rev-erbs are related to RORs, consistent with the fact that these receptors share the same response element[10,21]. Rev-erbs exhibit typically 58% and 37% amino acid identities with the DNA-binding and ligand-binding domains of RORs.

Drosophila E75 (NR1D3) was identified in a chromosomal walking strategy to isolate a puff that responds to ecdysone, located at position 75B on chromosome 3L[5]. The long gene that was cloned was shown to be a member of the nuclear receptor superfamily with a complex structure generating several isoforms. It exhibited 78% amino acid identity with Rev-erbα in the DBD and 38% in the LBD. The E75 gene isolated from *Manduca* exhibited 100% and 86% amino acid identities in the DBD and LBD of the *Drosophila* gene, respectively. A E75 gene was also cloned recently in the shrimp *Metapenaeus ensis*[18]. It exhibited 100% and 54% amino acid identities with the *Drosophila* E75 gene in the DBD and LBD.

Another *Drosophila* gene, called E78 (NR1E1) from its chromosomal location, is also closely related to Rev-erbs and E75[6]. As E75, this gene was found by a chromosomal walk in order to isolate the gene that is regulated in the early-late ecdysone inducible puff located at the 78C position on chromosome 3L. Again this is a long and complex gene generating several isoforms. The gene product is 70% and 27% identical with E75 in the DBD and LBD, respectively. In addition, it harbors 70% and 34% sequence identity with Rev-erbα in the same two domains, and 61% and 28% with RORs. These data clearly suggest that E78 is a member of the Rev-erb group. Nevertheless, its placement in phylogenetical trees is tedious, probably because of rapid evolutionary divergence, and it was located as a unique member of the group E of subfamily I in the nomenclature of nuclear receptors[22]. Since the shrimp E75 gene is clearly more closely related to E75 than to E78, we suggest that the E75/E78 split arose very early during protostomians evolution[18,21,22].

The same situation probably holds for the *Caenorhabditis* gene CNR14 (NR1G1), which was found by a PCR strategy. It may be viewed as a rapidly divergent member of the Rev-erb group[19]. In fact, depending on the phylogenetical procedure used, CNR14 and E78 appear as homologues but, since these relationships were weakly supported, CNR14 was placed as a unique member of group G in subfamily 1[21,22]. This gene exhibits 77% and 21% amino acid identities in the DBD and LBD, respectively, with the *Dopsophila* E78 gene[19]. The level of identity is lower with other members of the Rev-erb group (70–73% in the DBD and 18–19% in the LBD). It was also identified as SEX-1, a gene required for sex determination in *C. elegans*[22a].

DNA binding

The Rev-erbα gene product was shown to bind to DNA as a monomer to a new response element called the RevRE, which contains an AGGTCA motif linked in 5' to an A/T-rich sequence harboring the following consensus: A/T A A/T N T[23]. Later, it was shown by site-selection experiments as well as by the study of a natural RevRE found in the human Rev-erbα promoter that it is able to homodimerize on RevDR2 elements that are composed of one classic RevRE followed in 3' by an AGGTCA motif[24,25]. The two motifs are separated by two bases that are always a C and a T. The C-terminal extension of the DNA-binding domain has been shown to be necessary for the dimerization[24,26]. Reports of binding of Rev-erbα to other types of elements, such as DR4 elements were not confirmed by subsequent studies[27,28].

The binding specificity of Rev-erbβ is indistinguishable from the one of Rev-erbα. Both receptors are able to form monomeric or dimeric complexes depending on the response element[9,11,14,24,25].

It has been shown both for Rev-erbα and β that the A-box is implicated in the recognition of the 5' A/T-rich extension of the RevRE and confers the specificity for this element, as is the case for other receptors[9,23]. Mutational analysis has led to the suggestion that the A-box of Rev-erbs lies closer to the DNA-binding domain than for other receptors[9]. The resolution of the crystal structure of a dimer of Rev-erbα DBD on a Rev-DR2 element demonstrates that, in addition to the P-box interaction with the major groove of the DNA on the AGGTCA motifs, a second protein–DNA interface exists[29]. This surface is composed of the A/T-rich 5' extension of the response element, which is recognized by the C-terminal extension of the receptor beyond the core DBD. This region corresponds to the A-box defined by mutational studies, which is also called the GRIP box[29]. The dimerization between the two monomers involved the tip of the second zinc finger module of the upstream monomer and the GRIP box of the downstream one[29].

Partners

Rev-erbs are unable to interact with RXRs[14,23].

In insects, a functional negative interaction exists between the non-DNA-binding isoform of E75, called E75B and DHR3 (NR1F4) for the regulation of βFTZ-F1 (NR5A3). This interaction may imply a direct protein–protein contact between DHR3 and E75B, since E75B can recognize the promoter of βFTZ-F1 only when DHR3 is present[30].

The corepressor NCoR was found in a yeast two-hybrid screen as a partner for Rev-erbα[27]. It was shown that ectopic NCoR potentiates Rev-erb-mediated repression and that transcriptional repression of Rev-erb mutants correlates with their ability to bind NCoR[31]. The delineation of the region of Rev-erb that forms the interaction surface with NCoR was a subject of conflicting results[31-37]. The present consensus is that two regions of Rev-erbs, called CIR-1 and CIR-2 (for corepressor interacting region) that correspond to helices 3 and 11, respectively, are forming a unique NCoR interaction surface. The helix 1 contained in the hinge (D) domain of Rev-erbs is apparently not involved in direct contact with NCoR. A precise mutational analysis as well as modeling studies reveal that an hydrophobic surface containing residues from helix 3, the loop between

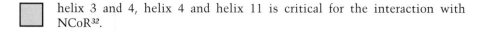

helix 3 and 4, helix 4 and helix 11 is critical for the interaction with NCoR[32].

Ligands

There is no ligand described for Rev-erbs and modeling studies of the ligand-binding domain suggest that the hydrophobic pocket found in liganded receptors is, in the case of Rev-erbα, full of densely packed side-chains of amino acids[32]. This suggests, but does not prove, that Rev-erbs are real orphan receptors.

Three-dimensional structure

The structure of a dimer of DNA-binding domains of Rev-erbα bound to a Rev-DR2 element has been solved[29]. This structure shows that two protein–DNA interaction regions exist: one between the P-box of the Rev-erbs monomer and the AGGTCA core motif, and the other between the C-terminal extension (GRIP box) and the 5′ A/T-rich extension of the site. Furthermore, this structure allows the interaction surface between the two Rev-erb monomers to be determined: the tip of the second zinc finger of the upstream monomer and the GRIP box of the downstream one[29].

There is no 3D structure available for the ligand-binding domain of Rev-erb but a modeling of this domain, based on the 3D structure of the RARγ ligand binding domain has been proposed[32].

Expression

Rev-erbα is expressed in a wide variety of tissues as a single mRNA of 3 kb[2,3]. Strong expression is observed in skeletal muscles, brown fat, liver, heart and brain, modest levels in pituitary and kidney, and very low levels in testis, lung and hypothalamus[2,3]. Rev-erbα was shown to be expressed in cell lines of the B lymphocyte lineage[38]. Consistent with earlier northern blot studies, the expression of a β-gal cassette inserted inside the Rev-erbα gene during knockout experiments suggest that expression starts in the heart at E10.5, extends to the eyes one day later and then increases during the second week of life[39,40]. In brain, a high expression in Purkinje cells of the cerebellum, olfactory granule cells, cerebral cortex and hippocampus was observed[39,41].

Rev-erbβ is found as two transcripts of 5.5 and 4 kb in rat and only one at 4.5 kb in human and mouse[9,11,12]. In all mammals examined, this gene is expressed in a wide variety of tissues, such as heart, brain, lung, liver, skeletal muscles and kidney[9–12]. Very weak expression was observed in spleen and testis. A positive regulation of Rev-erbβ expression by butylated hydroxytoluene was observed[9]. *In situ* hybridization experiments reveal a strong expression in the central nervous system in the cerebellar cortex as well as in the dentate gyrus and hippocampus[12]. In chicken, two transcripts of 4.5 and 3 kb were visualized with a very strong expression in spleen from 1-day-old chicks and no expression in other tissues[14]. During chicken development, *in situ* hybridization experiments show that Rev-erb β is expressed in the central and peripheral nervous system, including a prominent expression in notochord and floor plate as well as in blood islands[14].

It has been shown that Rev-erbα expression increases during adipocytes differentiation[42] and decreases during myogenic differentiation[43]. This down-regulation of Rev-erbα observed during myogenic differentiation correlates with the appearance of muscle-specific transcripts, such as myogenin and α-actin. A similar down-regulation of Rev-erbβ expression was observed during this process. It was shown functionally that Rev-erbα and β overexpression can suppress myogenic differentiation, suggesting these receptors play an active role in the regulation of this event[33]. These data also suggested that Rev-erbs may be able to regulate genes important for myogenic differentiation, such as p21[Cip1/Waf1], MyoD1 and myogenin[33,43].

A promoter of the human Rev-erbα gene was cloned[25]. Interestingly, the activity of this promoter is down-regulated by both Rev-erbα and β acting via a Rev-DR2 element located close to the start sites. The *in vivo* significance of this negative autoregulatory loop has been confirmed by the knockout experiments in which a burst of expression was observed in Rev-erbα[-/-] animals[39]. It has been shown that Rev-erbα expression is stimulated by fibrates, hypolipidemic drugs that act via the PPARα receptor[44]. Interestingly, the PPARα–RXR heterodimer directly regulate the human Rev-erbα promoter and do so *via* the same Rev-DR2 site on which Rev-erbα homodimers bind and repress its over-expression[45]. Indeed, competition between Rev-erbα homodimers, which repress transcription, and PPARα–RXR heterodimers, which activate transcription, for the occupancy of this site, has been shown. This suggests that a cross-talk between Rev-erbs and PPARs may exist on a subset of PPAR-regulated genes that contain Rev-DR2 elements in their promoters. Another similar example, the enoyl-CoA hydratase/3-hydroxyacyl-CoA dehydrogenase gene, has been also described[46]. The fact that PPARα may regulate Rev-erbα expression suggests that PPARγ may also be able to do so, and provides an explanation for the induction of Rev-erbα observed during adipocyte differentiation[42,45].

All these data suggest that the expression of Rev-erbα (and perhaps Rev-erbβ) is finely tuned. This has been confirmed by the observation that Rev-erbα expression is regulated along a circadian rhythm in rat liver[47]. The peak of Rev-erbα expression in this organ is around 16 hours. This regulation is conserved in other species, such as adult zebrafish as well as embryos[8]. In embryos, the rhythmic expression is developmentally regulated, since expression is first observed in the epiphysis at 48 hours post-fertilization (hpf), then in epiphysis and retina at 72 hpf, and finally in epiphysis, retina and optic tectum at 96 hpf. This has led to the suggestion that Rev-erbα is a clock-controlled gene[8]. This observation should be linked to the fact that RORs, and particularly RORβ, also exhibit a circadian regulation of expression and that RORβ knockout mice exhibit a circadian phenotype (see ROR). This suggests that one of the important functions of Rev-erbs and RORs could be linked to circadian rhythmicity.

In *Drosophila* it has been shown that E75 and E78 are both directly related by ecdysone[5,6]. The precise expression pattern during *Drosophila*

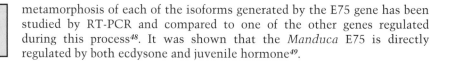

metamorphosis of each of the isoforms generated by the E75 gene has been studied by RT-PCR and compared to one of the other genes regulated during this process[48]. It was shown that the *Manduca* E75 is directly regulated by both ecdysone and juvenile hormone[49].

Activity and target genes

Although, Rev-erbα was first described as a constitutive transcriptional activator[23,27], it was rapidly made clear that both Rev-erbs are in fact transcriptional repressors[9–11,24,25,43]. This is consistent with the fact that Rev-erbs are devoid of a complete AF-2–AD region, since their ligand-binding domain is truncated at the C-terminal end compared with other receptors. Since Rev-erbs and RORs bind to the same half-site sequence, it was shown that they may regulate a common set of target genes with opposing biological activities[10,11]. The Rev-erb repressing activity is mediated by an interaction with the corepressor NCoR[31,32,37]. These reports propose that N-CoR interacts only with Rev-erb homodimers and not to Rev-erb monomers, suggesting that homodimers are alone able to repress transcription efficiently. Nevertheless, this was not confirmed by other authors, who show a clear Rev-erb-mediated repression through monomeric sites[25,50].

Several target genes on which Rev-erbs are able to repress transcription were described. It has been suggested that Rev-erbs may down-regulate the expression of CRBPI, p21[Cip1/Waf1], MyoD1 and myogenin, but this remains to be demonstrated by more detailed experiments[24,33,43]. The two first target genes to be described for which a direct regulation by Rev-erb was demonstrated were Rev-erbα itself and the N-myc gene[25,50]. In the Rev-erbα promoter, a typical RevDR2 element was observed, whereas N-myc is regulated by a half-site RevRE. In the latter case, the opposite effects of ROR and Rev-erb on transcription were deciphered, and it was suggested that ROR and Rev-erb may be linked to oncogenic transformation, since the deletion of their response element in N-myc increased the transforming potential of N-myc in a focus assay[38]. Other cases where an opposite effect of ROR and Rev-erb through the same DNA sequence was observed are the rat apoA1 and the enoyl-CoA hydratase/3-hydroxyacyl-CoA dehydrogenase gene[44,46,51]. The CYP4A6 encoding a member of the P450 fatty acid ω-hydroxylase family is also a target gene of Rev-erbα[28].

The cross-talk between Rev-erb and ROR may also exist in insects, since it was shown that E75B and DHR3 have opposite effects on the regulation of the βFTZ-F1 gene in *Drosophila*[30].

Knockout

The knockout of the Rev-erbα gene has been performed in mice[39]. Strikingly, the mice are viable and phenotypically normal, although the females exhibit a somehow reduced fertility. Fat tissues and muscles are normal and the transcription of the TRα2 isoform that overlaps the Rev-erbα gene appears normal. This suggests that the overlap has no major regulatory role *in vivo*. The only noticeable phenotype was observed in the cerebellum in which several abnormalities are present during the second post-natal week of life. The authors observed an alteration in the development of the Purkinje cells as well as a delay in proliferation and migration, and increased apoptosis in granule cells. This phenotype, which is much less severe than the RORα phenotype, nevertheless suggests that a cross-talk between the two genes exist during cerebellum development[39].

Genetic analysis in *Drosophila* indicate that E75 is required at a number of developmental stages, including both larval molting and metamorphosis (cited in reference 49).

Analysis of E78 mutants show that the gene is apparently dispensable for normal development but that it is required for a maximal puffing activity of a subset of late puffs, such as 63E and 82F[52].

Associated disease

There is no disease known to be associated with Rev-erbs in humans. However, the fact that Rev-erbα represses the transcription of the N-*myc* proto-oncogene may suggest a link with cancer[50].

Gene structure, promoter and isoforms

The gene structure of the Rev-erbα locus is complex and unique[2,3]. The last exon of the c-erbA-1 gene, which encodes the thyroid hormone receptor α gene, overlaps on 267 bp, the last exon of Rev-erbα. Interestingly, the last exon of c-erbA-1 is alternative and specific to an isoform of TRα, called TRα2, which acts as a dominant negative regulator of the normal receptor TRα1. It appears immediately that the c-erbA-1/Rev-erbα overlap offers a way to regulate the ratio finely between the α1 and α2 isoforms of TRα[2,3]. Indeed, it was shown both in cell culture and *in vivo* that high levels of Rev-erbα correlate with low levels of TRα2 relative to TRα1[14,38,53]. Furthermore, conditions that alter Rev-erbα expression levels often result in a reciprocal change in the ratio of TRα2 to TRα1[42,54]. The mechanisms underlying this regulatory effect have been studied first by showing *in vitro* that Rev-erbα sequences regulate TRα splicing[55]. More recently, it has been shown in cell cultures that Rev-erbα does indeed influence the TRα2/TRα1 ratio[56]. Nevertheless, the study of knockout Rev-erbα mice has not revealed a major effect of the deletion of Rev-erbα transcripts on the TRα2 level *in vivo*[39].

Beside the overlap, the Rev-erbα gene itself is a small gene of c. 6 kbp, consistent with its striking regulation along the circadian rhythm[2,3,57]. The location of the introns inside the DNA-binding domain is reminiscent of that observed for other members of subfamily 1: one intron lies between the two zinc fingers, two amino acids after the last Cys of the first finger, and the other is located inside the C-terminal extension (A-box) of the DNA-binding domain, i.e. just downstream from the core DNA-binding domain, a location conserved in all the superfamily.

In addition to this complex structure, the Rev-erbα gene is able to code for two different N-terminal isoforms that differ in their A/B region[2,3,7,57]. There is no available study on the distribution and functional differences between these isoforms that probably results from alternative promoter usage. The promoter of the human Rev-erbα gene governing the expression of the large isoform has been cloned[25]. The study of this promoter allowed to demonstrate a down-regulation by Rev-erbα and a stimulation by PPARα, through the same Rev-DR2 element[25,44,45].

Nothing is known about genomic organization or isoforms that may be present in the Rev-erbβ gene.

In contrast with its homologues in vertebrates, but as other regulatory genes implicated in the ecdysone response in *Drosophila*, the E75 gene is long (50 kbp) and complex, generating three isoforms, E75A, E75B and E75C[5]. E75B differs from E75A by the lack of the first zinc finger and is thus unable to bind to DNA. Nevertheless, this isoform may play an important regulatory role, since it was shown that it may impair the DHR3-mediated activation of the βFTZ-F1 gene. In addition, E75A, B and C have different expression patterns and ecdysone response during *Drosophila* metamorphosis[48].

As E75, E78 is a long (c. 35 kbp) and complex gene generating two isoforms, E78A and B. E78B encodes a truncated receptor that lacks the DNA-binding domain[6]. As for E75, the expression patterns of these isoforms are distinct: E78A is expressed during a brief interval in mid-pupal development, whereas E78B is predominantly expressed at puparium formation and immediately following E78A in pupae. In addition, E78B is directly inducible by ecdysone in late third-instar larvae[6].

Chromosomal location

The Rev-erbα gene, which overlaps with the TRα gene, is located on human chromosome 17 on q21 (see reference 58). The mouse Rev-erbα gene is located on chromosome 11.

Rev-erbβ is located on human chromosome 3 and, interestingly, is linked to the TRβ gene[58]. These two genes do not overlap as TRα and Rev-erbα, since they are separated by 1 Mb. Interestingly, this region also contains the RARβ gene. Since TRα and Rev-erbα are also linked to RARα, this argues for two ensembles of old synteny that was duplicated once early on during vertebrate evolution and which give rise to TRα, Rev-erbα and RARα, on one hand, and TRβ, Rev-erbβ and RARβ, on the other.

E75 and E78 are located on chromosome 3L at 75A9–B6 and 78C4–7, respectively[5,6].

Amino acid sequence for human Rev-erbα (NR1D1)

Accession number : M24898

```
  1  MTTLDSNNNTGGVITYIGSSGSSPSRTSPESLYSDNSNGSFQSLTQGCPTYFPPSPTGSLTQDPARSFGS
 71  IPPSLSDDGSPSSSSSSSSSSSSSFYNGSPPGSLQVAMEDSSRVSPSKSTSNITKLNGMVLLCKVCGDVAS
141  GFHYGVHACEGCKGFFRRSIQQNIQYKRCLKNENCSIVRINRNRCQQCRFKKCLSVGMSRDAVRFGRIPK
211  REKQRMLAEMQSAMNLANNQLSSQCPLETSPTQHPTPGPMGPSPPPAPVPSPLVGFSQFPQQLTPPRSPS
281  PEPTVEDVISQVARAHREIFTYAHDKLGSSPGNFNANHASGSPPATTPHRWENQGCPPAPNDNNTLAAQR
351  HNEALNGLRQAPSSYPPTWPPGPAHHSCHQSNSNGHRLCPTHVYAAPEGKAPANSPRQGNSKNVLLACPM
421  NMYPHGRSGRTVQEIWEDFSMSFTPAVREVVEFAKHIPGFRDLSQHDQVTLLKAGTFEVLMVRFASLFNV
491  KDQTVMFLSRTTYSLQELGAMGMGDLLSAMFDFSEKLNSLALTEEELGLFTAVVLVSADRSGMENSASVE
561  QLQETLLRALRALVLKNRPLETSRFTKLLLKLPDLRTLNNMHSEKLLSFRVDAQ 614
```

Amino acid sequence for human Rev-erbβ (NR1D2)

Accession number : L31785

```
   1  EVNAGGVIAYISSSSSASSPASCHSEGSENSFQSSSSSVPSSPNSSNSDTNGNPKNGDLANIEGILKNDR
  71  IDCSMKTSKSSAPGMTKSHSGVTKFSGMVLLCKVCGDVASGFHYGVHACEGCKGFFRRSIQQNIQYKKCL
 141  KNENCSIMRMNRNRCQQCRFKKCLSVGMSRDAVRFGRIPKREKQRMLIEMQSAMKTMMNSQFSGHLQNDT
 211  LVEHHEQTALPAQEQLRPKPQLEQENIKSSSPPSSDFAKEEVIGMVTRAHKDTFMYNQEQQENSAESMQP
 281  QRGERIPKNMEQYNLNHDHCGNGLSSHFPCSESQQHLNGQFKGRNIMHYPNGHAICIANGHCMNFSNAYT
 351  QRVCDRVPIDGFSQNENKNSYLCNTGGRMHLVCPMSKSPYVDPHKSGHEIWEEFSMSFTPAVKEVVEFAK
 421  RIPGFRDLSQHDQVNLLKAGTFEVLMVRFASLFDAKERTVTFLSGKKYSVDDLHSMGAGDLLNSMFEFSE
 491  KLNALQLSDEEMSLFTAVVLVSADRSGNRKNVNSVEALQETLIRALRTLIMKNHPNEASIFTKLLLKLPD
 561  LRSLNNMHSEELLAFKVHP 579
```

Amino acid sequence for *Drosophila* E75 (NR1D3)

Accession number : X51548

```
    1  MLMSADSSDSAKTSVICSTVSASMLAPPAPEQPSTTAPPILGVTGRSHLENALKLPPNTSVSAYYQHNSK
   71  LGMGQNYNPEFRSLVAPVTDLDTVPPTGVTMASSSNSPNSSVKLPHSGVIFVSKSSAVSTTDGPTAVLQQ
  141  QQPQQQMPQHFESLPHHHPQQEHQPQQQQQQHHLQHHPHPHVMYPHGYQQANLHHSGGIAVVPADSRPQT
  211  PEYIKSYPVMDTTVASSVKGEPELNIEFDGTTVLCRVCGDKASGFHYGVHSCEGCKGFFRRSIQQKIQYR
  281  PCTKNQQCSILRINRNRCQYCRLKKCIAVGMSRDAVRFGRVPKREKARILAAMQQSTQNRGQQRALATEL
  351  DDQPRLLAAVLRAHLETCEFTKEKVSAMRQRARDCPSYSMPTLLACPLNPAPELQSEQEFSQRFAHVIRG
  421  VIDFAGMIPGFQLLTQDDKFTLLKAGLFDALFVRLICMFDSSINSIICLNGQVMRRDAIQNGANARFLVD
  491  STFNFAERMNSMNLTDAEIGLFCAIVLITPDRPGLRNLELIEKMYSRLKGCLQYIVAQNRPDQPEFLAKL
  561  LETMPDLRTLSTLHTEKLVVFRTEHKELLRQQMWSMEDGNNSDGQQNKSPSGSWADAMDVEAAKSPLGSV
  631  SSTESADLDYGSPSSSQPQGVSLPSPPQQQPSALASSAPLLAATLSGGCPLRNRANSGSSGDSGAAEMDI
  701  VGSHAHLTQNGLTITPIVRHQQQQQQQQQIGILNNAHSRNLNGGHAMCQQQQQPHPQLHHHLTAGAARYRK
  771  LDSPTDSGIESGNEKNECKAVSSGGSSSCSSPRSSVDDALDCSDAAANHNQVVQHPQLSVVSVSPVRSPQ
  841  PSTSSHLKRQIVEDMPVLKRVLQAPPLYDTNSLMDEAYKPHKKFRALRHREFETAEADASSSTSGSNSLS
  911  AGSPRQSPVPNSVATPPPSAASAAAGNPAQSQLHMHLTRSSPKASMASSHSVLAKSLMAEPRMTPEQMKR
  981  SDIIQNYLKRENSTAASSTTNGVGNRSPSSSSTPPPSAVQNQQRWGSSSVITTTCQQRQQSVSPHSNGSS
 1051  SSSSSSSSSSSSSSSTSSNCSSSSASSCQYFQSPHSTSNGTSAPASSSSGSNSATPLLELQVDIADSAQP
 1121  LNLSKKSPTPPPSKLHALVAAANAVQRYPTLSADVTVTASNGGSSVGGGESGRQQQSAGECGLPQSGPER
 1191  RRAQGNAGGVRAGGGRWFYAEKWERQRLGVAVQRSRKQDHLERRELN 1237
```

Amino acid sequence for *Drosophila* E78 (NR1E1)

Accession number : U01087

```
   1  MDVYQIELEEQAQIRSKLLVETCVKHSSSEQQQLQVKQEDLIKDFTRDEEEQPSEEEAEEEDNEEDEEEE
  71  GEEEEEDEEEDEDEEALLPVVNFNANSDFNLHFFDTPEDSSTQGAYSEANSLESEQEEEKQTQQHQQQKQ
 141  HHRDLEDCLSAIEADPLQLLHCDDFYRTSALAESVAASLSPQQQQQRQHTHQQQQQQQQQQQHPGQQQHQ
 211  LNCTLSNGGGALYTISSVHQFGPASNHNTSSSSPSSSAAHSSPDSGCSSASSSGSSRSCGSSSASSSSSA
 281  VSSTISSGRSSNNSVVNPAATSSSVAHLNKEQQQQPLPTTQLQQQQQHQQQLQHPQQQQSFGLADSSSSN
 351  GSSNNNNGVSSKSFVPCKVCGDKASGYHYGVTSCEGCKGFFRRSIQKQIEYRCLRDGKCLVIRLNRNRCQ
 421  YCRFKKCLSAGMSRDSVRYGRVPKRSRELNGAAASSAAAGAPASLNVDDSTSSTLHPSHLQQQQQQHLLQ
 491  QQQQQQHQPQLPQHHQLQQQPHVSGVRVKTPSTPQTPQMCSIASSPSELGGCNSANNNNNNNNSSSGNA
 561  SGGSGVTSALLLWADTSNWWEAAWWEWRAWARMPTRSGVSRRLGGNGKRADRLRCHHVRVAGAPLNCSYT
 631  EELTRELMRRPVTVPQNGIASTVAESLEFQKIWLWQQFSARVTPGVQRIVEFAKRVPGFCDFTQDDQLIL
 701  IKLGFFEVWLTHVARLINEATLTLDDGAYLTRQLEILYDSDFVNALLNFANTLNAYGLSDTEIGLFSAMV
 771  LLASDRAGLSEPKVIGRARELVAEALRVQILRSRAGSPQALQLMPALEAKIPELRSLGAKHFSHLDWLRM
 841  NWTKLRLPPLFAEIFDIPKADDEL 864
```

References

1   Miyajima, N. et al. (1988) Nucl. Acids Res. 16, 11057–11074.
2   Miyajima, N. et al. (1989) Cell 57, 31–39.
3   Lazar, M.A. et al. (1989) Mol. Cell. Biol. 9, 1128–1136.
4   Feigl, G. et al. (1989) Nucl. Acids Res. 17, 7167–7178.
5   Segraves, W.A. and Hogness, D.S. (1990) Genes Dev. 4, 204–219.
6   Stone, B.L. and Thummel, C.S. (1993) Cell 75, 307–320.
7   Lazar, M.A. et al. (1990) DNA Cell Biol. 9, 77–83.
8   Delaunay, F. et al. (2000) Science 289, 297–300.
9   Dumas, B. et al. (1994) Mol. Endocrinol. 8, 996–1005.
10 Forman, B.M. et al. (1994) Mol. Endocrinol. 8, 1253–1261.
11 Retnakaran, R. et al. (1994) Mol. Endocrinol. 8, 1234–1244.
12 Enmark, E. et al. (1994) Biochem. Biophys. Res. Commun. 204, 49–56.
13 Pena de Ortiz, S. et al. (1994) Mol. Brain Res. 23, 278–283.
14 Bonnelye, E. et al. (1994) Cell Growth Diff. 5, 1357–1365.
15 Segraves, W.A. and Woldin, C. (1993) Insect Biochem. Mol. Biol. 23, 91–97.
16 Jindra, M. et al. (1994) Eur. J. Biochem. 221, 665–675.
17 Palli, S.R. et al. (1997) Dev. Genet. 20, 36–46.
18 Chan, S.M. (1998) FEBS Lett. 436, 395–400.
19 Kostrouch, Z. et al. (1995) Proc. Natl Acad. Sci. USA 92, 156–159.
20 Laudet, V. et al. (1992) EMBO J. 11, 1003–1013.
21 Laudet, V. (1997) J. Mol. Endocrinol. 19, 207–226.
22 Nuclear Receptor Nomenclature Committee (1999) Cell 97, 161–163.
22a Carmi, I. et al. (1998) Nature 396, 168–173.
23 Harding, H.P. and Lazar, M.A. (1993) Mol. Cell. Biol. 13, 3113–3121.
24 Harding, H.P. and Lazar, M.A. (1995) Mol. Cell. Biol. 15, 4791–4802.
25 Adelmant, G. et al. (1996) Proc. Natl Acad. Sci. USA 93, 3553–3558.
26 Terenzi, H. et al. (1998) Biochemistry 37, 11488–11495.
27 Spanjaard, R.A. et al. (1994) Mol. Endocrinol. 8, 286–295.
28 Hsu, M.H. et al. (1998) J. Biol. Chem. 273, 27988–27997.
29 Zhao, Q. et al. (1998) Mol. Cell 1, 849–861.
30 White, K.P. et al. (1997) Science 276, 114–117.
31 Zamir, I. et al. (1996) Mol. Cell. Biol. 16, 5458–5465.
32 Renaud, J.P. et al. (2000) Mol. Endocrinol. 14, 700–717.
33 Burke, L. et al. (1996) Nucl. Acids Res. 24, 3481–3489.
34 Downes, M. et al. (1996) Nucl. Acids Res. 24, 3490–3498.
35 Burke, L.J. et al. (1998) Mol. Endocrinol. 12, 248–262.
36 Downes, M. et al. (1996) Nucl. Acids Res. 24, 4379–4387.
37 Zamir, I. et al. (1997) Genes Dev. 11, 835–846.
38 Hastings, M.L. et al. (1997) Nucl. Acids Res. 25, 4296–4300.
39 Chomez, P. et al. (2000) Development 127, 1489–1498.
40 Janini, E.A. et al. (1992) Biochem. Biophys. Res. Commun. 184, 739–745.
41 Kainu, T. et al. (1996) Brain Res. 743, 315–319.
42 Chawla, A. and Lazar, M.A. (1993) J. Biol. Chem. 268, 16265–16269.
43 Downes, M. et al. (1995) Mol. Endocrinol. 9, 1666–1678.
44 Vu-Dac, N. et al. (1998) J. Biol. Chem. 273, 25713–25720.
45 Gervois, P. et al. (1999) Mol. Endocrinol. 13, 400–409.
46 Kassam, A. et al. (1999) J. Biol. Chem. 274, 22895–22900.

[47] Balsalobre, A. et al. (1998) Cell 93, 929–937.
[48] Huet, F. et al. (1993) Development 118, 613–627.
[49] Zhou, B. et al. (1998) Dev. Biol. 193, 127–138.
[50] Dussault, I and Giguère, V. (1997) Mol. Cell. Biol. 17, 1860–1867.
[51] Winrow, C.J. et al. (1998) J. Biol. Chem. 273, 31442–31448.
[52] Russell, S.R. et al. (1996) Genetics 144, 159–170.
[53] Bradley, D.J. (1989) Proc. Natl Acad. Sci. USA 86, 7250–7254.
[54] Lazar, M.A. et al. (1990) J. Biol. Chem. 265, 12859–12863.
[55] Munroe, S.H. et al. (1991) J. Biol. Chem. 266, 22083–22086.
[56] Hastings, M.L. et al. (2000) J. Biol. Chem. 275, 11507–11513.
[57] Laudet, V. et al. (1991) Nucl. Acids Res. 19, 1105–1112.
[58] Koh, Y.S. et al. (1999) Genomics 57, 289–292.

Names

The first member of the group to be discovered was in *Drosophila* and was called DHR3, for *'Drosophila* hormone receptor 3'. Following this nomenclature, the genes found in *Manduca* and in the nematode were called MHR3 and CNR3 (*Caenorhabditis* nuclear receptor 3), respectively, and, more recently, CHR3. In vertebrates, two members were found in the same period and were called RORα (for RAR-related orphan receptor) and RZRβ (for a new family 'Z' of potential retinoic acid binding receptors). Since RZRβ was wrongly described as a receptor for melatonin, its name tends to be avoided and replaced by RORβ.

Species	Other names	Accession number	References
Official name: NR1F1			
Human	RORα, RZRα, RORA	U04897	5
Mouse	RORα, RZRα	Z82994	53, 54
Mus caroli	RORα, RZRα	D45910	52
Mus cookii	RORα, RZRα	U22437	8
Official name: NR1F2			
Human	RZRβ, RORβ, RORB	Y08639	Unpub.
Rat	RZRβ, RORβ	L14610	3
Chicken	RZRβ, RORβ	Y08638	Unpub.
Official name: NR1F3			
Human	RORγ, TOR, RORC	U16997	6
Mouse	RORγ, TOR	U39071	7
Official name: NR1F4			
Drosophila	DHR3, DmHR3	M90806	1
Aedes	AHR3, AaHR3	U87543	Unpub.
Manduca	MHR3, MsHR3	X74566	2
Galleria	GHR3, GmHR3	U02621	56
Choristoneura	CHR3, CfHR3	U37528	55
Caenorhabditis	CNR3, CHR3	U13075	51

Isolation

The first members of this group to be discovered were the *Drosophila* and *Manduca* homologues, called DHR3 and MHR3, respectively[1,2]. Both were cloned by low-stringency screening with a E75 or a human RARα probe. Because of their expression pattern, it was immediately realized that both genes played an important role in the gene activation cascade that controls insect metamorphosis. Furthermore, in *Manduca* it was shown that the expression of MHR3 is regulated by ecdysone[2]. The *Manduca* and *Drosophila* sequences are strongly conserved (97% and 68% amino-acid identity in the DNA- and ligand-binding domains, respectively).

The first members of this group in vertebrates were identified by Michael Becker-André's and Vincent Giguere's laboratories[3–5]. Using highly degenerated PCR primers designed on the DNA-binding domain of known nuclear receptors, a signature for a new orphan, RZRα, was found by RT-PCR on human endothelial cells RNA[3]. The complete cDNA was then isolated using 5' and 3' RACE approaches. Independently, the same gene, called RORα was found from human retina and testis cDNA libraries[5]. Interestingly, several isoforms of this new member of the superfamily were described. RORα was shown to be strongly related to DHR3 (76% and 32% amino-acid identity in the DNA- and ligand-binding domains, respectively). In addition, these genes have in common a very long D domain harboring 31% amino-acid identity[3].

The RZRβ gene was discovered with the same RT-PCR strategy that was used for isolating RZRα. The gene was shown to be highly related to RORα (92% and 62% amino-acid identity in the C and E domains, respectively).

Finally, RORγ was also found by an RT-PCR strategy by Anton Jetten's laboratory and was later found independently by low-stringency screening of a murine T-cell cDNA library (hence the label TOR, for T-cell orphan receptor) with a fragment of RXRα cDNA[6,7]. RORγ exhibits 90% and 53% amino-acid identity with RORα, and 89% and 51% with RORβ in the DNA- and ligand-binding domains, respectively.

All these receptors shared several distinctive structural features: a very long D domain and a strongly conserved C-terminal extension (CTE) to the C domain, which is implicated in monomeric DNA binding. The RORs are related to the Rev-erbs (NR1D) and exhibit overall 58% and 37% amino-acid identities with the DNA- and ligand-binding domains of these orphan receptors.

DNA binding

RORs have unique DNA-binding properties. They bind to DNA as monomers on half-site elements with a 5' A/T-rich extension, which is identical to the response elements of the Rev-erb orphan receptors (T/A A/T T/A C A/T A/GGGTCA)[4,5]. The DNA-binding domain of ROR is in fact composed of three integrated parts: (1) the two classic zinc fingers that recognize the AGGTCA motif in the major groove of the DNA double helix; (2) the CTE that interacts with the adjacent minor groove and makes specific contacts with the 5' A/T-rich sequence in a manner similar to Rev-erb; and (3) the region N-terminal to the zinc finger that, in RORα, is distinct in the four different isoforms known. It has been shown that the four isoforms exhibit slightly different specificities for the precise DNA sequences that are recognized. Extensive mutational analysis of the CTE and N-terminal region by Vincent Giguere's laboratory has demonstrated that the N-terminal region makes specific contacts to the CTE, causing slight structural changes that regulate the interaction between the CTE and the 5' A/T-rich region[15–17]. In addition, the N-terminal region modifies the 130° angle that RORα induces by binding to its response element[17].

Although it has been reported that RORα may bind as a homodimer to Rev-DR2 sites (on which the Rev-erb orphan receptors also form efficient homodimers), it is now clear that these dimers are not energetically favored and there is no evidence that homodimer binding is favored over monomer binding in vitro[19,21]. The reason why RORα does not form cooperative homodimers in contrast to Rev-erb has been explored in detail[21]. It has been shown by mutational analysis that it is possible to transform a wild-type monomer binding RORα to a molecule able to homodimerize as efficiently as Rev-erb by the mutation of only four amino acids: three in the first zinc finger, and one in the second zinc finger. One of these positions contains an hydrophobic amino acid, whereas all the other homodimer-forming receptors of subfamily I harbor an amino acid with an aromatic side-chain. This allows a prediction of the monomer or a homodimer binding state of a new receptor to be made. This study showed that the transition between a monomer binding to an homodimer binding receptor may be extremely easy, providing a simple mechanism for receptor evolution[21].

Interestingly, it has been shown that mouse RORγ does not recognize the classical ROREs but binds with high affinity as a homodimer to DR4 and DR5 elements[11]. The situation is different for RORβ, which binds as a monomer to ROREs but activates transcription only when two properly spaced ROREs are present[4,15]. No clear homodimer formation was visualized for RORβ. A strict comparison of the DNA-binding specificities of the three ROR genes will be needed to achieve a better understanding of the significance of these differences.

In *Drosophila*, DHR3 was shown to bind to typical RORE[33]. It was later shown that it binds to three sites that harbor the structure of typical monomeric RORE upstream of the FTZ-F1 (NR5A3) gene[31]. In contrast, the *Caenorhabditis* CHR3 protein binds to DR5 elements, with much weaker binding being observed with half-site motifs or other direct repeats[32].

Partners

Apparently RORs are unable to interact with other members of the nuclear receptor superfamily, including Rev-erbs. Nevertheless, it has been shown that E75B a non-DNA-binding isoform generated from E75 gene, the *Drosophila* homologue of Rev-erb, is able to bind to the βFTZ-F1 promoter only if DHR3 is present, suggesting that a complex between E75B and DHR3 may exist[47].

It has been shown, in accordance with the ability of RORα to activate transcription, that these receptors interact with known coactivators such as TIF2 and the PPAR-binding protein (PBP), also known as TRAP220 or DRIP205, which is able to bind to PPAR, TR or VDR when their respective ligands are present[20]. Interestingly, in the case of RORα, this interaction is strong even in the absence of exogenous ligand, in contrast with the other receptors. It has been shown that TIF2 effectively enhances the transactivating properties of RORα. This interaction requires a strong activation domain that was mapped in the ROR ligand-binding domain[19]. Mutations in helix 3 and helix 12, as predicted from sequence analysis, impair both the interaction with TIF2 and the increase of the transactivation potential of RORα mediated by this coactivator. In contrast, a mutant in helix 12 retains the ability to interact with TRAP220, consistent with observations with other nuclear receptors, such as VDR.

Other authors report an interaction between RORα1 and the coactivator p300 as well as with the muscle-specific transcription factor MyoD1[22]. Using a dominant negative mutant, it has been shown that RORα1 is important for muscle differentiation. The interaction between RORα1 and

MyoD1 require the DNA-binding domain of RORα1, whereas the interaction with p300 is mediated by the E domain[22]. Interestingly MyoD1 was able to bind simultaneously to p300 and RORα1.

It has been shown, using GAL4 fusion proteins with various domains of RORα, that this receptor contains a repressive domain in the ligand-binding domain[19]. Consistent with this notion, an interaction was observed in GST pull-down experiments between RORα and the known corepressors NCoR and SMRT.

Recently, it has been shown that RORβ interacts with NIX1 (neural interacting factor X 1), a nuclear receptor cofactor, in an AF-2-dependent manner[63].

Ligands

The pineal hormone melatonin was claimed to be a ligand for RORα and β, and synthetic ligands were described such as the antiarthritic CGP52608 or the thiazolidinedione BRL49653[9-12,18] (see reference 23 for a review of melatonin receptors). Nevertheless, these provocative data, which came from only one research group, have not been confirmed by other studies[13-15]. Even if the expression pattern and knockout of RORβ may suggest a function in circadian rhythm consistent with a role as a melatonin receptor, the current view is that RORs are still orphan receptors in search of their ligand.

Three-dimensional structure

No 3D structure data are available.

Expression

RORα was found to be expressed as a series of mRNA transcripts. The predominant species is a c. 15 kb mRNA present in many organs (lung, muscle, brain, heart, peripheral blood leukocytes, spleen, liver, ovary, etc.) but smaller bands are also visible at 2.3 kb with a narrower distribution (leukocytes, testis, lung and liver)[3,29]. Other bands at 7.5, 5.5 and 2.0 kb are also visible, but with a weaker intensity. These may correspond to the four different isoforms described[5]. This gene is constitutively expressed during muscle differentiation. Interestingly, a dominant negative version of RORα represses myogenic differentiation[22]. RORα was found to be expressed in murine lens[14]. RORα is also expressed in rat intestinal epithelium and Caco-2 cells[25]. By detection of the β-gal activity in a heterozygous mice strain containing one allele in which part of the RORα gene was replaced by the β-gal enzyme, expression in various areas of the brain, including retinal ganglion cells, cerebellum, some nuclei in the thalamus and the suprachiasmatic nucleus, testis and skin was detected[26]. Strikingly, no expression was detected in other organs, such as liver, heart, spleen, lung or leukocytes, which have been found to contain RORα mRNAs[3,26,27]. RORα is up-regulated during the differentiation of 3T3-L1 cells into adipocytes as well as during osteogenic differentiation[28,57]. Consistent with a role of

RORα in cerebellum development, demonstrated by the analysis of the *staggerer* mouse mutant, which corresponds to a mutation in RORα (see below), an expression of RORα in Purkinje cells was demonstrated[29,30]. Recently, it was shown that RORα is expressed in cartilage[48].

In contrast, RORβ exhibits a more specific expression pattern, since it is uniquely expressed in brain. As RORα, most of the transcripts are found as a large 10 kb mRNA species and a minor fraction is detected at 2.4 kb. In spleen, a RORβ transcript of 1 kb is detected. Of note, the size of this transcript is too small to encode the peptide deduced from the brain-derived cDNA. The nature of this transcript is not known. The distribution of RORβ transcript was studied in rat both by competitive PCR and by *in situ* hybridization, and confirmed the specific expression in brain[9,49]. Strong expression was found in the pineal gland, hypothalamus and thalamus. A detailed study of the expression pattern in rat brain suggests a role in the processing of sensory information and circadian rhythm[49]. Other regions of RORβ expression were the eye, spinal cord or pituitary. In the eye, RORβ was shown to be expressed in retinal progenitor cells in the embryonic rat retina and to be important in regulating retinal progenitor proliferation, possibly via the homeobox gene, Chx10[24]. Interestingly, it has been shown that RORβ mRNA levels oscillate along a circadian rhythm in retina and in pineal gland in mouse, consistent with a role in the regulation of circadian rhythm[35].

Human RORγ was shown to be highly expressed in muscle as well as in thymus, testis, pancreas, prostate, heart and liver as a 3.2 kb mRNA[6]. In thymus, two other transcripts at 7.2 and 5.2 kb were also detected. In mouse, other authors found a predominant expression in thymus as in other tissues such as liver, heart, tongue, muscle, and diaphragm[7]. Notably, no expression was found in spleen or bone marrow. The major transcript is found at 2.5 kb but a 3.5 kb form is detected in several organs such as thymus. The same study reported the expression in a number of mammalian cell lines. A thymus-specific isoform of RORγ, called RORγt, is implicated in protection against apoptosis[59]. Interestingly, RORα and RORγ were shown to be up-regulated during differentiation of the 3T3-L1 cells in adipocytes just after PPARγ and C/EBPα[28]. Since Rev-erbα is also up-regulated during adipocyte differentiation, it is suspected that genes regulated through RORE are important in this differentiation process.

DHR3 is expressed as a series of at least three transcripts of 9, 7 and 5.5 kb, the 9 kb species being the only one detected during pupal development[1]. Peaks of DHR3 expression were detected at midembryogenesis, during each of the first two larval instar, at the end of the third larval instar extending into pre-pupal development and, finally, during pupal development[1,43]. The correspondence between these peaks and the six commonly accepted peaks in the ecdysone titer suggests that DHR3 transcription may be ecdysone induced. A wide expression during mid-embryogenesis and early metamorphosis was demonstrated by *in situ* hybridization[31]. An ecdysone regulation was also shown for the MHR3 gene in *Manduca sexta*[2]. It was demonstrated that MHR3 induction by ecdysone was largely independent of protein synthesis, suggesting a direct regulation by EcR. The direct regulation was proven by the identification of four EcREs present in the 5'-flanking region of the MHR3 gene[40]. These sites were regulated by the

EcRB1-USP1 heterodimer but not by the EcRB1-USP2, suggesting a specific role of each of the two isoforms of USP in *Manduca*.

In *Caenorhabditis*, the CHR3 protein is present in all blastomeres of the embryo from the two-cell stage up to approximately the 200-cell stage. This early presence suggests a maternal contribution that was demonstrated by *in situ* hybridization. Later, the expression became prominent in epidermal cells[32].

Activity and target genes

RORα and β were demonstrated to be constitutive transcriptional activators acting on ROREs[4,5]. Nevertheless, it has been suggested that the activity of RORβ is neuron specific and that this receptor is inactive in other cell types[15]. Nevertheless, an isoform RORβ2 differing in the A/B region is active in non-neuronal cells, suggesting that the AF-1 function may play a role in this differential ability[58]. Since the ROREs are identical to the RevREs, which are the response elements for the transcriptional repressors, Rev-erbs, it has been proposed that a cross-talk between RORs and Rev-erbs exists. This view suggests that these two groups of related receptors regulate with opposing activities an overlapping regulatory network[27,36]. Interestingly, RORγ, which apparently does not bind to the same element, is apparently unable to activate transcription, but may repress TR and RAR activities on DR4 and DR5 elements, respectively[11].

Recently, an elegant study has shown that the activity of RORα may be potentiated up to 20–30-fold by the Ca^{2+}/calmodulin-dependent protein kinase IV (CaMKIV), which plays a prominent role in the mediation of the Ca^{2+} response[41]. This striking effect is not due to a direct phosphorylation of RORα by CaMKIV, but rather involved an unknown coactivator that could be the direct target of the kinase.

The model of opposite action of RORs and Rev-erbs action on specific gene transcription has been verified in the case of N-*myc*[37]. This gene is up-regulated by RORα and down-regulated by Rev-erbβ through the same RORE element, located in the first intron of N-*myc*. Interestingly, mutations in the RORE increased the transforming potential of N-*myc* in the rat embryonic fibroblast transformation assay, suggesting that this element is effectively important for the regulation of N-*myc*, at least in cell cultures. This idea suggests that deregulation of RORα and/or Rev-erbβ could contribute to cell transformation.

The fact that RORs are transcriptional regulators has been confirmed by the identification of numerous target genes such as γF-crystallin on which a negative functional interference with the RXR–RAR heterodimer was demonstrated[24]. Other known target genes include laminin B1[38], apolipoprotein A1[25], mouse prosaposin[39], mouse oxytocin[44] and peroxisomal hydratase-dehydrogenase[13].

A repression domain was mapped in RORα and it was shown that this repression domain is more active in some cell types. In line with this observation, an interaction with NCoR and SMRT was demonstrated[19]. Nevertheless there are no known target genes clearly repressed by RORs. The only case, the mouse prosaposin, is relatively complex and the possible repressive effect of ROR may be due to a functional interference existing with Spl-type factors[39].

Contradictory results have been found concerning the functional relationship existing between RORs and liganded nuclear receptors, such as RAR and TR. Whereas several authors demonstrate a negative interference between these factors, possibly due to competition between identical binding sites[4,7,24,38], other authors have found opposite results. These authors described that RORα1 enhanced the transactivation by liganded TR on several response elements[42]. In addition the same authors have shown that RORα1 expression is stimulated by thyroid hormones, suggesting that positive feedback mechanisms exist between these two genes[30]. It has also been shown that RORα stimulates ligand-mediated transactivation by the PPAR–RXR heterodimer of the peroxisomal hydratase-dehydrogenase gene[13]. All these results emphasized the poorly understood and complex relationships between RORs and other members of the superfamily.

In *Drosophila*, it has been demonstrated that DHR3 is bound to numerous ecdysone-induced puffs in the polytene chromosome. One of these loci is the FTZ-F1 gene, which was demonstrated to be a target gene for DHR3[31,47]. Three DHR3 binding sites, with the typical structure of ROREs, were identified downstream from the start sites of an isoform of FTZ-F1, βFTZ-F1, the expression of which is activated by DHR3. Interestingly, the non-DNA-binding isoform E75B of the E75 (NR1D3) orphan receptor, which is a homologue of Rev-erbs, is able to inhibit the induction of βFTZ-F1 by forming a complex with DHR3 on the βFTZ-F1 promoter[47]. In contrast, it has been shown that DHR3 is sufficient to repress BR-C, E74A, E75A and E78B (NR1F1) expression, suggesting that it behaves as a transcriptional repressor for these genes. Unfortunately, the binding sites of DHR3 in the regulatory regions of these genes is not known. These data suggest that DHR3 acts as a switch that defines larval to pupal transition by arresting the early regulatory response to ecdysone (via down-regulation of BR-C, E74A, E75A and E78B) and inducing βFTZ-F1 in mid-prepupae[31,47].

Knockout

It has been shown that the *staggerer* mouse mutant, which has been known for a long time, corresponds to a deletion within the RORα gene that prevents translation of the LBD and possibly gives a dominant negative version[29]. The *staggerer* mutation abolishes the development of cerebellar Purkinje cells in a cell autonomous fashion, resulting in an immature morphology of these cells and a severe ataxia in these mice. These data suggest an essential role of RORα in the terminal differentiation of the Purkinje cells[29,53]. This has been confirmed by the experimental knockout of the RORα gene performed independently by two research teams[26,45]. Interestingly, although the absence of RORα causes the *staggerer* phenotype in the cerebellum, no apparent morphological effects on thalamus, hypothalamus and retina, regions in which RORα is expressed, were detected[26]. Similarly, testis and skin of RORα knockout mice display a normal phenotype. However the fur of these mice is significantly less dense than normal and grows poorly after being shaved[26]. It is anticipated that the careful study of these mice will reveal other abnormalities, for example, at the level of cholesterol metabolism[25].

Interestingly, the RORβ knockout mice also correspond to a known mouse mutant, now lost, called *vacillans*[35]. RORβ knockout animals display a typical duck-like gait, a transient inability for males to reproduce and severe disorganization of the retina that results in complete blindness. Furthermore, these mice exhibit a slightly abnormal circadian rhythm: although their rhythmicity is still entrained by the light–dark cycle, under constant conditions, they display an extended period of free-running rhythmicity, suggesting that this gene influences circadian rhythm[35].

The knockout of the RORγ gene has recently been published[50]. These mice lose the thymic expression of the antiapoptotic factor BcL-xL and thus the knockout animals exhibited a reduction in thymocyte survival. In addition, the gene was shown to be important for the development of lymphoid organs such as lymph nodes and Peyer's patches.

In *Drosophila*, the analysis of DHR3 mutants, as well as rescue experiments at different stages, reveals that this gene is required for embryogenesis, for the transition between pre-pupal and pupal stages and to the differentiation of adult structures during metamorphosis[34,46]. This phenotype is consistent with its role as a molecular switch, i.e. in ensuring that responses to the pre-pupal ecdysone pulse will be distinct from responses to the late larval pulse and thus that the animal progresses in an appropriate manner through the early stages of metamorphosis[34].

Associated disease

To date there is no disease associated with RORs in human. The phenotypes of the knockout mice may suggest that several nervous system abnormalities of humans may be linked to ROR deficiencies[29,35]. In addition, the fact that RORα activates the transcription of the N-*myc* proto-oncogene may suggest a link with cancer[37].

Gene structure, promoter and isoforms

The structure of the mouse RORγ gene has been determined[62]. The gene has a complex structure consisting of 11 exons spanning more than 21 kb. Each zinc finger is encoded by a different exon (3 and 4), and the position of the splice site between these two exons is identical to that of Rev-erb, RAR or TR[62]. The two mRNAs, of 2.3 and 3.0 kb, are derived from alternative polyadenylation. The structure of the RORα and RORβ genes are known but have not been described extensively[29,35,53].

Four isoforms are expressed from the RORα gene, RORα1–RORα4. These isoforms display different N-terminal domains, causing DNA-binding site preferences[5,16]. RORα1 binds to a large subset of ROREs, whereas the RORα2 isoform recognizes ROREs with a strict specificity and displays weaker transcriptional activity than RORα1. Of note, an exon encoding a functionally important region of the RORα2 isoform resides on the opposite strand of a cytochrome c-processed pseudogene[5]. Interestingly, these isoforms display different expression profiles: the thalamus has only RORα1 mRNA, RORα4 transcripts are predominant in leukocytes and skin; and RORα2 and RORα3 transcripts are exclusively detected in the testes; and in the remaining tissues, including the cerebellum, there is a mixture of RORα1 and RORα4[26].

A rat isoform of RORβ specifically expressed in the pineal gland and retina has been identified[58]. This isoform exhibits a different A/B region and binds to DNA in a more restricted set of ROREs than the regular isoform. In contrast to RORβ1, this RORβ2 isoform is able to activate transcription in non-neuronal cells, suggesting that this is caused by a different AF-1 function.

An N-terminal isoform of RORγ, RORγt, has been shown to be specifically expressed in the thymus. Ectopic expression of this isoform protects T-cell hybridomas from activation-induced cell death by inhibiting the up-regulation of the Fas ligand as well as IL2 production[59,61]. This isoform is up-regulated by signaling through the T-cell receptor[60].

Chromosomal location

RORα maps to a conserved region between human and mouse[8]. In human, it maps close to the PML gene in chromosome 15q21–q22, whereas in mouse it maps to chromosome 9[29]. RORβ is located on human chromosome 9 at position q22, a region synthenic to mouse chromosome 4, a location consistent with the location of the *vacillans* phenotype (see above)[35]. RORγ was mapped on human chromosome 1q21 and to mouse chromosome 3F2.1–2.2, a position synthenic to the human location[6,62]. DHR3 was mapped on the right arm of *Drosophila* chromosome 2R at position 46F[1]. CHR3 is located on the central part of chromosome 1 on cosmid C01H6 between *unc*-14 and *myo*-1[32].

Amino acid sequence for human RORα (NR1F1)

Accession number: U04897

```
  1  MESAPAAPDPAASEPGSSGADAAAGSRETPLNQESARKSEPPAPVRRQSYSSTSRGISVTKKTHTSQIEI
 71  IPCKICGDKSSGIHYGVITCEGCKGFFRRSQQSNATYSCPRQKNCLIDRTSRNRCQHCRLQKCLAVGMSR
141  DAVKFGRMSKKQRDSLYAEVQKHRMQQQQRDHQQQPGEAEPLTPTYNISANGLTELHDDLSNYIDGHTPE
211  GSKADSAVSSFYLDIQPSPDQSGLDINGIKPEPICDYTPASGFFPYCSFTNGETSPTVSMAELEHLAQNI
281  SKSHLETCQYLREELQQITWQTFLQEEIENYQNKQREVMWQLCAIKITEAIQYVVEFAKRIDGFMELCQN
351  DQIVLLKAGSLEVVFIRMCRAFDSQNNTVYFDGKYASPDVFKSLGCEDFISFVFEFGKSLCSMHLTEDEI
421  ALFSAFVLMSADRSWLQEKVKIEKLQQKIQLALQHVLQKNHREDGILTKLICKVSTLRALCGRHTEKLMA
491  FKAIYPDIVRLHFPPLYKELFTSEFEPAMQIDG   523
```

Amino acid sequence for human RORβ (NR1F2)

Accession number: Y08639

```
  1  MRAQIEVIPCKICGDKSSGIHYGVITCEGCKGFFRRSQQNNASYSCPRQRNCLIDRTNRNRCQHCRLQKC
 71  LALGMSRDAVKFGRMSKKQRDSLYAEVQKHQQRLQEQRQEQSGEAAERLARVYSSSISNGLSNLNNETSGT
141  YANGSVIDLPKSEGYYNVVSGQPSPDQSGLDMTGIKQIKQEPIYDLTSVPNLFTYSSFNNGQLAPGITMT
211  EIDRIAQNIIKSHLETCQYTMEELHQLAWQTHTYEEIKAYQSKSREALWQQCAIQITHAIQYVVEFAKRI
281  TGFMELCQNDQILLLLKSGCLEVVLVRMCRAFNPLNNTVLFEGKYGGMQMFKALGSDDLVNEAFDFAKNLC
351  SLQLTEEEIALFSSAVLISPDRAWLIEPRKVQKLQEKIYFALQHVIQKNHLDDETLAKLIAKIPTITAVC
421  NLHGEKLQVFKQSHPEIVNTLFPPLYKELFNPDCATACK   459
```

Amino acid sequence for human RORγ (NR1F3)

Accession number: U16997

```
  1 MDRAPQRQHRASRELLAAKKTHTSQIEVIPCKICGDKSSGIHYGVITCEGCKGFFRRSQRCNAAYSCTRQ
 71 QNCPIDRTSRNRCQHCRLQKCLALGMSRDAVKFGRMSKKQRDSLHAEVQKQLQQRQQQQEPVVKTPPAG
141 AQGADTLTYTLGLPDGQLPLGSSPDLPEASACPPGLLKASGSGPSYSNNLAKAGLNGASCHLEYSPERGK
211 AEGRESFYSTGSQLTPDRCGLRFEEHRHPGLGELGQGPDSYGSPSFRSTPEAPYASLTEIEHLVQSVCKS
281 YRETCQLRLEDLLRQRSNIFSREEVTGYQRKSMWEMWERCAHHLTEAIQYVVEFAKRLSGFMELCQNDQI
351 VLLKAGAMEVVLVRMCRAYNADNRTVFFEGKYGGMELFRALGCSELISSIFDFSHSLSALHFSEDEIALY
421 TALVLINAHRPGLQEKRKVEQLQYNLELAFHHHLCKTHRQSILAKLPPKGKLRSLCSQHVERLQIFQHLH
491 PIVVQAAFPPLYKELFSTETESPVGCPSDLEEGLLASPYGLLATSLDPVPPSPFSFPMNPGGWSPPALWK
560
```

Amino acid sequence for *Drosophila* DHR3 (NR1F4)

Accession number: M90806

```
  1 MYTQRMFDMWSSVTSKLEAHANNLGQSNVQSPAGQNNSSGSIKAQIEIIPCKVCGDKSSGVHYGVITCEG
 71 CKGFFRRSQSSVVNYQCPRNKQCVVDRVNRNRCQYCRLQKCLKLGMSRDAVKFGRMSKKQREKVEDEVRF
141 HRAQMRAQSDAAPDSSVYDTQTPSSSDQLHHNNYNSYSGGYSNNEVGYGSPYGYSASVTPQQTMQYDISA
211 DYVDSTTYEPRSTIIDPEFISHADGDINDVLIKTLAEAHANTNTKLEAVHDMFRKQPDVSRILYYKNLGQ
281 EELWLDCAEKLTQMIQNIIEFAKLIPGFMRLSQDDQILLLKTGSFELAIVRMSRLLDLSQNAVLYGDVML
351 PQEAFYTSDSEEMRLVSRIFQTAKSIAELKLTETELALYQSLVLLWPERNGVRGNTEIQRLFNLSMNAIR
421 QELETNHAPLKGDVTVLDTLLNNIPNFRDISILHMESLSKFKLQHPNVVFPALYKELFSIDSQQDLT
487
```

References

1 Koelle, M.R. et al. (1992) Proc. Natl Acad. Sci. USA 89, 6167–6171.
2 Palli, S.R. et al. (1992) Dev. Biol. 150, 306–316.
3 Becker-André, M. et al. (1993) Biochem. Biophys. Res. Commun. 194, 1371–1379.
4 Carlberg, C. et al. (1994) Mol. Endocrinol. 8, 757–770.
5 Giguère, V. et al. (1994) Genes Dev. 8, 538–553.
6 Hirose, T. et al. (1994) Biochem. Biophys. Res. Commun. 205, 1976–1983.
7 Ortiz, M.A. et al. (1995) Mol. Endocrinol. 9, 1679–1691.
8 Giguère, V. et al. (1995) Genomics 28, 596–598.
9 Becker-André, M. et al. (1994) J. Biol. Chem. 269, 28531–28534.
10 Wiesenberg, I. et al. (1995) Nucl. Acids Res. 23, 327–333.
11 Missbach, M. et al. (1996) J. Biol. Chem. 271, 13515–13522.
12 Wiesenberg, I. et al. (1998) Mol. Pharmacol. 53, 1131–1138.
13 Winrow, C.J. et al. (1998) J. Biol. Chem., 273, 31442–31448.
14 Tini, M. et al. (1995) J. Biol. Chem. 270, 20156–20161.
15 Greiner, E.F. et al. (1996) Proc. Natl Acad. Sci. USA 93, 10105–10110.
16 McBroom, L.D.B. et al. (1995) Mol. Cell. Biol. 15, 796–808.
17 Giguère, V. et al. (1995) Mol. Cell. Biol. 15, 2517–2526.
18 Schräder, M. et al. (1996) J. Biol. Chem. 271, 19732–19736.
19 Harding, H.P. et al. (1997) Mol. Endocrinol. 11, 1737–1746.
20 Atkins, G.B. et al. (1999) Mol. Endocrinol. 13, 1550–1557.
21 Moraitis, A.N. and Giguère, V. (1999) Mol. Endocrinol. 13, 431–439.

22 Lau, P. et al. (1999) Nucl. Acids Res. 27, 411–420.
23 Reppert, S.M. and Weaver, D.R. (1995) Cell 83, 1059–1062.
24 Chow, L. et al. (1998) Mech. Dev. 77, 149–164.
25 Vu-Dac, N. et al. (1997) J. Biol. Chem. 272, 22401–22404.
26 Steinmayr, M. et al. (1998) Proc. Natl Acad. Sci. USA 95, 3960–3965.
27 Forman, B.M. et al. (1994) Mol. Endocrinol. 8, 1253–1261.
28 Austin, S. et al. (1998) Cell Growth Diff. 9, 267–276.
29 Hamilton, B.A. et al. (1996) Nature 379, 736–739.
30 Koibuchi, N. and Chin, W.W. (1998) Endocrinology 139, 2335–2341.
31 Lam, G.T. et al. (1997) Development 124, 1757–1769.
32 Kostrouchova, M. et al. (1998) Development 125, 1617–1626.
33 Horner, M.A. et al. (1995) Dev. Biol. 168, 490–502.
34 Lam, G.T. et al. (1999) Dev. Biol. 212, 204–216.
35 André, E. et al. (1998) EMBO J. 17, 3867–3877.
36 Retnakaran, R. et al. (1994) Mol. Endocrinol. 8, 1234–1244.
37 Dussault, I and Giguère, V. (1997) Mol. Cell. Biol. 17, 1860–1867.
38 Matsui, T. (1996) Biochem. Biophys. Res. Commun. 220, 405–410.
39 Jin, P. et al. (1998) J. Biol. Chem. 273, 13208–13216.
40 Lan, Q. et al. (1999) Mol. Cell. Biol. 19, 4897–4906.
41 Kane, C.D. and Means, A.R. (2000) EMBO J. 19, 691–701.
42 Koibuchi, N. et al. (1999) Endocrinology 140, 1356–1364.
43 Huet, F. et al. (1995) Development, 121, 1195–1204.
44 Chu, K. et al. (1999) J. Mol. Endocrinol. 23, 337–346.
45 Dussault, I. et al. (1998) Mech. Dev. 70, 147–153.
46 Carney, G.E. et al. (1997) Proc. Natl Acad. Sci. USA 94, 12024–12029.
47 White, K.P. et al. (1997) Science 276, 114–117.
48 Bordji, K. et al. (2000) J. Biol. Chem. 275, 12243–12250.
49 Schaeren-Wiemers, N. et al. (1997) Eur. J. Neurosci. 9, 2687–2701.
50 Sun, Z. et al. (2000) Science 288, 2369–2373.
51 Kostrouch, Z. et al. (1995) Proc. Natl Acad. Sci. 92, 156–159.
52 Matsui, T. et al. (1995) Brain Res. Mol. Brain. Res. 33, 217–226.
53 Matysiak-Scholze, U. and Nehls, M (1997) Genomics 43, 78–84.
54 Medvedev, A. et al. (1996) Gene 181, 199–206.
55 Palli, S.R. et al. (1996) Insect Biochem. Mol. Biol. 26, 485–499.
56 Jindra, M. et al. (1994) Insect Biochem. Mol. Biol. 24, 763–773.
57 Meyer, T. et al. (2000) Proc. Natl Acad. Sci. USA 97, 9197–9202.
58 André, E. et al. (1998) Gene 216, 277–283.
59 He, Y.W. et al. (1998) Immunity 9, 797–806.
60 Villey, I. et al. (1999) Eur. J. Immunol. 29, 4072–4080.
61 He, Y.W. et al. (2000) J. Immunol. 164, 5668–5674.
62 Medvedev, A. et al. (1997) Genomics 46, 93–102.
63 Greiner, E.F. et al. (2000) Proc. Natl Acad. Sci. USA 97, 7160–7165.

Names

There is only one name for the ecdysone receptor, which was cloned in 1991[1]. It is called NR1H1 in the nomenclature for nuclear receptors[2]. It belongs to the same group as the vertebrate receptors LXRα (NR1H3), LXRβ (NR1H2) and FXR (NR1H4), which are also receptors for steroid hormones (oxysterols and bile acids, respectively).

Species	Other names	Accession number	References
Official name: NR1H1			
Drosophila melanogaster	DmEcR	M74078	1
Ceratitis capitata	CcEcR	AJ224341	108
Lucilia cuprina	LcEcR	U75355	12
Aedes aegypti	AaEcR	U02021	5
Aedes albopictus	AaEcR	AF210733	109
Chironomus tentans	CtEcR	S60739	6
Bombyx mori	BmEcR	L35266	3
Manduca sexta	MsEcR	U19812	4
Precis coenia	PcEcR	AJ251809	Unpub.
Bicyclus anynana	BaEcR	AJ251810	Unpub.
Heliothis virescens	HvEcR	Y09009	13
Choristoneura fumiferana	CfEcR	U29531	11
Tenebrio molitor	TmEcR	Y11533	7
Locusta migratoria	LmEcR	AF049136	110
Celuca pugilator	UpEcR	AF034086	8,9
Amblyomma americanum	AamEcR	AF020186	10

Isolation

The *Drosophila* EcR cDNA was identified by David Hogness' laboratory in a screening of the *Drosophila* genome for members of the nuclear receptor superfamily that was carried out with a probe encompassing the DNA-binding domain of the E75 orphan receptor gene[1]. An orthologous EcR gene was identified in numerous other insect species either by PCR with degenerated oligonucleotides or by low stringency screening with a *Drosophila* EcR probe[4-7,11-13] (see also references 16, 17, 20 and 67 for reviews). In addition, EcR was also found in two other types of arthropods, namely a crustacean, the fiddler crab *Celuca pugilator*, and a chelicerate, the Ixodid tick *Amblyomma americana*[8-10] (see references 14 and 15 for reviews).

Overall, the various EcRs exhibit 84–92% sequence identity in their DNA-binding domain and 60–70% in their ligand-binding domain. This level of sequence identity is particularly low when EcRs from Diptera or Lepidoptera are compared with their homologues in other insects and even in arthropods[10,15,18]. This suggests that, for unknown reasons, the selective pressure acting on EcR may have been modified specifically for Dipteras and Lepidopteras. A similar, even more spectacular, effect has been observed for USP (NR2B4), the heterodimeric partner of EcR[18,19].

There is still no published EcR gene outside the arthropods. The *C. elegans* genome is apparently devoid of an EcR homologue, although the effects of ecdysteroids were described in nematodes[38,39].

The *Drosophila* EcR exhibits 63–70% sequence identity in the DBD with LXRs and FXR, the other members of the NR1H group. In the LBD, the amino-acid identity level is 41% for LXRs and 34% for FXR. EcR is also distantly related to PXR, CAR (group NR1I) and DHR96 (NR1J1). In the DBD, it exhibits 44–52% sequence identity with these receptors and 20–32% in the LBD.

DNA binding

Ecdysone response elements (EcREs) were characterized even before the cloning of the *Drosophila* EcR gene. A 23 bp sequence from the promoter of the ecdysone-inducible *hsp27* gene has been shown to function as an EcRE in that it confers 20-fold ecdysone inducibility on an ecdysone non-responsive promoter[21]. This element has a palindromic structure and was shown to contain four direct and inverted repeats based on the consensus sequence AGGTCA. It was also shown that the sequence of the repeats as well as their spacing are important for ecdysone inducibility[22]. Similar elements were found in other ecdysone-responsive genes[23]. These results have thus supported the notion of a common origin of the hormone response elements of vertebrates and arthropods, even before the cloning of EcR[22].

When the EcR gene was cloned it was shown to bind to the EcRE from the *hsp27* gene[1]. Later, it was shown that EcR is unable to bind DNA alone but heterodimerizes with USP (NR2B4), the homologue of the 9-*cis* retinoic acid RXR in vertebrates[24,25].

In general, natural EcREs are imperfect palindromes composed of two half-site sequences with the consensus sequence AGGTCA separated by a unique central base pair[23,27–28]. However, EcREs composed of direct repeats have also been described[29,31,32]. By *in vitro* selection of the DNA-binding site, it was shown that the preferred target sequence of the EcR–USP heterodimer is in fact a perfect palindromic sequence in which the classical core sequence is linked at 5′ to a G to give the consensus GAGGTCA[30]. The two repeats are separated by a unique central base, which is always a pair A/T. Whereas EcR–USP showed a pronounced binding polarity on DR4, with USP in the 5′ position (as RXR in the RXR–RAR heterodimer), no such polarity was found for the core recognition motif[30]. It was also observed that the EcR–USP heterodimer is able to bind to direct repeats with a lower affinity[30,32]. When several elements were tested in competition experiments, the following order of decreasing affinity was observed: PAL1 > DR4 > DR5 > PAL0 > DR2 > DR1 > *hsp27* = DR3 > DR0, where DR represents direct repeats and PAL palindromic elements[30].

Partners

EcR forms an obligatory heterodimer with USP and is unable to bind alone either as a monomer or homodimer to DNA[24,25]. It has been shown that RXR may replace USP as a partner for EcR, suggesting that the heterodimerization properties of USP and RXR are strongly conserved. Nevertheless EcR from different species do not have the same ability to dimerize with RXR[36]. In addition to allowing DNA binding on EcRE, the heterodimerization is critical for efficient hormone binding, an effect that is unique in the nuclear receptor superfamily[25,26].

Studies using the two-hybrid system have demonstrated that the product of the Seven-up gene (SVP; NR2F3, the homologue of COUP-TFs in *Drosophila*) can interact with EcR[40]. SVP can also reduce the ability of EcR and USP to transactivate through the hsp27 EcRE, although it is unable to bind to this sequence. These observations suggest that SVP may form an inactive heterodimer with EcR reducing the level of EcR-USP, the functional ecdysone receptor in the cell (reviewed in reference 41).

It has been shown that EcR actively represses transcription in the absence of ligand and that the pulse of ecdysone triggers a spectacular increase of the transcriptional activity of the receptor[34]. The repressive effect of the EcR–USP heterodimer is at least in part explained by its interaction with SMRTER, which is a corepressor that is structurally divergent but functionally similar to the vertebrate corepressors NCoR and SMRT[35]. SMRTER mediates repression by interacting with Sin3A, a repressor known to form a complex with the histone deacetylase Rpd3/HDAC. Interestingly, a mutant of EcR unable to interact with SMRTER, induces developmental abnormalities and lethality when introduced in *Drosophila*, suggesting that the SMRTER–EcR interactions are critical for EcR function *in vivo*[35].

EcR can also interact with heat-shock proteins, which appear to be required for the DNA-binding activity of the EcR–USP heterodimer[100]. It has been shown that the heterodimer needs a molecular chaperone-containing heterocomplex that contains Hsp83 (the *Drosophila* homologue of vertebrate Hsp90), Hsc70, Hip, Hop, KKBP52 and p23. Interestingly, Hsp90 and Hsc70 are required for EcR activity *in vivo* and a direct physical interaction between these two proteins has been demonstrated *in vitro*. This study clearly shows that the interactions between nuclear receptors and heat-shock proteins are not specific to classic steroid receptors (NR3A group) but can also be of importance for other receptor types[100].

Ligands

The EcR produced in *Drosophila* cells is able to bind directly to ecdysteroids[1]. The first experiments were performed with labeled iodoponasterone A and demonstrate a specific binding that was abrogated by incubation of cell extracts with an anti-EcR specific antibody[1]. Furthermore, these experiments reveal that EcR may confer ecdysone responsiveness to ecdysone-resistant cells[1]. The ability of several compounds to trigger EcR transcriptional activity was tested after transient transfection of an EcR cDNA in mammalian cells. It was shown that neither α-ecdysone, 20-hydroxyecdysone nor polypodine B act as agonists for EcR in this system, whereas muristerone A, and to a lesser extent, ponasterone A were active[37]. The lack of effect of 20-hydroxyecdysone was surprising since it is a natural ligand of EcR. This may be due to the replacement of USP by RXR in the mammalian cell assay. Half-maximal stimulation of reporter activity was observed at 0.5 mM of muristerone A. It was subsequently shown that efficient ecdysone binding required both EcR and USP[26,33] (see also reference 100 for a recent discussion of this aspect).

It has to be emphasized that a variety of ecdysteroids exist *in vivo* and that the most active compounds are not necessarily identical in different species (see Wang[101] for references). Study of the interaction between physiological ecdysteroids and their receptors has been hampered by the relatively low affinity of receptor–ligand interaction. It was found that ponasterone A had K_d values in the nanomolar range (0.8 nM for *Drosophila*, 2.8 nM for *Aedes* and 3 nM for *Choristoneura*; see Dhadialla[42] and Wang[101] for references). The K_d for tebufenozide a synthetic compounds for the same receptors were 336, 28 and 0.5 nM, respectively. These data suggest that the large sequence variability existing inside the EcR LBD may result in substantial differences in the binding of various ligands, either natural or synthetic. The basis of these differences were explored in the case of *Aedes* and *Drosophila* receptors and it was shown that the differential effect of ligands was linked to the EcR part of the heterodimer, and not USP. In the LBD the region encompassing helices 9 and 10 previously implicated in dimerization was shown to be important for this effect. A single amino acid differences in *Aedes* and *Drosophila* (Tyr 611 in *Drosophila*) was shown to be critical for ligand specificity[101].

The EcR is a target for the development of a new generation of insecticides. The idea behind the use of these compounds is to promote inappropriately the metamorphosis by activation of the ecdysone response. Several molecules with ecdysteroid activities such as the non-steroidal agonists bisacylhydrazines are currently under development. Interestingly, some of these compounds, such as tebufenozide and RH-2485, are very selectively toxic to lepidopterans, suggesting that it may be possible to develop pest-specific insecticides[42].

The strong activity of EcR in the presence of its ligand and its repressive action in the apo-form has led several researchers to use it as a conditional expression system in mammalian cells and transgenic mice[99].

Three-dimensional structure

No 3D structure is yet available for either the DBD or LBD of EcR.

Expression

In *Drosophila* the first expression data revealed a predominant 6 kb transcript that was detectable at varying levels during all of development except at 0–3 h, when a faint 3.7 kb band was observed[1]. The most striking feature of this expression pattern is the intense peak of EcR mRNA during the end of the third instar larvae[1]. This expression level is weaker in pre-pupae and declines during pupal life. With specific antibodies it was shown that the EcR protein is widely distributed throughout the embryo. In late third instar and pre-pupal tissues, EcR was detected in imaginal discs, the fat body, tracheae and salivary gland. Staining was also observed in all larval tissues examined: the central nervous system, gut, cells associated with cuticular structure, etc[1]. The protein was localized in the nucleus, even in the absence of ecdysone[1]. This dynamic expression pattern should nevertheless be studied in the light of the three isoforms generated from the EcR gene[46] (see below; reviewed in reference 49).

The EcR gene expression is directly induced by ecdysone[45]. EcR belongs, as E74B, to the class I transcripts that are induced in mid-third-instar larvae in response to low but increasing titers of ecdysone. As the hormone concentration peaks in late third-instar larvae, these transcripts are repressed and the class II mRNAs (E74A, E75A and E75B) are induced[45,55].

The effect of ecdysone and the expression of EcR has been studied in detail in several *Drosophila* organs that are subjected to change during metamorphosis. This has been the case for the central nervous system in which a subset of neuronal populations that specifically expressed high levels of EcR-A are fated to die when the titer of ecdysone declines when the adults emerge from the pupae[57]. It has also been shown in other neurons that the appearance of the two receptor isoforms correlates with different types of ecdysone response: EcR-A predominates when cells are undergoing maturational responses, whereas EcR-B1 predominates during proliferative activity or regressive responses. This complex behavioral pattern of neurons in response to the same increase in ecdysone titer illustrates very well the complexity of the response of target tissues to the pulse of hormone concentration[48,58]. Other tissues in which the effect of ecdysone, the expression of EcR and/or its function in regulating metamorphosis were shown are: the ovary, in which EcR and USP regulates the timing and progression of morphogenesis[59] as well as egg-chamber development[66]; the eye, in which furrow progression is under the control of the hormone[60–62]; the cuticle[56,63,68]; or the muscles[64,65] (see reference 67 for an extensive review).

The expression pattern of the various orthologues of EcR cloned in insect and other arthropods also exhibits a dynamic pattern with a relatively wide expression in numerous organs and direct regulation by ecdysone[3–13,43,44]. Of note, in the tobacco hornworm *Manduca sexta*, EcR gene expression is under the control of both 20-hydroxyecdysone (20E) and juvenile hormone (JH). It has been shown that 20E up-regulates EcR-B1 and EcR-A expression, whereas JH modulates this induction in a complex manner[56].

Activity and target genes

EcR is a repressor without ligand and a transcriptional activator when ligand is added[34]. The transcriptional repression has been explained at least in part by the interaction with SMRTER, a functional homologue of the vertebrate corepressors NCoR and SMRT[35]. In addition, EcR has been shown to be able to dimerize with other nuclear receptors, such as SVP[40]. It is believed that this interaction inhibits the DNA-binding activity of EcR and thus inactivates it. It is clear that EcR can also repress some genes, such as βFTZ-F1 (NR5A3) but the molecular basis for this effect is still not understood[69,72,73].

The molecular effects of ecdysone on gene expression have been well studied in salivary gland in which ecdysone regulates the puffing patterns of the polytene chromosome. Specific patterns arise as the animal progresses through its development, and the changes in these patterns have been well documented (reviewed in reference 41). A small set of early puffs can be directly and rapidly induced in culture using ecdysone, and these match the early puffs formed at the end of the third larval instar. Protein synthesis is then required for both early puff regression and late puff formation. These observations led to the Ashburner model, in which ecdysone triggers a regulatory cascade by activating the early puffs, which can induce the late puffs as well as use a feedback mechanism to repress their own puffing[50,51]. Genes cloned from the early puffs include the Broad-complex, E74 and E75, which all encode transcription factors, E75 (NR1D3) being a homologue of the Rev-erb orphan receptors[52-54]. The early genes like the early puffs have been shown to be directly and transiently induced by ecdysone[45,55,75,87,88] and to be essential for proper entry into metamorphosis[84-86] (see reference 51 for a review).

Beside E75, many other orphan members of the nuclear receptor superfamily are ecdysone-regulated genes and their regulatory network was studied by a number of research teams, including the laboratory of Carl Thummel. This is the case for βFTZ-F1 (NR5A3), the expression of which is repressed by ecdysone and is thus expressed during mid-prepupae when the ecdysone titer is low[69,70,78] (reviewed in reference 41). The expression of βFTZ-F1 is activated by another orphan receptor DHR3 (NR1F4)[72,73]. DHR3, which regulates βFTZ-F1 expression, is itself an ecdysone-regulated gene both in Drosophila and Manduca[73,76,77]. Two other orphan receptors expressed throughout third-instar larval and pre-pupal development and regulated by ecdysone are DHR78 (NR2D1) and DHR96 (NR1J1), which are homologues of TR2/4 (NR2C1 and 2) and VDR/CAR (NR1I1–NR1I4), respectively[71]. Interestingly, the DHR96 gene product binds the hsp27 EcRE, suggesting a possible interference with EcR–USP signaling. Recent results have led to the suggestion that DHR78 may control EcR expression and may be a receptor for an unknown hormone that is produced during mid-third instar[74]. All these genes have in common to be orphan members of the nuclear receptor superfamily and long genes bearing complex transcriptional units generating several isoforms that are differentially regulated. This complex pattern undoubtedly accounts for the generation of the complex response of target tissues to the increase in only one known hormone, 20-hydroxyecdysone. In the mosquito Aedes, the partner of EcR, USP, has been shown to be regulated by ecdysone. Strikingly, the two N-terminal isoforms of USP are differentially regulated by EcR in fat body cultures: USP-A is inhibited, whereas USP-B is activated by ecdysone[102]. The expression of the USP-2 isoform of Manduca also correlates with the ecdysone titer, whereas USP-1 transcription complements that of USP-2[76].

Apart from members of the nuclear receptor superfamily, many other EcR target genes were cloned. As already mentioned, the first was hsp27[21], the EcRE of which is now the canonical EcR target sequence. Other known target genes, for which in many cases a consensus EcRE was found, are: β3 tubulin[79]; the Eip28/29 and Eip40 genes, which where identified by virtue of their immediate response to ecdysone in cell cultures and which are interesting models for tissue-specific regulation by this hormone[23,80]; the Broad-complex gene, which generate an extremely complex set of transcripts and also accounts for the tissue-specific response to ecdysone[54,81], the Fbp1 gene regulated by ecdysone in the fat body tissue[27,82]; or the E63-1 gene that corresponds to the early puff 63F and which encodes a calcium-binding protein[83]. Intermolt genes that are the target of EcR were also identified, among which are ng-1, ng-2 and Sgs-4[89,90]. More recently, late genes, and among them genes implicated in apoptosis and that play a role during metamorphosis have been cloned (see reference 91 for a review). Interesting examples of these late genes are an ecdysone-inducible *Drosophila* caspase, DRONC[92], or L82, a gene corresponding to a late puff and which is conserved in nematodes and yeast[93]. The gene diap2, which encodes a death inhibitor, is repressed by the ecdysone cascade, whereas *reaper* (rpr) and *head involution defective* (hid), two death activators, are induced[96]. A new family of ecdysone-regulated zinc-finger proteins, encoded by the *crooked legs* gene was identified and shown to be required for leg and wing morphogenesis[94,106]. It has been proposed that the ecdysone-regulated crooked-leg proteins may in turn regulate the EcR gene itself as well as other class I transcripts such as E74B[94]. More recently, L63, the gene corresponding to the 63E late puff, was identified as a cyclin-dependent kinase (CDK)[104]. Other interesting recent examples of ecdysone-regulated genes can be found in references 103, 105 and 107.

Our view of the transcriptional response to ecdysone is likely to be completely revised by two major technological advances: the sequencing of the whole *Drosophila* genome, which provides information on all the known genes in this organism and on their regulatory regions, and the DNA microarrays that allow study of the expression of a large number of transcripts in a given physiological situation. A first step in this direction was made by David Hogness's laboratory at which microarrays were used to study the expression of several thousand *Drosophila* genes during metamorphosis[95]. Studies of this type, focused on genes directly regulated by ecdysone, will allow more complete understanding of the ecdysone regulatory hierarchy.

Knockout

A large number of mutant alleles are available for EcR[47,48]. Those within the common region are generally lethal during embryonic development, although the defect responsible for this lethality has not yet been identified. Recently, a conditional rescue system has allowed the phenotype of EcR null mutants during molting and metamorphosis to be studied[97]. This study revealed that EcR is required for hatching at each larval molt and for the initiation of metamorphosis. The study of a temperature-sensitive mutant affecting the three isoform, EcR[A483T] has also allowed the role of EcR in oogenesis to be studied[98].

Isoform-specific mutants were also generated. EcRB1 mutants die at the onset of metamorphosis, while animals null for both EcR-B1 and B2 mostly die as second instar larvae, with a few surviving to the end of the third instar. EcRA and EcRB2 mutants have not yet been identified. The EcR-B1 and B2 mutants exhibit molting defects, resulting in many mutants carrying two sets of larval mouthhooks and two cuticles from different larval instar. In addition, mutants that are rescued after early larval lethality are defective in wandering behavior during which the wild-type animals leave the food in search of a place to undergo metamorphosis. None of the external morphological changes associated with the onset of metamorphosis have occurred, indicating that these animals have failed to initiate metamorphosis[47,48]. Interestingly, the same type of defects are observed in USP mutants rescued to overcome the early lethality.

In the mutants, the larval salivary gland and midgut are not destroyed as in the wild type[47]. EcR-B1 animals show only limited midgut imaginal cell proliferation and, ultimately, the adult gut failed to form in these animals[47]. The dramatic remodeling of the larval nervous system observed during normal metamorphosis is also disrupted in EcR-B1 and B2 null mutants[48]. The proliferation of the histoblast nests, the cells which eventually form the adult abdominal epidermis, is also blocked in EcR-B1 mutants[47]. The imaginal disc of the EcR-B1 mutants do evert and partially elongate, but disc development is blocked earlier in EcR-B1 and B2 animals, with the discs everting inside the animals (reviewed in reference 49).

Another well-studied effect of ecdysone is the regulation of the puffing patterns of the polytene chromosomes of the larval salivary gland and expression of the early genes directly regulated by ecdysone[51]. In EcR-B1 mutants it has been observed that the intermolt puffs (puffs triggered by a low titer pulse of ecdysone at mid-third instar) form normally but then failed to regress while neither the early nor the late puffs are formed. These results have led to the idea that EcR regulates late third instar events such as wandering behavior and the onset of metamorphosis, but not the mid-third instar events that prepare the animal for metamorphosis (see reference 49 for a review).

Associated disease

Not applicable.

Gene structure, promoter and isoforms

The *Drosophila* EcR gene can generate three different isoforms called A, B1 and B2[46]. All three isoforms share a common DNA- and ligand-binding domain but each has a unique N-terminus. EcR-A is generally expressed in the tissues that will eventually grow and differentiate into adult structures, while EcR-B1 is expressed in those tissues fated to die during metamorphosis. EcR-B2 expression pattern is not yet known. These differential expression patterns suggest that different EcR isoforms may be involved in regulating the diverse roles of ecdysone during *Drosophila* metamorphosis.

At the functional level, all three EcR isoforms recognize the EcRE and binds to USP. An AF-1 function has not been studied in detail in these isoforms.

Chromosomal location

The EcR gene was mapped to the cytological position 42A on the third *Drosophila* chromosome.[1]

Amino acid sequence for *Drosophila* EcR (NR1H1)

Accession number: M74078

```
  1 MKRRWSNNGGFMRLPEESSSEVTSSSNGLVLPSGVNMSPSSLDSHDYCDQDLWLCGNESGSFGGSNGHGL
 71 SQQQQSVITLAMHGCSSTLPAQTTIIPINGNANGNGGSTNGQYVPGATNLGALANGMLNGGFNGMQQQIQ
141 NGHGLINSTTPSTPTTPLHLQQNLGGAGGGGIGGMGILHHANGTPNGLIGVVGGGGGVGLGVGGGGVGGL
211 GMQHTPRSDSVNSISSGRDDLSPSSSLNGYSANESCDAKKSKKGPAPRVQEELCLVCGDRASGYHYNALT
281 CEGCKGFFRRSVTKSAVYCCKFGRACEMDMYMRRKCQECRLKKCLAVGMRPECVVPENQCAMKRREKKAQ
351 KEKDKMTTSPSSQHGGNGSLASGGGQDFVKKEILDLMTCEPPQHATIPLLPDEILAKCQARNIPSLTYNQ
421 LAVIYKLIWYQDGYEQPSEEDLRRIMSQPDENESQTDVSFRHITEITILTVQLIVEFAKGLPAFTKIPQE
491 DQITLLKACSSEVMMLRMARRYDHSSDSIFFANNRSYTRDSYKMAGMADNIEDLLHFCRQMFSMKVDNVE
561 YALLTAIVIFSDRPGLEKAQLVEAIQSYYIDTLRIYILNRHCGDSMSLVFYAKLLSILTELRTLGNQNAE
631 MCFSLKLKNRKLPKFLEEIWDVHAIPPSVQSHLQITQEENERLERAERMRASVGGAITAGIDCDSASTSA
701 AAAAAQHQPQPQPQPQPSSLTQNDSQHQTQPQLQPQLPPQLQGQLQPQLQPQLQTQLQPQIQPQPQLLPV
771 SAPVPASVTAPGSLSAVSTSSEYMGGSAAIGPITPATTSSITAAVTASSTTSAVPMGNGVGVGVGVGGNV
841 SMYANAQTAMALMGVALHSHQEQLIGGVAVKSEHSTTA  878
```

References
1 Koelle, M.R. et al. (1991) Cell 67, 59–77.
2 Nuclear Receptor Nomenclature Committee (1999) Cell 97, 161–163.
3 Swevers et al. (1995) Insect Biochem. Mol. Biol. 25, 857–866.
4 Fujiwara, H. et al. (1995) Insect Biochem. Mol. Biol. 25, 845–856.
5 Cho, W.L. et al. (1995) Insect Biochem. Mol. Biol. 25, 19–27.
6 Imhof, M.O. et al. (1993) Insect Biochem. Mol. Biol. 23, 115–124.
7 Mouillet, J–F. et al. (1997) Eur. J. Biochem. 248, 856–863.
8 Durica, D. and Hopkins, P.M. (1996) Gene 171, 237–241.
9 Chung, A.C.K. et al. (1998) Mol. Cell. Endocrinol. 139, 209–227.
10 Guo, X. et al. (1998) Insect Biochem. Mol. Biol. 27, 945–962.
11 Kothapalli, R. et al. (1995) Dev. Genet. 17, 319–330.
12 Hannan, G.N. and Hill, R.J. (1997) Insect Biochem. Mol. Biol. 27, 479–488.
13 Martinez, A. et al. (1999) Insect Biochem. Mol. Biol. 29, 915–930.
14 Durica, D.S. et al. (1999) Am. Zool. 39, 758–773.
15 Palmer, M.J. et al. (1999) Am. Zool. 39, 747–757.
16 Riddiford, L.M. et al. (1999) Am. Zool. 39, 736–746.
17 Raikhel, A.S. et al. (1999) Am. Zool. 39, 722–735.
18 Zelus, D. et al. (submitted).
19 Hayward, D.C. et al. (1999) Dev. Genes Evol. 209, 564–571.
20 Henrich, V.C. and Brown, N.E. (1995) Insect Biochem. Mol. Biol. 25, 881–897.
21 Riddihough, G. and Pelham, H.R.B. (1987) EMBO J. 6, 3729–3734.
22 Martinez, E. et al. (1991) EMBO J. 10, 263–268.
23 Cherbas, L. et al. (1991) Genes Dev. 5, 120–131.
24 Yao, T.P. et al. (1992) Cell 71, 63–72.
25 Thomas, H.E. et al. (1993) Nature 362, 471–475.
26 Yao, T.P. et al. (1993) Nature 366, 476–479.
27 Antoniewski, C. et al. (1994) Mol. Cell. Biol. 14, 4465–4474.
28 Lehmann, M. et al. (1997) Mech. Dev. 62, 15–27.
29 D'Avino, P.P. et al. (1995) Mol. Cell. Endocrinol. 113, 1–9.

[30] Vögtli, M. et al. (1998) Nucl. Acids Res. 26, 2407–2414.
[31] Antoniewski, C. et al. (1996) Mol. Cell. Biol. 16, 2977–2986.
[32] Horner, M.A. et al. (1995) Dev. Biol. 168, 490–502.
[34] Dobens, L. et al. (1991) Mol. Cell. Biol. 11, 1846–1853.
[35] Tsai, C–C. et al. (1999) Mol. Cell 4, 175–186.
[36] Suhr, S.T. et al. (1998) Proc. Natl Acad. Sci. USA 95, 7999–8004.
[37] Christopherson, K.S. et al. (1992) Proc. Natl Acad. Sci. USA 89, 6314–6318.
[38] Sluder, A.E. et al. (1999) Genome Res. 9, 103–120.
[39] Barker, G.C. and Rees, H.H. (1990) Parasitol. Today 6, 384–387.
[40] Zelhof A.C. et al. (1995) Mol. Cell. Biol. 15, 6736–6745.
[41] Thummel, C.S. (1995) Cell, 83, 871–877.
[42] Dhadialla, T.S. et al. (1998) Annu. Rev. Entomol. 43, 545–569.
[43] Jindra, M. et al. (1996) Dev. Biol. 180, 258–272.
[44] Chung, A.C.K. et al. (1998) Gen. Comp. Endocrinol. 109, 375–389.
[45] Karim, F.D. and Thummel, C.S. (1992) EMBO J. 11, 4083–4093.
[46] Talbot , W.S. et al. (1993) Cell 73, 1323–1337.
[47] Bender, M. et al. (1997) Cell 91, 777–788.
[48] Schubiger, M. et al. (1998) Development 125, 2053–2062.
[49] Hall, B.L. (1999) Am. Zool. 39, 714–721.
[50] Ashburner, M. (1990) Cell 61, 1–3.
[51] Thummel, C.S. (1996) Trends Genet. 1, 306–310.
[52] Burtis, K.C. et al. (1990) Cell 61, 85–99.
[53] Segraves, W.A. and Hogness, D.S. (1990) Genes Dev. 4, 204–219.
[54] DiBello, P.R. et al. (1991) Genetics 129, 385–397.
[55] Karim, F.D. and Thummel, C.S. (1991) Genes Dev. 5, 1067–1079.
[56] Hiruma, K. et al. (1999) Dev. Genes Evol. 209, 18–30.
[57] Robinow, S. et al. (1993) Development 119, 1251–1259.
[58] Truman, J.W. et al. (1994) Development 120, 219–234.
[59] Hodin, J. and Riddiford, L.M. (1998) Dev. Genes Evol. 208, 304–317.
[60] Brennan, C.A. et al. (1998) Development 125, 2653–2664.
[61] Champlin, D.T. and Truman, J.W. (1998) Development 125, 2009–2018.
[62] Champlin, D.T. and Truman, J.W. (1998) Development 125, 269–277.
[63] Krämer, B. and Wolbert, P. (1998) Dev. Genes Evol. 208, 205–212.
[64] Bayline, R.J. et al. (1998) Dev. Genes Evol. 208, 369–381.
[65] Champlin, D.T. et al. (1999) Dev. Genes Evol. 209, 265–274.
[66] Buszczak, M. et al. (1999) Development 126, 4581–4589.
[67] Gilbert, J.R. et al. (eds) (1996) Metamorphosis: Postembryonic Reprogramming of Gene Expression in Amphibian and Insect Cells. San Diego: Academic Press.
[68] Charles, J–P. et al. (1992) Eur. J. Biochem. 206, 813–819.
[69] Woodward, C.T. et al. (1994) Cell 79, 607–615.
[70] Sun, G.C. et al. (1994) Dev. Biol. 162, 426–437.
[71] Fisk, G.J. and Thummel, C.S. (1995) Proc. Natl Acad. Sci. USA 92, 10604–10608.
[72] White, K.P. et al. (1997) Science 276, 114–118.
[73] Lam, G.T. et al. (1997) Development 124, 1757–1769.
[74] Fisk, G.J. and Thummel, C.S. (1998) Cell 93, 543–555.
[75] Zhou, B. et al. (1998) Dev. Biol. 193, 127–138.

[76] Lan, Q. et al. (1999) Mol. Cell. Biol. 19, 4897–4906.
[77] Lam, G.T. et al. (1999) Dev. Biol. 212, 204–216.
[78] Broadus, J. et al. (1999) Mol. Cell 3, 143–149.
[79] Bruhat, A. et al. (1990) Nucl. Acids Res. 18, 2861–2867.
[80] Andres, A.J. and Cherbas, P. (1992) Development 116, 865–876.
[81] von Kalm, L. et al. (1994) EMBO J. 13, 3505–3516.
[82] Brodu, V. et al. (1999) Mol. Cell. Biol. 19, 5732–5742.
[83] Andres, A.J. and Thummel, C.S. (1995) Development 121, 2667–2679.
[84] Fletcher, J.C. and Thummel, C.S. (1995) Genetics 141, 1025–1035.
[85] Fletcher, J.C. and Thummel, C.S. (1995) Development 121, 1411–1421.
[86] Fletcher, J.C. and Thummel, C.S. (1995) Development 121, 1455–1465.
[87] Huet, F. et al. (1995) Development 118, 613–627
[88] Huet, F. et al. (1993) Development 121, 1195–1204.
[89] Lehmann, M. and Koeller, G. (1995) EMBO J. 14, 716–726.
[90] Crispi, S. et al. (1998) J. Mol. Biol. 275, 561–574.
[91] Rodriguez, A. et al. (1998) Am. J. Human Genet. 62, 514–519.
[92] Dorstyn, L. et al. (1999) Proc. Natl Acad. Sci. USA 96, 4307–4312.
[93] Stowers, R.S. et al. (1999) Dev. Biol. 213, 116–130.
[94] D'Avino, P.P. and Thummel, C.S. (1998) Development 125, 1733–1745.
[95] White, K.P. et al. (1999) Science 286, 2179–2184.
[96] Jiang, C. et al. (2000) Mol. Cell 5, 445–455.
[97] Li, T.R. and Bender, M. (2000) Development 127, 2897–2905.
[98] Carney, G.E. and Bender, M. (2000) Genetics 154, 1203–1211.
[99] No, D. et al. (1996) Proc. Natl. Acad. Sci. USA 93, 3346–3351.
[100] Arbeitman, M.N. and Hogness, D.S. (2000) Cell 101, 67–77.
[101] Wang, S.F. et al. (2000) Mol. Cell. Biol. 20, 3870–3879.
[102] Wang, S.F. et al. (2000) Dev. Biol. 218, 99–113.
[103] Pecasse, F. et al. (2000) Dev. Biol. 221, 53–67.
[104] Stowers, R.S. et al. (2000) Dev. Biol. 221, 23–40.
[105] Ahmed, A. et al. (1999) Proc. Natl Acad. Sci. USA 96, 14795–14800.
[106] D'Avino, P.P. and Thummel, C.S. (2000) Dev. Biol. 220, 211–224.
[107] Matsuoka, T. and Fujiwara, H. (2000) Dev. Genet. Evol. 210, 120–128.
[108] Verras, M. et al. (1999) Eur. J. Biochem. 265, 798–808.
[109] Jayachandran, G. and Fallon, A.M. et al. (2000) Arch. Insect Biochem. Physiol. 43, 87–96.
[110] Saleh, D.S. et al. (1998) Mol. Cell. Endocrinol. 143, 91–99.

Names

Two closely related LXR genes are known. Since they were found by different research groups at the same time the nomenclature is a little confusing. The following table should help the reader.

Species	Other names	Accession number	References
Official name: NR1H2			
Human	UR, NER, LXRβ	U07132	1, 2
Mouse	RIP15, LXRβ	U09419	3
Rat	OR-1, UR, LXRβ	U14533	1, 4
Official name: NR1H3			
Human	LXRα	U22662	5
Mouse	LXRα	AJ132601	31, 32
Rat	RLD-1, LXRα	U11685	6

Isolation

The NR1H2 gene was characterized at the same time by four groups in different mammalian species. Shinar et al. used a RT-PCR-based strategy to isolate a sequence, called NER, from the human Saos-2/B10 osteoblastic cell line[2]. Song et al. first characterized the clone from rat using a rat vagina cDNA library and a mixture of oligonucleotide probes corresponding to highly conserved regions of the C domain of nuclear receptors[1]. The clone was called UR (ubiquitous receptor) because of its very large expression pattern. The same strategy was used by Jan-Åke Gustafsson's laboratory in a screen for orphan receptors (OR-1: orphan receptor-1) using a rat liver cDNA library[4]. The human sequence was also characterized by Song et al. in a PC3 prostate cancer cell line cDNA library using the rat UR as a probe[1]. Finally, in a two-hybrid screen for RXR-interacting proteins, David Moore's laboratory isolated from a mouse liver library a cDNA, called RIP15, corresponding to the same gene[3]. A processed but truncated LXRβ pseudogene was recently found in the mouse genome (accession number: MMU133116)[31].

The NR1H3 gene was first characterized by Magnus Pfahl's group by screening of a rat liver-derived cDNA library using a probe encoding the DNA-binding domain of the v-erbA probe[6]. The clone was called RLD-1 (rat liver derived-1). It was shown that the product encoded by this gene was able to bind to DR4 elements as heterodimers with RXR. Independently, David Mangelsdorf's team isolated the human cDNA by low-stringency screening of a human liver cDNA library with a probe encompassing the RARα DNA-binding domain[5]. The sequence, called LXRα (liver X receptor) exhibits a close relationship with NR1H2 and with the ecdysone receptor. Interestingly, it was shown that, on DR4 elements LXRα may shift RXR from a silent partner to a retinoid responsive receptor, suggesting that the LXR–RXR heterodimer may define a new retinoid signaling pathway.

In the DNA-binding domain, LXRs display 76% amino-acid identities with FXR (NR1H4), 67% with EcR (NR1H1) and a range of 44–59% with members of the NR1I group (PXR, CAR). In the ligand-binding domain, the

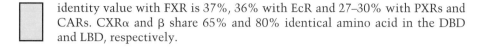

identity value with FXR is 37%, 36% with EcR and 27–30% with PXRs and CARs. CXRα and β share 65% and 80% identical amino acid in the DBD and LBD, respectively.

DNA binding

The literature has convincingly shown that LXRα can bind to DNA on DR4 elements as heterodimers with RXR. LXR alone appears to be unable to bind to DNA. No direct binding of the RXR–LXR heterodimer on DR3 or DR5 elements could be detected[6], whereas very weak competition by cold DR5 and DR3 could be observed after binding of RXR–LXR to a DR4 element[5]. Willy et al. defined an element called the LXRE derived from the 5'-regulatory region of the mouse mammary tumor virus long terminal repeat[5]. This element has the structure of a DR4 but the sequences of the two repeated motifs are quite divergent: GGTTTA and AGTTCA for the 5' and 3' motifs, respectively. Apfel et al. also reported that the sequence of the spacer is important for high-affinity binding of RLD-1 to DNA, since changing a CGAA spacer to a GGCC spacer decreases the binding of an RLD-1-RXR heterodimer but not the binding of RXR-TRα[6]. From all these data it is clear that the LXR–RXR heterodimer will only bind to a subset of DR-4, whereas the *in vivo* relevance of these experiments remains to be established.

LXRβ exhibits very similar DNA-binding activities on DR4 elements in association with RXR. Song et al. detected a weak direct binding of UR on a DR5 probe but clearly the affinity for the DR4 appears to be much stronger[1]. No binding on palindromic elements was detected by Teboul et al., whereas Seol et al. reported specific binding on IRO and IR1 elements[3,4]. The same authors reported the binding of the RIP15–RXR heterodimer to DR5, DR4 and DR2 with comparable affinities and no binding on DR0, DR1 and DR3. The reasons for these discrepancies are unclear. A site selection assay using gel shift and PCR result in the discovery of a new target sequence for LXRβ[26]. This element is a DR1 with a spacer containing an obligatory A. This element is specific to the LXRβ–RXR heterodimer and it was shown that LXRα does not recognize it.

Recently, a new DNA-binding site for LXRα was discovered upstream of the renin and c-*myc* genes[32]. This element, called CNRE (TCTCACAG) is responsible for the cAMP regulation of these two genes and bears a weak homology with cAMP response elements (TGACGTCA). In contrast to previous work, it was shown that LXRα recognizes it as a monomer.

Partners

RXRs are partners of all members of the NR1H group. Interestingly, it has been shown by coimmunoprecipitation that OR-1 can form a heterodimer with RXR in the absence of DNA[4]. Within the heterodimer bound to a direct repeat element, RXR occupies the 5' half-site of the response element and LXRα the 3' one[8]. This specific orientation does not exist for the RXR–LXRβ heterodimer bound to the DR1s element found by site-selection assay[26].

Interestingly, the RXR–LXR heterodimer exhibits unique transcriptional properties that are shared by both OR-1 and LXRα[8,10]. Unlike other nuclear receptors, in the absence of ligand, the RXR–LXR heterodimer does not associate with a corepressor protein and instead displays a dimer-induced transcriptional activity. Upon addition of either an RXR or an LXR ligand the heterodimer undergoes a conformational change that results in coactivator recruitment and an increase in transactivation. When ligands for each receptor are both present, the effect is synergistic. In addition, ligand binding by RXR allows the AF-2 region of LXR to recruit a coactivator by a dimerization-mediated conformational change. The mechanism by which the activation potential of one receptor is enabled by the binding of its partner is apparently unique to the RXR–LXR heterodimer.

Using the yeast two-hybrid system, it has been proposed that PPARα may interact with LXRα[9]. This complex is not formed on DNA and LXRα does not form a specific complex with PPARα on direct repeats or on natural PPREs. However, LXRα may inhibit the binding of the PPARα–RXR to PPREs, suggesting that it may modulate the PPAR signaling pathways in the cell. The *in vivo* significance of these findings is unclear but it is interesting to note that both LXRα and PPARα are expressed in liver.

LXRβ interacts with the coactivator SRC-1 as well as with RIP140[30]. The requirement for the binding of these two molecules is distinct and their binding to LXRβ is competitive. Interestingly, the dimerization of LXRβ with RXR is sufficient to recruit SRC-1, suggesting that the constitutive activation state of the heterodimer is mediated by SRC-1[30].

Ligands

Using concentrated lipid extracts from breeding-bull testis, Janowski et al. found that oxysterols enhance the transactivation mediated by the LXRα–RXR heterodimer[7,12]. Remarkably, only a specific group of oxysterols activated LXRα. A screen of several compounds revealed that the strongest LXRα activator is the naturally occurring compound 22(R)-hydroxycholesterol (22(R)-HC). Other molecules, such as 20(S)HC, 24-HC, 25-HC, 7α-HC or FF-MAS, are also potent activators. The EC_{50} for these compounds is around 1.5 μM. That these molecules are indeed ligands was demonstrated by direct binding assays[16]. The three most potent ligands found were 24(S),25-epoxycholesterol (K_d = 70 nM), 24(S)-HC (K_d = 220 nM) and 22(R)-HC (K_d = 380 nM). These ligands bind to the receptors at concentrations that occur *in vivo*. Furthermore, structural requirements of ligands for LXRα and OR-1 have been described[16]. These studies revealed that position-specific monooxidation of the sterol side-chain is required for LXR high-affinity binding and activation. In addition, introduction of an oxygen on the sterol B-ring results in a ligand that can only recognize LXRα and not OR-1[16]. The activation of LXRα by its ligand can apparently be further enhanced by phosphorylation events[19].

The activators of LXRα are precursors of steroid hormones and bile acids. It has been proposed that LXRα may act as a sensor of specific sterols, such as 22(R)-OH or 7α-HC, to regulate the metabolic pathways of steroid hormones and bile acids[7]. Such a 'feedforward' mechanism is consistent with the high expression of this receptor in liver (see references 13, 14 and 22 for detailed reviews). This has been substantiated by the results of the inactivation of the gene in mouse (see below) as well as by the observation that distinct products of the mevalonate pathway can either stimulate or inhibit (e.g. geranylgeraniol) LXRα activity in cultured cells[17,18]. Recently, nonsteroidal LXR-selective agonists were described[34]. When injected into mice and hamsters, these compounds induce lipogenesis genes, resulting in increased plasma triglyceride and phospholipid levels in both species.

Three-dimensional structure

No 3D structure of the DNA-binding or ligand-binding domain is available to date.

Expression

LXRβ exhibits a very large expression pattern that justifies its name 'ubiquitous receptor'[1]. This gene is expressed as a 2.3 kb mRNA in a wide variety of tissues (heart, liver, kidney, brain, testis, ovary, adrenal, uterus, prostate, vagina, lung, spleen) with comparable levels. A broad expression is also observed in cell lines: prostate carcinoma PC-3 and LNCaP cells, skin fibroblasts, B-cells (RPMI 1788, WEHI-231, etc.), and osteoblastic Saos-2 cells[1,2]. The protein is also ubiquitously expressed as a 50 kDa nuclear form.

LXRα in human is predominantly expressed in the liver as a 1.9 kb mRNA[1]. Other human tissues exhibiting LXRα/RLD-1 expression are kidney and spleen as well as intestine at a lower level[1]. In rat, the highest expression level was found in the pituitary, spleen, fat tissues and liver, and lower levels were observed in a number of tissues including testes, prostate and skin. Apparently, the expression level may differ in the same tissues from different animals, since the rat kidney exhibits a very low (if any) expression, whereas the human kidney supports a high expression level of LXRα. The expression of LXRα appears to decline together with that of the RXRs and PPARα during acute-phase response in rodent liver[27]. A recent report has shown that LXRα expression is regulated in liver by fatty acids and that the mouse LXRα promoter is a direct target of PPARα in liver[29].

Little is known on the expression of these two genes during embryonic development. Northern blot analysis of total mouse embryo mRNA at various stages performed by Willy et al. have shown that low levels of LXRα can be detected from embryonic day 13.5 and continue to increase until day 18.5[5]. Teboul et al. noted a prominent expression of OR-1 in the thymus, brown fat, spleen, gastrointestinal tract, salivary gland, thyroid gland, pituitary and retina in rat fetus at 21 days of development[4]. Moderate levels were observed in brain and cerebellum, in perichondrium around developing bones, as well as in liver, lung, heart and skin. Low levels were detected in skeletal muscles. According to these authors the strongest expression level in adult rats was in adrenal cortex, lymph nodes and prostate. Kainu et al. described the expression pattern of OR-1 in rat brain

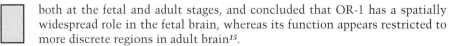

both at the fetal and adult stages, and concluded that OR-1 has a spatially widespread role in the fetal brain, whereas its function appears restricted to more discrete regions in adult brain[15].

Activity and target genes

The first target gene clearly characterized is the cholesterol 7α-hydroxylase (CYP7A) gene, which encodes a rate-limiting enzyme in the catabolism of cholesterol to bile acids[12] (see reference 33 for a review). This regulation occurs through a specific LXRE element found in the promoter of this gene. Interestingly, this LXRE is a much stronger response element for LXRα than for LXRβ, further substantiating the idea that both receptors have distinct physiological functions. In line with these findings, the transcription of the CYP7A gene is down-regulated in LXRα knockout mice[18]. LXR also has an opposite effect to that of FXR (which is a bile acid receptor) on this gene suggesting that both receptors control the rate-limiting step of cholesterol degradation to bile acids in opposing ways[23, 24].

Since several other genes (SREBP1 and 2, HMG CoA reductase, HMG CoA synthase, etc.) are deregulated in these mice, it is likely that LXRα regulates many other genes related to cholesterol metabolism. It has recently been shown that the SREBP-1c gene, but not the SREBP-1a or SREBP-2 is a direct target of LXRα and β through a DR4 element, which binds the RXR–LXR heterodimer[35]. Other P450 genes of the CYP2, CYP3 and CYP4 families could also be LXR target genes[21, 24]. More recently, the human White gene and its mouse homologue ABC8, both members of the ATP-binding cassette transporter superfamily, have been shown to be regulated by LXRβ through a specific DR1s element[25]. The human ABC1 promoter is also controlled by LXRα and LXRβ[28,36,37]. The regulation of these genes by LXRs is believed to play a major role in cellular cholesterol efflux (see reference 33 for a review). This transport of cholesterol is abnormal in several human diseases such as Tangier disease or sitosterolemia[33,37,38]. The renin and c-*myc* genes were shown recently to be regulated by cAMP via the interaction of LXRα with a cAMP-responsive site in their promoter[32].

Knockout

The LXRα gene has been knocked out in mice[18]. Resulting animals exhibit no obvious phenotype when reared under standard laboratory conditions. Nevertheless, when fed with a cholesterol-rich diet, the LXR[−/−] mice fail to show a compensatory increase in bile acids synthesis and excretion. As a result, these animals accumulate massive amounts of cholesterol esters in their livers, which results in chronic hepatomegaly. As stated above, these mice do not regulate the CYP7A gene transcription, suggesting that oxysterols directly regulate bile acid synthesis and thus cholesterol degradation through this rate-limiting enzyme (see references 13 and 14 for detailed reviews).

The LXRβ null mice have not been published yet. Obviously, the functions of LXRβ and LXRα are distinct, since LXRβ is apparently unable to compensate for LXRα function in the liver. Peet et al. quote unpublished data suggesting that LXRβ null mice are viable with no obvious phenotype[13].

Associated disease

Despite its important role in regulating cholesterol metabolism, up to now no disease has been directly linked to the LXRα gene. No alteration of the amount of LXRα mRNA has been found in the adipose tissue of obese patients or in patients with non-insulin-dependent diabetes mellitus (NIDDM). LXRs play important role in ABC transporters, such as ABC1, ABC5 and ABC8, which are involved in cellular cholesterol efflux. These ABC transporters are directly implicated in syndromes linked to cholesterol transport, such as Tangier disease and sitosterolemia (see reference 33 for a review). It is believed that LXR specific agonist may have important medical applications in the regulation of cholesterol homeostasis.

Gene structure, promoter and isoforms

The mouse LXRα gene is organized in nine exons that encompass the translated portion of the gene[18]. Exons 1 and 2 encode the A/B domain, exon 3 encodes the whole C domain whereas exons 4–9 encode the D and E domains. A characterization of the mouse LXRα promoter was performed[29,31]. It was shown that this promoter is directly regulated by PPARα consistent with the *in vivo* regulation of LXRα expression by fatty acids in the liver[29].

As LXRα, the LXRβ promoter harbors no obvious TATA box, but contains NFκB and Ets binding sites that suggest a possible role in the hematopoietic and immune system[31].

No isoforms have been described for these genes.

Chromosomal location

The chromosomal location of LXRβ has been described using FISH in human[11]. This gene is located on chromosome 19q13.3, a region that contains many other genes such as prostate-specific antigen (PSA) or luteinizing hormone β. The LXRβ gene is not close to any other nuclear receptor and its chromosomal location does not suggest any association with human disorders. The precise chromosomal location of LXRα is unknown but the gene has been found on chromosome 11 in the complete sequence of the human genome.

Amino acid sequence for human LXRβ (NR1H2)

Accession number: U07132

```
  1  MSSPTTSSLDTPLPGNGPPQPGAPSSSPTVKEEGPEPWPGGPDPDVPGTDEASSACSTDWVIPDPEEEPE
 71  RKRKKGPAPKMLGHELCRVCGDKASGFHYNVLSCEGCKGFFRRSVVRGGARRYACRGGGTCQMDAFMRR
141  KCQQCRLRKCKEAGMREQCVLSEEQIRKKKIRKQQQQESQSQSQSPVGPQGSSSSASGPGASPGGSEAGS
211  QGSGEGEGVQLTAAQELMIQQLVAAQLQCNKRSFSDQPKVTPWPLGADPQSRDARQQRFAHFTELAIISV
281  QEIVDFAKQVPGFLQLGREDQIALLKASTIEIMLLETARRYNHETECITFLKDFTYSKDDFHRAGLQVEF
351  INPIFEFSRAMRRLGLDDAEYALLIAINIFSADRPNVQEPGRVEALQQPYVEALLSYTRIKRPQDQLRFP
421  RMLMKLVSLRTLSSVHSEQVFALRLQDKKLPPLLSEIWDVHE       463
```

Amino acid sequence for human LXRα (NR1H3)

Accession number: U22662

```
  1  MSLWLGAPVPDIPPDSAVELWKPGAQDASSQAQGGSSCILREEARMPHSAGGTAGVGLEAAEPTALLTRA
 71  EPPSEPTEIRPQKRKKGPAPKMLGNELCSVCGDKASGFHYNVLSCEGCKGFFRRSVIKGAHYICHSGGHC
141  PMDTYMRRKCQECRLRKCRQAGMREECVLSEEQIRLKKLKRQEEEQAHATSLPPRRSSPPQILPQLSPEQ
211  LGMIEKLVAAQQQCNRRSFSDRLRVTPWPMAPDPHSREARQQRFAHFTELAIVSVQEIVDFAKQLPGFLQ
281  LSREDQIALLKTSAIEVMLLETSRRYNPGSESITFLKDFSYNREDFAKAGLQVEFINPIFEFSRAMNELQ
351  LNDAEFALLIAISIFSADRPNVQDQLQVERLQHTYVEALHAYVSIHHPHDRLMFPRMLMKLVSLRTLSSV
421  HSEQVFALRLQDKKLPPLLSEIWDVHE    447
```

References

1 Song, C. et al. (1994) Proc. Natl Acad. Sci. USA 91, 10809–10813.
2 Shinar, D.M. et al. (1994) Gene 147, 273–276.
3 Seol, W. et al. (1995) Mol. Endocrinol. 9, 72–85.
4 Teboul, M. et al. (1995) Proc. Natl Acad. Sci. USA 92, 2096–2100.
5 Willy, P.J. et al. (1995) Genes Dev. 9, 1033–1045.
6 Apfel, R. et al. (1995) Mol. Cell. Biol. 14, 7025–7035.
7 Janowski, B.A. et al. (1996) Nature 383, 728–731.
8 Willy, P.J. and Mangelsdorf, D.J. (1997) Genes Dev. 11, 289–298.
9 Miyata, K.S. et al. (1996) J. Biol. Chem. 271, 9189–9192.
10 Wiebel, F.F. and Gustafsson, J.A. (1997) Mol. Cell. Biol. 17, 3977–3986.
11 LeBeau, M.M. (1995) Genomics 26, 166–168.
12 Lehman, J.M. (1997) J. Biol. Chem., 272, 3137–3140.
13 Peet, D.J. (1998) Curr. Opin. Genet. Dev., 8, 571–575.
14 Accad, M. and Farese R.V. (1998) Curr. Biol. 8, R601–R604.
15 Kainu, T. et al. (1996) J. Mol. Neurosci. 7, 29–39.
16 Janowski, B.A. et al. (1999) Proc. Natl Acad. Sci. USA 96, 266–271.
17 Forman, B.M. et al. (1997) Proc. Natl Acad. Sci. USA 94, 10588–10593.
18 Peet, D.J. et al. (1998) Cell 93, 693–704.
19 Huang, C-J. et al. (1998) Biochem. Biophys. Res. Commun. 243, 657–663
20 Auboeuf, D. et al. (1997) Diabetes 46, 1319–1327.
21 Waxman, D.J. (1999) Arch. Biochem. Biophys. 369, 11–23.
22 Wolf, G. (1999) Nutr. Rev. 57, 196–198.
23 Wang, H. et al. (1999) Mol. Cell 3, 543–553.
24 Russell, D.W. (1999) Cell 97, 539–542.
25 Venkateswaran, A. et al. (2000) J. Biol. Chem. 275, 14700–14707.
26 Feltkamp, D. et al. (1999) J. Biol. Chem. 274, 10421–10429.
27 Beigneux, A.P. et al. (2000) J. Biol. Chem. 275, 16390–16399.
28 Costet, P. et al. (2000) J. Biol. Chem. 275, 28240–28245.
29 Tobin, K.A. et al. (2000) Mol. Endocrinol. 14, 741–752.
30 Wiebel, F.F. et al. (1999) Mol. Endocrinol. 13, 1105–1118.
31 Alberti, S. et al. (2000) Gene 243, 93–103.
32 Tamura, K. et al. (2000) Proc. Natl Acad. Sci. USA 97, 8513–8518.
33 Allayee, H. et al. (2000) Science 290, 1709–1711.
34 Schultz, J.R. et al. (2000) Genes Dev. 14, 2831–2838.
35 Repa, J.J. et al. (2000) Genes Dev. 14, 2819–2830.
36 Vankateswaran, A. et al. (2000) Proc. Natl Acad. Sci. USA 97, 12097–12102.
37 Repa, J.J. et al. (2000) Science 289, 1524–1529.
38 Berge, K.E. et al. (2000) Science 290, 1771–1775.

Names

FXR is closely related to the LXR group and was included in this group in the Nomenclature for Nuclear Receptors. Only one FXR group member has been described to date and the possibility remains that it is a third member of the LXR group that has experienced a strong sequence divergence due to the acquisition of distinct functional properties. The three different names given to the human, mouse and rat orthologues reflect their characterization by three different laboratories. FXR is sometimes called BAR since it is a bile acid receptor.

Species	Other names	Accession number	References
Official name: NR1H4			
Human	HRR-1	U68233	Unpub.
Mouse	RIP14, BAR	U09416 to 418	1
Rat	FXR	U18374	2

Isolation

The FXR gene was isolated in rat by Cary Weinberger's group by low-stringency screening of a rat liver cDNA library with an oligonucleotide corresponding to the highly conserved P-box of the nuclear receptor superfamily (TCEGCK(G/V)FF)[2]. This clone was called FXR (farnesoid X receptor) because it was shown to be activated by farnesoid compounds (see below). The mouse orthologue was isolated at the same time by Seol et al. as an RXR-interacting partner and called RIP14[1]. The human orthologue, HRR-1, was isolated by Papetti et al. and the sequence deposited in as a public database, but it has not been published.

In the DNA-binding domain, FXR displays 76% amino-acid identities with LXRs (NR1H2 and NR1H3), 70% with EcR (NR1H1) and a range of 44–55% with members of the NR1I group (PXR, CAR). In the ligand-binding domain, the identity value with LXRs is 37%, 34% with EcR, and 27–30% with PXRs and CARs.

DNA binding

FXR binds to DNA only as a heterodimer with RXR. Apparently it binds to at least two distinct response elements: a DR5, such as the one found upstream of the RARβ gene (βRARE)[1] and palindromic elements, called IR1 (for inverted repeats separated by 1 bp), such as the EcRE, found upstream of the *Drosophila hsp27* gene[1,2]. It was also found to bind to IR0, DR2 and DR5 elements though with a lower affinity[1]. The various isoforms found in mouse do not bind DNA with the same affinity[1]. Recently, a selection of binding site by gel shift and PCR revealed that the preferred sequence is a IR1 but that other types of motifs containing two AGGTCA sequences were also recognized[13]. The importance of the core sequence, the spacing nucleotide and the flanking nucleotides for DNA binding by FXR–RXR was systematically evaluated in this study.

Partners

FXR is an obligate partner of the RXR and was in fact identified by Seol et al. by this property in a two-hybrid screen for RXR-interacting proteins[1]. Its interaction potential with other nuclear receptors is still unknown. The interaction with RXR does not require DNA binding.

Using a two-hybrid system in CV-1 cells, SRC-1 was shown to bind to FXR in the presence of both RXR and the FXR ligand, chenodeoxycholic acid[6-8].

Ligands

FXR was shown to be transcriptionally activated by farnesol metabolites such as farnesol itself, juvenile hormone III and, to a lesser extent, farnesal, farnesyl acetate, farnesoic acid or geranylgeraniol[2]. These compounds activate FXR only in the presence of RXR. Interestingly, it is not activated by squalene, cholesterol nor by methoprene, a synthetic analogue of juvenile hormone. A synergistic effect was observed when specific ligands for RXR and farnesoid compounds were tested in combination on the RXR–FXR heterodimer. The EC_{50} for the most active compound are around 5–50 μM a value that is apparently comparable to the *in vivo* concentration of these compounds (0.5-10 μM) that can be estimated from the Michaelis constant of cellular enzymes that use them as substrates. No direct binding of the activating compounds to FXR has yet been described, suggesting that the physiological ligand may be a derivative of farnesoid metabolism. This pattern of activation of FXR by farnesyl pyrophosphate metabolite, a central molecule in the mevalonate pathway that gives rise to cholesterol, carotenoids or farnesoids, has led Cary Weinberger to suggest that FXR may be part of the signaling pathway for transcriptional control of cholesterol metabolism[3].

This hypothesis has recently been demonstrated by a series of three papers showing that bile acids are physiological ligands for FXR[6-8]. These analyses demonstrate that several bile acids are potent and selective activators of FXR-mediated transcription and that these compounds activate FXR at concentrations that are found in liver and intestinal cells. Using a fluorescence-based peptide-binding assay, the direct binding to FXR of chenodeoxycholic acid (but not other bile acids, such as cholic acid or lithocholic acid, in contrast to the activation profile) was demonstrated[8]. In fact, the affinity of FXR for chenodeoxycholic acid is in the micromolar range (4.5 μM), suggesting that the affinity of FXR for its ligand is not high. The direct binding of chenodeoxycholic acid to FXR is further confirmed by CARLA-type assays in which the binding of the coactivator SRC-1 to FXR is shown to be dependent on the presence of chenodeoxycholic acid[6-8]. Furthermore, these papers also show that FXR is a global regulator of bile acid metabolism (see references 9 and 10 for reviews). FXR represses the synthesis of these compounds (via a feedback mechanism) in the liver via the regulation of the cholesterol 7α-hydroxylase (CYP7A) gene, which is the rate-limiting enzyme for bile acid synthesis[6,7]. Furthermore, FXR activates (in presence of its ligand) the expression of the IBABP (ileal bile acid binding protein), which is responsible for the transport of bile acids from the gut to the liver, suggesting that it activates the recycling of these compounds[7, 9].

Interestingly, it has also been shown that FXR can be activated by retinoids such as all-*trans* retinoic acid or the synthetic retinoid TTNPB[14]. In the FXR–RXR heterodimer, the effects of all-*trans* RA and 9-*cis* RA (which binds the RXR partner) are additive. Of note, no evidence of direct binding of these retinoids on FXR was found[14].

Three-dimensional structure

No 3D structure of the DNA- or ligand-binding domain is available to date.

Expression

The expression of FXR (as a 2.3 kb mRNA in rodents) is relatively restricted to the liver, the intestinal villi, the renal tubules and the adrenal cortex[2,3]. Of note, activators of FXR have recently been shown to accelerate the development of the fetal epidermal permeability barrier[5]. This could suggest that FXR may be expressed in the epidermal cells of the fetus during development.

Activity and target genes

Two genes, cholesterol 7α-hydroxylase (CYP7A1) and IBABP, which are implicated in bile acid biosynthesis and recycling, respectively, are target genes of FXR[7,11,12]. The CYP7A1 enzyme controls the rate-limiting step of the catabolism of cholesterol into bile acids in the liver. The element by which the CYP7A1 gene is down-regulated by FXR was mapped and it was shown that the FXR–RXR heterodimer cannot bind to this sequence suggesting an indirect regulation[12]. Indeed, it has been shown that the down-regulation of CYP7A1 gene by bile acid is accomplished by three nuclear hormone receptors: FXR, SHP and LRH-1[17,18] (reviewed in reference 15). Binding of bile acids to FXR induces the transcription of the SHP gene through an IR-1 element. Increased levels of SHP induce CYP7A1 gene repression, since SHP physically interacts with LRH-1, which is an activator of the CYP7A1 promoter through a monomeric half-site sequence. It is interesting to remember that the CYP7A1 gene may be up-regulated by LXR (NR1H3) in the presence of its ligand 24(S),25-epoxycholesterol (see reference 9 for a review). As discussed above, this induction requires LRH-1 and this requirement explains the tissue-specific action of oxysterols on CYP7A1[17,18]. It has been independently shown that FXR can suppress LXR-mediated transactivation on a synthetic LXRE element[6]. This repression can be relieved by LXR ligands. Taken together, these data reinforce the notion that LXR and FXR together control cholesterol homeostasis in the liver[9]. Other genes, such as sterol 27-hydroxylase (CYP27) and sterol-12α hydroxylase (CYP12), may be target genes regulated in an opposite manner by LXR and FXR[6].

The mouse IBABP promoter contains a unique inverted repeat of the

AGGTCA core sequence with a 1 bp spacer (IR1) located 142 upstream of the transcriptional start site. This element mediates the positive regulation by FXR in presence of chenodeoxycholic acid. The same types of result were found on the human IBABP gene[11].

A recent study has suggested three other target genes for FXR by database search, using a consensus IR1 element determined by a detailed analysis of the preferred target sequence of the FXR–RXR heterodimer[13]. The three genes, phospholipid transfer protein (PLTP), phenylethanolamine N-methyltransferase (PNMT) and carnitine palmitoyltransferase-II (CPT-II) are all expressed in liver, kidney, adrenal and/or gut tissues in which FXR expression has been observed[13]. The IR1 element found in their promoters fits with the consensus IR1 determined by functional studies.

Knockout

Mice lacking the FXR developed normally and are fertile[16]. In contrast to normal mouse, these animals display elevated bile acid levels in the serum as well as increased cholesterol and triglycerides in the serum and in the liver. The lipoprotein profile of these mice in the serum is reminiscent of a proatherogenic situation. In addition to these features, the FXR$^{-/-}$ mice display decreased fecal bile acid excretion owing to decreased expression of hepatic bile acid transport protein. In these animals there is no repression of CYP7A1 expression by bile acids nor induction of IBABP as in wild-type animals. Thus, when fed with a large amount of cholic acid, the mutant mice display an impaired ability to deal with excess dietary bile acids. Under these conditions, the FXR$^{-/-}$ animals lose their adipose tissue and body weight, and later develop hepatoxicity owing to the accumulation of bile salts. The study of the FXR knockout mice clearly confirms the important role of FXR in bile acid and lipid homeostasis (see reference 15 for a review).

Associated disease

No disease has been associated with FXR yet.

Gene structure, promoter and isoforms

The mouse FXR gene exhibits at least two isoforms that differ in the N-terminal A/B region and in the hinge domain[1]. These isoforms exhibit different DNA-binding abilities.

The gene structure of FXR is not known.

Chromosomal location

The mouse FXR gene has been mapped to chromosome 10[4]. Coincidentally, three genes encoding farnesyl pyro-phosphate synthase (Fpsl: FPP synthase-related loci) have been mapped on the same mouse chromosome near the FXR gene[3, 4].

Amino acid sequence for rat FXR (NR1H4)

Accession number: U18374

```
  1 MNLIGPSHLQATDEFALSENLFGVLTEHAAGPLGQNLDLESYSPYNNVQFPQVQPQISSSSYYSNLGFYP
 71 QQPEDWYSPGLYELRRMPTESVYQGETEVSEMPVTKKPRMAASSAGRIKGDELCVVCGDRASGYHYNALT
141 CEGCKGFFRRSITKNAVYKCKNGGNCVMDMYMRRKCQDCRLRKCREMGMLAECLLTEIQCKSKRLRKNVK
211 QHADQTVNEDSEGRDLRQVTSTTKLCREKTELTVDQQTLLDYIMDSYSKQRMPQEITNKILKEEFSAEEN
281 FLILTEMATSHVQILVEFTKRLPGFQTLDHEDQIALLKGSAVEAMFLRSAEIFNKKLPAGHADLLEERIR
351 KSGISDEYITPMFSFYKSVGELKMTQEEYALLTAIVILSPDRQYIKDREAVEKLQEPLLDVLQKLCKIYQ
421 PENPQHFACLLGRLTELRTFNHHHAEMLMSWRVNDHKFTPLLCEIWDVQ 469
```

References

1 Seol, W. et al. (1995) Mol. Endocrinol. 9, 72–85.
2 Forman, B.M. et al. (1995) Cell 81, 687–693.
3 Weinberger, C. (1996) Trends. Endocrinol. Metab. 7, 1–6.
4 Kozak, C.A. et al. (1996) Mamm. Genom. 7, 164–165.
5 Hanley, K. et al. (1997) J. Clin. Invest. 100, 705–712.
6 Wang, H. et al. (1999) Mol. Cell 3, 543–553.
7 Makishima, M. et al. (1999) Science 284, 1362–1365.
8 Parks, D.J. et al. (1999) Science 284, 1365–1368.
9 Russell, D.W. (1999) Cell 97, 539–542.
10 Gustafsson, J-A. (1999) Science 284, 1285–1286.
11 Grober, J. et al. (1999) J. Biol. Chem. 274, 29749–29754.
12 Chiang, J.Y.L. et al. (2000) J. Biol. Chem. 275, 10918–10924.
13 Laffitte, B.A. et al. (2000) J. Biol. Chem. 275, 10638–10647.
14 Zavacki, A.M. et al. (1997) Proc. Natl Acad. Sci. USA 94, 7909–7914.
15 Chawla, A. et al. (2000) Cell 103, 1–4.
16 Sinal, C.J. et al. (2000) Cell 102, 731–744.
17 Lu, T.T. et al. (2000) Mol. Cell 6, 507–515.
18 Goodwin, B. et al. (2000) Mol. Cell 6, 517–526.

Names

The vitamin D receptor (VDR, NR1I1) is a steroid receptor related to the PXR (NR1I2) and CARs (NR1I3 and NR1I4) that were first described as orphan receptors. It binds 1,25-dihydroxyvitamin D_3 (1,25$(OH)_2$-D_3, also known as calcitriol), a hormone involved in calcium homeostasis[1]. VDR is known from a relatively narrow range of species and only in vertebrates. Molecular phylogeny suggests that DHR96 (NR1J1) is its orthologue in *Drosophila* and that PXR and CARs may be considered as paralogous genes[2]. Interestingly, together with EcR (NR1H1), LXR (NR1H2 and 3) and FXR (NR1H4), and with only the exception of DHR96 (NR1J1), the ligand of which remains unknown, all these receptors bind steroids, suggesting that they form a second steroid receptor group distinct from the classic receptors (ER, GR, MR, AR, PR, NR3A and 3C groups). The human vitamin D receptor is a protein 427 amino acid long with a very short A/B region of only 21 amino acids.

Species	Other names	Accession number	References
Official name: NR1I1			
Human	VDR	J03258	4
Mouse	VDR	D31969	6
Rat	VDR	J04147	3
Chicken	VDR	AF011356	1, 5
Quail	VDR	U12641	5
Xenopus	VDR	U91846	149
Zebrafish	VDR	AF164512	Unpub.
Japanese flounder	VDRa	AB037674	150
Japanese flounder	VDRb	AB037673	150

Isolation

The first cDNA encoding a vitamin D receptor was isolated using a monoclonal antibody raised against the purified protein and a chicken intestinal cDNA expression library[1]. The resulting clone was a partial cDNA encompassing strong sequence identity with the v-*erbA* oncogene as well as chicken ER and PR, showing that VDR was a new member of the nuclear receptor superfamilly. It was immediately clear that this cDNA is only distantly related to the classic steroid receptors, such as ER, PR or GR. This cDNA was able to hybridize to a rat transcript that substantially increases after exposure of cells to 1,25$(OH)_2$-D_3 in accordance with the known behavior of VDR[1]. A complete quail cDNA was cloned later on by another group using the rat cDNA as a probe[5]. A complete rat cDNA was also cloned using an expression library derived from rat kidney and a set of three monoclonal antibodies[3]. The product derived from this cDNA shows high affinity binding to 1,25$(OH)_2$-D_3 but not to estradiol, progesterone or other steroids. The human cDNA was isolated using the chicken cDNA that was first isolated[4].

In the DBD, VDR exhibits 59% amino-acid identity with *Drosophila* DHR96, 56% with human PXR and CAR, and 44% with LXRs, FXR and EcR. In the LBD, it is more closely related to PXR (41% amino acid

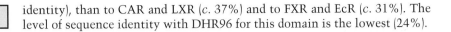

identity), than to CAR and LXR (c. 37%) and to FXR and EcR (c. 31%). The level of sequence identity with DHR96 for this domain is the lowest (24%).

DNA binding

Soon after its isolation, the VDR gene product was shown to bind DNA through the C domain, which contains zinc fingers[7]. Analysis of 1,25(OH)$_2$-D$_3$ target genes, such as the human osteocalcin gene, shows VDREs containing three half-sites oriented as inverted and direct repeats[36]. In the case of the mouse osteopontin gene, two direct repeat elements were found[37]. This was confirmed by the establishment of the 3-4-5 rule that revealed that VDR preferentially bound to direct repeat elements spaced by three nucleotides[34,35]. Detailed study of the interaction of VDR DBD to DNA clearly shows its marked preference for direct repeats when compared to palindromic elements, in contrast to other steroid receptors, such as ER or PR[8,12]. In solution, the VDR DBD is able to form strong homodimers with a dyad symmetry, whereas natural VDREs are direct repeat elements[14]. As for TR or RAR, VDR can bind to DNA in the absence of a ligand[11]. The addition of the hormone 1,25(OH)$_2$-D$_3$ results in an increased affinity of the receptor for DNA.

As for TR, it was realized relatively early on that VDR needs a partner for efficient DNA binding[9,10,59,86]. This partner has been identified as RXR, which is very important in helping to discriminate the correct spaced direct repeat (here DR3) for efficient DNA binding[38–42]. Nevertheless, several studies consistently report that VDR can bind to DNA as homodimers. This has been shown to occur on the DR3 element found in the osteopontin gene[8]. This has been directly verified by *in vitro* selection of binding sites using EMSA and PCR[18]. These experiments reveal that VDR homodimers can recognize DR3 elements of the GGTTCA core motif (such as that found in the osteopontin gene) but not those containing the classic AGGTCA motif, whereas the RXR–VDR heterodimers can bind to both types[18,19]. Other studies have suggested that VDR may bind as homodimers to direct repeats spaced by six nucleotides, to palindromes without spacing or to inverted palindrome with 12 nucleotide spacing[15,20], (see reference 16 for comments). However, these results are controversial (see reference 65 for a review).

An interesting cross-talk between VDR and AP-1 signaling has been observed on the osteocalcin gene that is widely used as a model of 1,25(OH)$_2$-D$_3$ regulation[26,43]. It has been shown that the VDRE of the osteocalcin gene may be recognized by the *fos–jun* heterodimer. Another VDRE element containing an AP-1 binding site has been characterized in the alkaline phosphatase gene promoter[26,43]. Occupancy of the site by *fos–jun* suppress both basal level and 1,25(OH)$_2$-D$_3$-enhanced osteocalcin gene transcription suggesting that both signaling pathways play antagonistic biological functions (see reference 44 for a review).

In the absence of hormone, VDR fused at either the N- or C-terminal side with the GFP is predominantly nuclear but with a significant cytoplasmic presence[126]. The addition of 1,25(OH)$_2$-D$_3$ promoted the nuclear import of cytoplasmic VDR in a few hours. A short segment of 20 amino acids in the hinge region enabled cytoplasmic GFP-tagged protein to translocate to the

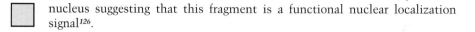

nucleus suggesting that this fragment is a functional nuclear localization signal[126].

Partners

RXRs are the major partners for vitamin D receptor since by heterodimerizing with VDR they increase their DNA-binding affinity and select the correct spacing of direct repeat elements[9,10,38–42]. Mutagenesis experiments of the VDR DBD has revealed some of the region implicated in VDR dimer binding to DNA[14]. The T-box of VDR, just C-terminal to the second zinc finger has been shown to be important for heterodimeric DNA binding[21]. The dimerization interface inside the LBD has also been mapped[13,22,76]. The region at the end of helix 3 has been shown to be vital for heterodimerization with RXR as well as for transcriptional activation[13,22,76,138,148]. In addition the region of helix 10 also known as the ninth heptad repeat has been implicated in RXR heterodimerization via mutagenesis experiments[23].

Other nuclear receptors, such as TRs or RARs have been proposed to form heterodimers with VDR but these findings were not confirmed by other groups, and it is now believed that, as the TR–RAR heterodimer, these complexes play very little role if any *in vivo*[17,24,25,42,147].

A direct physical interaction was found between c-*jun* and VDR in the context of the negative regulation by liganded VDR of the GM-CSF gene[107]. The promoter of this gene contains a VDRE adjacent to an AP-1 site and it has been shown that VDR stabilizes the binding of a *fos–jun* heterodimer to the AP-1 site through a direct interaction with c-*jun*. This interaction apparently led to an inhibitory effect on c-*jun* transactivation function. Thus, VDR has multiple mechanisms to cross-talk with AP-1 signaling[26,43,44,107].

VDR interacts with several classes of protein, which mediate its transcriptional activity (see reference 156 for a review). VDR specifically binds to several members of the basic transcriptional apparatus, such as TFIIB or the TBP-associated factor, $TAF_{II}55$[27,30,146]. The distal half of the VDR LBD is not required for interaction with TFIIB, whereas the interaction with $TAF_{II}55$ requires the region encompassing helices H3–H5. The interaction between $TAF_{II}55$ and VDR is ligand-dependent[30], in contrast to the VDR–TFIIB, which appears to be independent of the presence of the ligand. In the case of the VDR–TFIIB interaction, a synergistic effect of both factors on transcription was observed when experiments were carried out in P19 embryonal carcinoma cells, whereas in NIH3T3 TFIIB, expression with VDR led to specific dose-dependent repression of transcription[27]. This suggests that this interaction may be functionally modulated by a cell-specific factor, the identity of which remains unknown. Another TAF, $TAF_{II}28$, also interacts in a ligand-reversible manner with VDR, as well as with TR[136].

Several proteins were shown to play a role as coactivator in VDR-mediated transcription. This is the case for the members of the p160 family, SRC-1 and TIF2, which interact with VDR in a ligand-dependent fashion through the C-terminal AF-2 AD region in helix H12[28,29,33,165]. An interaction also occurs between VDR and CBP[29,138]. Mutations in either helix 3 (K246) or helix 12 (L417 and E420) disrupt the interaction between VDR and coactivators, such as SRC-1, AIB-1 and CBP[138,148]. This interaction results in the enhancement of VDR-mediated transactivation[29,138]. Interestingly, SRC-1 interacts with Smad3 a downstream member of the TGFβ signaling pathway[32]. Smad3 itself can act as a VDR coactivator, establishing a molecular basis for cross-talk between TGFβ and vitamin D signaling pathways. Of note, the Smad7 protein, which is known to play an inhibitory role on TGFβ signaling, inhibits the Smad3–VDR interaction[140].

The interaction between VDR and p160 coactivators may be differentially affected by the use of various vitamin D analogues. For example, OCT (22-oxa-1a,25-dihydroxyvitamin D_3) induced interaction of VDR with TIF2, but not SRC-1 or AIB-1, the third member of the p160 family[33]. This is interesting because this compound, although a vitamin D analogue, does no have the same biological activity as $1,25(OH)_2$-D_3 (see below). VDR binds to a ligand- and AF-2-dependent manner to the DRIP complex, which mediates ligand-dependent enhancement of transcription by VDR and TR in cell-free transcription assays[117]. The DRIP complex is similar to the TRAP or Mediator complexes and, among others, contains the TRAP220 factor, also called DRIP205[118,119] which directly interact with VDR and other nuclear receptors (see reference 120 for a review). A member of this complex TRAP100 does not bind to nuclear receptors but, by its association to TRAP220, is able to coactivate TR- and VDR-mediated transactivation[31]. The TRAP220–VDR interaction is dependent on two LXXLL motifs that are present on TRAP220[137]. VDR exhibits a strong AF-2-dependent preference for the second motif, in contrast to other receptors such as RXR, suggesting that a single TRAP220 molecule binds to the RXR–VDR heterodimer. VDR is also able to bind to another coactivator called NCoA-62 which is related to BX42, a factor implicated in ecdysone-stimulated gene expression in *Drosophila*[135]. NCoA-62 strongly coactivates VDR-mediated transactivation. While it also interacts with other receptors, such as RAR or RXR, it displays a modest effect on ligand-dependent transactivation controlled by these receptors[135].

The interaction between apo-VDR and corepressors, such as NCoR and SMRT, was shown to be weaker than the interaction with apo-TR or apo-RAR, in accordance with the relatively weak transcriptional repression exerted by apo-VDR[127]. Other authors report a VDRE-dependent repression of the 25-hydroxyvitamin D_3-24-hydroxylase (CYP24) promoter by apo-VDR[128]. These authors found a strong interaction of VDR with a composite domain of NCoR containing the two interaction domains, ID-I and ID-II[128].

Ligands

The natural ligand of VDR is 1,25-dihydroxyvitamin D_3, a hormone involved in calcium homeostasis[54,124]. Expression of the cloned VDR cDNA into COS cells led to the production of a receptor, which binds $1,25(OH)_2$-D_3 with high affinity (K_d 5.2×10^{-11} M)[4]. This value is similar to the one found by biochemical purification of the receptor from a number of different sources. Several authors have purified the receptor or the isolated ligand-binding domain following overexpression in either bacterial cells or using the baculovirus system[45–47]. This has led to a highly purified protein preparation that binds $1,25(OH)_2$-D_3 with an affinity similar to the native source.

The binding of the ligand has important effects on the vitamin D receptor. It has been shown that ligand binding increases the stability of the protein by increasing its half-life from 2 to 6 hours[48]. This may explain at least in part the induction of VDR by $1,25(OH)_2$-D_3, which is not observed at the mRNA level. In addition, the binding of $1,25(OH)_2$-D_3 modulates the DNA binding of VDR[49]. In solution, apo-VDR exists as a monomer and can bind DNA as a homodimer to the osteopontin gene VDRE. The binding of the ligand reduces the amount of homodimer bound to DNA and favors the formation of the VDR–RXR heterodimer[49]. Of note, the ligand for RXR, 9-*cis* RA, decreases VDR-RXR binding to the VDRE and increases RXR homodimer formation[49] (see reference 144 for a discussion on the effects of 9-*cis* RA on RXR DNA binding). A complex cross-talk exists between VDR and RXR ligands that differentially affects the ability of their cognate receptors to bind to the osteopontin VDRE[50].

In addition to its role in maintaining calcium homeostasis in the body, it has been shown that $1,25(OH)_2$-D_3 has a much wider scope of activity, which includes inhibition of cellular proliferation, induction of differentiation and immunosuppression (see references 54–56, 60 and 65 for reviews). Thus $1,25(OH)_2$-D_3 is now used for many therapeutic applications. Nevertheless, the use of $1,25(OH)_2$-D_3 has been hampered by its dose-dependent induction of hypercalcemia. This has prompted numerous groups to search for $1,25(OH)_2$-D_3 analogues that may have antiproliferative, prodifferentiation and immunosuppressive activities while being less calcemic than $1,25(OH)_2$-D_3[52,125]. Examples of such analogues are MC-903 (1α,24S-dihydroxy-22-ene-25,26,27-cyclopropyl-vitamin D_3), various 16-ene molecules or 22-oxa-$1,25(OH)_2D_3$, which exhibits significantly higher activity than $1,25(OH)_2$-D_3 to induce differentiation but has little of its calcemic activity in animal studies (see references 51, 54, 57, 60 and references therein). Other compounds such as 20-epi analogues are considered as promising, since their antiproliferative potency is several levels of magnitude higher than $1,25(OH)_2$-D_3 but their calcemic activity is comparable with $1,25(OH)_2$-D_3. These compounds are potent inducers of DRIP coactivator complex binding to liganded-VDR[169]. The mechanism of action of some of these analogues has been addressed and a correlation was observed between their ability to allow VDR–RXR heterodimerization and their level of transcriptional activation[51,52]. The study of other compounds has shown that the A ring of $1,25(OH)_2$-D_3 is critical for its calcemic activity as well as for its high-affinity binding to the receptor[53,54,124]. Analogues modified in the A ring have a modified ability to interact with the AF-2 helix, and an inverse relationship between their potency and their interaction with AF-2 residues was noticed[53]. A more recent paper has noted that OCT induced interaction of VDR with certain coactivators of the p160 family, such as TIF2, but not with others, such as SRC-1 or AIB-1[33]. This indeed suggests that OCT acts via the conformation of helix 12[33]. Recently, non-steroidal analogues of vitamin D were described that also exhibit less calcemic activity than $1,25(OH)_2$-D_3[61,62]. In summary, analogues of $1,25(OH)_2$-D_3 have at least two mechanisms of action: via the regulation of the VDR–RXR complex formation, and via the repositioning of the AF-2 helix.

A specific antagonist of $1,25(OH)_2$-D_3 binding to VDR was recently described and characterized[170,171]. This molecule, called TEI-9647, a 1α-25 lactone analogue of $1,25(OH)_2$-D_3, binds more strongly to VDR than $1,25(OH)_2$-D_3 but is unable to promote coactivator binding. In contrast, TEI-9647 decreases the VDR–RXR and the VDR–SRC-1 interactions[171]. In accordance with these in vitro findings, TEI-9647 inhibits the activity of several VDR-responsive promoters, such as the rat 24-hydroxylase gene promoter, and do not induce the differentiation of HL-60 cells, in contrast to $1,25(OH)_2$-D_3[170].

Three-dimensional structure

No structure of the VDR DBD is yet available. In contrast, the structure of the human VDR LBD has been determined by X-ray crystallography[58]. In the N-terminal part of the LBD, the VDR exhibits a non-conserved loop (from residues 165–215 of human VDR) for which no biological function has been found. The presence of this loop may explain the difficulties that were encountered to crystallize the LBD since it is poorly structured. Thus this loop was removed from the LBD and the structure of the mutant LBD was determined after crystallization in the presence of $1,25(OH)_2$-D_3. The overall structure had a strong similarity to the structure of liganded RARγ LBD. The activation helix 12 was found in an agonist position. The ligand lies within a hydrophobic pocket with the A ring oriented toward the C-terminus of helix H5 and the 25-hydroxyl group close to helices H7 and H11. The distance separating the 25-hydroxyl and the 1-hydroxyl group in the A ring is 13 Å. Some residues found mutated in vitamin D-resistant rickets, such as Arg 271 in helix 5 or His 305 and Ile 314 (mutated toward Leu, Gln and Ser, respectively[63,64]), are in direct contact with the ligand. Others, such as Arg 391, which is mutated in Cys in rickets[64], affect the dimerization interface of the receptor. The ligand-binding cavity of VDR is larger (697 Å^3) than that of ER, PR or RARγ, and the ligand occupies only 56% of this cavity. Modelling of ligand docking was carried out with some $1,25(OH)_2$-D_3 analogues, such as MC903 or EB1089 (1α25-dihydroxy-22,24diene-24,26,27-trihomovitamin D_3), which exhibit different side-chains connected to the D ring. These ligands can be accommodated in the binding pocket relatively easily[58]. In some cases, such as 20-epi analogues, the ligand form additional contacts with the binding cavity, which may further stabilize the helix H12, in accordance with functional studies[53]. These compounds are highly efficient in promoting VDR–DRIP complex interaction[169]. Undoubtedly, this structure will allow us to design new agonists of VDR with more specific biological effects[58].

Expression

According to its three main functions[54,56,60], stimulation of intestinal calcium and phosphorus absorption, mediation of bone remodeling and conservation of minerals in the kidney, the VDR gene is expressed in chicken intestine and kidney as two transcripts of 2.6 and 3.2 kb[1]. In brain only the larger form was found. The VDR protein was detected in the shell gland of immature, laying and molting hens[66]. In human a unique 4.4 kb transcript was observed in several cell lines: T47D (breast cancer), ATCC 407 (intestine), and HL-60[4]. Expression of VDR was also observed in 3T3-L1 pre-adipocytes, consistent with the role of $1,25(OH)_2$-D_3 in preventing adipocyte differentiation[74]. Skin is a major target tissue for $1,25(OH)_2$-D_3[121]. Growth and differentiation of the predominant cell type in the skin, the keratinocyte, are regulated by $1,25(OH)_2$-D_3. In line with the strong effect of $1,25(OH)_2$-D_3, VDR is expressed and functionally active in human skin *in vivo* as well as in human keratinocytes[75,122,123]. VDR was found to be expressed in bone and has an important role in the differentiation of bone cells as well as in calcium metabolism in bone[141,143]. VDR was found in osteoblasts, osteocytes as well as in osteoclasts[157,158,166,167]. It has also been detected in dental tissue[166]. VDR was also detected in the nervous system and *in situ* hybridization experiments demonstrate that rat oligodendrocytes express it[151]. An expression of VDR was also detected in the rat olfactory epithelium as well as throughout the brain in rat embryos[153,154]. Spinal cord, dorsal root and other ganglia also exhibited positive staining for VDR. During late development the level of expression strongly decreases[154]. VDR is also present in bone marrow, thymus, small intestine as well as numerous other tissues (see reference 156 for a review).

The regulation of the abundance of VDR is an important mechanism that modulates cell responsiveness to $1,25(OH)_2$-D_3. Several hormones such as $1,25(OH)_2$-D_3 itself, glucocorticoids, estrogens, retinoic acid and parathyroid hormone (PTH) have been shown to regulate the abundance of VDR, which also appears to be linked to the rate of cell proliferation (see references 54 and 60 for reviews). In accordance with this observation, it has been shown that mitogens, such as EGF or IGF-1, increase VDR mRNA and protein levels via a cAMP signaling pathway[69,70]. In contrast, activation of protein kinase C with basic FGF results in a significant down-regulation of VDR expression[71].

VDR protein expression has been shown to be up-regulated by $1,25(OH)_2$-D_3[67,68]. In fibroblast and intestine cells, this effect was shown to be the result of the stabilizing effect of the ligand on the VDR protein and was not observed at the mRNA level[48,72,73]. The same effect was noticed in keratinocytes in which it was shown that $1,25(OH)_2$-D_3 blocks ubiquitin/proteasome-mediated degradation of VDR[75]. In contrast, in 3T3-L1 pre-adipocytes the up-regulation of VDR expression by $1,25(OH)_2$-D_3 was shown to occur at the transcript level since a 4–5-fold increase of VDR mRNA was noticed after $1,25(OH)_2$-D_3 treatment[74]. Such a positive effect of $1,25(OH)_2$-D_3 on VDR mRNA was also found by other authors[67,145]. It is possible that both effects (i.e. regulation at the transcriptional level and protein stabilization) may be important in different tissues and/or physiological regulation (see reference 65 for a review).

Activity and target genes

VDR is a transcriptional activator. It exhibits a short A/B region (21 amino acids for the human VDR but several N-terminal variants, for which no functional characterization is available, were recently described[172]) with no AF-1 function, since deletion of these residues does not alter the functional capacity of the receptor[9,59]. An activation function was mapped in the LBD and was shown to be dependent on the 12 C-terminal amino acids[29,76,148]. The region of helix 3 is also important for transcriptional activation[138,148]. Mutations of amino acids in (K246) or 12 (L417 or E420) inhibit VDR transcriptional activity as well as its interactions with coactivators[138]. Thus VDR possesses an AF-2 activity that is similar to the one of other nuclear receptors (see reference 156 for a review).

VDR may also activate transcription in the absence of ligand as in the case of the human osteocalcin gene[81]. This effect was linked to the phosphorylation of the receptor. Some evidence does indeed suggest that VDR may be phosphorylated and that $1,25(OH)_2$-D_3 stimulates the phosphorylation of the receptor[7,78,160]. In transfected osteoblastic ROS 17/2.8 cells, the receptor is phosphorylated in the N-terminal part of the LBD[80]. Consistent with this observation, human VDR was shown to be phosphorylated by casein kinase II at Ser 208, resulting in a potentiation of its transcriptional activity[79]. Interestingly, phosphorylation may be able to activate VDR in the absence of ligand[77,81]. Human VDR can be phosphorylated by protein kinase C on Ser 51, a residue adjacent to the P-box crucial for its transactivation function[159]. Using the osteocalcin gene promoter as a model system it was shown that treatment with okadaic acid, a phosphatase inhibitor or with dopamine, led to a significant induction of VDR transcriptional activity[77].

Among the numerous VDR target genes the most extensively studied is undoubtedly the osteocalcin gene, which is regulated by $1,25(OH)_2$-D_3 in osteoblasts both *in vitro* and *in vivo*[36,82,83]. A VDRE has been located 460–440 nucleotides upstream of the transcriptional start site within the osteocalcin gene promoter[36,84-86]. Study of this response element reveals that AP-1 is also able to recognize it and that AP-1 and VDR have antagonistic effects on osteocalcin expression[26,43,44]. In addition $1,25(OH)_2$-D_3 treatment induces the formation of protein–DNA complexes that do not contain VDR in the TATA and CAAT box of the osteocalcin promoter[87]. It was thus suggested that interactions occur between VDRE-, CAAT- and TATA-protein–DNA complexes. The VDRE of the rat osteocalcin gene contains three AGGTCA-like motifs that form a DR3 and a palindromic element (ctGGGTGAatgAGGACAttacTGACCG). Mutagenesis of this element has revealed that it is the DR3 motif and not the palindrome that confers $1,25(OH)_2$-D_3-responsiveness[88,97]. In accordance with this result, the sequence of the VDRE of the human gene also contains a DR3 motif[36,65,84]. The human gene promoter is regulated *in vivo* by $1,25(OH)_2$-D_3, since transgenic mice expressing this promoter linked to the CAT gene exhibit a reduced CAT activity in calvaria femur and brain when a vitamin D deficiency was experimentally created[89]. In rat, full $1,25(OH)_2$-D_3-inducibility requires sequences downstream of the VDRE consensus site[93,97]. This promoter also contains other sites for transcription factors such as CBFA1, the factor instrumental for osteoblast differentiation, C/EBP or ATF[91,92]. Interestingly, mutation of the CBFA sites abolishes 1,25(OH)$_2$-D$_3$-responsiveness of the promoter[92]. In contrast to the rat and human case, the mouse osteocalcin gene, despite containing a conserved VDRE, is not regulated by 1,25(OH)$_2$-D$_3$ either *in vitro* or in transgenic mice (see reference 90 and references therein). The basis of this difference is not known but should be related to the region outside the VDRE since this element is present in mouse osteocalcin genes[90].

Another well-known VDR target gene is the osteopontin gene which is regulated by 1,25(OH)$_2$-D$_3$ via a DR3 element in which the two core motifs exhibit the GGTTCA consensus sequence[8,12,37]. In bones, the rat calbindin-9k[94] and mouse and chicken calbindin-D$_{28k}$ genes[95,96] that encode calcium-binding proteins and are expressed in intestine and kidney are also regulated by VDR. The characterization of the calbindin-D$_{28k}$ VDRE has led to some controversy and the precise mechanisms of 1,25(OH)$_2$-D$_3$ regulation of this gene are still unclear[104]. Among other noticeable VDR target genes in osteoblasts it is interesting to cite the c-*fos* gene, which contains in its promoter a complex element containing three direct repeats of AGGTCA motifs separated by 7 and 20 bp as well as a CTF/NF-1 binding site[99]. In liver and kidney, VDR regulates the 25-hydroxyvitamin D$_3$-24-hydroxylase (CYP24), a key enzyme in the degradation of 1,25(OH)$_2$-D$_3$, suggesting that 1,25(OH)$_2$-D$_3$ can up-regulate its own degradation[98,129]. The promoter of the rat CYP24 gene contains two VDREs that are classic DR3 elements[131]. Interestingly, there is a functional cooperation between RXR–VDR bound to the downstream VDRE and members of the ETS family of transcription factors, such as Ets1 and Ets2, bound on a nearby Ets binding site[130].

A number of VDR target genes has been described in myeloid cells consistent with the role of this hormone in macrophage and osteoclast differentiation. One of the first of these target genes was the chicken carbonic anhydrase II gene, which is a VDR target gene in erythroid cells[108,109] as well as in osteoclasts that derived from monocytic cells[141,142]. In order to understand the role of 1,25(OH)$_2$-D$_3$ in monocyte/macrophage differentiation, Leonard Freedman's laboratory has applied a differential screen for direct VDR target genes induced during U937 cells differentiation. This screen led to the identification of HoxA10 and p21[Waf1/Cip1] as direct VDR target gene[106,110]. Both genes strongly decreased the proliferation of these cells and increase their differentiation[106,110]. In addition, it has been shown that 1,25(OH)$_2$-D$_3$ increases HIV-1 replication in monocytes[105]. The GM-CSF gene is a negative target of liganded VDR in myeloid cells (see above)[107]. The avian $\alpha_v\beta_3$ integrin promoter is regulated by VDR in monocytes and osteoclasts, and contains a classic VDRE[157,158].

VDR can inhibit the expression of several genes such as the IL2 gene[161], the calcitonin gene[100], the PTH gene[101,163], the parathyroid hormone-related peptide (PTHrP)[102,164] or the renal 25-hydroxyvitamin D$_3$, 1α-hydroxylase, which catalyses the metabolic activation of 25-hydroxyvitamin D$_3$ into 1,25(OH)$_2$-D$_3$[155]. The human PTH gene promoter contains a VDRE, which exhibits only one AGGTCA motif that is recognized by VDR alone or by VDR–RXR, depending on the cell lines that are used to produce nuclear extracts[101]. In contrast, the avian PTH gene, which is also negatively regulated by VDR contains a DR3 element[103]. Mutagenesis of this element reveals that two specific positions seems to account for its ability to promote VDR repression[103]. The rat PTHrP promoter contains a unique half-site as the human PTH promoter and a DR3 element as the avian PTH promoter[102]. Both elements appear to interact with VDR. Concerning the calcitonin gene it is believed that negative $1,25(OH)_2$-D_3 regulation may be due to an interference with a cAMP-induced enhancer[100]. The GM-CSF gene is also down-regulated by liganded VDR through competition with the transcription factor NFAT as well as by a direct interaction between VDR and c-Jun that results in an inhibition of the transcriptional activity of c-Jun[107]. The repression of the IL2 gene by VDR is caused by the direct inhibition of the formation of an NFATp/AP1 complex that positively regulates the IL2 gene promoter[161].

Knockout

The knockout of the VDR gene has been done at Shigeaki Kato's laboratory and gives rise to embryos that develop without obvious abnormalities until weaning, even if the expression of some VDR target genes, such as calbindin-D_{9k} or osteopontin, was reduced before that time[111]. The level of osteocalcin gene expression is not reduced in these animals. After weaning, the mutant embryos exhibit growth retardation, alopecia, hypocalcemia and infertility. Analysis of the bones reveal growth retardation with loss of bone density[111,168]. The mice also show a flat face and a short nose. These features are typical of the human syndrome of type II vitamin D-dependent rickets in which mutations of the VDR gene have been identified (see below). Nevertheless, in contrast to human patients, the knockout mice died approximately 15 weeks after birth[111]. These animals also exhibit uterine hypoplasia with impaired folliculogenesis. Interestingly, experimentally induced vitamin D deficiency in mice does not induce alopecia and uterine abnormalities. A 10-fold increase of serum levels of $1,25(OH)_2$-D_3 was observed in VDR$^{-/-}$ mice older than 4 weeks, consistent with the observed up-regulation of 24-hydroxylase by VDR[98]. No abnormalities in the immunological cell population was noticed in the VDR$^{-/-}$ mice. The heterozygous mice do not show obvious abnormalities and have a normal bone density. This is striking because in humans, polymorphisms in the VDR gene are related to differences in bone mineral density (see below). Together these results establish the critical role of VDR in growth, bone formation and female reproduction in the post-weaning stage[111]. The skin abnormalities found in the VDR$^{-/-}$ mice are strikingly similar to those found in mice bearing the specific RXRα mutation in keratinocytes obtained using the Cre-ER^T recombinase method[139]. This suggests that the *in vivo* partner of VDR in skin is RXRα and that these two receptors together control epidermal keratinocyte proliferation and differentiation.

Associated disease

Mutations of the VDR gene were found in human patients that exhibit hypocalcemic rickets[63,64,112,113]. This rare autosomal recessive disorder results from target organ resistance to the action of 1,25(OH)$_2$-D$_3$ (see reference 156 for an extensive review). In the first case in which a mutation was found, the receptor was shown to display normal binding to 1,25(OH)$_2$-D$_3$ but reduced DNA interaction. Consistent with this observation in two families, mutations were found in the DBD, respectively, in the tip of the first and second zinc fingers[112]. This results in a receptor that has a markedly decreased interaction with DNA. Following these first cases, other mutations were found, including the mutation of a Tyr codon to a stop codon in exon 7, resulting in a truncated protein with a large part of the LBD missing. The resulting receptor does not bind the ligand and does not transactivate a reporter gene construct[113]. Point mutations in the LBD giving rise to receptors with reduced affinity for the ligand were also described[63,64]. The link between VDR and hypocalcemic rickets was largely confirmed by the analysis of the phenotype of VDR$^{-/-}$ mice[111].

Interestingly, a relationship was found in humans between bone density and a polymorphism in the VDR gene[114,115]. First, it was demonstrated that it is possible to predict circulating osteocalcin levels knowing the polymorphism present for VDR in a given patient[115]. Then the relationship was extended to bone density[114]. It was found that the genotype associated with lower bone density was overrepresented in post-menopausal women with bone densities lower than controls. It is nevertheless clear that interaction with environmental factors, particularly dietary calcium intake and other genetic factors, have to be considered (see references 56 and 156 for a review).

A genetic association between VDR and hyperparathyroidism exists[132]. Liganded VDR is able to repress the expression of the PTH gene[101,103]. An excess of the b allele of VDR, which was associated with impaired VDR function, was found in patients with hyperparathyroidism suggesting that impaired VDR function may induce an excess of PTH production in these patients[132].

Gene structure, promoter and isoforms

The human VDR gene is composed of 11 exons and spans approximately 75 kb of genomic DNA[133]. The DBD is encoded by two exons with the intron lying between the two zinc fingers, 4 amino acids downstream from the last Cys of the first finger. Three unique isoforms are produced by the differential splicing of the two non-coding exons called 1B and 1C (see reference 156 for a review). The coding part of the receptor is contained in exons 2–9. The human promoter is located in a GC-rich region upstream of exon 1A. In the small intestine, it is directly regulated by the caudal-related homeodomain protein Cdx-2, which binds to an A/T-rich element within the promoter[162]. The mouse promoter as its human counterpart exhibits numerous Sp1 sites but no TATA box[173]. In the human, the structure of the 5′ part of the gene is more complex than anticipated because it contains several alternative exons that can generate three N-terminal variants of VDR[172]. In addition, a tissue-specific distal promoter was found in humans. This promoter controls the formation of unique transcripts in tissues involved in calcium regulation[172].

In addition to the human N-terminal variants[172], an isoform of VDR, rVDR1, generated by retention of intron 8 was found in rat[134]. The corresponding transcript is expressed at a lower level than the regular transcript. This results in a protein that lacks the C-terminal 86 amino acids of the bona fide receptor, which are replaced by 19 unrelated amino acids. The rVDR1 isoform does not bind 1,25(OH)$_2$-D$_3$ and does not heterodimerize with RXR, but can form heterodimers with the regular VDR on consensus VDRE and can act as a dominant negative inhibitor of VDR-mediated transactivation[134]. It is striking that the allelism in the human VDR gene reported to be linked to bone density was also found in the intron 8, suggesting that it may affect rVDR1[114,115].

Chromosomal location

The human VDR gene is located on chromosome 12cen–q12 and the rat gene on chromosome 7[116,152].

Amino acid sequence for human VDR (NR1I1)

Accession number: J03258

```
  1  MEAMAASTSLPDPGDFDRNVPRICGVCGDRATGFHFNAMTCEGCKGFFRRSMKRKALFTCPFNGDCRITK
 71  DNRRHCQACRLKRCVDIGMMKEFILTDEEVQRKREMILKRKEEEALKDSLRPKLSEEQQRIIAILLDAHH
141  KTYDPTYSDFCQFRPPVRVNDGGGSHPSRPNSRHTPSFSGDSSSSCSDHCITSSDMMDSSSFSNLDLSEE
211  DSDDPSVTLELSQLSMLPHLADLVSYSIQKVIGFAKMIPGFRDLTSEDQIVLLKSSAIEVIMLRSNESFT
281  MDDMSWTCGNQDYKYRVSDVTKAGHSLELIEPLIKFQVGLKKLNLHEEEHVLLMAICIVSPDRPGVQDAA
351  LIEAIQDRLSNTLQTYIRCRHPPPGSHLLYAKMIQKLADLRSLNEEHSKQYRCLSFQPECSMKLTPLVLE
421  VFGNEIS    427
```

References

[1] McDonnell, D.P. et al. (1987) Science 235, 1214–1217.

[2] Escriva, H. et al. (2000) BioEssays 22, 717–727.

[3] Burmester, J.K. et al. (1988) Proc. Natl Acad. Sci. USA 85, 1005–1009.

[4] Baker, A.R. et al. (1988) Proc. Natl Acad. Sci. USA 85, 3294–3298.

[5] Elaroussi, M.A. et al. (1994) Proc. Natl Acad. Sci. USA 91,11596–11600.

[6] Kamei, Y. et al. (1995) Gene 152, 281–282.

[7] Mc Donnell, D. et al. (1989) Mol. Endocrinol. 3, 635–644.

[8] Freedman, L.P. and Towers, T.L. (1991) Mol. Endocrinol. 5, 1815–1826.

[9] Sone, T. et al. (1991) Mol. Endocrinol. 5, 1578–1586.

[10] Ross, T.K. et al. (1992) Proc. Natl Acad. Sci. USA 89, 256–260.

[11] Ross, T.K. et al. (1993) Proc. Natl Acad. Sci. USA 90, 9257–9260.

[12] Nishikawa, J.I. et al. (1993) J. Biol. Chem. 268, 19739–19743.

[13] Rosen, E.D. et al. (1993) J. Biol. Chem. 268, 11534–11541.

[14] Towers, T.L. et al. (1993) Proc. Natl Acad. Sci. USA 90, 6310–6314.

[15] Carlsberg, C. et al. (1993) Nature 361, 657–660.

[16] Green, S. (1993) Nature 361, 590–591.
[17] Schräder, M. et al. (1994) Nature 370, 382–386.
[18] Freedman, L.P. et al. (1994) Mol. Endocrinol. 8, 265–273.
[19] Nishikawa, J.I. et al. (1994) Nucl. Acids Res. 22, 2902–2907.
[20] Schräder, M. et al. (1995) Mol. Cell. Biol. 15, 1154–1161.
[21] Hsieh, J.C. et al. (1995) Biochem. Biophys. Res. Commun. 215, 1–7.
[22] Whitfield, G.K. et al. (1995) Mol. Endocrinol. 9, 1166–1179.
[23] Nakajima, S. et al. (1994) Mol. Endocrinol. 8, 159–172.
[24] Schräder, M. et al. (1994) J. Biol. Chem. 269, 5501–5504.
[25] Schräder, M. et al. (1993) J. Biol. Chem. 268, 17830–17836.
[26] Owen, T.A. et al. (1990) Proc. Natl Acad. Sci. USA 87, 9990–9994.
[27] Blanco, J.C.G. et al. (1995) Proc. Natl Acad. Sci. USA 92, 1535–1539.
[28] Gill, R.K. et al. (1998) Mol. Endocrinol. 12, 57–65.
[29] Castillo, A.I. et al. (1999) Mol. Endocrinol. 13, 1141–1154.
[30] Lavigne, A.C. et al. (1999) Mol. Cell. Biol. 19, 5486–5494.
[31] Zhang, J. and Fondell, J.D. (1999) Mol. Endocrinol. 13, 1130–1140.
[32] Yanagisawa, J. et al. (1999) Science 283, 1317–1321.
[33] Takeyama, K.I. et al. (1999) Mol. Cell. Biol. 19, 1049–1055.
[34] Umesono, K. et al. (1991) Cell 65, 1255–1266.
[35] Näär, A.M. et al. (1991) Cell 65, 1267–1279.
[36] Kerner, S.A. et al. (1989) Proc. Natl Acad. Sci. USA 86, 4455–4459.
[37] Noda, M. et al. (1990) Proc. Natl Acad. Sci. USA 87, 9995–9999.
[38] Yu, V.C. et al. (1991) Cell 67, 1251–1266.
[39] Kliewer, S.A. et al. (1992) Nature 355, 446–449.
[40] Zhang, X.K. et al. (1992) Nature 355, 441–446.
[41] Leid, M. et al. (1992) Cell 68, 377–395.
[42] MacDonald, P.N. et al. (1993) Mol. Cell. Biol. 13, 5907–5917.
[43] Schüle, R. et al. (1990) Cell 61, 497–504.
[44] Schüle, R. and Evans, R.M. (1991) Trends Genet. 7, 377–381.
[45] Ross, T.K. et al. (1991) Proc. Natl Acad. Sci. USA 88, 6555–6559.
[46] Nakajima, S. et al. (1993) Biochem. Biophys. Res. Commun. 197, 478–485.
[47] Craig, T.A. and Kumar, R. (1996) Biochem. Biophys. Res. Commun. 218, 902–907.
[48] Arbour, N.C. et al. (1993) Mol. Endocrinol. 7, 1307–1312.
[49] Cheskis, B. and Freedman, L.P. (1994) Mol. Cell. Biol. 14, 3329–3338.
[50] Lemon, B.D. and Freedman, L.P. (1996) Mol. Cell. Biol. 16, 1006–1016.
[51] Cheskis, B. et al. (1995) Mol. Endocrinol. 9, 1814–1824.
[52] Zhao, X.Y. et al. (1997) Mol. Endocrinol. 11, 366–378.
[53] Peleg, S. et al. (1998) Mol. Endocrinol. 12, 525–535.
[54] Bouillon, R. et al. (1995) Endocrine Rev. 16, 200–257.
[55] Lemire, J.M. (1992) J. Cell. Biochem. 49, 26–31.
[56] Ferrari, S. et al. (1998) Trends Endocrinol. Metab. 9, 259–265.
[57] Abe, J. et al. (1989) Endocrinology 124, 2645–2647.
[58] Rochel, N. et al. (2000) Mol. Cell 5, 173–179.
[59] Sone, T. et al. (1991) J. Biol. Chem. 266, 23296–23305.
[60] De Luca, H. and Zierold, C. (1998) Nutr. Rev. 56, 54–74.
[61] Boehm, M.F. et al. (1999) Chem. Biol. 6, 265–275.
[62] Verstuyf, A. et al. (1998) J. Bone Min. Res. 13, 549–558.

63 Kristjansson, K. et al. (1993) J. Clin. Invest. 92, 12–16.
64 Whitfield, G.K. et al. (1996) Mol. Endocrinol. 10, 1617–1631.
65 Christakos, S. et al. (1996) Biochem. J. 316, 361–371.
66 Yoshimura, Y. et al. (1997) Gen. Comp. Endocrinol. 108, 282–289.
67 Mangelsdorf, D.J. (1987) Proc. Natl Acad. Sci. USA 84, 354–358.
68 Strom, M. et al. (1989) Proc. Natl Acad. Sci. USA 86, 9770–9773.
69 Krishnan, A.V. and Feldman, D. (1991) J. Bone Min. Res. 6, 1099–1107.
70 Krishnan, A.V. and Feldman, D. (1992) Mol. Endocrinol. 6, 198–206.
71 Krishnan, A.V. and Feldman, D. (1991) Mol. Endocrinol. 5, 605–612.
72 Wiese, R.J. et al. (1992) J. Biol. Chem. 267, 20082–20086.
73 Santiso–Mere, D. et al. (1993) Mol. Endocrinol. 7, 833–839.
74 Kamei, Y. et al. (1993) Biochem. Biophys. Res. Commun. 193, 948–955.
75 Li, X.Y. et al. (1999) Mol. Endocrinol. 13, 1686–1694.
76 Jin, C.H. et al. (1996) Mol. Endocrinol. 10, 945–957.
77 Matkovits, T. and Christakos, S. (1995) Mol. Endocrinol. 9, 232–242.
78 Brown, T.A. and De Luca, H.F. (1990) J. Biol. Chem. 265, 10025–10029.
79 Jurutka, P.W. et al. (1996) Proc. Natl Acad. Sci. USA 93, 3519–3524.
80 Jones, B.B. et al. (1991) Mol. Endocrinol. 5, 1137–1146.
81 Darwish, H.M. et al. (1993) Biochim. Biophys. Acta 1167, 29–36.
82 Lian, J.B. et al. (1989) Proc. Natl Acad. Sci. USA 86, 1143–1147.
83 Demay, M.B. et al. (1990) Proc. Natl Acad. Sci. USA 87, 369–373.
84 Morrison, N.A. et al. (1989) Science 246, 1158–1161.
85 Markose, E.R. et al. (1990) Proc. Natl Acad. Sci. USA 87, 1701–1705.
86 Liao, J. et al. (1990) Proc. Natl Acad. Sci. USA 87, 9751–9755.
87 Bortell, R. et al. (1992) Proc. Natl Acad. Sci. USA 89, 6119–6123.
88 Demay, M.B. et al. (1992) Mol. Endocrinol. 6, 557–562.
89 Kesterson, R.A. et al. (1993) Mol. Endocrinol. 7, 462–467.
90 Sims, N. et al. (1997) Mol. Endocrinol. 11, 1695–1708.
91 Bidwell, J.P. et al. (1993) Proc. Natl Acad. Sci. USA 90, 3162–3166.
92 Javed, A. et al. (1999) Mol. Cell. Biol. 19, 7491–7500.
93 Sneddon, W.B. et al. (1997) Mol. Endocrinol. 11, 210–217.
94 Darwish, H.M. and De Luca, H.F. (1992) Proc. Natl Acad. Sci. USA 89, 603–607.
95 Cancela, L. et al. (1992) Mol. Endocrinol. 6, 468–475.
96 Gill, R.K. and Christakos, S. (1993) Proc. Natl Acad. Sci. USA 90, 2984–2988.
97 Terpening, C.M. et al. (1991) Mol. Endocrinol. 5, 373–385.
98 Zierold, C. et al. (1994) Proc. Natl Acad. Sci. USA 91, 900–902.
99 Candeliere, G.A. et al. (1996) Mol. Cell. Biol. 16, 584–592.
100 Peleg, S. et al. (1993) Mol. Endocrinol. 7, 999–1008.
101 Mackey, S.L. et al. (1996) Mol. Endocrinol. 10, 298–305.
102 Falzon, M. (1996) Mol. Endocrinol. 10, 672–681.
103 Koszewski, N.J. et al. (1999) Mol. Endocrinol. 13, 455–465.
104 Ferrari, S. et al. (1994) Mol. Endocrinol. 8, 173–181.
105 Skolnik, P.R. et al. (1991) Proc. Natl Acad. Sci. USA 88, 6632–6636.
106 Rots, N.Y. et al. (1998) Mol. Cell. Biol. 18, 1911–1918.
107 Towers, T.L. et al. (1999) Mol. Cell. Biol. 19, 4191–4199.
108 Billecocq, A. et al. (1990) Proc. Natl Acad. Sci. USA 87, 6470–6474.
109 Lomri, A. and Baron, R. (1992) Proc. Natl Acad. Sci. USA 89, 4688–4692.

[110] Liu, M. et al. (1996) Genes Dev.10, 142–153.

[111] Yoshizawa, T. et al. (1997) Nature Genet. 16, 391–396.

[112] Hughes, M.R. et al. (1988) Science 242, 1702–1705.

[113] Ritchie, H.H. et al. (1989) Proc. Natl Acad. Sci. USA 86, 9783–9787.

[114] Morrison, N.A. et al. (1994) Nature 367, 284–287.

[115] Morrison, N.A. et al. (1992) Proc. Natl Acad. Sci. USA 89, 6665–6669.

[116] Szpirer, J. et al. (1991) Genomics 11, 168–173.

[117] Rachez, C. et al. (1998) Genes Dev. 12, 1787–1800.

[118] Rachez, C. et al. (1999) Nature 398, 824–828.

[119] Yuan, C.X. et al. (1998) Proc. Natl Acad. Sci. USA 95, 7939–7944.

[120] Kingston, R.E. (1999) Nature 398, 199–200.

[121] Berger, U. et al. (1988) J. Clin. Invest. Metab. 67, 607–613.

[122] Li, X.Y. et al. (1997) J. Invest. Dermatol. 108, 506–512.

[123] Kang, S. et al. (1997) J. Invest. Dermatol. 108, 513–518.

[124] Norman, A.W. et al. (1971) Science 173, 51–54.

[125] Bikle, D.D. (1992) Endocrinol. Rev. 13, 765–784.

[126] Michigami, T. et al. (1999) J. Biol. Chem. 274, 33531–33538.

[127] Tagami, T. et al. (1998) Biochem. Biophys. Res. Commun. 253, 358–363.

[128] Dwiveddi, P.P. et al. (1998) J. Mol. Endocrinol. 20, 327–335.

[129] Ohyama, Y. et al. (1994) J. Biol. Chem. 269, 10545–10550.

[130] Dwiveddi, P.P. et al. (2000) J. Biol. Chem. 275, 47–55.

[131] Kerry, D.M. et al. (1996) J. Biol. Chem. 271, 29715–29721.

[132] Carling, T. et al. (1995) Nature Med. 1, 1309–1311.

[133] Miyamoto, K. et al. (1997) Mol. Endocrinol. 11, 1165–1179.

[134] Ebihara, K. et al. (1996) Mol. Cell. Biol. 16, 3393–3400.

[135] Baudino, T.A. et al. (1998) J. Biol. Chem. 273, 16434–16441.

[136] Mengus, G. et al. (2000) J. Biol. Chem. 275, 10064–10071.

[137] Ren, Y. et al. (2000) Mol. Cell. Biol. 20, 5433–5446.

[138] Jiménez–Lara, A.M. and Aranda, A. (1999) J. Biol. Chem. 274, 13503–13510.

[139] Li, M. et al. (2000) Nature 407, 633–636.

[140] Yanagi, Y. et al. (1999) J. Biol. Chem. 274, 12971–12974.

[141] Quélo, I. et al. (1998) J. Biol. Chem. 273, 10638–10646.

[142] Quélo, I. et al. (2000) Biochem. Biophys. Res. Commun. 271, 481–491.

[143] Roodman, G.D. (1996) Endocrine Rev. 17, 308–332.

[144] Chen, Z.P. et al. (1994) J. Biol. Chem. 269, 25770–25776.

[145] Mohonen, A. et al. (1991) Biochim. Biophys. Acta 1088, 111–118.

[146] MacDonald, P.N. et al. (1995) J. Biol. Chem. 270, 4748–4752.

[147] Thompson, P.D. et al. (1999) J. Cell. Biochem. 75, 462–480.

[148] Kraichely, D.M. et al. (1999) J. Biol. Chem. 274, 14352–14358.

[149] Li, Y.C. et al. (1997) Endocrinology 138, 2347–2353.

[150] Suzuki, T. et al. (2000) Biochem. Biophys. Res. Commun. 270, 40–45.

[151] Baas, D. et al. (2000) Glia 31, 59–68.

[152] Taymans, S.E. et al. (1999) J. Bone. Miner. Res. 14, 1163–1166.

[153] Glaser, S.D. et al. (1999) Cell. Mol. Neurobiol. 19, 613–624.

[154] Veenstra, T.D. et al. (1998) Brain Res. 804, 193–205.

[155] Takeyama, K.I. et al. (1997) Science 277, 1827–1830.

[156] Malloy, P.J. et al. (1999) Endocrine Rev. 20, 156–188.

[157] Cao, X. et al. (1993) J. Biol. Chem. 268, 27371–27380.

[158] Medhora, M.M. et al. (1993) J. Biol. Chem. 268, 1456–1461.
[159] Hsieh, J.C. et al. (1991) Proc. Natl Acad. Sci. USA 88, 9315–9319.
[160] Hilliard IV, G.M. et al. (1994) Biochemistry 33, 4300–4311.
[161] Alroy, I. et al. (1995) Mol. Cell. Biol. 15, 5789–5799.
[162] Yamamoto, H. et al. (1999) J. Bone Min. Res. 14, 240–247.
[163] Russell, J. et al. (1999) J. Bone Min. Res. 14, 1828–1837.
[164] Nishishita, T. et al. (1998) J. Biol. Chem. 273, 10901–10907.
[165] Chen, S. et al. (2000) J. Biol. Chem. 275, 15039–15048.
[166] Davideau, J.L. et al. (1996) Endocrinology 137, 3577–3585.
[167] Mee, A.P. et al. (1996) Bone 18, 295–299.
[168] Amling, M. et al. (1999) Endocrinology 140, 4982–4987.
[169] Yang, W. and Freedman, L.P. (1999) J. Biol. Chem. 274, 16838–16845.
[170] Miura, D. et al. (1999) J. Biol. Chem. 274, 16392–16399.
[171] Ozomo, K. et al. (1999) J. Biol. Chem. 274, 32376–32381.
[172] Crofts, L.A. et al. (1998) Proc. Natl Acad. Sci. USA 95, 10529–10534.
[173] Jehan, F. and DeLuca, H.F. (1997) Proc. Natl Acad. Sci. USA 94, 10138–10143.

Names

The nomenclature of this group is especially confusing because orthologues were given different names by different laboratories. The first described sequence in this group was cloned in *Xenopus* in 1994 and was called xONR1 for *Xenopus* orphan nuclear receptor[1]. The same *Xenopus* sequence was described later on and called BXR[2]. A related mouse gene was identified as a pregnane activated receptor and called PXR (for pregnane X receptor)[3]. Later, a closely related human gene was identified by Ron Evans's laboratory as a steroid- and xenobiotic-sensing nuclear receptor and called SXR (for steroid and xenobiotic receptor)[4], whereas it was also found by Steve Kliewer's laboratory and called PXR[5]. The human gene was also called PAR (for pregnane-activated receptor) by Anders Berkenstam's group in Sweden[6]. The relationships between the human, mouse and *Xenopus* sequences are not completely resolved and to date it has been difficult to establish clearly whether they are orthologues or paralogues. For the reasons detailed below, we considered them as orthologues.

This gene clearly belongs to the NR1I group of receptors that also contains the vitamin D receptor (NR1I1) and the CAR receptors (NR1I3 and NR1I4). In this book, these various receptors are presented separately in order to discuss their functional specificities in more detail. Their relationships may nevertheless be of more significance than anticipated.

Species	Other names	Accession number	References
Official name: NR1I2			
Human	SXR, PXR, PAR	AF061056	4–6
Mouse	PXR	AF031814	3
Rat	PXR	AF151377	7, 12
Rabbit	PXR	AF188476	12
Xenopus	xONR1, BXR	X75163	1, 2

Isolation

Xenopus NR1I2 was identified from a *Xenopus* stage 13 embryonic cDNA library screened with the coding region of the human vitamin D receptor[1]. It is effectively more closely related to VDR than to any other receptor (71% and 40% sequence identity in the DBD and LBD, respectively). Since it was unable to bind vitamin D and its ligand was not known, it was described as an orphan receptor. The same gene was also described by Ron Evans's laboratory after a screen for maternally expressed nuclear hormone receptors[2].

The mouse sequence called PXR was isolated after an *in silico* screen of public EST databases. This search yielded a clone from a mouse liver library that had homology with a variety of LBDs of nuclear hormone receptor. Screening of a cDNA library with the corresponding sequence gave rise to mouse NR1I2, which was called PXR[3].

Three research groups isolated a human gene strongly related to the mouse PXR sequence[4–6]. Interestingly, the conclusions of these groups differ on the homologous status of the human and mouse sequences. Because mouse and human sequences exhibit only 95% and 73% identity in their DBD and LBD respectively, Ron Evans's group suggested that both could be paralogous (i.e. may represent the α and β subtypes of a novel group of receptors)[4]. Their conclusion was reinforced by the observation that both receptors exhibit different pharmacological properties (see below). Nevertheless, a human sequence more related to mouse PXR was not isolated. In contrast, Steve Kliewer's laboratory suggested that both genes are true orthologues because their DBDs are strongly conserved, explaining the divergence of the LBD by a rapid evolutionary divergence. The very similar expression pattern of both genes and their activation profile by a common set of compounds supported this conclusion[5]. Finally, Bertilsson et al. isolated PAR using a computational search for ESTs based on hidden Markov models[6]. In the absence of firmer data (such as chromosomal localization, genomic organization etc.) and given the lack of any proof of the existence of two paralogues, we follow the conclusion of Steve Kliewer's laboratory[3,5] and we consider the human and mouse sequences as orthologues of the same NR1I2 gene. The rat homologue exhibits 100% and 96% sequence identity in the DBD and LBD, respectively, with the mouse PXR, suggesting that these two sequences are clearly orthologous[7,12]. The view that the human, mouse and rat genes are orthologues is reinforced by the observation that the rabbit sequence is also divergent sharing in the ligand binding domain only 76% amino acid identity with the rat PXR, 78% with the mouse and 82% with the human[12]. Since it is no more strongly related to either human or mouse, it is clear that all these sequences are orthologues evolving at high speed. This is confirmed by the strong conservation of the DBD of these receptors (92–100% amino-acid identities). The divergence of the LBD correlates with a striking difference in the types of compounds recognized by each type of receptor[12].

The same question of orthology stands for the *Xenopus* sequence that exhibits 73% and 52% identity in the DBD and LBD, respectively, with human PXR. These low values may suggest that the *Xenopus* and mammal sequences belong to two different paralogous genes. Nevertheless, given the low sequence identity existing within mammals, these low values may be also explained by a rapid evolutionary rate of this receptor across all vertebrates. The significance of such an accelerated speed may be interesting to correlate to the variety of compounds bound by the receptor and to the apparent species specificity of some of these compounds (such as PCN or corticosterone; see below).

In the DBD, PXRs display 55% amino-acid identities with CARs (NR1I3 and NR1I4), 63% with DHR96 (NR1J1), 55% with VDR (NR1I1) and a range of 51–55% with members of the NR1H group (LXR, FXR, EcR). In the LBD, the identity value with CARs is between 45 and 50%, 23% with DHR96, 41% with VDR and 24–29% with LXRs, FXR and EcR.

DNA binding

The first descriptions of PXR have suggested that this receptor binds to DNA as a heterodimer with RXR on a DR3 element, the same element as the phylogenetically related VDR[1]. This was confirmed later, on a variety of synthetic and natural DR3 elements both for the mouse and human homologues[3,5]. Nevertheless, other reports described a different picture for both the human and *Xenopus* sequences with a DR4 element being the more efficient, relatively weak binding being observed on DR3 and DR5[2,4]. The precise reasons for these discrepancies are not clear, although it has to be emphasized that the various laboratories did not use the same core sequences: the binding on DR4 was observed on the RARE of the RARβ gene encompassing AGTTCA and not AGGTCA as a core motif. It is possible that the sequence of the core elements influences the preferred spacing of the heterodimer. In any case transcriptional activations and direct DNA binding on natural elements were reported both for DR3 and DR4 elements, and even ER6 elements, suggesting that PXR may exhibit a promiscuous binding to different response elements[3-6].

Partners

RXR is the only heterodimerization partner described to date for PXR. No DNA binding of PXR is observed in the absence of RXR[1-6].

It has been shown by *in vitro* pull down assay that SRC-1 is able to recognize the liganded PXR[3,5]. The binding was observed using only a fragment of SRC-1 fused to the GST protein and no information for the *in vivo* significance of this interaction is available to date. An interaction between mouse PXR bound to phthalic acid and RIP140 was also observed[16].

Ligands

When the *Xenopus* NR1I2 gene was described, the fact that it was more closely related to the vitamin D receptor than to any others led the authors to test whether it bound 1α,25-dihydroxyvitamin D3[1]. It was clearly shown that this is not the case. Later, it was shown that the mouse homologue can be activated by a variety of pregnane derivatives, either synthetic or natural[3]. The most potent synthetic compounds were dexamethasone *t*-butylacetate, 6,16α-dimethyl pregnenolone and pregnenolone 16α-carbonitrile (PCN). Remarkably, it was found that PXR was activated both by agonists (dexamethasone) and antagonists (RU486) of the glucocorticoid receptor. The most active natural compounds were pregnenolone, 17α-hydroxypregnenolone, progesterone, 17α-hydroxyprogesterone and 5β-pregnane-3,20-dione, i.e. pregnane derivatives. All these compounds were active at the micromolar range. Importantly, androgens, estrogens or naturally occurring glucocorticoids such as cortisol and cortisosterone were inactive. This latter finding is surprising given the fact that both dexamethasone and RU486 activate PXR. The direct binding of some of these compounds to mouse PXR was demonstrated by a CARLA-type assay using a GST–PXR fusion protein and a fragment of the coactivator protein SRC-1[3,13]. The mouse PXR also binds endocrine-disrupting chemicals, such as phthalic acid and nonylphenol, that were previously believed to act exclusively through estrogen receptors[16].

The human homologue was also found to be activated by a wide variety of compounds and steroid derivatives[2,5,6,13]. Interestingly, the activation profile of the various compounds on the human receptor differs from the one of the mouse receptor. For example, whereas rifampicin (an antibiotic) is an efficient activator of human PXR, it activates only moderately its mouse homologue. PCN has exactly the opposite effect as also clotrimazole (an antimycotic) and lovistatin (an antihypercholesterolemic drug). Ron Evans's laboratory proposed that human PXR plays the role of a steroid sensor that is recognized with low affinity and broad specificity by a wide variety of compounds that should be eliminated by the cytochrome P450 genes, which are *in vivo* targets of PXR (see below). Direct binding of rifampicine, dexamethasone *t*-butylacetate and clotrimazole to human PXR was demonstrated by a CARLA-type assay identical to that performed on the mouse receptor[5,13].

From all the currently available data it is thus clear that PXR are rapidly evolving nuclear receptors binding a wide variety of compounds. Their specificity strongly change from one species to another, in accordance with the strong evolutionary divergence of the ligand-binding domain[12,13].

The *Xenopus* homologue, described as BXR, is also able to bind a ligand. Interestingly, it exhibits a radically different family of ligands[4]. The ligand of *Xenopus* NR1I2 was searched based on the assumption that it would be colocalized with its receptor during *Xenopus* development. By fractionating and purification of embryonic extracts, benzoates such as 4-amino butyl benzoate (4-ABB) or 3-amino ethyl benzoate (3-AEB) were demonstrated to be activators of *Xenopus* NR1I2. The most potent activator was 4-ABB and it directly binds to the receptor with a K_d (determined by Skatchard analysis) of 332 nM. The authors proposed that *Xenopus* NR1I2 and benzoate control a previously unsuspected vertebrate signaling pathway.

Three-dimensional structure

No 3D structure is available.

Expression

The mammalian PXR are all expressed at high levels in liver and more moderately in the intestine (colon and small intestine) as a family of transcripts from 9.0 to 2.5 kb[2,3,5,6]. Depending on the authors, low levels of expression in stomach and kidney were also observed. In human, the prominent transcript is 2.6 kb with two less abundant transcripts at 4.3 and 5 kb. *In situ* hybridization performed on human embryos at 10 weeks of development revealed that, in the intestine, the expression is limited to the cells of the mucosal layer[6]. In day 18 mouse embryo the same findings were reported with an expression in the epithelium of the intestine as well as in the liver[3]. Apparently, the expression pattern in rat is similar[7]. The human PXR is expressed in both normal and neoplastic breast tissue, and its level does not differ between the tumor and the adjacent normal tissue. However, a significant inverse relationship was found between the expression level of PXR and the estrogen receptor status[8].

The *Xenopus* homologue is expressed in the unfertilized egg, suggesting it is a maternal RNA, and its level remains relatively constant until gastrulation[2]. It persists thereafter at a much reduced level until the tadpole stage. At the tailbud stage, the expression is restricted to the hatching gland, whereas in the adult it is expressed in the brain as well as in the gonads[14].

Activity and target genes

PXR is clearly a transcriptional activator but it is apparently strictly dependent on the presence of both its partner RXR and its ligand[2-7]. Numerous target genes all belonging to the detoxifying enzymes of the cytochrome P450 family were found (reviewed in reference 9). PXR binds to a DR3 element found in the promoters of rodent CYP3A1 and CYP3A2 genes and activates transcription through these elements[3]. In human, a computer search identified CYP2A1, CYP2A2, CYP2C1, CYP2C6, CYP3A1, CYP3A2, P450 oxydoreductase and UDP-glucuronosyltransferase as candidate target genes[4]. It was shown the PXR can activate transcription on artificial reporter constructs containing the direct repeat or inverted repeat elements present on the promoters of these genes. The human CYP3A4 was also demonstrated to be a direct target gene as the human CYP3A7[5,6,10]. The rat CYP3A23 gene is also regulated by PXR through a DR3 element[11]. Recently, it was shown that PXR and the related receptors CAR both regulates CYP3A and CYP2B genes. This provides a molecular mechanism for how toxic drugs may regulate several CYP genes as well as provides an explanation for drug–drug interactions[18].

Knockout

The targeted disruption of the mouse PXR gene abolishes the induction of CYP3A gene by xenobiotics and other inducers such as dexamethasone and pregnenolone-16α-carbonitrile[17,18]. In accordance with these findings a transgenic mouse expressing an activated form of human PXR (PXR sequence fused to VP16 activation domain) under the control of the liver-specific albumin promoter causes a constitutive upregulation of CYP3A and enhanced protection against xenobiotic compounds. By introducing the human PXR into the mouse PXR$^{-/-}$ mice, Ron Evans' laboratory generate 'humanized' transgenic mice that are responsive to human-specific inducers such as the antibiotic rifampicin[17].

Associated disease

There is no known associated disease with PXR. Nevertheless, it was shown recently that hypericum extracts, which are used widely for the treatment of depression but that are known to induce CYP3A4 gene expression in the liver, contain hyperforin, which was shown to be a ligand for human PXR[15]. The wide spectrum of molecules that activates PXR clearly suggests that this receptor will have an important role to play in medicine.

Gene structure, promoter and isoforms

Two isoforms (PXR.1 and PXR.2) of the mouse PXR gene have been described[3]. These isoforms differ by an insertion of 41 amino acids in the N-terminal part of the LBD, which is present in mouse PXR.1 and absent in mouse PXR.2[3]. Interestingly, the mouse PXR.2 isoform displayed a much more restricted activation profile than the PXR.1 isoform. On the synthetic compounds, only dexamethasone *t*-butylacetate activated PXR.2 efficiently. For natural compounds only 5β-pregnane-3,20-dione was found to activate PXR.2 transcriptional activity.

In humans, two different isoforms of the gene were described giving rise to two proteins with different A/B regions[6]. This may suggest that the corresponding transcripts are generated by the use of two alternative promoters. The two isoforms were activated by the same range of compounds.

The genomic organization of the gene has still not been described in the literature.

Chromosomal location

The chromosomal location is not known. The gene has been tentatively located on chromosome 3 in the sequence of the human genome.

Amino acid sequence for human PXR (NR1I2)

Accession number: AF061056

```
  1  MEVRPKESWNHADFVHCEDTESVPGKPSVNADEEVGGPQICRVCGDKATGYHFNVMTCEGCKGFFRRAMK
 71  RNARLRCPFRKGACEITRKTRRQCQACRLRKCLESGMKKEMIMSDEAVEERRALIKRKKSERTGTQPLGV
141  QGLTEEQRMMIRELMDAQMKTFDTTFSHFKNFRLPGVLSSGCELPESLQAPSREEAAKWSQVRKDLCSLK
211  VSLQLRGEDGSVWNYKPPADSGGKEIFSLLPHMADMSTYMFKGIIISFAKVISYFRDLPIEDQISLLKGAA
281  FELCQLRFNTVFNAETGTWECGRLSYCLEDTAGGFQQLLLLEPMLKFHYMLKKLQLHEEEYVLMQAISLFS
351  PDRPGVLQHRVVDQLQEQFAITLKSYIECNRPQPAHRFLFLKIMAMLTELRSINAQHTQRLLRIQDIHPF
421  ATPLMQELFGITGS 434
```

Amino acid sequence for mouse PXR (NR1I2)

Accession number: AF031814

```
  1  MRPEESWSRVGLVQCEEADSALEEPINVEEEDGGLQICRVCGDKANGYHFNVMTCEGCKGFFRRAMKRNV
 71  RLRCPFRKGTCEITRKTRRQCQACRLRKCLESGMKKEMIMSDAAVEQRRALIKRKKREKIEAPPPGGQGL
141  TEEQQALIQELMDAQMQTFDTTFSHFKDFRLPAVFHSGCELPEFLQASLLEDPATWSQIMKDRVPMKISL
211  QLRGEDGSIWNYQPPSKSDGKEIIPLLPHLADVSTYMFKGVINFAKVISYFRDLPIEDQISLLKGATFEM
281  CILRFNTMFDTETGTWECGRLAYCFEDPNGGFQKLLLLDPLMKFHCMLKKLQLHKEEYVLMQAISLFSPDR
351  PGVVQRSVVDQLQERFALTLKAYIECSRPYPAHRFLFLKIMAVLTELRSINAQQTQQLLRIQDSHPFATP
421  LMQELFSSTDG 431
```

Amino acid sequence for *Xenopus* xONR1 (NR1I2)

Accession number: X75163

```
  1  MWKVQETLVLEEEEEEEDASNSCGTGEDEDDGDPKICRACGDRATGYHFNAMTCEGCKGFFRRAVKRNLR
 71  LSCPFQNSCVINKSNRRHCQACRLKKCLDIGMRKELIMSDAAVEQRRALIKRKHKLTKLPPTPPGASLTP
141  EQQHFLTQLVGAHTKTFDFNFTFSKNFRPIRRSSDPTQEPQATSSEAFLMLPHISDLVTYMIKGIISFAK
211  MLPYFKSLDIEDQIALLKGSVAEVSVIRFNTVFNSDTNTWECGPFTYDTEDMFLAGFRQLFLEPLVRIHR
281  MMRKLNLQSEEYAMMAALSIFASDRPGVCDWEKIQKLQEHIALTLKDFIDSQRPPSPQNRLLYPKIMECL
351  TELRTVNDIHSKQLLEIWDIQPDATPLMREVFGSPE 386
```

References

1 Smith, D.P. et al. (1994) Nucl. Acids Res. 22, 66–71.
2 Blumberg, B. et al. (1998) Genes Dev. 12, 1269–1277.
3 Kliewer, S.A. et al. (1998) Cell 92, 73–82.
4 Blumberg, B. et al. (1998) Genes Dev. 12, 3195–3205.
5 Lehmann, J.M. et al. (1998) J. Clin. Invest. 102, 1016–1023.
6 Bertilsson, G. et al. (1998) Proc. Natl Acad. Sci. USA 95, 12208–12213.
7 Zhang, H. et al. (1999) Arch. Biochem. Biophys. 368, 14–22.
8 Dotzlaw, H. et al. (1999) Clin. Cancer Res. 5, 2103–2107.
9 Waxman, D.J. (1999) Arch. Biochem. Biophys. 369, 11–23.
10 Pascussi, J-M. (1999) Biochem. Biophys. Res. Commun. 260, 377–381.
11 Schuetz, E.G. et al. (1998) Mol. Pharmacol. 54, 1113–1117.
12 Jones, S.A. et al. (2000) Mol. Endocrinol. 14, 27–39.
13 Moore, L.B. et al. (2000) J. Biol. Chem. 275, 15122–15127.
14 Heath, L.A. et al. (2000) Int. J. Dev. Biol. 44, 141–144.
15 Moore, L.B. et al. (2000) Proc. Natl Acad. Sci. USA 97, 7500–7502.
16 Masuyama, H. et al. (2000) Mol. Endocrinol. 14, 421–428.
17 Xie, W. et al. (2000) Nature 406, 435–439.
18 Xie, W. et al. (2000) Genes Dev. 14, 3014–3023.

CAR

Names

The first gene of this group was found in humans and was first called MB67 and later CARα[1]. A cDNA from mouse was then identified and called CARβ[2]. The relationship between these two sequences are unclear and is discussed below. The two genes are robustly placed inside the NR1I group of receptors that also contains the vitamin D receptor (NR1I1) and the pregnane receptor (NR1I2). In this book, these various receptors are presented separately in order to discuss their functional specificities in more detail. Their relatedness may nevertheless be of more significance than anticipated.

Species	Other names	Accession number	References
Official name: NR1I3			
Human	MB67, hCAR, CARα	Z30425	1
Official name: NR1I4			
Mouse	mCAR, CARβ	AF009327	2
Official name: NR1I group			
Chicken	CXR	AF276753	11

Isolation

The CARα (NR1I3) gene was isolated by David Moore's laboratory by screening of a human cDNA library with degenerate oligonucleotides based on the sequence of the P-box region of the RAR, TR and orphan receptors[1]. Since this receptor exhibits a constitutive activity, it was subsequently called CAR (constitutive actived receptor)[2].

A mouse-related gene, NR1I4, called mCAR or CARβ, was found by screening of a mouse liver cDNA library with a CARα probe[2]. The identity between CARα and CARβ in the DBD and LBD is 88%, and 72%, respectively. These values, which are extremely low between two putative orthologues in human and mouse for the nuclear receptor superfamily (e.g. the values for the rapidly evolving PPARα gene are around 93%), prompted the authors to propose that they derived from two different paralogous genes. We follow the authors in this interpretation but to date no direct proof (such as the isolation of human CARβ or mouse CARα) is available. We thus cannot rule out the possibility that the two genes are real orthologues that have experienced an extremely rapid evolution. Nevertheless, the recent identification of a chicken receptor closely related to the CARs and called CXR may challenge this view[11]. In a phylogenetic tree this CXR appears to be equally distant to CARα and CARβ, suggesting that these two genes may be real orthologues in mammals. This is also confirmed by the analysis of the sequence of the human genome. In this case, the official nomenclature of this group will have to be modified.

Both CARα and CARβ have in common a number of functional and structural peculiarities. For example, they both contain a very short A/B region (8 AA for CARα, 18 AA for CARβ). In the DBD, CARs display 55% amino-acid identities with PXR (NR1I2), 54% with VDR (NR1I1), 46%

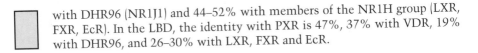

with DHR96 (NR1J1) and 44–52% with members of the NR1H group (LXR, FXR, EcR). In the LBD, the identity with PXR is 47%, 37% with VDR, 19% with DHR96, and 26–30% with LXR, FXR and EcR.

DNA binding

CARα and CARβ were first demonstrated to bind specifically to a subset of retinoid response elements having a DR5 organization[1,2]. For example, it was shown to bind to the RAREβ element, which is present upstream of the RARβ gene or to the DR5 element of the alcohol deshydrogenase (ADH3) gene[1]. No binding was observed on the RAREs present in the ApoA1, CRBP1 or MCAD promoters, suggesting that only some RAREs may bind CARα. No binding on palindromic or on DR4 elements was reported in these studies[1,2]. The binding was shown to be strictly dependent on the heterodimerization with RXR.

Nevertheless, the situation appears to be more complex, since by searching the nuclear receptors that are able to mediate the phenobarbital (PB) sensitivity of Cyp2B genes in human, rat or mouse, the Masahiko Negishi's group have observed the binding of the CAR–RXR heterodimer on a DR4 element (NR1 site) present in the phenobarbital-responsive enhancer of the mouse CYP2B10 and human CYP2B6 genes[3,4]. Furthermore, binding to an everted repeat with a 6 bp spacer element (ER-6) located in the human CYP3A4 gene was also described[4]. An interaction with a variety of elements, including a complex direct-repeat element found in the MMTV promoter, was also found in a recent study by David Moore's laboratory, suggesting that CARs are relatively versatile in their selection of target sequences[8].

Thus, the precise rules governing CAR binding to DNA are still poorly understood. The respective influences of the sequence of the spacer, the spacing or the structure of the binding sites have to be carefully tested before any definitive conclusion can be taken.

Partners

RXR is an obligatory partner for DNA binding. The polarity of the heterodimer on the response element is not known[1,2].

The coactivator SRC-1 was shown by a CARLA-type assay to bind constitutively to the apo-form of CAR. The binding of the ligands, androstanol and androstenol, displace this binding[5]. No other coactivator is known.

Ligands

By a systematic screening for molecules able to modulate CARβ transcriptional activity, it was found that, surprisingly, two androstane metabolites, androstenol and androstanol were able to inhibit CARβ transcriptional activity[5]. The EC$_{50}$ of these compounds is around 400 nM. This value is higher than the levels of androstenol reported in the circulation in adult male but it has to be emphasized that the physiology and activity of these compounds are poorly known, and that more studies will be necessary before concluding on their role as natural CARβ ligands. Interestingly, the inhibition is specific for these two compounds, since other closely related molecules, including DHEA, dihydrostestosterone, testosterone or androsterone, are inactive in the same test. Androstenol is known as a human pheromone but the CARβ structure–activity profile differs from that of the pheromonal response, suggesting that CARβ is not the pheromone receptor. It was later shown that these compounds have the same effects on the closely related CARα[9].

The most surprising finding is of course that the two compounds inhibit and not activate CARβ transcriptional activity. The fact that this inhibition is induced by the direct binding of the compounds to CARβ LBD is strongly suggested by the reproduction of the effect using GAL4 chimeras with CARβ LBD, in both mammalian and yeast cells. Furthermore, by a CARLA-type assay, *in vitro*, CARβ LBD binds in the apo-form to the coactivator SRC-1 (explaining its constitutive activity without ligand) but releases this coactivator upon binding of androstenol. Since no radioactive compound is available, no direct binding assay has been performed yet, but these data collectively indicate that androstenol and androstanol bind directly to CARβ and shut off its activity. The question, discussed in reference 6, is whether the two molecules represent inverse agonists (i.e. molecules having the opposite effect to an agonist) or if they are an antagonist (i.e. molecules acting against an agonist). The resolution of such a question, and the interesting issues regarding the mechanism of action of these inhibiting ligands await the resolution of the 3D structure of CARβ LBD in the apo- and the holo-form.

The negative effects of androstenol have been independently reproduced in a study on the regulation of the human CYP2B6 gene by phenobarbital[4]. Interestingly, the same study indicates that a variety of other compounds, such as phenobarbital, TCPOBOP or methoxychlor, activate (and do not inhibit) CARβ, suggesting that, as for the related receptor PXR, it may be more versatile than anticipated. In addition, it has been demonstrated that, in contrast to HepG2 cells, *in vivo* binding of CARβ to the CYP2B6 promoter occurs only after induction by phenobarbital[3]. This has been studied in more detail, and it has been shown that CARβ is in the cytoplasm in untreated liver, and phenobarbital induce a nuclear translocation of the protein by an okadaic-acid-sensitive mechanism[7]. Interestingly, androstenol does not inhibit the nuclear localization of CARβ even at high concentration. This and other observations suggest that the nuclear accumulation of CARβ in the nucleus is not ligand dependent.

The positive effect of TCPOBOP on CARβ activity can be explained by the demonstration that this compound is an agonist ligand for CARβ[8]. The observed EC_{50} of 20 nM is consistent with the low concentration of this compound that is required to stimulate CYP gene expression in primary hepatocytes. The direct interaction of TCPOBOP with CARβ was suggested by three lines of evidence: (1) TCBOPOB increases CARβ transcriptional activity; (2) it stimulates coactivator binding; and (3) mutations designed to block ligand binding block the inhibitory effects of androstanes and the stimulatory effect of TCPOBOP. Interestingly, these mutations do not abrogate the constitutive activity of CARβ, suggesting that this activity is really ligand independent. The current model is thus that CAR are constitutively active receptors that can be further activated or repressed by different types of compounds[8].

These findings were independently found by Steven Kliewer's team, which systematically compared the activators for CARs and PXR[9]. These authors found that TCPOBOP activates both human PXR and CARβ, but not CARα or mouse PXR, further stressing the importance of the species differences for these receptors. It was also observed that clotrimazole is an antagonist of CARα but not CARβ. Another interesting observation was that all four receptors share 5β-pregnane-3,20-dione as an activator. Finally, it has also been demonstrated by fluorescence energy transfer (FRET) assays that steroids or clotrimazole are bona fide CAR ligands[9].

Three-dimensional structure

No 3D structure is available.

Expression

Human CARα is expressed as a series of distinct mRNA species: a 1.4–1.7 kb unresolved cluster, a 2.1 kb mRNA, a third species at 2.9 kb and a very weakly expressed 7 kb transcript also being visible at long exposure[1]. The major expression site is the liver. Low levels of a 3.0 kb transcript are also observed in heart and muscles with even lower levels in kidney and lung. Kidney also expresses a 1.6 kb transcript at low levels[1].

Mouse CARβ also expressed several transcripts with a broader band of 1.3–1.7 kb and additional species at 3.0, 4.0 and 5.7 kb[2]. The major expression site is the liver. Hybridization with a 5'UTR probe indicates that the various species do not correspond to variants with distinct N-terminal sequences as frequently observed in other NRs. This also excludes some of the observed bands being the result of a cross-hybridization with a related gene. A 3'UTR probe suggests that alternative polyadenylation may be responsible for at least some of these different transcript species. As discussed below, two isoforms of CARβ exist and this may also explain some of the different transcripts.

The number and size of the transcripts observed in human and mouse are also difficult to interpret since the relationships between CARα and CARβ (whether they are orthologues or paralogues) are not clear.

The chicken CXR gene is expressed as 2.0 and 1.4 kb transcripts in tissues where drug induction of CYP genes occurs namely, liver, small intestine and colon[11].

Activity and target genes

Both CARα and CARβ are constitutively active in both mammalian and yeast cells[1,2]. The transcriptional activity of CARs is inhibited by androstenol and androstanol, and is activated by TCPOBOP as discussed above[5,8].

The RARβ and the alcohol dehydrogenase genes have been proposed as CAR target genes[1,2]. These constitutively active receptors may be responsible for the low levels of RARβ expression observed in situations in which retinoids are not present and thus cannot induce the expression of the gene. Nevertheless, no complete study has yet addressed this idea.

Several cytochrome P450-encoding genes, such as mouse CYP2B10, human CYP2B6 genes or human CYP3A4, are target genes for CARs[3,4]. It has been shown that CYP3A and CYP2B are regulated by both CAR and PXR, suggesting that xenobiotics can activate multiple CYP genes through these receptors, resulting in more efficient protection against the harmful effects of toxic compounds[10]. More recently, it has been shown that CARβ interacts with the MMTV promoter[8]. The chicken CXR gene regulates the CYP2H1 gene through a phenobarbital-responsive element[11].

Knockout

No knockout studies have been described in the literature.

Associated disease

There is no known associated disease with CAR.

Gene structure, promoter and isoforms

In mouse, CARβ gives rise to two isoforms, one long (mCAR1) and one shorter in the C-terminal (mCAR2)[2]. The structure of the mouse gene is known: it is split into nine exons and exon 8, corresponding to the central part of the LBD, is split out in the short mCAR2 isoform. Exon 2 encodes the short A/B region and the first zinc finger, exon 3 the second zinc finger, exon 4 the hinge and exons 5–9 the ligand binding domain. The 5′ end of the mouse CARβ cDNA has been mapped by primer extension.

Chromosomal location

The chromosomal locations of CARα and CARβ are not known.

Amino acid sequence for human CARα (NR1I3)

Accession number: Z30425

```
  1 MASREDELRNCVVCGDQATGYHFNALTCEGCKGFFRRTVSKSIGPTCPFAGSCEVSKTQRRHCPACRLQK
 71 CLDAGMRKDMILSAEALALRRAKQAQRRAQQTPVQLSKEQEELIRTLLGAHTRHMGTMFEQFVQFRPPAH
141 LFIHHQPLPTLAPVLPLVTHFADINTFMVLQVIKFTKDLPVFRSLPIEDQISLLKGAAVEICHIVLNTTF
211 CLQTQNFLCGPLRYTIEDGARVGFQVEFLELLFHFHGTLRKLQLQPEPEYVLLAAMALFSPDRPGVTQRDE
281 IDQLQEEMALTLQSYIKGQQRRPRDRFLYAKLLGLLAELRSINEAYGYQIQHIQGLSAMMPLLQEICS
348
```

Amino acid sequence for mouse CARβ (NR1I4)

Accession number: AF009327

```
  1 MTAMLTLETMASEEEYGPRNCVVCGDRATGYHFHALTCEGCKGFFRRTVSKTIGPICPFAGRCEVSKAQR
 71 RHCPACRLQKCLNVGMRKDMILSAEALALRRARQAQRRAEKASLQLNQQQKELVQILLGAHTRHVGPLFD
141 QFVQFKPPAYLFMHHRPFQPRGPVLPLLTHFADINTFMVQQIIKFTKDLPLFRSLTMEDQISLLKGAAVE
211 ILHISLNTTFCLQTENFFCGPLCYKMEDAVHAGFQYEFLESILHFHKNLKGLHLQEPEYVLMAATALFSP
281 DRPGVTQREEIDQLQEEMALILNNHIMEQQSRLQSRFLYAKLMGLLADLRSINNAYSYELQRLEELSAMT
351 PLLGEICS 358
```

References

1 Baes, M. et al. (1994) Mol. Cell. Biol. 14, 1544–1552.
2 Choi, H-S. et al. (1997) J. Biol. Chem. 272, 23565–23571.
3 Honkakoski, P. et al. (1998) Mol. Cell. Biol. 18, 5652–5658.
4 Sueyoshi, T. et al. (1999) J. Biol. Chem. 274, 6043–6046.
5 Forman, B.M. et al. (1998) Nature 395, 612–615.
6 Picard, D. (1998) Nature 395, 543–544.
7 Kawamoto, T. et al. (1999) Mol. Cell. Biol. 19, 6318–6322.
8 Tzameli, I. et al. (2000) Mol. Cell. Biol. 20, 2951–2958.
9 Moore, L.B. et al. (2000) J. Biol. Chem. 275, 15122–15127.
10 Xie, W. et al. (2000) Genes Dev. 14, 3014–3023.
11 Handschin, C. et al. (2000) Proc. Natl Acad. Sci. USA 97, 10769–10774.

Names

DHR96 and DAF-12 are protostome nuclear receptors closely related to the VDR (NR1I1), PXR (NR1I2), CAR (NR1I3 and NR1I4) vertebrate receptors[1,2]. Since these receptors are variable, their evolutionary distances are large and they were classified in a separate group inside subfamily 1. There is little doubt that DHR96 is indeed the orthologue of the vertebrate genes[3]. It is clear the closest homologue of DAF-12 is DHR96[4]. Since other uncharacterized nematode receptors closely related to this group, such as NHR-8 and NHR-48, exist in the genome sequence of *Caenorhabditis*[5], its precise relationship with other nuclear receptors should await a complete pylogenetical analysis. The name DHR96 stands for *Drosophila* hormone receptor, whereas DAF-12 was characterized as a Dauer larva mutant in *Caenorhabditis*.

Species	Other names	Accession number	References
Official name: NR1J1			
Drosophila	DHR96	U36792	3
Drosophila grimshawi	DHR96	U85656	Unpub.
Official name: NR1J group			
Caenorhabditis	DAF-12	AF136238	4
Strongyloides stercoralis	DAF-12	Not available	4
Caenorhabditis	NHR-8	AF083226	4, 5
Caenorhabditis	NHR-48	Z79604	4, 5

Isolation

The *Drosophila* DHR96 gene was identified by a rapid cloning and screening strategy designed to identify new members of the NHR superfamily that are expressed during the onset of metamorphosis[1]. This screen was essentially based on the construction of a library of short sequence signatures for nuclear receptors that was based on PCR with degenerated primers.

DAF-12 was identified as a mutant in the development of a specialized diapause stage of *Caenorhabditis* development, the Dauer larva, which forms when environmental conditions became unfavorable. There are several alleles of this gene, some of which are Dauer constitutive mutants in which Dauer larvae always form irrespective of the environmental conditions and some which are Dauer-defective, i.e. in which there is a bypass of the Dauer larvae even in unfavorable conditions[6]. In addition, the mutants exhibit a modestly shortened life-span extension as well as heterochronic changes in the formation of some larval stages in gonadal and extragonadal tissues[4,6]. Even before its cloning, the gene was placed in a cascade by genetic experiments. It was shown that DAF-12 acts downstream of lin-14 but upstream of lin-28[6]. The gene was cloned by transposon-insertion mutagenesis in screens for Dauer-defective mutants[4]. The sequence of the genomic DNA flanking the Tc1 insertion in mutants with DAF-12 phenotypes exhibits sequence identity to the DBD of nuclear receptors. The corresponding cosmid was fully sequenced by the *Caenorhabditis* genome project and was shown to rescue the phenotype. Using cosmid fragments as probes, several cDNAs were isolated from a mixed-stage nematode cDNA library[4].

NHR-8 and NHR-48 were found in the complete sequence of the *Caenorhabditis* genome. A cDNA is available for NHR-8 but not for NHR-48, the sequence of which is thus only putative since it was predicted from the genomic sequence by Genefinder[5].

In the DBD, DHR96 displays 60% sequence identity with VDR (NR1I1), 58% with PXR (NR1I2) and 53% with CARs. In the LBD, the identity value with VDR is 23%, 21% with PXR and 18% with CARs.

In the DBD, DAF-12 exhibits 96% sequence identity with its orthologue from another nematode species, *Strongyloides stercoralis*, and 67% with NHR-8, 69% with DHR96, 58% with PXR, 50% with CARs and 53% with VDR. DAF-12, NHR-8, NHR-48 and DHR96 contain 13 contiguous conserved residues that comprise the DNA recognition helix including the P-box sequence CESCKA, which is unique inside the superfamily. Within the LBD, the level of identity is lower: 64% with DAF-12 from *Strongyloides stercoralis*, 40% with DHR96, 34% with NHR-8, and only 20% with vertebrate receptors PXR, CAR and VDR. Of note, the helix 12 of DAF-12 containing the AF-2 AD region is well conserved with other nuclear receptors.

DNA binding

DHR96 was shown to bind the ecdysone response element (EcRE) found in the *hsp*27 gene[1]. This sequence is a palindromic arrangement of the sequences AGTGCA and GGTTCA. This element can also bind the EcR–USP heterodimer, suggesting a possible competition between these receptors. This is of particular interest since DHR96 expression is regulated by ecdysone (see below). No binding was found on DR3, DR4, DR5 elements or on palindromic HRE, or on monomeric half-sites[1].

No DNA binding data are available for DAF-12, NHR-8 or NHR-48.

Partners

There are no known partners for DHR96 and DAF-12. Of note, the heterodimerization properties of DHR96 with USP (NR2B4) were not tested. Given the fact that all the vertebrate homologues of DHR96 need USP to bind to DNA, it is likely that DHR96 also requires USP.

Ligands

There are no known ligands. Given the fact that all the vertebrate homologues of DHR96 and DAF-12 bind steroid-derived compounds, it is possible that DHR96 and DAF-12 also bind such compounds but this remains a speculation. Note that the identity level between these receptors in the LBD is not high (20–30%).

Three-dimensional structure

There is no known 3D structure.

Expression

DHR96 temporal pattern expression was studied by northern blot from *Drosophila* third-instar larvae and pre-pupae[3]. DHR96 encodes two transcripts of 2.8 and 0.6 kb, and is expressed throughout third-instar larval and pre-pupal development with distinct increases in abundance at 106 hours after egg laying. The 0.6 kb transcript, which can be detected with a probe encompassing the DBD, is present at the beginning of the third larval instar and is down-regulated at 94 hours. Interestingly, it was shown that DHR96 expression is regulated by ecdysone in cultures of isolated larval organs. Only the 2.8 kb transcript of DHR96 can be detected at the beginning of the culture and its level increases during the culture for up to 8 hours of treatment.

The DAF-12 expression pattern was determined by the fusion of the GFP gene with the exon 1 of the gene[4]. The transgenic construct was based on a large cosmid in order to contain the entire DAF-12 regulatory region. DAF-12 is expressed widely in target tissues from embryo to adult but is up-regulated during midlarval stages. In the adult, expression persists in the nervous system and somatic gonads (consistent with several mutant alleles that exhibit gonadal phenotypes), two tissues that regulate adult longevity. Expression was also found in the epidermis, intestine and musculature. The analysis of the mutant phenotype and the expression pattern of the gene suggest that it regulates the advance from the second to the third and later larval-stage fates and mediates the choice between alternate third-stage programs, Dauer diapause or reproductive development. Since it is a nuclear receptor related to vertebrate ligand-activated receptors, it was proposed that DAF-12 integrates hormonal signals in cellular targets to coordinate these major life history traits[4].

No expression data are available for NHR-8 and NHR-48. The only available information is that NHR-48 is an expressed gene, since ESTs corresponding to the genomic sequence were isolated[5].

Activity and target genes

There was no functional characterization of either DHR96, DAF-12, NHR-8 or NHR-48 gene products. Since their vertebrates homologues are transcriptional activators and their AF-2 AD region is conserved, it is likely that these four genes are also activators but this remains to be experimentally tested. No target genes were studied.

Knockout

The phenotype of the DHR96 mutant has still not been described.

The DAF-12 mutant was studied extensively[4,6]. Six classes of allele exist based on separable heterochronic and Dauer phenotypes. Interestingly, the molecular lesions of each class are clearly distinct. Class 2 and 3 alleles are all mutants in the DBD, whereas class 1, 4, 5 and 6 are mutants in the LBD. Only one mutant lies in the hinge region[4]. For example, class 3 mutants are Dauer defective and present impenetrant heterochronic phenotypes. Most of the class 3 alleles are inside the DBD, some of the mutations affecting critical residues known to be important for the integrity of the DBD, such as P-box sequence or conserved cysteines. It is thus likely that class 3 mutants result in a protein with strongly impaired DNA-binding activity. Genetic analysis suggests that these class 3 mutants are loss of function null alleles. In contrast, the unique class 2 mutant, which also exhibits a Dauer-defective and moderate heterochronic phenotype, contains a deletion in exon 3 that creates a stop codon upstream of the DBD. Of note, the unique mutant class with a Dauer-constitutive phenotype (class 6) exhibits mutation of the same amino acid within the helix 3 of the LBD (R564).

Associated disease

See the entries devoted to the vertebrate homologues of these genes.

Gene structure, promoter and isoforms

The gene structure of DHR96 has not been published, although it is now available on the complete sequence of the *Drosophila* genome (see flybase for details). There is no known isoform of DHR96. The promoter of this gene has not been characterized.

The DAF-12 gene consists of 17 exons spanning 21 kb[4]. A 14.9 kb intron separates exons 2 and 3, whereas exons 3–17 are densely packed within 4.7 kb. The two zinc fingers are on separate exons (5 and 6), the intron lying 13 amino acids after the last cysteine of the first zinc finger. The LBD is encoded by exons 13–17.

Three different isoforms, 12A1, 12A3 and 12B, were found for DAF-12[4]. The 12A and 12B isoforms differed at their 5' ends, since the 12B isoform starts at exon 13 giving rise to a 267 amino acid product that does not contain a DBD. 12A1 and 12A3 are identical except that 12A3 has an internal portion of exon 12 spliced out, removing 16 amino acids. The size of these two products are thus 753 and 737 amino acids, respectively.

The NHR-8 gene contains nine exons, with DBD encoded by exons 2 and 3, and LBD by exons 6–9[4,5]. The gene is *c.* 3 kbp shorter than DAF-12. The NHR-48 gene contains 16 exons with the DBD encoded by a unique exon (5) and LBD by exons 13–16. The gene is approximately the same size as DAF-12[4,5].

Several isoforms also exist for NHR-8 and NHR-48[4]. The NHR-8 isoforms differ in their 5' ends: isoform 8A1 starts at exon 1, isoform 8A2 inside this exon and isoform 8B at exon 2. For NHR-48 at least five isoforms exist that differ in their 5' end, as well as by the removal of exon 10 or the use of an alternate splice donor site in exon 7.

Chromosomal location

DHR96 is at position 96B14-15 on chromosome 3R[3].

DAF-12 maps to *Caenorhabditis* chromosome X, as its close homologue NHR-48. NHR-8 is on chromosome IV[4,5].

Amino acid sequence for *Drosophila* DHR96 (NR1J1)

Accession number: U36792

```
  1 MSPPKNCAVCGDKALGYNFNAVTCESCKAFFRRNALAKKQFTCPFNQNCDITVVTRRFCQKCRLRKCLDI
 71 GMKSENIMSEEDKLIKRRKIETNRAKRRLMENGTDACDADGGEERDHKAPADSSSSNLDHYSGSQDSQSC
141 GSADSGANGCSGRQASSPGTQVNPLQMTAEKIVDQIVSDPDRASQAINRLMRTQKEAISVMEKVISSQKD
211 ALRLVSHLIDYPGDALKIISKFMNSPFNALTVFTKFMSSPTDGVEIISKIVDSPADVVEFMQNLMHSPED
281 AIDIMNKFMNTPAEALRILNRILSGGGANAAQQTADRKPLLDKEPAVKPAAPAERADTVIQSMLGNSPPI
351 SPHDAAVDLQYHSPGVGEQPSTSSSHPLPYIANSPDFDLKTFMQTNYNDEPSLDSDFSINSIESVLSEVI
421 RIEYQAFNSIQQAASRVKEEMSYGTQSTYGGCNSAANNSQPHLQQPICAPSTQQLDRELNEAEQMKLREL
491 RLASEALYDPVDEDLSALMMGDDRIKPDDTRHNPKLLQLINLTAVAIKRLIKMAKKITAFRDMCQEDQVA
561 LLKGGCTEMMIMRSVMIYDDDRAAWKVPHTKENMGNIRTDLLKFAEGNIYEEHQKFITTFDEKWRMDENI
631 ILIMCAIVLFTSARSRVIHKDVIRLEQNSYYYLLRRYLESVYSGCEARNAFIKLIQKISDVERLNKFIIN
701 VYLNVNPSQVEPLLREIFDLKNH   723
```

Amino acid sequence for *Caenorhabditis* DAF-12 (NR1J group).

Accession number: AF136238

```
  1 MGTNGGVIAEQSMEIETNENPDKVEEPVVRRKRVTRRRHRRIHSKNNCLTPPNSDDDPQMSTPDDPVIHS
 71 PPSIGAAPGMNGYHGSGVKLEESSGACGSPDDGLLDSSEESRRRQKTCRVCGDHATGYNFNVITCESCKA
141 FFRRNALRPKEFKCPYSEDCEINSVSRRFCQKCRLRKCFTVGMKKEWILNEEQLRRRKNSRLNNTGTCNK
211 RSQPGNQQSPQGPNQQPHLSPHHPGVAIYPPQPQRPLTINPMDNQMMHHMQANRPNAMPQLISPPGAQPY
281 PLTSPVGSSASDSPPNRSLTMMHHNGEKSPDGYDPNIMAHRAPPPSFNNRPKMDSGQVVLSTEEYKQLLSR
351 IPGAQVPGLMNEEEPINKRAAYNCNGHPMPAETTPPYSAPMSDMSLSRHNSTSSGTEKNHMTHSTVSAIP
421 GNSAQNHFDIASFGMGIVTATGGGDAAEEMYKRMNMFYENCIQSALDSPENQEPKPQEAMIPKEEYMTPT
491 HGFQYQSDPYQVPPAERNINYQLNAAELKALDAVREAFYGMDDPMEQGRQMQSFLKANKTPADIMNIMDV
561 TMRRFVKVAKGVPAFREVSQEGKFSLLKGGMIEMLTVRGVTRYDASTNSFKTPTIKGQNVSVNVDDMFAK
631 LNANAQAQKAKCLEFFGFFDEEIKKNELAVYLVMLAVLFSVRSDPPMNENDVRIVTERHNHFMSLLNRYL
701 ESLFGEQARRIFERIPKALGLLNEIARNAGMLFMGTVRSGEAEELPGEFFKIK   753
```

References

1 Laudet, V. (1997) J. Mol. Endocrinol. 19, 207–226.
2 Escriva, H. (2000) BioEssays 22, 717–727.
3 Fisk, F.M. and Thummel, C.S. (1995) Proc. Natl Acad. Sci. USA 92, 10604–10608.
4 Antebi, A. et al. (2000) Genes Dev. 14, 1512–1527.
5 Sluder, A.E. et al. (1999) Genome Res. 9, 103–120.
6 Antebi, A. et al. (1998) Development 125, 1191–1205.

HNF4

Names

HNF4 (hepatocyte nuclear factor 4) was cloned by Frances Sladek in James Darnell's laboratory as a nuclear factor expressed in liver[1]. The *Drosophila* homologue was called dHNF4[2]. Later, two paralogues were found in human and *Xenopus* and were called HNF4β and HNF4γ. The first member of this group is thus now called HNF4α[3,7]. These genes form the group A of subfamily 2 in the nomenclature[4].

Species	Other names	Accession number	References
Official name: NR2A1			
Human	HNF4, HNF4α	X76930	6
Mouse	HNF4, HNF4α	D29015	8, 9
Rat	HNF4, HNF4α	D10554	1, 5
Xenopus	HNF4, HNF4α	Z37526	10
Official name: NR2A2			
Xenopus	HNF4β	Z49827	7
Official name: NR2A3			
Human	HNF4γ	Z49826	3
Mouse	HNF4γ	AJ242626	78
Official name: NR2A4			
Drosophila	dHNF4	U70874	2
Bombyx	BmHNF4	UP3843X	11
Aedes aegypti	AaHNF4	AF059027	76

Isolation

The HNF4 gene was cloned as a liver-enriched transcription factor as were the three preceding HNFs[1]. It was previously known as a factor detected in crude liver extracts binding to a DNA element required for the transcription of the transthyretin (TTR) and the apolipoprotein CIII genes[12]. HNF4 was biochemically purified using this DNA-binding property. A partial amino-acid sequence of the purified protein was obtained. From these peptide sequences, PCR primers were derived and a partial cDNA clone was obtained that was shown to belong to a new member of the nuclear receptor superfamily[1,13]. Mouse, human, *Xenopus* and *Drosophila* HNF4 were cloned using the rat cDNA under low-stringency screening procedures[2,6,8–10].

HNF4β was obtained by screening a *Xenopus* liver cDNA library with a rat HNF4α probe. Several clones were clearly shown to be orthologues of HNF4α[10], whereas others were suggested to represent a new gene that was called HNF4β[7]. The overall identity between *Xenopus* HNF4α and β is 67%, with 94% in the DNA-binding domain and 72% in the LBD. The A/B and D region are more divergent as the F domain (25% identity only).

HNF4γ was isolated using a probe for *Xenopus* HNF4β in the hope of finding the orthologue of HNF4β in mammals[3]. However, from the observed sequence, it was concluded that it represents a third member of the HNF4 group in vertebrates that was thus called HNF4γ. This gene exhibits a very long A/B domain (378 amino acids), and 92% and 79% amino-acid identity with the DBD and LBD of human HNF4α, respectively.

The reason HNF4β and HNF4γ, which were both identified in Gerhart Ryffel's laboratory, are not considered as orthologues is based on the relatively low relatedness between them, the presence of a long A/B region in human HNF4γ, as well as on their different expression patterns (see below). Nevertheless, the definitive proof that the two homologues effectively correspond to different genes will be the discovery of an orthologue of HNF4β in human and a orthologue of HNF4γ in *Xenopus*. Without such arguments one cannot exclude that *Xenopus* HNF4β and human HNF4γ are two orthologues that, for an unknown reason, evolved very rapidly, as for example the PPARs (NR1C) or the PXRs (NR1I). Of note the complete sequence of the human genome does not contain a third HNF4 gene (M. Robinson and V. Laudet, in press).

All HNF4 group members contain a C-terminal F domain of various lengths clearly separable from the ligand-binding domain. This is a relatively unusual characteristic in nuclear receptors. In addition, they all share a common P-box with the sequence CDGCKG, which is specific to this group. In the phylogeny, HNF4 is distantly related to the RXR group (NR2B), inside subfamily II.

DNA binding

The rat HNF4α gene was cloned by virtue of its DNA-binding property on a regulatory element of the rat transthyretin gene[1]. This element encompasses what we define now as a divergent DR1 element[12]. The resulting cDNA product was shown to bind specifically to this sequence as well as to a similar element found in the apolipoprotein CIII gene (see reference 13 for a review).

Following the cloning of HNF4, a number of target genes were found and were shown to contain DR1 elements in their promoters. This is the case for the hepatocyte nuclear factor 1 (HNF1)[14], or apolipoproteins B, AII or CIII[15,16]. Other target genes are reviewed in reference 17 and will be discussed below. It has been found that HNF4α binds to DNA as an exclusive homodimer[18].

Drosophila HNF4, as well as human HNF4γ and *Xenopus* HNF4β, binds to the same DR1 element as HNF4α[2,3,7].

Partners

HNF4 binds to DNA as an exclusive homodimer and does not form a heterodimer with either RXR, RAR, TR or ER[18]. In solution, the wild-type protein exists as an homodimer and it has been demonstrated that a very strong homodimerization interface exists in the LBD of HNF4[18-20]. Strikingly, the DBD of HNF4 alone is able to bind to DNA as a heterodimer with RXR, suggesting that the strong homodimerization interface that exists in the LBD impairs the latent heterodimerization ability of the DBD[19]. There is no other known partner for HNF4.

It has been shown that the classic isoform of HNF4α, HNF4α1, is able to interact physically and functionally with the coactivators GRIP1, SRC1, P300 and CBP in the absence of any exogenous ligands[22,63]. Interestingly, the F domain is able to obscure an AF-2-independent binding site for GRIP1, consistent with the notion that this domain acts as a transcriptional inhibition domain in the context of the full-length protein[23,65]. In accordance with this finding, HNF4α2, an isoform of HNF4α containing a 10 amino-acid insertion in the middle of the F domain, abrogates the inhibition of coactivator binding[22]. This study also suggests that there is a physical contact between the F domain and the LBD of HNF4α. It has been shown that the isolated AF-1 region of HNF4 interacts with multiple components of the basal transcription complex, such as TAF$_{II}$31, TAF$_{II}$80, TFIIH-p62, ADA2 or CBP[67]. The *in vivo* relevance of these interaction remains to be established. The interaction of HNF4 with CBP has recently been confirmed, since it was shown that CBP acetylates HNF4 in lysine residues of the nuclear localization sequence[66]. This acetylation is required for the proper nuclear retention of HNF4.

Ligands

Recently, a ligand has been proposed for HNF4α[21]. This report proposed that long-chain fatty acids directly modulate the transcriptional activity of HNF4α by binding as their acyl-CoA thioesters to the LBD of HNF4α. This binding may modulate positively or negatively the transcriptional activity of HNF4α, depending on the length and degree of saturation of the fatty acid: poly- or monounsaturated acyl-CoAs inhibited the constitutive activity of HNF4α, whereas different saturated acyl-CoAs, such as palmitoyl-CoA or steroyl-CoA, activated or inhibited it, respectively. This regulatory effect may be achieved by a regulation of the homodimerization and/or DNA-binding activity of HNF4. Nevertheless, this spectacular discovery has not been independently confirmed and the weak activation or repression observed led to some skepticism of its physiological relevance. We thus still consider HNF4 as an orphan receptor, whereas the possibility remains that a derivative of fatty acids may be a natural ligand of this protein. Of note, the interaction between HNF4α and coactivators such as GRIP1 is observed without any exogenous ligand, both *in vivo* and *in vitro*[22].

Three-dimensional structure

No 3D structure is yet available for either the DBD or LBD of HNF4.

Expression

The rat HNF4α was shown to be expressed as a 4.5 kb mRNA in liver, kidney and intestine, but not in other organs[1]. A weaker transcript at 2.3 kbp was also observed in liver and kidney. In humans, HNF4 was shown to be expressed in hepatoma cell lines, such as HepG2[6]. It was also shown to be expressed in pancreatic islet cells and insulinoma cells[56]. In mouse, extensive *in situ* hybridization studies were performed. The transcript was found in a limited number of tissues: liver, kidney, intestine, stomach and skin[8,9]. The mRNA appears in the primary endoderm at E4.5, was restricted to the primary differentiation of the visceral yolk sac from E5.5 to E8.5 and was found in liver, kidney (in mesonephric tubules), pancreas, stomach, and intestine from E8.5 until birth[8,9]. Later, an expression was detected in the metanephric tubules of the kidney. In *Xenopus*, HNF4α is expressed first as a maternal mRNA and protein in the fertilized egg, then as a gradient from animal to vegetal pole in the cleaving embryo, and then in the embryonic regions that will give the liver and the kidney[7,10].

This promoter of HNF4α contains an HNF1-binding site, which is required but not sufficient for liver specific expression. Distal elements located 5.5–6.5 kb upstream of the start sites are important to achieve the correct expression pattern of the mouse HNF4 gene[24]. Recently, it was shown that HNF4α is under the transcriptional control of the HNF3 family members[45]. Control of HNF4α expression by HNF6 and HNF1β has been demonstrated[72,75]. It was also shown that, when the gene encoding the serine/threonine kinase receptor ALK2 was inactivated in mouse, the expression of HNF4α was reduced, suggesting that the pathway governed by ALK2 ultimately reaches the HNF4α promoter[73]. Similar observations were made for the BMP signaling pathway[68].

As HNF4α, the *Xenopus* HNF4β protein is a maternal factor during early embryogenesis and is then present in nearly equal amounts during embryogenesis[7]. The distribution of the transcripts, assayed by RT-PCR, was much more dynamic. In the adult, the protein is found in a much larger set of organs: liver and kidney (as HNF4α), but also stomach, intestine, ovary, testis and lung[7].

The HNF4γ gene is also expressed differentially when compared to the human HNF4α[3]. HNF4γ was detected in a *c.* 6.0 kb species in pancreas, kidney, small intestine and testis, but, strikingly, not in liver and only very weakly in colon. In general, HNF4γ is more than 10-fold less abundant than HNF4α[3].

Interestingly, the *Drosophila* dHNF4 gene is expressed as a doublet at 4.6 and 3.3 kb in tissues homologous to the ones in which HNF4α is expressed[2], i.e. in midgut, fat bodies and malpighian tubules, suggesting that the function of the gene is conserved in most if not all metazoans. Interestingly, a uniformly distributed strong signal was observed in the very early embryo (0–2 hours of development). The signal then declines, becomes barely visible and then surges again at 8 hours of development, being localized in the endodermal cells that correspond to the posterior midgut primordium. The mRNA is expressed in ovaries, indicating that the early transcript corresponds to a maternal mRNA[2]. In the mosquito *Aedes*, several isoforms exhibit different expression patterns[76].

Activity and target genes

HNF4 is a constitutive transcriptional activator and, as discussed above, interacts with coactivators[1,17,22,63].

The HNF4α proteins are phosphorylated on Ser and Thr residues, and it has been proposed that this phosphorylation may modulate the DNA-binding activity of the protein[25]. In addition, it has also been shown that HNF4 is phosphorylated on Tyr residues and that this phosphorylation is important for its DNA-binding activity and, consequently, for its transactivation potential[26]. Thus, despite the fact that HNF4 is a constitutive transcriptional activator, it is clear that its activity can be modulated by post-translational mechanisms, independently of any ligand binding.

HNF4 has been shown to play a significant role in the regulation of many of the various genes expressed in the liver, both those that are destined for the bloodstream and those that remain in the liver. The variety of HNF4 target genes makes it very difficult to identify any unifying theme between them. Following the review in reference 17, we can tentatively classify them in to the following groups:

1 Apolipoproteins CIII[1], B[15,16], AII[15], A1[28,32].
2 Clotting factors such as factor IX[27], factor VII[35].
3 Other serum proteins such as α1-microglobulin[34] or α1antitrypsin[41], as well as serine protease inhibitors[39].
4 Genes involved in fatty acid metabolism, such as medium-chain acyl-CoA dehydrogenase (MCAD)[47].
5 Cytochrome P450.
6 Genes involved in amino-acid metabolism, such as the tyrosine amino transferase (TAT) gene[30].
7 Genes involved in glucose metabolism, such as L-pyruvate kinase[29], the phosphoenolpyruvate carboxykinase (PEPCK)[33,36,40] or the Aldolase[71] genes.
8 Genes linked to retinoid signaling, such as the cellular retinol binding protein II[48].
9 Viral genes such as the enhancer I of hepatitis B virus[31].
10 Other HNF genes, such as HNF1 (see below and reference 17 for a review), HNF6[42], alphafetoprotein[70] or the human angiotensinogen[69] genes.

Other genes not classified in this scheme are the prolactine receptor[37] and, in *Xenopus*, the activin A, which is instrumental in mesoderm formation[38]. On all these genes HNF4 regulates transcription through DR1 elements that are often divergent when compared to the consensus. In several cases, HNF4 serves as an accessory factor for the induction by other stimuli, such as glucocorticoids, in the case of the PEPCK gene[36].

The regulation of HNF1 by HNF4 and the way by which specific liver expression of their target genes is achieved were extensively studied. Since neither HNF4 nor other factors, such as HNF1, HNF3 or C/EBP, are specifically expressed in the liver, it was proposed that a regulatory hierarchy controlled hepatic-specific gene expression. These factors function in unique, often synergistic, combinations to stimulate cell-specific transcription. In this view, a master regulator of the liver phenotype should be located upstream of the cascade. In was shown that HNF4 is located at the top of this cascade, since it regulates HNF1α expression directly[14,42]. The importance of this regulation was shown by the fact that HNF4 can rescue HNF1α expression in dedifferentiated hepatomas, suggesting that HNF4 is indeed at the top of the transcriptional hierarchy[42,49]. It was further shown that the HNF1α promoter can be activated by HNF1α by an autoregulatory loop, and that HNF1α and HNF4 synergize in activating the HNF1 promoter[43,44]. The regulation of HNF4 by HNF1 has also been observed in *Xenopus*[74]. More recently, it was shown that this HNF4/HNF1α cascade was under the control of HNF3 family members[45]. The importance of the relationship between HNF4 and HNF1α was well demonstrated by their common implication in the same disease – type 2 diabetes (see below).

Knockout

Mice lacking the HNF4 gene die very early on in embryogenesis: at E7.5 the embryos are clearly abnormal and generally are found dead at E9.5[50]. It was shown that the gene was essential for the completion of gastrulation: the knockout animals initiate gastrulation as evidenced by production of cells expressing markers of nascent mesoderm, such as Brachyury, but they fail to produce cells expressing late mesoderm markers[50]. At this stage, in wild-type embryos, HNF4 expression is restricted to the extraembryonic visceral endoderm with no HNF4 detected in embryonic tissues[9]. This suggests that the correct HNF4 expression in visceral endoderm has a critical function in maintaining gastrulation. This model was tested by performing tetraploid rescue of HNF4$^{-/-}$ embryos and by showing that HNF4 was required for normal expression of several secreted factors (alphafetoproteins, apolipoproteins, retinol binding protein, transthyretin and transferrin) by the visceral endoderm and was essential for normal hepatocyte differentiation[51,77].

In *Drosophila*, a deletion of the chromosomal region spanning the dHNF4 locus was characterized[2]. In these animals, the maternal mRNA expression was not affected but the zygotic expression was abolished. After stage 10, i.e. when the mRNA transcript appears in the embryos, visible defects were clearly noticed in the midgut, malpighian tubules and salivary glands. This suggests that dHNF4 in *Drosophila* is effectively primordial for gut formation as its expression pattern suggested.

Associated disease

HNF4 has been associated to hemophilia B Leyden because it transcriptionally regulates the blood coagulation factor IX[27]. It has been shown that this disease may be caused by a single point mutation in the factor IX promoter that disrupts a HNF4 binding site, resulting in a strong decrease in factor IX production[27].

More recently, HNF4α was associated with a form of type 2 diabetes, maturity-onset diabetes of the young (MODY1)[52]. A family with MODY1 was shown to contain a heterozygous nonsense mutation at position 268 of the protein (Q268X) that segregates with the disease. The mutation gives rise to a truncated product that contains a normal DBD but only a partial LBD. Strikingly, the resulting protein neither binds to DNA alone nor binds it as a dimer with the wild-type protein[54]. In contrast, the mutant protein is capable of dimerizing in solution with the wild-type protein. Transient transfection assays have revealed that this protein does not display a dominant negative activity and exhibits an abnormal subcellular localization[54]. Other mutations were found: R154X, F75delT and K99fsdelAA that give rise to truncated products, or missense mutations, such as G115S, R217W, located in the T-box, V255M E276Q or V393I[54,55,64]. In general, these mutations do not give rise to dominant negative mutants but impair to various degrees the function of the protein[55,62]. Interestingly, HNF1α itself was found mutated in another class of type 2 diabetes, MODY3[57], reinforcing the importance of the transcriptional link between HNF4α and HNF1α. It is believed that mutations in HNF4α might cause MODY through impaired HNF1α gene function. This connection is revealed by the disruption of the HNF4 binding site in the HNF1 promoter that has been identified in an Italian family with MODY[61].

Gene structure, promoter and isoforms

The HNF4α gene consists of 12 exons spanning c. 30 kbp, among which several correspond to alternative exons[8,24]. The gene structure is unique in the nuclear receptor superfamily in that the second zinc finger is encoded by two exons with an intron lying in the middle of the D-box[8].

The promoter of HNF4α has been characterized in mouse and was shown to be GC rich with no visible TATA box[8,9,24]. A region sufficient to drive liver-specific expression of HNF4 was identified using transgenic mice.

The structure of HNF4β and HNF4γ is unknown.

The HNF4α gene is alternatively spliced and may generate up to seven different isoforms. HNF4α1 is the initially identified transcript[1]; HNF4α2 is a splice variant that contains a larger exon 9, resulting in 10 extra amino acids in the F domain and has been identified in rat[5], mouse[59] and human[3,6]. HNF4α3, initially called HNF4-C, has been identified in human liver and exhibits a shorter and distinct C-terminus[60]. HNF4α4, which is found in human kidney and liver, contains two additional exons in the N-terminal region[3,53]. The fifth splice variant, HNF4α5, carries the same two N-terminal exons as HNF4α4, together with the 10 extra amino acids in the F domain found in HNF4α2[3,53]. HNF4α6 combines the same two N-terminal exons as HNF4α4 with the C-terminus found in HNF4α3[53]. The last, HNF4α7, is expressed in a tissue-specific manner in embryonic stem cells, probably from a new tissue-specific promoter and is characterized by an alternative first exon[58]. It is believed that each of these isoforms have unique properties, allowing a precise regulation of HNF4 activity. It has been shown that HNF4α4 has a lower transactivation potential than HNF4α2[3] and that HNF4α7 is a weaker activator than HNF4α1[58]. It has also been shown that the insertion of 10 amino acids in the F domain found in HNF4α2 and HNF4α5 increases the strength of the interaction with the coactivator GRIP1, resulting in a protein exhibiting a stronger transactivation potential[22].

Xenopus HNF4β also displays alternative splicing, and at least three isoforms generated by this gene are known[7]. The HNF4β2 and β3 isoforms contain additional 5' UTR exons, which seem to affect RNA stability. The protein products encoded by these three isoforms are identical.

No isoforms of human HNF4γ or *Drosophila* HNF4 gene have been described[2,3]. Interestingly, three isoforms differing in their N-terminal parts were cloned in the mosquito *Aedes*. Two of them exhibit different A/B region and thus possibly different AF-1 activity, whereas the third has the entire A/B region and DBD deleted, suggesting that it may behave as an inhibitor. These three isoforms exhibit different tissue- and stage-specific expression patterns[76].

Chromosomal location

HNF4α is located on human chromosome 20 at position q12–q13.1[3,52], a position associated with MODY1, whereas HNF4γ is located on human chromosome 8 in an unknown position[3]. The mouse HNF4γ gene is at a synthenic position[78]. The *Drosophila* HNF4 gene is located on the left arm of chromosome 2 in the region 29E[2].

Amino acid sequence for human HNF4α (NR2A1)

Accession number: X76930

```
  1  MDMADYSAALDPAYTTLEFENVQVLTMGNDTSPSEGTNLNAPNSLGVSALCAICGDRATGKHYGASSCDG
 71  CKGFFRRSVRKNHMYSCRFSRQCVVDKDKRNQCRYCRLKKCFRAGMKKEAVQNERDRISTRRSSYEDSSL
141  PSINALLQAEVLSRQITSPVSGINGDIRAKKIASIADVCESMKEQLLVLVEWAKYIPAFCELPLDDQVAL
211  LRAHAGEHLLLGATKRSMVFKDVLLLGNDYIVPRHCPELAEMSRVSIRILDELVLPFQELQIDDNEYAYL
281  KAIIFFDPDAKGLSDPGKIKRLRSQVQVSLEDYINDRQYDSRGRFGELLLLLPTLQSITWQMIEQIQFIK
351  LFGMAKIDNLLQEMLLGGSPSDAPHAHHPLHPHLMQEHMGTNVIVANTMPTHLSNGQMCEWPRPRGQAAT
421  PETPQPSPPGASGSEPYKLLPGAVATIVKPLSAIPQPTITKQEVI  465
```

Amino acid sequence for *Xenopus* HNF4β (NR2A2)

Accession number : Z49827

```
  1  MDMPDYTETLDSSYTMLEFDSIRVLPSNTEIITVETASPGLLNNGINSFCAICGDRATGKHYGASSCDGC
 71  KGFFRRSVRKNHVYACRFSRQCIVDKDKRNQCRYCRLRKCFRAGMKKEAVQNERDRISMRRSSYEDNGSL
141  SINVLTQAEAMVHQYSPVSPVHSSDISMKKVASISDVCESMKQQLLLLVEWAKYIPAFCELPLDDQVALL
211  RAHAGAHLLLGVAKRSLPYKDFLLLGNDFIMPMHCPELEIARVPCRILDELVKPLREIQIDDNEYVCLKA
281  IIFFDPDCKGLSDQTKVKNMRFQVQVNLEDYINDRQFDSRGRFSDILLLLPPLQSITWQMIEQVQFAKLF
351  GVARIDSLLQELLLGGTTMDGGQYINSGHSSLNLDLLPGPTVHSHNLHSVIHTVSSLSPETSPPTNSTSE
421  DYKMNTATVSSIPLMQRTVIAKKEIL  446
```

Amino acid sequence for human HNF4γ (NR2A3)

Accession number : Z49826

```
  1  MNTTDNGVNCLCAICGDRATGKHYGASTCDGCKGFFRRSIRKSHIYSCRFSRQCVVDKDKRNQCRYCRLR
 71  KCFRAGMKKEAVQNERDRISTRRSTFDGSNIPSINTLAQAEVRSRQISVSSPGSSTDINVKKIASIGDVC
141  ESMKQQLLVLVEWAKYIPAFCELPLDDQVALLRAHAGEHLLLGATKRSMMYKDILLLGNNYVIHRNSCEV
211  EISRVANRVLDELVRPFQEIQIDDNEYACLKAIVFFDPDAKGLSDPVKIKNMRFQVQIGLEDYINDRQYD
281  SRGRFGELLLLLPTLQSITWQMIEQIQFVKLFGMVKIDNLLQEMLLGGASNDGSHLHHPMHPHLSQDPLT
351  GQTILLGPMSTLVHADQISTPETPLPSPPQGSGQEQYKIAANQASVISHQHLSKQKQL  408
```

Amino acid sequence for *Drosophila* HNF4 (NR2A4)

Accession number : U70874

```
  1 MHADALASAYPAASQPHSPIGLALSPNGGGLGLSNSSNQSSENFALCNGNGNAGSAGGGSASSGSNNNNS
 71 MFSPYNNLNGSGSGTNSSQQQLQQQQQQQSPTVCAICGDRATGKHYGASSCDGCKGFFRRSVRKNHQYTC
141 RFARNCVVDKDKRNQCRYCRLRKCFKAGMKKEAVQNERDRISCRRTSNDDPDPGNGLSVISLVKAENESR
211 QSKAGAAMEPNINEDLSNKQFASINDVCESMKQQLLTLVEWAKQIPAFNELQLDDQVALLRAHAGEHLLL
281 GLSRRSMHLKDVLLLSNNCVITRHCPDPLVSPNLDISRIGARIIDELVTVMKDVGIDDTEFACIKALVFF
351 DPNAKGLNEPHRIKSLRHQILNNLEDYISDRQYESRGRFGEILLILPVLQSITWQMIEQIQFAKIFGVAH
421 IDSLLQEMLLGGELADNPLPLSPPNQSNDYQSPTHTGNMEGGNQVNSSLDSLATSGGPGSHSLDLEVQHI
491 QALIEANSADDSFRAYAASTAAAAAAAVSSSSSAPASVAPASISPPLNSPKSQHQHQQHATHQQQQESSY
561 LDMPVKHYNGSRSGPLPTQHSPQRMHPYQRAVASPVEVSSGGGGLGLRNPADITLNEYNRSEGSSAEELL
631 RRTPLKIRAPEMLTAPAGYGTEPCRMTLKQEPETGY  666
```

References

1 Sladek, F.M. et al. (1990) Genes Dev. 4, 2353–2365.
2 Zhong, W. et al. (1993) EMBO J. 12, 537–544.
3 Drewes, T. et al. (1996) Mol. Cell. Biol. 16, 925–931.
4 Nuclear Receptor Nomenclature Committee (1999) Cell 97, 161–163.
5 Hata, S. et al. (1992) Biochim. Biophys. Acta 1131, 211–213.
6 Chartier, F. et al. (1994) Gene 147, 269–272.
7 Holewa, B. et al. (1997) Mol. Cell. Biol. 17, 687–694.
8 Taraviras, S. et al. (1994) Mech. Dev. 48, 67–79.
9 Duncan, S.A. et al. (1994) Proc. Natl Acad. Sci. USA 91, 7598–7602.
10 Holewa, B. et al. (1996) Mech. Dev. 54, 45–57.
11 Swevers, L. et al. (1998) Mech. Dev. 72, 3–13.
12 Costa, R.H. et al. (1989) Mol. Cell. Biol. 9, 1415–1425.
13 Sladek, F.M. and Darnell, J.E. (1992) Curr. Opin. Genet. Dev. 2, 256–259.
14 Tian, J.M. and Schibler, U. (1991) Genes Dev. 5, 2225–2234.
15 Ladias, J.A. et al. (1992) J. Biol. Chem. 267, 15849–15860.
16 Metzger, S. et al. (1993) J. Biol. Chem. 268, 16831–16838.
17 Sladek, F.M. (1994) In Transcriptional Regulation of Liver-Specific Genes (Tronche, F. and Yaniv, M. ed.) R.G. Landes Company, Austin, TX USA.
18 Jiang, G. et al. (1995) Mol. Cell. Biol. 15, 5131–5143.
19 Jiang, G. et al. (1997) Mol. Cell. Biol. 17, 6546–6554.
20 Jiang, G. and Sladek, F.M. (1997) J. Biol. Chem. 272, 1218–1225.
21 Hertz, R. et al. (1998) Nature 392, 512–516.
22 Sladek, F.M. et al. (1999) Mol. Cell. Biol. 19, 6509–6522.
23 Hadzopoulou-Cladaras, M. et al. (1997) J. Biol. Chem. 272, 539–550.
24 Zhong, W. et al. (1994) Mol. Cell. Biol. 14, 7276–7284.
25 Jiang, G. et al. (1997) Arch. Biochem. Biophys. 340, 1–9.
26 Ktistaki, E. et al. (1995) Proc. Natl Acad. Sci. USA 92, 9876–9880.
27 Reijnen, M.J. et al. (1992) Proc. Natl Acad. Sci. USA 89, 6300–6303.
28 Chen, J. et al. (1993) Nucl. Acids Res. 21, 1205–1211.
29 Diaz-Guerra, M-J.M. et al. (1993) Mol. Cell. Biol. 13, 7725–7733.
30 Nitsch, D. et al. (1993) Genes Dev. 7, 308–319.
31 Garcia, A.D. et al. (1993) J. Virol. 67, 3940–3950.
32 Fuernkranz, H.A. et al. (1994) Nucl. Acids Res. 22, 5665–5671.

33 Yanuka-Kashles, O. et al. (1994) Mol. Cell. Biol. 14, 7124–7133.
34 Rouet, P. et al. (1995) Nucl. Acids Res. 23, 395–404.
35 Erdmann, D. and Heim, J. (1995) J. Biol. Chem. 270, 22988–22996.
36 Hall, R.K. et al. (1995) Proc. Natl. Acad. Sci. USA 92, 412–416.
37 Moldrup, A. et al. (1996) Mol. Endocrinol. 10, 661–671.
38 Weber, H. et al. (1996) Development 122, 1975–1984.
39 Rollini, P. and Fournier, R.E.K. (1999) Proc. Natl Acad. Sci. USA 96, 10308–10313.
40 Wang, J-C. et al. (1999) Mol. Endocrinol. 13, 604–618.
41 Bulla, G.A. (1999) Nucl. Acids Res. 27, 1190–1197.
42 Kuo, C.J. et al. (1992) Nature 355, 457–461.
43 Miura, N. et al. (1993) Nucl. Acids Res. 21, 3731–3736.
44 Kritis, A. et al. (1993) Nucl. Acids Res. 21, 5882–5889.
45 Duncan, S.A. et al. (1998) Science 281, 692–695.
46 Lahuna, O. et al. (2000) Mol. Endocrinol. 14, 285–294.
47 Carter, M.E. et al. (1993) J. Biol. Chem. 268, 13805–13810.
48 Nakshatri, H. and Chambon, P. (1994) J. Biol. Chem. 269, 890–902.
49 Griffo, G. et al. (1993) J. Cell Biol. 121, 887–898.
50 Chen, W.S. et al. (1994) Genes Dev. 8, 2466–2477.
51 Duncan, S.A. et al. (1997) Development 124, 279–287.
52 Yamagata, K. et al. (1996) Nature 384, 458–460.
53 Furuta, H. et al. (1997) Diabetes 46, 1652–1657.
54 Sladek, F.M. et al. (1998) Diabetes 47, 985–990.
55 Lausen, J. et al. (2000) Nucl. Acids Res. 28, 430–437.
56 Miquerol, L. et al. (1994) J. Biol. Chem. 269, 8944–8951.
57 Yamagata, K. et al. (1996) Nature 384, 455–458.
58 Nakhei, H. et al. (1998) Nucl. Acids Res. 26, 497–504.
59 Hata, S. et al. (1995) Biochim. Biophys. Acta 1260, 55–61.
60 Kritis, A.A. et al. (1996) Gene 173, 275–280.
61 Gragnoli, C. et al. (1997) Diabetes 46, 1648–1651.
62 Stoffel, M. and Duncan, S.A. (1997) Proc. Natl Acad. Sci. USA 94, 13209–13214.
63 Wang, J.C. et al. (1998) J. Biol. Chem. 273, 30847–30850.
64 Hani, E. et al. (1998) J. Clin. Invest. 101, 521-526.
65 Iyemere, V.P. et al. (1998) Nucl. Acids Res. 26, 2098–2104.
66 Soutoglou, E. et al. (2000) Molec. Cell 5, 745–751.
67 Green, V.J. et al. (1998) J. Biol. Chem. 273, 29950–29957.
68 Coucouvanis, E. and Martin, G.R. (1999) Development 126, 535–546.
69 Yanai, K. et al. (1999) J. Biol. Chem. 274, 34605–34612.
70 Magee, T.R. et al. (1998) J. Biol. Chem. 273, 30024–30032.
71 Gregori, C. et al. (1998) J. Biol. Chem. 273, 25237–25243.
72 Morrisey, E.E. et al. (1998) Genes Dev. 12, 3579–3590.
73 Gu, Z. et al. (1999) Development 126, 2551–2561.
74 Nastos, A. et al. (1998) Nucl. Acids Res. 26, 5602–5608.
75 Coffinier, C. et al. (1999) Development 126, 4785–4794.
76 Kapitskaya, M.Z. et al. (1998) J. Biol. Chem. 273, 29801–29810.
77 Li, J. et al. (2000) Genes Dev. 14, 464–474.
78 Taraviras, S. et al. (2000) Biochim. Biophys. Acta 1490, 21–32.

Names

The first member of the RXR group was identified by Keiko Ozato's laboratory as a mouse factor that binds to a regulatory element of the major histocompatibility complex (MHC) class I genes, called region II[1]. It was thus called H-2RIIBP for H-2 region II binding protein. Independently, Ron Evans's laboratory identified a new orphan receptor, which responded specifically to retinoids and was thus called RXR for retinoid X receptor[2]. This name very well described one important function of this receptor and is now used by all researchers in this field, and the clone initially found by Ron Evans's laboratory is now called RXRα. H-2RIIBP was later shown to be the mouse homologue of the RXRβ gene[4]. The same gene was also found in rat by Michael Rosenfeld's laboratory and called RCoR-1 for RAR coregulator[4]. A third gene, RXRγ, was found in chicken[3] as well as in mammals[5,6].

An homologue for RXR was found in *Drosophila* by several groups. Ron Evans's laboratory found that this gene (first called XR2C) was in fact the one mutated in the ultraspiracle (USP) *Drosophila* mutant[7]. Two other groups found the same gene as a regulator of the s15 chorion gene and called it CF1 (for chorion factor 1)[8] and, as a new nuclear receptor, for its cytological position 2C1[9]. RXR homologues were found in a wide variety of metazoans. Since the USP genes from dipterans and lepidopterans appear to be extremely divergent compared to those of other insects and metazoans, the name USP tends to be specifically used in dipterans and lepidopterans, whereas the others are simply called RXR[10,11,21,23].

Species	Other names	Accession number	References
Official name: NR2B1			
Human	RXRα	X52773	2
Mouse	RXRα	X66223	5
Rat	RXRα	L06482	255
Chicken	RXRα	not avail.	220
Xenopus	RXRα	L11446	14
Zebrafish	RXRα	U29940	16
Official name: NR2B2			
Human	RXRβ	M84820	6
Mouse	RXRβ, H-2RIIBP	X66224	5
Rat	RXRβ, RCoR-1	M81766	4
Xenopus	RXRβA	X87366	147
Xenopus	RXRβB	S73269	15
Zebrafish	RXRβA, RXRδ	U29941	16
Zebrafish	RXRβB, RXRε	U29942	16
Official name: NR2B3			
Human	RXRγ	U38480	5
Mouse	RXRγ	X66225	5
Chicken	RXRγ	X58997	3
Xenopus	RXRγ	L11443	14
Zebrafish	RXRγ	U29894	16
Official name: NR2B4			
Drosophila	USP, Ultraspiracle, 2C1, CF1, XR2C	X53417	7
Aedes	USP, Ultraspiracle, AaUSP	AF210734	256, 195
Chironomus	USP, Ultraspiracle, CtUSP	AF045891	198

Choristoneura	USP, Ultraspiracle, CfUSP	AF016368	199
Bombyx	USP, Ultraspiracle, BmUSP	U06073	18
Manduca	USP, Ultraspiracle, MsUSP	U44837	19
Apis	USP, AmUSP	AF263459	Unpub.
Tenebrio	USP, TmUSP	AJ251542	20
Locusta	RXR	AF136372	10
Amblyomma	RXR1, AaRXR1	AF035577	21
Amblyomma	RXR2, AaRXR2	AF035578	21
Celuca	RXR, UpRXR	AF032983	22
Polyandrocarpa	RXR, PmRXR	AB030318	257
Branchiostoma	RXR	not avail.	202
Schistosoma	RXR-1	AF094759	24
Schistosoma	RXR-2, SmRXR	AF158102	25, 26
Tripedalia	RXR	AF091121	29

Isolation

H-2RIIBP, now referred to as RXRβ, was isolated by its ability to bind the regulatory elements of the MHC class I genes in the mouse[1]. An expression library of mouse liver was screened with double-stranded oligonucleotides that encompass the regulatory elements that are critical for MHC class I genes transcription, which harbor a AGGTCA motif[1]. The RXRα gene was found by low-stringency screening of a human and liver cDNA library with a cDNA fragment encompassing the human RARα DBD[2]. A chicken gene, now known as RXRγ, was also found by low-stringency screening with a RARα probe and was later found to be the first version of a third gene, RXRγ[3]. The three genes were also found in mouse[5], xenopus[14,15] and zebrafish[16], in which a fourth gene was also found. This fourth gene, called RXRε, is the result of a specific duplication of RXRβ gene that arose in the teleost fishes lineage[16,17].

Interestingly, the RXRs were also discovered as factors that are able to enhance the binding to DNA of several members of the nuclear receptor superfamily, such as TR, RAR or VDR (see references 12 and 13 for reviews). It was well known that nuclear factors were necessary to achieve high-affinity DNA binding of these receptors and several groups have attempted to isolate such factors. One of the strategies was a sequential screening of expression libraries using both RARE and RARα, in order to find proteins that are able to bind to both the DNA element recognized by RARs and RARα itself[4]. This clever strategy yielded RXRβ, which was effectively shown to be able to enhance the DNA binding of RAR as well as TR and VDR. Another approach was the biochemical purification of an activity able to help DNA binding of RARγ to an RARE. The resultant protein turned out to be human RXRβ[6].

The USP gene was found, like RXRα, by low-stringency screening of a cDNA library with a human probe[7]. It was also cloned as a regulator of the s15 chorion gene in *Drosophila*[8] as well as through a search for nuclear receptors in *Drosophila* using consensus oligonucleotide probe[9]. Many homologues of USP were found in insects, mainly in dipterans (*Aedes*[195], *Chironomus*[198]) and lepidopterans (*Bombyx*[18], *Manduca*) either by low-stringency screening or by PCR[18,19]. These sequences were found to be related to the *Drosophila* USP. USP homologues in other insects (*Locusta* or the beetle *Tenebrio*) as well as in arthropods (the tick *Amblyomma* or the fiddler crab *Celuca*) are less numerous and were found to be distant from *Drosophila* USP and, strikingly, more related to RXR (see references 23 and 30 for reviews)[10,20-22]. In the tick, two genes resulting from a specific gene duplication[21,23]. The same situation was found in a flat worm *Schistosoma mansoni*[24-26] in which one of the copy appears to be more divergent (see reference 27 for a review). Homologues of RXR were also found in early metazoans such as the Cnidarians (see reference 30 for a review). A short RXR signature was identified by PCR in *Hydra vulgaris*[28], whereas a full-size cDNA that generates a protein apparently able to bind 9-cis retinoic acid was described in the diploblast *Tripedalia cystophora*[29].

Identity levels between the three vertebrate RXR genes are high: 88–95% in the DBD and 89–93% in the LBD, the two most closely related being RXRα and γ. The *Drosophila* USP gene is identical at 82–85% and 49–51% with the three human paralogues in the DBD and LBD, respectively. The USP genes from insects outside dipterans and lepidopterans are much more closely related to vertebrates RXR in their LBD (c. 70%)[11,23].

DNA binding

The basis of the isolation of RXRβ was its ability to bind on a specific region of the MHC class I genes regulatory elements[1]. This region was shown to contain an AGGTCA core motif and may also contain an additional degenerated core motif (GGGTGG) with a 1 bp spacing with the consensus one. The same study reports that RXRβ was able to bind to an ERE but not to other elements, such as the growth hormone gene TRE or a GRE from the TAT gene. These results apparently contrast with that obtained by Ron Evans's laboratory during the characterization of human RXRα, since it was shown that the product of this gene can bind a palindromic TRE[2]. In fact, it was shown later that baculovirus-produced RXRβ is able to bind to the ERE as well as to some, but not all, TREs and RAREs[31]. It was shown later that RXR produced in reticulocyte lysates is able to form strong homodimers on the DR1 element and more weakly on DR4 and DR5 elements based on the AGGTCA motif[37]. The binding on direct repeats based on the AGTTCA motif was much weaker[37]. A systematic study of the DNA-binding abilities of pure and functionally homogenous recombinant RXRα (lacking the A/B region) produced in bacteria, baculovirus or in COS cells was done[32]. This study revealed that it binds to DR1, DR4 and DR5 elements as a dimer and more weakly to DR0 and DR3 as a monomer[32]. *In vivo* occupancy of RXR binding sites, such as the ones in the regulatory regions of the MHC class I genes or in the RARβ2 promoter, was demonstrated in tissues in which these genes are strongly expressed and was found to be hormone-dependent[38,208].

One major breakthrough in the nuclear receptor field was the observation that RXR is a common partner for numerous nuclear receptors, such as TR, RAR and VDR, and that each heterodimer can recognize a distinct type of response element[4,6,34-36,52]. For example, the RXR–VDR heterodimer binds to a direct repeat of the AGGTCA core motif with a 3 bp spacing (DR3), whereas RXR–TR binds to a DR4 and RXR–RAR to a DR5 (reviewed in references 12 and 13 ; see below). The spacing was shown to be not the only determinant for the specificity of the binding, since the sequence of the core motif itself, the sequence of the spacer or that of the flanking nucleotides also plays an important role in this process (see references 42 and 44 among others, and reference 13 for a review). Many other nuclear receptors have been shown to interact with RXRs, including some orphan receptors (see below). The interaction of RXR not only increases the DNA-binding efficiency of its partner but it also alters the binding-site repertoires of RAR or TR. This is mainly dependent on the heterodimerization interface, which is present inside the DBDs of these receptors[37,39,40]. The dimerization interface, which is present in the LBD, further stabilizes but does not change the binding repertoire dictated by the DBDs. The various heterodimers can bind to a wide variety of response elements, including direct repeats, palindromes and inverted palindromes[37,42]. This suggests that the DBD of these receptors, including their common partner, is rotationally flexible with respect to the LBD. Since RXR was found to be located in the 5' position of the various direct repeats, this suggests that the flexibility existing between the DBD and the LBD allowed the heterodimer to adapt its DNA-binding interface to each type of response element[37,39,40,44]. This flexibility was further demonstrated by the observation that, on DR1, heterodimers of RAR and RXR can bind to DNA in a reverse polarity (i.e. 5'RAR–RXR-3')[41]. Interestingly, these heterodimers cannot activate transcription even in the presence of the cognate ligands and thus they inhibit the transcription of RXR homodimers through these elements[41]. Each type of complex of a given receptor, such as TR (monomer, homodimer or heterodimer), contacts different nucleotides within the response elements, suggesting that the precise interaction with DNA is different (see reference 43 as an example). The same notion has also been demonstrated by the analysis of the bending induced by the various types of complexes of RAR, TR and TR on the response elements[47].

It has been proposed that the ligand, 9-cis retinoic acid promotes the binding of homodimers of RXR to DR1 elements[33]. These results led to the suggestion that 9-cis retinoic acid may play an important role in modifying the equilibrium between monomeric, homodimeric and heterodimeric receptors[33,53]. Nevertheless, this effect was not observed using purified RXR preparation and it was suggested that it may be a consequence of a conformational stabilization of improperly folded in vitro synthesized proteins[32]. The study of the effect of RXR-specific ligands on the stability of RXR tetramers (see below) reveals that ligand binding led to tetramer dissociation, giving rise to a large number of RXR molecules that can form either homodimers or heterodimers[201,203]. Thus the modulation of protein–protein interaction by the ligand may alter the magnitude of a transcriptional response.

In insects it was shown that USP forms a strong and stable heterodimer with the ecdysone receptor, EcR (NR1H1), and that this heterodimer binds to the described EcRE, such as that found upstream of the *hsp27* gene[48,49]. Experiments using *in vitro* selection of the DNA-binding site revealed that the preferred target sequence of the EcR–USP heterodimer is a perfect palindromic sequence based on GAGGTCA[50]. The two repeats are separated by a unique central base, which is always A or T. It was also observed that the EcR–USP heterodimer is able to bind to direct repeats with a lower affinity[50,51]. This is in accordance with reports showing that several EcREs found in genes regulated by the EcR–USP heterodimer are direct repeats (see reference 45 and the EcR chapters for examples). In accordance with these findings, the mosquito EcR–USP heterodimer is able to bind to palindromic elements as well as, although less avidly, to direct repeats[46]. Direct repeats in which the two half-sites are separated by 11–13 bp were also found to be recognized by this EcR–USP heterodimer.

Partners

RXR is the major heterodimeric partner of numerous nuclear receptors. The first, for which an interaction with RXR was shown to be required for high-affinity DNA binding, were RAR, TR and VDR[4,6,34–36,52,58]. Other partners for RXR include PPARs[57,74], LXRs[67–70], FXR[66], PXR[75–79], CAR[59,60] and two members of the NGFIB group, NGFIB and NURR1[71,72] (see the relevant chapters for an in-depth discussion of each case). COUP-TF has been described as a partner for RXR but this remains controversial[56,58,61–64]. An interaction between SHP and RXR has been found, but it cannot be compared to a real heterodimer formation, since it involved the AF-2 region of RXR, which recognizes an LXXLL motif within SHP in contrast to the other heterodimers (see the relevant chapter)[65]. USP has been found to form heterodimer with EcR[48,49] as well as with DHR38[73,197]. USP is not only important for efficient DNA binding of EcR but also for its ability to bind its ligand, 20-hydroxy-ecdysone[194] (see the EcR chapter). It has been suggested by phylogenetic analysis that to be an efficient partner for RXR a nuclear receptor should belong to two of the six known subfamilies, mainly subfamily I and IV[30,81]. This does mean that all members of these subfamilies interact with RXR, since it is clear that, for example, Rev-erb genes that belong to subfamily I are unable to form an heterodimer with RXR[80].

In most cases, the partner of RXR does not exhibit a marked preference for one of the three RXR subtypes. It has even been found that RXR may replace USP as a partner for EcR, suggesting that the heterodimerization properties of USP and RXR are strongly conserved[55]. USP is also able to heterodimerize with TRβ to transactivate the human Apo-AII promoter[196]. Nevertheless, in some cases, such as the RXR–VDR heterodimer bound to the rat 24-hydroxylase promoter, a preference for one RXR subtype over another (in this case RXRα and RXRγ over RXRβ) was detected[54].

The regions of RXR implicated in heterodimer formation have been mapped. The DBD as well as the LBD contains dimerization interfaces[4,6,35,36,52]. It is now clear that the DBD dictates the type of response element bound by each heterodimer, whereas the LBD interface only reinforces the strength of the interaction[37,39,40,86]. The interaction interfaces within the DBD and LBD have been precisely mapped by structure–function studies. Within the LBD it was known that the regions referred to as heptad repeats, and in particular the ninth heptad, close to helix 10, are important for dimerization of nuclear receptors[88]. It was not clear whether this region plays a role in heterodimer formation between RXR and other nuclear receptors. Deletion experiments have shown that the C-terminus of RXR, which contains the ninth heptad, is required for heterodimer formation[6,36,52]. Further analysis clearly emphasized the importance of that region for the dimerizing activity of RXR[82,85–87]. Interestingly, detailed mutagenesis of this region in RXRα has allowed separation of the homodimer and heterodimer formation[85]. The interaction between the LBDs of TR and RXR is increased by binding of TR- and/or RXR-specific ligands on their respective receptors[260].

Within the DBD, the D-box of the second zinc finger is specifically required for heterodimerization with TR and RAR on DR4 and DR5 elements, respectively[83]. Interestingly, the interaction surface between RAR and RXR on a DR2 element involved the second zinc finger of RXR with the exclusion of the D-box. This suggests that the interface between RAR and RXR within the DBD is not the same when the heterodimer is bound on a DR2 or on a DR5 element[84]. The same region, i.e. the second zinc finger of RXR, as well as the T-box region, are implicated in the formation of RXR homodimers on DR1 elements[84,165].

The importance of the heterodimerization between RXRs and their partners has been shown by genetic studies (see below) as well as by the study of dominant negative RXRs in transient transfection. Mainly two types of dominant negative mutants were generated: (1) mutants in which the DBD was deleted and which form non-functional dimers with RARs[89]; and (2) mutants deleted of their extreme C-terminus (AF-2 function), which dimerize with RARs and compete with bona fide RXR–RAR heterodimers for binding to their cognate DNA elements[90]. In both cases, these mutants abrogate retinoic acid gene regulation, suggesting that the RXR function is critical for RAR signaling[89,90].

RXR is not only able to form homodimers on DR1 elements, it can also form stable tetramers[91,92]. This was observed in RXR target genes, such as the cellular retinol-binding protein II (CRBP-II) gene that contains numerous weak half-site sequences that cooperatively recruit RXR monomers to form a functional tetramer as well as other high-order structures[91]. This ability to form tetramer is apparently isoform specific since RXRβ cannot form these complexes. Tetramers of apo-RXRα were also observed in solution and apparently occur predominantly at a protein concentration of about 70 nM[32,92]. Interestingly, ligand binding induces dissociation of these tetramers to smaller species, suggesting that tetramers represent a repressed stored version of RXR that is available for either homodimer or heterodimer formation[93,201,203]. The region of the protein implicated in tetramer formation is located in helix 11 in which two consecutive Phe residues were shown to be critical for tetramer formation[94]. The conformation of the tetramerization domain, but not of the dimerization domain in helix 10, is apparently modified upon ligand binding, explaining why tetramers are disrupted in the holo-receptor. These data have recently been confirmed by the resolution of the three-dimensional structure of a tetramer of RXRα LBD either in the apo-form or complexed with antagonists[262]. These structures reveal that the RXR LBD tetramer forms a compact, disk-shaped complex, consisting of two symmetrical dimers that are packed along helix 3 and 11. In each monomer the helix 12, containing the AF-2 domain, protrudes away from the core domain and spans into the coactivator binding site in the adjacent monomer of the symmetrical dimer. This configuration physically excludes the binding of coactivators and suggests that an autorepression mechanism is mediated by the helix 12 within the tetramer.

RXR interacts with other proteins that are required to mediate its transcriptional activity. This is the case for proteins that participate to the pre-initiation complex, such as the TATA-binding protein, TBP. The AF-2 AD region of RXR, in the extreme C-terminal part of the protein, interacts with TBP in a ligand-independent manner[95]. RXRs also interact with TBP-associated factors (TAFs), such as TFIIB, $TAF_{II}110$ and $TAF_{II}28$, in a ligand- and AF-2-dependent manner[95,100,189,204]. $TAF_{II}28$, potentiates the AF-2-mediated transcriptional activation by RXR but not through a direct physical interaction[100]. Other interactions of RXR with nuclear proteins involve the Oct1 transcription factor[258] and the poly(ADP-ribose) polymerase (PARP)[259]. In both cases the interaction surface is localized in the DBD of RXRα and results in the inhibition of the transcriptional activation of target promoters by the heterodimer.

Given its pivotal role in enhancing the transcriptional abilities of many nuclear receptors, it is not surprising that RXR interacts with a number of coactivators. Among these are the members of the p160 family of transcriptional coactivators SRC-1[105], TIF2[106] and ACTR[107]. All of these factors interact with RXR in a ligand- and AF-2-dependent manner (see references 108 and 109 for reviews). Given the fact that the different partners of an heterodimer can have distinct but overlapping abilities to bind coactivators and that coactivators contain several receptor interaction domains, the stoichiometry of the interactions between coactivators and heterodimers is complex[97,110–113]. Two molecules of TIF2 can bind in a cooperative manner to an heterodimer between RXR and TR even in the presence of only one ligand[111]. How the receptors present in the heterodimer interpret the presence of ligand is still a matter of discussion and of strong research effort[108,109,115,183]. The interaction surface between RXR and coactivators of the p160 family is composed of an hydrophobic surface that is formed by the helix 12 as well as by residues in helices 3 and 5 of the LBD[114,183]. This has been confirmed by crystallographic analysis (see below).

RXR interacts in a ligand-dependent manner with numerous other coactivators, such as TRAP220[98], a member of the TRAP/DRIP/SMCC complex that plays an important role in transcriptional activation by nuclear receptors (see reference 99 for a review). The interaction between RXR and the cointegrators CBP and P300 that participate in the p160 coactivator complex has also been described[118,119]. Other molecules that interact with RXR and may play a role in mediating its transcriptional effects are RIP140[120], TIF1[121,122], SUG1[122,123], NSD1[124], TLS[125], PSU1[126], NRIF3[127], ASC-2[128] and RAP250[129]. Interestingly, IκBβ proteins were shown to interact directly with RXR through the AF-2 region and to inhibit 9-*cis*-RA-mediated transactivation[206]. Bcl3, another IκB family member, also interacts with RXR through two different regions of RXR[205]. In accordance with its ability also to interact with TFIIB, TBP and TFIIA, Bcl3 behaves as a coactivator of RXR.

No interaction was described between RXR and corepressors, suggesting that RXR homodimers in the absence of ligand have a weak repressing activity[116,117]. No interaction was found between RXR and NCoR, whereas a very weak and probably insignificant one was observed with SMRT[101,102]. This is due to the helix 12 of RXR that masks a functional but latent NCoR binding site that is present within RXRs[96]. It is believed that the heterodimerization, of RXR with TR, unmasks the NCoR binding site of RXR, suggesting that RXR directly participates in repression[96]. The persistence of the interaction between RAR and NCoR in the presence of a ligand explains why the RAR–RXR heterodimer bound on the DR1 element represses transcription, whereas it activates when bound on DR5 (see reference 104 for a review)[103]. Several lines of evidence suggest nevertheless that RXR is critical for the repressing activity of RXR–RAR or RXR–TR heterodimers (discussed in reference 96).

Ligands

When it was discovered, RXRα was immediately suspected to be a receptor for an unknown retinoid, since it can be activated by a high concentration of all-*trans* retinoic acid without directly interacting with this compound[2]. A systematic search for a natural ligand for RXR led to the discovery that 9-*cis* retinoic acid, an isomerization product of all-*trans* RA is a high-affinity ligand for RXRα as well as for the two other subtypes, RXRβ and RXRγ[5,166,167]. It was further shown that, among several stereoisomers of all-*trans* RA, including 7-*cis*, 9-*cis*, 11-*cis* and 13-*cis*, only 9-*cis* RA displays a high-affinity binding to RXRα[168]. It was further demonstrated that RXR cannot bind all-*trans* RA, whereas RARs bind both all-*trans* RA and 9-*cis* RA[168]. The K_d value for the binding of 9-*cis* RA to RXR is around 10 nM[166,167], whereas the K_d for RAR is 0.2–0.7 nM[168]. Interestingly, two of the four subtypes of RXR known in zebrafish, RXRδ and RXRε, which result from a fish-specific duplication of the RXRβ gene, are unable to bind to 9-*cis* RA and are transcriptionally inactive[16].

Since 9-*cis* RA can bind to both RXR and RAR, it was critical to find synthetic compounds that only recognize RXR in order to decipher the role played by this receptor and its ligand activation[174]. The most potent of these compounds, such as SR11217 and SR11237, activate RXR at 100 nM. Interestingly, these two compounds were found to activate only RXR homodimers and not the RAR–RXR heterodimer. Other groups have found similar ligands, such as LG100268[187] and LGD1069[188], which specifically bind to RXR and activate RXR homodimer-dependent transcription. All these RXR-specific compounds are called 'rexinoids'[178]. A compound that is an antagonist toward an RXR homodimer but functions as an agonist of a specific RXR heterodimer have been described[189]. These types of compounds were used to understand better by what kind of mechanism a given gene was activated by retinoids. For example, in F9 cells, the RARβ gene was not activated by SR11237, whereas the rat growth hormone gene was induced by this compound, suggesting that RXR plays an active role in its transcription[175]. Even complex physiological processes can be scrutinized using this kind of analysis. The effect of retinoids in promoting granulocytic differentiation and apoptosis of HL60 cells is due to an RAR-dependent effect on differentiation and to an RXR-dependent effect on apoptosis[176]. Some of the RXR-specific compounds are promising tools for the treatment of metabolic diseases, such as diabetes and obesity (see reference 178 for an example). RXRs can also be activated by several types of natural compounds, such as the insect growth regulator, methoprene[177] or, in the case of the RXRα mutant F318A, by oleic acid[162,181]. Recently, docosahexaenoic acid, a long-chain polyunsaturated fatty acid that is highly enriched in mammalian brain, has been proposed as a ligand of vertebrate RXR[263]. This striking finding has reactivated the discussion on the physiological significance of 9-cis retinoic acid binding to RXR.

It has been proposed that 9-cis RA promotes the binding to DNA of RXR homodimers and thus shifts the equilibrium inside the cells from RXR engaged in heterodimers, which can participate in various hormonal responses to RXR engaged in homodimers[33,53]. This effect has nevertheless not been reproduced using purified RXR and its reality thus appears dubious[32].

Using the specific ligands, it was possible to understand the relationships better existing between each partner in the heterodimer. This is particularly interesting in the case of heterodimers containing two liganded receptors, since the various types of heterodimers behave differently with the addition of only one or both ligands (see references 213 and 214 for examples, and references 215 and 216 for reviews). Several lines of evidence, including the in vivo genetic evidence described below, suggest that RXR is transcriptionally active[137,146,210]. The study of a constitutively active RXR mutant F318A suggests that different types of partner may exert different effects, from inhibition to synergism, on RXR transcriptional activity[181]. This and other lines of evidence derived from the study of differentiation of myeloid cells by retinoids and synthetic analogues of retinoids[180,184] led to the concept of RXR subordination[181]. This concept suggests that the structure of the RXR–RAR heterodimer is incompatible with the existence of an active RXR AF-2 interaction surface for coactivators. This subordination occurs independently of corepressor binding and is relieved when a ligand to RAR induces an allosteric change in conformation allowing the generation of an active surface for coactivator binding in RXR[181]. Both ligands have clear synergistic effects[210,211]. Another model called the 'phantom ligand effect' suggests that the ligand-induced transcriptional activity is exerted entirely by the RAR partner[182]. This model impies that the function of agonist-bound RXR would be limited to an allosteric increase of the affinity of coactivators to RAR and thus that RXR by itself is transcriptionally silent[41,71,174,240]. This model probably accounts for the relationships between RXR and LXR, since ligand binding by RXR allows the AF-2 region of LXR to recruit a coactivator by a dimerization-mediated conformational change[191]. The same model was suggested for the PPAR–RXR heterodimer since the mutation of the AF-2 region of PPARγ extinguishes the activity of the heterodimer liganded either by PPAR- or RXR-specific ligands, whereas the mutation of the AF-2 domain of RXR has no effect[213]. Nevertheless, in the case of the RXR–RAR heterodimer, a growing body of evidence suggests that target genes of the RAR–RXR heterodimer can be activated by both subunits (see reference 212 as an example). This discussion is thus not yet resolved.

Since little was known of the *in vivo* function of 9-*cis* RA, several groups have characterized its biological functions in a variety of model systems. Like all-*trans* RA, 9-*cis* RA was found to be a potent inducer of digit-pattern formation in the chick wing bud[172]. Using this system, it was shown that 9-*cis* RA can regulate endogenous gene expression, suggesting that ligand-bound RXR can mediate retinoid signaling *in vivo*[170]. Endogenous 9-*cis* retinoic acid was detected in a number of situations *in vivo*. In *Xenopus* embryo it was localized primarily within the anterior and posterior (but not middle) dorsal regions of the embryo[173]. RXR-selective ligands produce malformation along the anterior–posterior axis in *Xenopus* embryos and regulates endogenous genes such as HoxA1 and Xlim-1[209]. It is also produced and released by the wound epidermis in the Urodele *Notophtalmus viridescens*[171]. The fact that RXR is active *in vivo*, suggesting that it may be liganded with 9-*cis* RA, has been demonstrated using a reporter transgenic mouse line. RXR was found to be active in the developing spinal cord, which was already known as an important place for retinoid synthesis[146]. Rexinoid treatment of diabetic and obese mice led to sensitization to insulin, suggesting that, together with PPARγ, RXR is a promising target in metabolic disease[178]. This is also the case for other diseases, such as cancer, since it was shown that differentiation of all-*trans* RA-resistant t(15:17) leukemic promyeloblasts can be achieved by a RAR-independent RXR signaling[190].

The *Drosophila* homologue of RXR, USP, has long been considered as an orphan receptor. Thus, the claim that juvenile hormone may be a ligand for USP[185,186] was thus considered with skepticism. Nevertheless, the recent finding that a ligand occupies a hydrophobic pocket inside the LBD of the *Heliothis* USP suggests that USP may not be an orphan receptor and that a ligand remains to be identified for this receptor[164]. Since insect USPs are extremely variable, it is possible that some insect USPs have lost their ability to bind this unknown ligand, whereas others have kept it[10,11,21,23].

Three-dimensional structure

The structure of the DNA-binding domain of RXRα in solution has been determined by NMR. The general architecture of this DBD is identical to that of other nuclear receptors, such as ER and GR, with two perpendicular oriented α-helices among which one, containing the P-box sequence, is exposed for DNA binding[156]. However the RXR DBD contains a third α-helix immediately after the second zinc finger in the regions known as the T-box. This helix has been shown by structural and mutational analysis to mediate protein–protein and protein–DNA interactions required for heterodimeric binding[156,157,165]. The important role of the third helix for RXR interaction with DNA has been confirmed by the resolution of the structure of RXR and TR bound to a DR5 element[157]. This structure also confirms the mutational analysis that defines the interface that exists between RXR and TR, and reveals that steric hindrance plays an important role in the discrimination of the correct spacing between direct repeats by the various heterodimers[83,84].

The structure of the LBD of RXRα in the absence of ligand (apo-RXRα) was the first structure of the LBD of a nuclear receptor to be solved[158]. The RXRα LBD was crystallized as an homodimer. Two α-helices H10 and, to a lesser extent H9, form the dimerization interface, consistent with the mutational studies that identify helix 10, which contains the ninth heptad repeat as the critical region for dimerization (see above and reference 158). The loop between H8 and H9 also plays a role in the dimer interface together with H11 and the loop between H9 and H10. The AF-2 AD region is part of the last helix, now called helix H12, that extends from the core of the LBD. Interestingly, in this conformation it impairs the binding of coactivators in the hydrophobic cleft and thus locks the apo-receptor in the inactive state (see references 159 and 160 for reviews). More recently, the structure of the RXRα LBD complexed with its ligand, 9-*cis* retinoic acid, was solved and compared to the apo-RXRα structure[161]. There was no contact between the ligand and helix H12 that nevertheless forms a lid to the hydrophobic pocket as in the case of the holo-RARγ LBD. The hydrophobic pocket itself is formed by residues conserved in other RXRs located on helices H3, H5, H7 and H11 as well as on the β-turn. Several amino acids, such as Ile 268, Cys 269, Ala 271, Ala 272, Phe 313, Arg 316, Leu 326, Ala 327, Cys 432 and His 435 among others, directly interact with the ligand. The fact that most of the residues that are in contact with the ligand are conserved in the three RXR subtypes may explain why, in contrast to RARs, no isoform-specific ligands have been found. The crystal structure of a heterodimer between RXR–RAR LBDs has also been determined[162]. Interestingly, it was not the wild-type RXR that was used but a constitutively active mutant, RXRα F318A[181], whereas the RAR moiety was complexed with an antagonist. Surprisingly, a ligand, a fatty acid in C18 oleic acid was observed inside the RXR LBD! It was shown that this fatty acid can regulate RXRα F318A transcriptional activity but not that of the wild-type RXRα. This suggests that the presence of oleic acid accounts for the constitutive activity of RXRα F318A. This finding is striking given that a polyunsaturated fatty acid has been found to bind and activate to mammalian RXRs[263]. This structure also allowed the dimerization interface between RAR and RXR in the LBD to be studied. This interface was shown to be similar to that found in RXR homodimers, involving residues from helices H7, H9, H10 and H11 as well as loops 8–9 and 9–10. The precise comparison of the surface contributed by each helix in the dimerization interface in the case of the homodimer and heterodimer shows that, in both cases H9 and H10 contribute to more than 75% of the total surface. Some differences in the relative involvement of other secondary structures are, however, observed and may explain the existence of RXR mutants that discriminate between homodimer and heterodimer formation (see above). This analysis also allowed a clearer understanding of the rules governing the specificity of heterodimer formation between nuclear receptors. For example, within the receptors that interact with RXR, the conservation of interface residues is significantly higher than that of the entire LBD. These data have also been confirmed by the resolution of the structure of the LBDs of RXRα complexed to 9-*cis* RA and holo-PPARγ forming a heterodimer together with a coactivator peptide[163]. In this structure, the heterodimer interface is composed of conserved motifs in PPARγ and RARα that formed a coiled coil along helix 10 with additional charge interactions from helices 7 and 9. This structure also shed light on the permissive activation of the PPARγ–RARα heterodimer by 9-*cis* RA. The structure of the RXRα LBD tetramer has recently been solved[262]. This structure, discussed above, reveals the tetramer interface and shows how an autorepression mechanism involving helix 12 maintains the tetramers in an inactive state.

Recently, the crystal structure of the *Heliothis* USP LBD was solved[164]. As in the case of the RXRα F318A mutant, a ligand, phosphatidylglycerol or phosphatidylethanolamine, which induces an unusual antagonist conformation, was found buried in the structure. Important structural differences with RXRα LBD were observed. This concerned the loop connecting helices H1 and H3, which appears to be a key element in the USP LBD structure in its antagonistic conformation. Indeed this loop sterically impairs the H12 helix containing the AF-2 function causing it to have an antagonist conformation. One important observation made in this structure is that the USP LBD contains a large binding cavity that is similar in shape to the RXR hydrophobic pocket. This suggests that a natural ligand of USP may exist.

Expression

Northern blot experiments revealed that rat RXRα is expressed as a 4.8 kb mRNA in liver as well as in muscle, lung and kidney[2]. Low levels of expression were found in intestine, heart and spleen, whereas no expression was detected in brain, adrenal, pituitary and testes. In mouse embryo the expression of the 5.6 kb RXRα transcript was expressed from E10.5 to birth at a fairly constant level[5]. Mouse RXRβ is expressed as a 2.4 kb transcript at high levels in brain and thymus, and at a lower level in spleen and liver[1]. In embryos, two transcripts of 3.4 and 2.8 kb were detected throughout embryogenesis[5]. In rat, RNAase protection experiments confirm the expression in spleen and reveal the presence of a transcript in muscle, lung, heart and kidney, but not in liver, gut and pancreas[4]. By northern blot two transcripts of 3.0 and 2.7 kb were detected in adult rat tissues[5]. RXRγ was found as two transcripts of 2.0 and 2.5 kb in adult rat tissues. Interestingly, these two transcripts do not have the same expression pattern: the 2.0 kb transcript was found in adrenal, heart, kidney, liver and muscle, whereas the larger one was found in brain and at a lower level in lung[5].

In situ hybridization was also performed during mouse development in order to delineate the expression pattern of each RXR gene[5,217]. The RXRα transcript is found predominantly in the epithelia of the digestive system, such as the villi of the small intestine or the epithelia of the colon, the skin and the liver. Expression is low in the central nervous system and in the skeleton. RXRα was found expressed in the limb during chicken development[220]. RXRβ was even more widespread in most tissues of the embryo, including the central nervous system. This ubiquitous expression has also been found in *Xenopus*[15]. RXRγ displays the highest restricted pattern of the three receptors. The most intense signal was found in the corpus striatum in the brain as well as in the pituitary. Transcripts were also detected in the myogenic lineage, i.e. in myotomes and subsequently in various muscles, including those of the face and the limb. Earlier in development a signal was observed in the ventral horn of the spinal cord. This expression in the peripheral nervous system was also observed in chicken[3]. The myogenic expression was confirmed by other workers that observed that the onset of RXRγ expression in muscle coincides with the differentiation of limb myoblasts[193]. Of note, in the central nervous system, the expression of the RXR proteins differs notably from the picture observed by *in situ* hybridization, suggesting an important post-transcriptional regulation of RXR expression[218].

RXR gene expression in F9 cells is regulated by all-*trans* and 9-*cis* retinoic acid, and both compounds have the same effect on the various genes with no synergism when they are added simultaneously[219]. The level of RXRα and of one of the RXRγ transcripts (the large one at 3.8 kb) increases with retinoid treatment, whereas the amount of the small RXRγ transcript (2.3 kb) decreases. RXRβ mRNA levels are unaffected by retinoids.

USP has been shown to be expressed as a unique 2.4 kb mRNA throughout the life cycle, including a high level in adult[7]. At the pre-metamorphic stage, the highest level of USP transcript and protein was observed during the late third larval instar[221]. Ecdysone has complex effects on USP gene expression. In mosquito, in which two N-terminal USP isoforms exist, it exerts opposite effects on each of them[102]. In the budworm *Choristoneura*, USP expression follows ecdysteroid titers, suggesting a positive regulation[199]. In *Manduca* a decrease of one of the USP isoforms is observed after ecdysone treatment[222]. The *Tenebrio* USP protein is phosphorylated by 20-hydroxyecdysone treatment, suggesting that post-transcriptional modifications may also play an important role in the ecdysone response.

Activity and target genes

As discussed above, RXRs are weak repressors[116,117] and the precise role of their transcriptional activation function in the context of the various heterodimers has been questioned (see reference 215 for a review). Analysis of the transactivation functions have led to the conclusion that RXRs contain two autonomous activation functions: AF-1 in the A/B region, and AF-2 in the LBD[235,238]. The AF-1 function of RXRγ is located in the 103 first residues and function in a ligand-independent manner[238]. Since the A/B region is often variable in various isoforms because of alternative splicing, it is anticipated that the various RXR isoforms will have different cell-type- and promoter-specific activities. The AF-2 function is separable from the ligand-binding pocket and is located in the extreme C-terminus of RXRs[179,235,237]. It functions in a ligand-dependent manner and it is dependent on the nature of the response element on which the RAR–RXR heterodimer is bound[237]. Given the complex allosteric regulation existing between each partner (see above), it is believed that the AF-2 function can be different in each type of heterodimer.

RXR is also active in yeast[241–243] as well as in *in vitro* transcription assays[239,240]. The first *in vitro* transcription experiments were conducted in the context of the RARα/RXRα heterodimer using a DR5 RARE as a response element[240]. They have been useful to demonstrate clearly that the ligand induces a conformational change in the C-terminal part of the receptor. In this assay the RXR AF-2 function was found to be inactive and this finding has fuelled the controversy on the role of RXR transcriptional activation (see above)[240]. This assay was also useful to observe the inhibitory effect of chromatin of transcriptional activity and the ability of liganded heterodimer to relieve this repression[239]. The addition of the p300 coactivator further enhances the transcription. A subordination of RXR to the RAR activity was observed in this assay. In yeast, both RXRα and RARα were found to be transcriptionally active[242]. The RAR–RXR and TR–RXR heterodimers are active in a ligand-dependent manner on a variety of response elements[241,243].

RXRα is a phosphoprotein that is a target of the MAP kinase pathway. The phosphorylation of RXRα at a consensus site for MAP kinase at Ser 260 results in attenuated transactivation by its partner, VDR[234]. Other authors have confirmed the phosphorylation of RXRα on specific sites of the A/B and D/E regions by MAP kinases, such as JNK or MKK4[236]. This phosphorylation also inhibits the transcriptional activity of RXRα in a Tyr 249-dependent manner[236]. The AF-1 and AF-2 functions of RXRγ can be regulated by cAMP-dependent protein kinase[238].

Given the fact that RXR is an obligate heterodimeric partner for many nuclear receptors, the number of its potential target genes is enormous. These are presented in each specific entry for TR, RAR, VDR, PPAR as well as EcR. Recent findings emphasized the importance of RXR in gene regulation as a partner of LXR and FXR[252]. Below we will only indicate those who seem physiologically interesting and those that are targets of RXR homodimers. The cellular retinol-binding protein type II (CRBPII), which plays an important role in the retinoic acid synthesis pathway, is up-regulated by all-*trans* retinoic acid in the presence of RXR but not of RAR[246]. The CRBPII promoter contains five DR1 elements[246] that are recognized by RXR that can form specific oligomers[91]. Another enzyme implicated in the storage of all-*trans* retinoic acid, mouse cellular retinoic acid binding protein II (CRABPII), can be regulated by RAR–RXR via two direct repeat elements that are DR2 and DR1. This promoter is responsive to both 9-*cis* RA and all-*trans* RA as well as to HNF4 and COUP-TFs[245,248]. The human gene is also regulated by RAR–RXR but via a DR5 element present far upstream of the proximal promoter[249]. This suggests that a distinct type of RARE can confer a similar physiological response to homologous genes. Another enzyme, P450RAI also known as CYP26, is implicated in retinoic acid metabolism and is regulated by the RARγ–RXRα heterodimer in F9 cells[247]. An enzyme implicated in vitamin D metabolism, 24-OH-hydroxylase, can be regulated *in vivo* by rexinoid in an RXR-specific manner[250]. This clearly confirms that RXR is a bona fide hormone receptor. Several other target genes are interesting to mention, such as the dopamine D2 receptor, which is regulated by retinoic acid via a RARE present in its promoter[251]. This finding is interesting, since the knockout of the RXRβ gene led to mice with impaired dopamine signaling in which this receptor is down-regulated[153]. The alphafetoprotein gene promoter has been shown to respond to RXR[253]. The case of negative regulation of the TSH-β gene by TR via a negative TRE is interesting since liganded RXR antagonizes the inhibitory action of TR[254].

Knockout

The knockout of all the members of the RXR group have been performed in the mouse (see reference 144 for a review). The inactivation of the RXRα gene is lethal during embryogenesis[130,131]. The most prominent defect of these animals is a hypoplastic development of the ventricular chamber of the heart as well as ocular malformation. Both defects were observed in the vitamin A deficiency syndrome, supporting the idea that RXRα is involved in retinoid signaling *in vivo*[155]. Nevertheless, the limbs of these animals appear normal despite the fact that retinoid signaling has been shown to play an important role in limb development. This is probably due to redundancy with other retinoid receptors. Interestingly, treatment of embryos with vitamin A, which normally induces teratogenic effects in limbs, has no effects in RXRα$^{-/-}$ mice, implicating RXRα in the teratogenic effects of retinoic acid[132]. RXRα also plays a role in retinoic acid-induced cleft palate[135]. RXRα is also required for the correct differentiation of retinoic-acid treated F9 cells[140]. The heart phenotype in RXRα mutant mice has been scrutinized by several teams. The genes that are regulated by RXRα during heart development have been identified by differential display[133]. The majority of these genes are implicated in the generation of energy and metabolism. This is consistent with the observation that mitochondria are more numerous in the myocytes of RXRα mutant heart than in their wild-type littermates. The tissue-specific knockout of RXRα in the heart gives rise to normal embryos without congenital heart abnormalities[134]. This suggests that the defect observed in RXRα null mice is non-cell autonomous and that interaction between cardiomyocytes and other cell types for which RXRα function is required takes place during development[134,136]. Other tissue-specific mutations involving RXRα have been performed in hepatocytes[138] and skin[139]. The hepatocyte-specific mutation reveals that RXRα is a crucial partner for many other nuclear receptors, such as PPARα (NR1C1), CARβ (NR1I4), PXR (NR1I2), LXR (NR1H2 and 3) and FXR (NR1H4), which cannot activate their target genes efficiently when it is inactivated[138]. The skin-specific mutation of RXRα has been performed using an inducible Cre-ErT recombinase that is only active in the presence of tamoxifen expressed under the control of genes expressed in specific layers of the skin. This allows the specific ablation of the RXRα gene in a specific location when animals received tamoxifen. This elegant method, combined with crosses with other mice mutants for RXRβ and γ, allowed the respective roles of RXRα, β and γ to be deciphered. It showed that it is RXRα that plays the major role during skin development and that some redundancy exists between RXRα and β[139]. Transgenic mice using a dominant negative mutant of RXRα have also implicated this gene in skin development[145]. Nevertheless, the results obtained with dominant negative mutants that may have a broad inhibiting ability are more difficult to interpret than those obtained by conditional knockout.

In order to understand whether the transcriptional activity of RXRα is required for its function, or if its dimerization properties are the only important role of RXRα, mice in which the AF-2 function of RXRα was destroyed were generated[137]. Since these animals recapitulate many of the defects of the RXRα null mice, this experiment clearly demonstrates that AF-2 function is critical for the function of RXRα *in vivo*.

The mutation of RXRβ in the mouse led to 50% embryonic lethality[141]. Strikingly, the mutant mice that survive appear normal, but the males are sterile and exhibit testicular defects and abnormal spermatogenesis. In addition, these mice exhibit locomotor deficiency related to a dysfunction of the dopamine signaling pathway[153]. Expression of dopamine receptor was reduced in the ventral striatum of the mutant mice. The mouse mutant for RXRγ appears normal[154]. Nevertheless, it has recently been observed that these mice exhibit a syndrome of resistance to thyroid hormones which is in accordance with the expression of RXRγ in the thyrotrope cells of the anterior pituitary[207]. Since all three types of mutants were constructed in Pierre Chambon's laboratory, in addition to RAR mutants, numerous combinations of mutants were generated[131,137,139,142,143,153,154]. Strikingly, mice in which RXRβ and RXRγ have been inactivated and which are heterozygous for the mutation of RXRα are viable[154]. The differentiation of F9 cells by retinoic acid has also been studied in the context of these various combination of mutants[192]. This led to the observation that distinct RAR–RXR heterodimers have different roles in the control of target genes in F9 cells. All these results demonstrate unambiguously that the RAR–RXR heterodimer transduces *in vivo* the retinoid signal. The analysis of the phenotype of the various mice suggest that the main partner of RARs being apparently the RXRα gene[143]. The double RXRβ–RXRγ animals have the same defects as the RXRβ animals[153], whereas the RXRα–RXRβ mutants die earlier than the RXRα−/− mutants, suggesting that RXRβ may have some overlapping role with RXRα during development, notably in heart[155].

Despite the fact that its role in development was clearly established by knockout experiments and the importance of its AF-2 function was illustrated by the phenotype of the RXR ΔAF-2 mice, the role of activation of RXRα by its ligand, 9-*cis* retinoic acid *in vivo* is still a matter of controversy. By constructing mice that allow to test for the presence of activated RXR *in vivo*, using a RXRLBD–GAL4DBD fusion transgenic construct and a UAS-Lacz reporter gene, Thomas Perlmann's laboratory strongly suggests, but does not formally prove, that ligand activation play an important role in RXR function[146]. Activation of RXR was detected in spinal cord in a pattern that suggests a role of RXR in the maturation of motor neurons. This is consistent with experiments of ectopic overexpression of RXRs done in *Xenopus*, which suggest a role for these receptors in neurogenesis[147].

The function of USP has been studied in *Drosophila* by the analysis of the phenotype of mutant flies (see reference 148 for a review). These analyses strongly support a role for the EcR–USP heterodimer as a functional ecdysteroid receptor as the onset of metamorphosis functioning both as a repressor and an activator[151,152]. These analyses also demonstrate that USP may have functions that are independent of those of EcR during *Drosophila* development. The analysis of USP mutants reveals that the gene is required for multiple functions in a wide variety of tissues at different stages of development[149]. This is the case for eye morphogenesis[149,150], as well as in the somatic and germline tissues of adult females and during egg shell morphogenesis[149].

Associated disease

Through its heterodimeric interaction with many other nuclear receptors, RXR may play a role in a wide variety of diseases. Indeed, synthetic retinoids ligands for RXR are developed for the treatment of many diseases, including psoriasis, acne or cancer and, more recently, for metabolic diseases, such as obesity and diabetes (see reference 200 for a review). The case of diabetes appears particularly promising, since rexinoids have been shown to reduce insulin resistance in diabetic and obese mice[178]. Detailed comparison of the effects of rexinoid, PPARγ ligands or both has allowed a better understanding of their respective physiological actions[169]. Rexinoids also have an effect on the terminal differentiation of human liposarcoma cells that is also induced by PPARγ ligands[244]. The important potential of rexinoids for the treatment of metabolic diseases is substantiated by the observation that rexinoid-treated mice exhibit marked changes in cholesterol balance. This is due at least in part to the regulation of absorption and efflux of cholesterol by RXR via heterodimer formation with LXRs and FXR[252]. The existence of a RAR-independent RXR signaling pathway that allows the differentiation of cells from patients with acute promyelocytic leukemia suggests that RXR may also be a potential target in cancer research[190]. Other observations suggest that RXRs may play an important part in some diseases. For example, the observation that RXRβ[-/-] mice exhibit defects in dopamine signaling pathway renders possible a role for this receptor in Parkinson's disease and schizophrenia[153]. Likewise, the fact that RXRγ[-/-] mice display a thyroid hormone resistance syndrome should prompt the study of this gene in patients with resistance syndromes but with no mutations in TRβ[207].

Gene structure, promoter and isoforms

The genomic organization and the promoter of the RXRα gene have not been described up to now. Three isoforms of mouse RXRα differing in the A/B domain are known[232]. Two of these isoforms (RXRα2 and α3) encode an identical protein that lacks 28 N-terminal amino acid residues that are part of the AF-1 function when compared to the RXRα1. RXRα2 and α3 are specifically expressed in testis and their expression is strongly up-regulated in this tissue at puberty.

The human gene encoding RXRβ has been entirely sequenced[226]. It is composed of 10 exons spanning 6.5 kb of genomic DNA. The genomic organization of the mouse RXRβ gene has also been determined[231]. The DBD is split into two exons in a highly unusual way when compared to other nuclear receptors since the intron lies inside the first zinc finger, six amino acids after the second Cys residue. Variants of RXRβ differing in the A/B region were identified in human and mouse[228,231]. No functional characterization of these isoforms was performed. In the mouse the expression of the two isoforms, one being 72 amino acid longer in the N-terminal, has been shown to be slightly different[231]. These mouse isoforms arose from alternative promoter usage, both of which are GC-rich. Sequence analysis of the human promoter region reveals a GC-rich region without a TATA box[226]. Interestingly, an isoform containing only a four-amino-acid insertion within the LBD has been detected in rodent and human[229]. This isoform, called RXRβ2E, does not bind 9-cis retinoic acid with high affinity and did not activate transcription[233]. It is still able to form heterodimers with TR, VDR, RAR or LXR. It cannot mediate the ligand-dependent function of RXR but can still perform its ligand-independent ones.

The genomic organization of the mouse RXRγ gene was determined[230]. The gene spans at least 50 kb and has 11 exons including two alternative first exons that led to the prediction that this gene should have at least two different promoters. The unusual position of the intron lying inside the DBD coding region is conserved. The exon–intron organization of RXRγ is not identical to that of RXRβ, since it contains an additional intron in the N-terminal part of the LBD[230,231]. Such differences between paralogous members of the nuclear receptor superfamily is very unusual. Two N-terminally different isoforms of RXRγ with different expression patterns were found. RARγ1, which is 125 amino acids longer than RXRγ2 is expressed in brain and muscle, whereas RXRγ2 is highly expressed in both cardiac and skeletal muscles and very little in liver[230]. Both are able to activate transcription. In chicken the promoter controlling the expression of the RXRγ2 isoform has been characterized[261]. This promoter allows the photoreceptor-specific expression of RXRγ2 in the developing retina.

The *Drosophila* USP gene has no intron and its promoter has not been described[7-9]. By contrast, many of the USP homologues cloned in other species such as mosquito[102,195] or moth[19] do contain isoforms, mainly in the N-terminal region, and have a more complex genomic organization. In many cases, these isoforms exhibit different expression patterns as well as different regulation by ecdysone. In *Manduca*, a differential effect of USP-1

or USP-2 implicated in heterodimer with EcR, on the regulation of the MHR3 (NR1F4), were observed[227]. The USP gene was thus considerably simplified in *Drosophila*.

Chromosomal location

The human RXRα gene is on chromosme 9q34.3[224,225], whereas the mouse gene is located on chromosome 2 near the centromere[223]. The human RXRβ gene is on chromosome 6p21.3[225] and its mouse homologue on the H-2 region of chromosome 17[223]. The human RXRβ gene has been entirely sequenced to test for its implication in a disorder of ossification of the spine[226]. The human RXRγ gene is on chromosome 1 band q22–q23[225] and the mouse RXRγ is located close to the Pbx gene on the distal part of chromosome 1[223].

In *Drosophila*, USP is located in the 2C1–3 position on the distal portion of the X-chromosome[9].

Amino acid sequence for human RXRα (NR2B1)

Accession number: X52773

```
  1  MDTKHFLPLDFSTQVNSSLTSPTGRGSMAAPSLHPSLGPGIGSPGQLHSPISTLSSPINGMGPPFSVISS
 71  PMPGPHSMSVPTTPTLGFSTGSPQLSSPMNPVSSSEDIKPPLGLNGVLKVPAHPSGNMASFTKHICAICGD
141  RSSGKHYGVYSCEGCKGFFKRTVRKDLTYTCRDNKDCLIDKRQRNRCQYCRYQKCLAMGMKREAVQEERQ
211  RGKDRNENEVESTSSANEDMPVERILEAELAVEPKTETYVEANMGLNPSSPNDPVTNICQAADKQLFTLV
281  EWAKRIPHFSELPLDDQVILLRAGWNELLIASFSHRSIAVKDGILLATGLHVHRNSAHSAGVGAIFDRVL
351  TELVSKMRDMQMDKTELGCLRAIVLFNPDSKGLSNPAEVEALREKVYASLEAYCKHKYPEQPGRFAKLLL
421  RLPALRSIGLKCLEHLFFFKLIGDTPIDTFLMEMLEAPHQMT  462
```

Amino acid sequence for human RXRβ (NR2B2)

Accession number: M84820

```
  1  MSWAARPPFLPQRHAAGQCGPVGVRKEMHCGVASRWRRRRPWLDPAAAAAAVAGGEQQTPEPEPGEAGR
 71  DGMGDSGRDSRSPDSSSPNPLPQGVPPPSPPGPPLPPSTAPTLGGSGAPPPPPMPPPPLGSPFPVISSSM
141  GSPGLPPPAPPGFSGPVSSPQINSTVSLPGGGSGPPEDVKPPVLGVRGLHCPPPPGGPGAGKRLCAICGD
211  RSSGKHYGVYSCEGCKGFFKRTIRKDLTYSCRDNKDCTVDKRQRNRCQYCRYQKCLATGMKREAVQEERQ
281  RGKDKDGDGEGAGGAPEEMPVDRILEAELAVEQKSDQGVEGPGGTGGSGSSPNDPVTNICQAADKQLFTL
351  VEWAKRIPHFSSLPLDDQVILLRAGWNELLIASFSHRSIDVRDGILLATGLHVHRNSAHSAGVGAIFDRV
421  LTELVSKMRDMRMDKTELGCLRAIILFNPDAKGLSNPSEVEVLREKVYASLETYCKQKYPEQQGRFAKLL
491  LRLPALRSIGLKCLEHLFFFKLIGDTPIDTFLMEMLEAPHQLA  533
```

Amino acid sequence for human RXRγ (NR2B3)

Accession number: U38480

```
  1  MYGNYSHFMKFPAGYGGSPGHTGSTSMSPSAALSTGKPMDSHPSYTDTPVSAPRTLSAVGTPLNALGSPY
 71  RVITSAMGPPSGALAAPPGINLVAPPSSQLNVVNSVSSSEDIKPLPGLPGIGNMNYPSTSPGSLVKHICA
141  ICGDRSSGKHYGVYSCEGCKGFFKRTIRKDLIYTCRDNKDCLIDKRQRNRCQYCRYQKCLVMGMKREAVQ
211  EERQRSRERAESEAECATSGHEDMPVERILEAELAVEPKTESYGDMNMENSTNDPVTNICHAADKQLFTL
281  VEWAKRIPHFSDLTLEDQVILLRAGWNELLIASFSHRSVSVQDGILLATGLHVHRSSAHSAGVGSIFDRV
351  LTELVSKMKDMQMDKSELGCLRAIVLFNPDAKGLSNPSEVETLREKVYATLEAYTKQKYPEQPGRFAKLL
421  LRLPALRSIGLKCLEHLFFFKLIGDTPIDTFLMEMLETPLQIT  463
```

Amino acid sequence for *Drosophila* USP (NR2B4)

Accession number: X53417

```
1    MDNCDQDASFRLSHIKEEVKPDISQLNDSNNSSFSPKAESPVPFMQAMSMVHVLPGSNSASSNNNSAGDA
71   QMAQAPNSAGGSAAAAVQQQYPPNHPLSGSKHLCSICGDRASGKHYGVYSCEGCKGFFKRTVRKDLTYAC
141  RENRNCIIDKRQRNRCQYCRYQKCLTCGMKREAVQEERQRGARNAAGRLSASGGGSSGPGSVGGSSSQGG
211  GGEGGVSGGMGSGNGSDDFMTNSVSRDFSIERIIEAEQRAETQCGDRALTFLRVGPYSTVQPDYKGAVSA
281  LCQVVNKQLFQMVEYARMMPHFAQVPLDDQVILLKAAWIELLIANVAWCSIVSLDDGGGGGGGGLGHDGS
351  FERRSPGLQPQQLFLNQSFSYHRNSAIKAGVSAIFDRILSELSVKMKRLNLDRRELSCLKAIILYNPDIR
421  GIKSRAEIEMCREKVYACLDEHCRLEHPGDDGRFAQLLLRLPALRSISLKCQDHLFLFRITSDRPLEELF
491  LEQLEAPPPPGLAMKLE  507
```

References

1 Hamada, K. et al. (1989) Proc. Natl Acad Sci. USA 86, 8289–8293.
2 Mangelsdorf, D.J. et al. (1990) Nature 345, 224–229.
3 Rowe, A. et al. (1991) Development 111, 771–778.
4 Yu, V.C. et al. (1991) Cell 67, 1251–1266.
5 Mangelsdorf, D.J. et al. (1992) Genes Dev. 6, 329–344.
6 Leid, M. et al. (1992) Cell 68, 377–395.
7 Oro, A.E. et al. (1990) Nature 347, 298–301.
8 Shea, M.J. et al. (1990 Genes Dev. 4, 1128–1140.
9 Henrich, V.C. et al. (1990) Nucl. Acids Res. 18, 4143–4148.
10 Hayward, D.C. et al. (1999) Dev. Genes Evol. 209, 564–571.
11 Zelus, D. et al. submitted.
12 Laudet, V. and Stéhelin, D. (1992) Curr. Biol. 2, 293–295.
13 Glass, C.K. (1994) Endocrine Rev. 15, 391–407.
14 Blumberg, B. et al. (1992) Proc. Natl Acad. Sci. USA 89, 2321–2325.
15 Marklew, S. et al. (1994) Biochim. Biophys. Acta 1218, 267–272.
16 Jones, B.B. et al. (1995) Mol. Cell. Biol. 15, 5226–5234.
17 Gongora, R. et al. (1998) Immunogenetics 48, 141–143.
18 Tzertzinis, G. et al. (1994) J. Mol. Biol. 238, 479–486.
19 Jindra, M. et al. (1997) Insect Mol. Biol. 6, 41–53.
20 Nicolai, M. et al. (2000) Insect Mol. Biol. 9, 241–249.
21 Guo, X. et al. (1998) Mol. Cell. Endocrinol. 139, 45–60.
22 Chung, A.C. et al. (1998) Mol. Cell. Endocrinol. 139, 209–227.
23 Palmer, M.J. et al. (1999) Am. Zool. 39, 747–757.
24 Freebern, W.J. et al. (1999) J. Biol. Chem. 274, 4577–4585.
25 Freebern, W.J. et al. (1999) Gene 233, 33–38.
26 de Mendonça, R.L. et al. (2000) Eur. J. Biochem. 267, 3208–3219.
27 de Mendonça, R.L. et al. (2000) Parasitol. Today 16, 233–240.
28 Escriva, H. et al. (1997) Proc. Natl Acad. Sci. USA 94, 6803–6808.
29 Kostrouch, Z. et al. (1998) Proc. Natl Acad. Sci. USA 95, 13442–13447.
30 Escriva, H. et al. (2000) BioEssays 22, 717–727.
31 Marks, M.S. et al. (1992) Mol. Endocrinol. 6, 219–230.
32 Chen, Z.P. et al. (1994) J. Biol. Chem. 269, 25770–25776.
33 Zhang, X.K. et al. (1992) Nature 358, 587–591.
34 Bugge, T.H. et al. (1992) EMBO J. 11, 1409–1418.
35 Kliewer, S.A. et al. (1992) Nature 355, 446–449.
36 Zhang, X.K. et al. (1992) Nature 355, 441–446.

[37] Mader, S. et al. (1993) EMBO J. 12, 5029–5041.
[38] Dey, A. et al. (1992) Mol. Cell. Biol. 12, 3590–3599.
[39] Perlmann, T. et al. (1993) Genes Dev. 7, 1411–1422.
[40] Kurokawa, R. et al. (1993) Genes Dev. 7, 1423–1435.
[41] Kurokawa, R. et al. (1994) Nature 371, 528–531.
[42] Mader, S. et al. (1993) J. Biol. Chem. 268, 591–600.
[43] Ikeda, M. et al. (1994) Endocrinology 135, 1628–1638.
[44] Predki, P.F. et al. (1994) Mol. Endocrinol. 8, 31–39.
[45] Antoniewski, C. et al. (1996) Mol. Cell. Biol. 16, 2977–2986.
[46] Wang, S.F. et al. (1998) J. Biol. Chem. 273, 27531–27540.
[47] Lu, X.P. et al. (1993) Mol. Cell. Biol. 13, 6509–6519.
[48] Yao, T.P. et al. (1992) Cell 71, 63–72.
[49] Thomas, H.E. et al. (1993) Nature 362, 471–475.
[50] Vögtli, M. et al. (1998) Nucl. Acids Res. 26, 2407–2414.
[51] Horner, M.A. et al. (1995) Dev. Biol. 168, 490–502.
[52] Marks, M.S. et al. (1992) EMBO J. 11, 1419–1435.
[53] Lehmann, J.M. et al. (1993) Mol. Cell. Biol. 13, 7698–7707.
[54] Kephart, D.D. et al. (1996) Mol. Endocrinol. 10, 408–419.
[55] Suhr, S.T. et al. (1998) Proc. Natl Acad. Sci. USA 95, 7999–8004.
[56] Butler, A.J. and Parker, M.G. (1995) Nucl. Acids Res. 23, 4143–4150.
[57] Tontonoz, P. et al. (1994) Nucl. Acids Res. 22, 5628–5634.
[58] Berrodin, T.J. et al. (1992) Mol. Endocrinol. 6, 1468–1478.
[59] Baes, M. et al. (1994) Mol. Cell. Biol. 14, 1544–1552.
[60] Choi, H-S. et al. (1997) J. Biol. Chem. 272, 23565–23571.
[61] Cooney, A.J. et al. (1993) J. Biol. Chem. 268, 4152–4160.
[62] Kliewer, S.A. et al. (1992) Proc. Natl Acad. Sci. USA 89, 1448–1452.
[63] Widom, R.L. et al. (1992) Mol. Cell. Biol. 12, 3380–3389.
[64] Tran, P. et al. (1992) Mol. Cell. Biol.12, 4666–4676.
[65] Lee, Y.K. et al. (2000) Mol. Cell. Biol. 20, 187–195.
[66] Seol, W. et al. (1995) Mol. Endocrinol. 9, 72–85.
[67] Teboul, M. et al. (1995) Proc. Natl Acad. Sci. USA 92, 2096–2100.
[68] Willy, P.J. and Mangelsdorf, D.J. (1997) Genes Dev. 11, 289–298.
[69] Wiebel, F.F. and Gustafsson, J.A. (1997) Mol. Cell. Biol. 17, 3977–3986.
[70] Feltkamp, D. et al. (1999) J. Biol. Chem. 274, 10421–10429.
[71] Forman, B.M. et al. (1995) Cell 81, 541–550.
[72] Perlmann, T. and Jansson, L. (1995) Genes Dev. 9, 769–782.
[73] Fisk, G.J. and Thummel, C.S. (1995) Proc. Natl Acad. Sci. USA 92, 10604–10608.
[74] Kliewer, S.A. et al. (1992) Nature 358, 771–774.
[75] Smith, D.P. et al. (1994) Nucl. Acids Res. 22, 66–71.
[76] Blumberg, B. et al. (1998) Genes Dev. 12, 1269–1277.
[77] Kliewer, S.A. et al. (1998) Cell 92, 73–82.
[78] Blumberg, B. et al. (1998) Genes Dev. 12, 3195–3205.
[79] Lehmann, J.M. et al. (1998) J. Clin. Invest. 102, 1016–1023.
[80] Bonnelye, E. et al. (1994) Cell Growth Diff. 5, 1357–1365.
[81] Laudet, V. (1997) J. Mol. Endocrinol. 19, 207–226.
[82] Rosen, E.D. et al. (1993) J. Biol. Chem. 268, 11534–11541.
[83] Zechel, C. et al. (1994) EMBO J. 13, 1414–1424.
[84] Zechel, C. et al. (1994) EMBO J. 13, 1425–1433.

[85] Zhang, X.K. et al. (1994) Mol. Cell. Biol. 14, 4311–4323.
[86] Perlmann, T. et al. (1996) Mol. Endocrinol. 10, 958–966.
[87] Lee, S.K. et al. (1998) Mol. Endocrinol. 12, 325–332.
[88] Forman, B.M. et al. (1990) Mol. Endocrinol. 4, 1293–1302.
[89] Minucci, S. et al. (1994) Mol. Cell. Biol. 14, 360–372.
[90] Blanco, J.C.G. et al. (1996) Genes Cells 1, 209–221.
[91] Chen, H. and Privalsky, M.L. (1995) Proc. Natl Acad. Sci. USA 92, 422–426.
[92] Kersten, S. et al. (1995) Proc. Natl Acad. Sci. USA 92, 8645–8649.
[93] Kersten, S. et al. (1997) J. Biol. Chem 272, 12771–12777.
[94] Kersten, S. et al. (1997) J. Biol. Chem. 272, 29759–29768.
[95] Schulman, I.G. et al. (1995) Proc. Natl Acad. Sci. USA 92, 8288–8292.
[96] Zhang, J. et al. (1999) Mol. Cell. Biol. 19, 6448–6457.
[97] Westin, S. et al. (1998) Nature 395, 199–202.
[98] Yuan, C.X. et al. (1998) Proc. Natl Acad. Sci. USA 95, 7939–7944.
[99] Lemon, B.D. and Freedman, L.P. (1999) Curr. Opin. Genet. Dev. 9, 499–504.
[100] May, M. et al. (1996) EMBO J. 15, 3093–3104.
[101] Hörlein, A.J. et al. (1995) Nature 377, 397–404.
[102] Chen, J.D. and Evans, R.M. (1995) Nature 377, 454–457.
[103] Kurokawa, R. et al. (1995) Nature 377, 451–454.
[104] Perlmann, T. and Vennström, B. (1995) Nature 377, 387–388.
[105] Onate, S.A. et al. (1995) Science 270, 1354–1357.
[106] Voegel, J.J. et al. (1996) EMBO J. 15, 3667–3675.
[107] Chen, H. et al. (1997) Cell 90, 569–580.
[108] McKenna, N.J. et al. (1999) Endocrine Rev. 20, 321–344.
[109] Glass, C.K. and Rosenfeld, M.G. (2000) Genes Dev. 14, 121–141.
[110] Ding, X.F. et al. (1998) Mol. Endocrinol. 12, 302–313.
[111] Leers, J. et al. (1998) Mol. Cell. Biol. 18, 6001–6013.
[112] Voegel, J.J. et al. (1998) EMBO J. 17, 507–519.
[113] McInerney, E.M. et al. (1998) Genes Dev. 12, 3357–3368.
[114] Feng, W. et al. (1998) Science, 280, 1747–1749.
[115] Jekayumar, M. et al. (1997) Mol. Endocrinol. 11, 755–767.
[116] Schulman, I.G. et al. (1996) Mol. Cell. Biol. 16, 3807–3813.
[117] Zhang, J. et al. (1997) Mol. Cell. Biol. 17, 6887–6897.
[118] Chakravarti, D. et al. (1996) Nature 383, 99–103.
[119] Kamei, Y. et al. (1996) Cell 85, 403–414.
[120] L'Horset, F. et al. (1996) Mol. Cell. Biol. 16, 6029–6036.
[121] Le Douarin, B. et al. (1995) EMBO J. 14, 2020–2033.
[122] Vom Baur, E. et al. (1996) EMBO J. 15, 110–124.
[123] Lee, J.W. et al. (1995) Nature 374, 91–94.
[124] Huang, N. et al. (1998) EMBO J. 17, 3398–3412.
[125] Powers, C.A. et al. (1998) Mol. Endocrinol. 12, 4–18.
[126] Gaudon, C. et al. (1999) EMBO J. 18, 2229–2240.
[127] Li, D. et al. (1999) Mol. Cell. Biol. 19, 7191–7202.
[128] Lee, S.K. et al. (1999) J. Biol. Chem. 274, 34283–34293.
[129] Caira, F. et al. (2000) J. Biol. Chem. 275, 5308–5317.
[130] Sucov, H.M. et al. (1994) Genes Dev. 8, 1007–1018.
[131] Kastner, P. et al. (1994) Cell 78, 987–1003.

[132] Sucov, H.M. et al. (1995) Development 121, 3997–4003.
[133] Ruiz-Lozano, P. et al. (1998) Development 125, 533–544.
[134] Chen, J. et al. (1998) Development 125, 1943–1949.
[135] Nugent, P. et al. (1999) Int. J. Dev. Biol. 43, 567–570.
[136] Tran, C.M. and Sucov, H.M. (1998) Development 125, 1951–1956.
[137] Mascrez, B. et al. (1998) Development 125, 4691–4707.
[138] Wan, Y.J.Y. et al. (2000) Mol. Cell. Biol. 20, 4436–4444.
[139] Li, M. et al. (2000) Nature 407, 633–636.
[140] Clifford, J. et al. (1996) EMBO J. 15, 4142–4155.
[141] Kastner, P. et al. (1996) Genes Dev. 10, 80–92.
[142] Wendling, O. et al. (1999) Proc. Natl Acad. Sci. USA 96, 547–551.
[143] Kastner, P. et al. (1997) Development 124, 313–326.
[144] Kastner, P. et al. (1995) Cell 83, 859–869.
[145] Feng, X. et al. (1997) Genes Dev. 11, 59–71.
[146] Solomin, L. et al. (1998) Nature 395, 398–402.
[147] Sharpe, C.R. and Goldstone, K. (1997) Development 124, 515–523.
[148] Kozlova, T. and Thummel, C.S. (2000) Trends Endocrinol. Metab. 11, 276–280.
[149] Oro, A.E. et al. (1992) Development 115, 449–462.
[150] Zelhof, A.C. et al. (1997) Development 124, 2499–2506.
[151] Hall, B.L. and Thummel, C.S. (1998) Development 125, 4709–4717.
[152] Schubiger, M. and Truman, J.W. (2000) Development 127, 1151–1159.
[153] Krezel, W. et al. (1998) Science 279, 863–867.
[154] Krezel, W. et al. (1996) Proc. Natl Acad. Sci. USA 93, 9010–9014.
[155] Kastner, P. et al. (1997) Development 124, 4749–4758.
[156] Less, M.S. et al. (1993) Science 260, 1117–1121.
[157] Rastinejad, F. et al. (1995) Nature 375, 203–211.
[158] Bourguet, W. et al. (1995) Nature 375, 377–382.
[159] Moras, D. and Gronemeyer, H. (1998) Curr. Opin. Cell Biol. 10, 384–391.
[160] Weatherman, R.V. et al. (1999) Annu. Rev. Biochem. 68, 559–581.
[161] Egea, P.F. et al. (2000) EMBO J. 19, 2592–2601.
[162] Bourguet, W. et al. (2000) Mol. Cell 5, 289–298.
[163] Gampe, R.T. et al. (2000) Mol. Cell 5, 545–555.
[164] Billas, I.M.L. et al. (2000) J. Biol. Chem. (in press).
[165] Wilson, T.E. et al. (1992) Science 256, 107–110.
[166] Levin, A.A. et al. (1992) Nature 355, 359–361.
[167] Heyman, R.A. et al. (1992) Cell 68, 397–406.
[168] Allenby, G. et al. (1993) Proc. Natl Acad. Sci. USA 90, 30–34.
[169] Lenhart, J.M. et al. (1999) Diabetologia 42, 545–554.
[170] Lu, H.C. et al. (1997) Development 124, 195–203.
[171] Viviano, C.M. et al. (1995) Development 121, 3753–3762.
[172] Thaller, C. et al. (1993) Development 118, 957–965.
[173] Kraft, J.C. et al. (1994) Proc. Natl Acad. Sci. USA 91, 3067–3071.
[174] Lehmann, J.M. et al. (1992) Science 258, 1944–1946.
[175] Davis, K.D. et al. (1994) Mol. Cell. Biol. 14, 7105–7110.
[176] Nagy, L. et al. (1995) Mol. Cell. Biol. 15, 3540–3551.
[177] Harmon, M.A. et al. (1995) Proc. Natl Acad. Sci. USA 92, 6157–6160.
[178] Mukherjee, R. et al. (1997) Nature 386, 407–410.

[179] Leng, X. et al. (1995) Mol. Cell. Biol. 15, 255–263.
[180] Chen, J.Y. et al. (1996) Nature 382, 819–822.
[181] Vivat, V. et al. (1997) EMBO J. 16, 5697–5709.
[182] Schulman, I.G. et al. (1997) Genes Dev. 11, 209–308.
[183] Mouchon, A. et al. (1999) Mol. Cell. Biol. 19, 3073–3085.
[184] Johnson, B.S. et al. (1999) Mol. Cell. Biol. 19, 3372–3382.
[185] Jones, G. and Sharp, P.A. (1997) Proc. Natl Acad. Sci. USA 94, 13499–13503.
[186] Jones, G. and Jones, D. (2000) Insect Biochem. Mol. Biol. 30, 671–679.
[187] Boehm, M.F. et al. (1995) J. Med. Chem. 38, 3146–3155.
[188] Boehm, M.F. et al. (1994) J. Med. Chem. 37, 2930–2941.
[189] Lala, D. et al. (1996) Nature 383, 450–453.
[190] Benoit, G. et al. (1999) EMBO J. 18, 7011–7018.
[191] Willy, P.J. and Mangelsdorf, D.J. (1997) Genes Dev. 11, 289–298.
[192] Chiba, H. et al. (1997) Mol. Cell. Biol. 17, 3013–3020.
[193] Georgiadis, P. and Brickell, P.M (1997) Dev. Dynamics 210, 227–235.
[194] Yao, T.P. et al. (1993) Nature 366, 476–479.
[195] Kapitskaya, M. et al. (1996) Mol. Cell. Endocrinol. 121, 119–132.
[196] Hatzivassiliou, E. et al. (1997) Biochemistry 36, 9221–9231.
[197] Sutherland, J.D. et al. (1995) Proc. Natl Acad. Sci. USA 92, 7966–7970.
[198] Vogtli, M. et al. (1999) Insect Biochem. Mol. Biol. 29, 931–942.
[199] Perera, S.C. et al. (1998) Dev. Genet. 22, 169–179.
[200] Thacher, S.M. et al. (2000) Curr. Pharm. Des. 6, 25–58.
[201] Dong, D. and Noy, N. (1998) Biochemistry 37, 10691–10700.
[202] Escriva, H. et al. (submitted).
[203] Chen, Z. et al. (1998) J. Mol. Biol. 275, 55–65.
[204] Leong, G.M. et al. (1998) J. Biol. Chem. 273, 2296–2305.
[205] Na, S.Y. et al. (1998) J. Biol. Chem. 273, 30933–30938.
[206] Na, S.Y. et al. (1998) J. Biol. Chem. 273, 3212–3215.
[207] Brown, N.S. et al. (2000) J. Clin. Invest. 106, 73–79.
[208] Bhattacharyya, N. et al. (1997) Mol. Cell. Biol. 17, 6481–6490.
[209] Minucci, S. et al. (1996) Proc. Natl Acad. Sci. USA 93, 1803–1807.
[210] Minucci, S. et al. (1997) Mol. Cell. Biol. 17, 644–655.
[211] Roy, B. et al. (1995) Mol. Cell. Biol. 15, 6481–6487.
[212] Botling, J. et al. (1997) J. Biol. Chem. 272, 9443–9449.
[213] Schulman, I.G. et al. (1998) Mol. Cell. Biol. 18, 3483–3494.
[214] Forman, B.M. et al. (1995) Cell 81, 541–550.
[215] Leblanc, B.P. and Stunnenberg, H.G. (1995) Genes Dev. 9, 1811–1816.
[216] Chambon, P. (1996) FASEB J. 10, 940–954.
[217] Dolle, P. et al. (1994) Mech. Dev. 45, 91–104.
[218] Krezel, W. et al. (1999) Neuroscience 89, 1291–1300.
[219] Wan, Y.J. et al. (1994) Exp. Cell Res. 210, 56–61.
[220] Pombo Seleiro, E.A. et al. (1995) Roux's Arch. Dev. Biol. 204, 244–249.
[221] Henrich, V.C. et al. (1994) Dev. Biol. 165, 38–52.
[222] Lan, Q. et al. (1997) Insect Mol. Biol. 6, 3–10.
[223] Hoopes, C.W. et al. (1992) Genomics 14, 611–617.
[224] Jones, K.A. et al. (1993) Ann. Hum. Genet. 57, 195–201.
[225] Almasan, A. et al. (1994) Genomics 20, 397–403.
[226] Numasawa, T. et al. (1999) J. Bone Miner. Res. 14, 500–508.

227 Lan, Q. et al. (1999) Mol. Cell. Biol. 19, 4897–4906.

228 Fleichhauer, K. et al. (1992) Nucl. Acids Res. 20, 1801.

229 Epplen, C.E. and Epplen, J.T. (1992) Mamm. Genome 3, 472–475.

230 Liu, Q. and Linney, E. (1993) Mol. Endocrinol. 7, 651–658.

231 Nagata, T. et al. (1994) Gene 142, 183–189.

232 Brocard, J. et al. (1996) Biochem. Biophys. Res. Commun. 229, 211–218.

233 Fujita, A. and Mitsuhashi, T. (1999) Biochem. Biophys. Res. Commun. 255, 625–630.

234 Solomon, C. et al. (1999) J. Clin. Invest. 103, 1729–1735.

235 Nagpal, S. et al. (1993) EMBO J. 12, 2349–2360.

236 Lee, H.Y. et al. (2000) J. Biol. Chem. 275, 32193–32199.

237 Durand, B. et al. (1994) EMBO J. 13, 5370–5382.

238 Dowhan, D.H. and Muscat, G.E.O. (1996) Nucl. Acids Res. 24, 264–271.

239 Dilworth, F.J. et al. (1999) Proc. Natl Acad. Sci. USA 96, 1995–2000.

240 Valcarel, R. et al. (1994) Genes Dev. 8, 3068–3079.

241 Hall, B.L. et al. (1993) Proc. Natl Acad. Sci. USA 90, 6929–6933.

242 Heery, D.M. et al. (1993) Proc. Natl Acad. Sci. USA 90, 4281–4285.

243 Heery, D.M. et al. (1994) Nucl. Acids Res. 22, 726–731.

244 Tontonoz, P. et al. (1997) Proc. Natl Acad. Sci. USA 94, 237–241.

245 Nakshatri, H. and Chambon, P. (1994) J. Biol. Chem. 269, 890–902.

246 Mangelsdorf, D.J. et al. (1991) Cell 66, 555–561.

247 Abu-Abed, S.S. et al. (1988) J. Biol. Chem. 273, 2409–2415.

248 Durand, B. et al. (1992) Cell 71, 73–85.

249 Astrom, A. et al. (1994) J. Biol. Chem. 269, 22334–22339.

250 Allegretto, E.A. et al. (1995) J. Biol. Chem. 270, 23906–23909.

251 Samad, T.A. et al. (1997) J. Biol. Chem. 94, 14349–14354.

252 Repa, J.J. et al. (2000) Science 289, 1524–1529.

253 Liu, Y. and Chiu, J.F. (1994) Nucl. Acids Res. 22, 1079–1086.

254 Cohen, O. et al. (1995) J. Biol. Chem. 270, 13899–13905.

255 Gearing, K.L. et al. (1993) Proc. Natl Acad. Sci. USA 90, 1440–1444.

256 Jayachandran, G. and Fallon, A.M. (2000) Arch. Insect Biochem. Physiol.43, 87–96.

257 Kamimura, M. et al. (2000) Dev. Growth Differ. 42, 1–8.

258 Kakizawa, T. et al. (1999) J. Biol. Chem. 274, 19103–19108.

259 Miyamoto, T. et al. (1999) Mol. Cell. Biol. 19, 2644–2649.

260 Kakizawa, T. et al. (1997) J. Biol. Chem. 272, 23799–23804.

261 Ameixa, C. and Brickell, P.M. (2000) Biochem J. 347, 485–490.

262 Gampe, R.T. et al. (2000) Genes Dev. 14, 2229–2241.

263 Mata de Urquiza, A. et al. (2000) Science 290, 2140–2144.

Names

These two closely related genes were identified by a number of research groups and therefore bear numerous names. TR2 was the first one to be proposed for NR2C1, whereas several names were given to NR2C2 when it was discovered in 1994. To emphasize its relationship to TR2, we used TR4. These two orphan receptors are closely related to the TLL (NR2E) group in subfamily II. The unique homologue known in sea urchin and *Drosophila* that belongs to the NR2D group of nuclear receptors is only distantly related to TR2 and TR4, and clusters with them with low bootstrap values[41].

Species	Other names	Accession number	References
Official name: NR2C1			
Human	TR2, TR2-11	M29960	1, 2
Mouse	TR2	U28265	8
Xenopus	xDOR2	AF013295	11
Axolotl	aDOR1	AF008302	10
Official name: NR2C2			
Human	TR4, TAK	L27586	3, 4
Mouse	TR2R1	U11688	5
Rat	TR4	L27513	4
Rabbit	TR4	not available	12
Official name: NR2D1			
Sea urchin	SpSHR2	U38281	9
Drosophila	DHR78, XR78E/F	U36791	6, 7
Tenebrio	THR6, TmHR78	AJ005765	40

Isolation

TR2 (NR2C1) was one of the first orphan receptors identified in 1988[1]. It was isolated by low-stringency screening of a human testis cDNA library with a 41 bp oligonucleotide probe corresponding to the DBD of GR, PR and ER[1]. The isolated clone, called TR2-5 was later shown to be an alternative splice variant of the regular transcript encoding an orphan receptor with a complete LBD (see below). This full-size isoform is called TR2-11[2]. Other orthologues of TR2 were found in the frog *Xenopus laevis*[10] as well as in the salamander *Ambystoma mexicanum* (axolotl)[11]. The *Ambystoma* sequence was identified from a cDNA library of axolotl neurulas under low-stringency screening with a newt RARβ cDNA fragment as a probe[11].

TR4 (NR2C2) was identified by three research groups in 1994 by RT-PCR with degenerate primers on lymphoblastoma Raji cells or on supraoptic nucleus mRNAs (3,4), or by low-stringency screening of a mouse brain cDNA library with a probe encoding the DBD of the COUP-TF orphan receptor[5]. TR4 is the most closely related receptor to TR2, since it harbors an overall 64% sequence identity with TR2 (36% in the A/B region, 80% in the DBD and 66% in the LBD). A rabbit homologue was recently described by the screening of a rabbit heart cDNA library with a degenerate oligonucleotide corresponding to the DBD of nuclear receptors[12].

Recently, a homologue to both TR2 and TR4, called SpSHR2 (NR2D1), was identified in sea urchin by low-stringency screening of a genomic library with a human COUP-TF-I probe encompassing the C domain. It harbors *c.* 75% identity with TR2 and TR4 in the DBD and 55% in the LBD.

The unique *Drosophila* homologue, DHR78 (NR2D1), was identified by a rapid cloning and screening strategy designed to identify new members of the NHR superfamily that are expressed during the onset of metamorphosis[6]. It was also found by another team by low-stringency screening of a *Drosophila* genomic library using a mouse RXRβ DBD probe[7]. It harbors *c.* 72% amino acid identity with TR2 and TR4 in the DBD and only 22% in the LBD. A homologue was also found in the mealworm *Tenebrio molitor*[40].

DNA binding

TR2 and TR4 are able to bind to direct repeats of the AGGTCA motif[13,14]. TR4 exhibits a preference for DR1, DR3, DR4 and DR5, and recognizes these elements as a homodimer[14]. In a site selection analysis, it was shown that TR4 exhibits the greatest activity for DR1 elements to which it binds as a homodimer[21]. It does not bind to a half-site motif and has a very low affinity for palindromic or inverted repeat elements. Interestingly, the transcriptional activity of TR4 on different direct repeats can be opposite[20] (see below).

DHR78, the *Drosophila* homologue of TR2 and TR4, binds to motifs containing two AGGTCA half-sites arranged as either direct or palindromic repeats[6]. It has also been shown that DR1 is the most efficient DNA-binding site for DHR78, as for TR4. No binding was observed on half-site sequences[7]. Since the ecdysone receptor (NR1H1) can bind to direct repeats, the competition between EcR and DHR78 for the occupancy of these sites was tested. Indeed, it was shown that DHR78 inhibits ecdysone signaling and that this effect depends on the DNA-binding site used[7].

Partners

It was shown by several teams that TR2 and TR4 cannot form heterodimers with RXR[14,21]. In contrast, TR2 and TR4 interact together and form heterodimers[16].

An interaction between TR4 and the androgen receptor was recently demonstrated[16]. AR functions as a repressor to down-regulate TR4 target genes by preventing the binding of TR4 to its target sequence. Interestingly, the opposite is also correct, since TR4 represses AR-mediated gene activation. Three isolated domains of AR interact with TR4: the N-terminal A/B region, the C domain and the LBD. This renders difficult the comprehension of the mechanism of this interaction. The domains of TR4 involved in the interaction are unknown.

Recently, an interaction between the A/B domain of TR2 and the cAMP-response element binding protein CREMτ was demonstrated[36]. This interaction occurs in the RARβ2 promoter, which is activated by TR2. This activation is further enhanced by cAMP, suggesting that the CREMτ–TR2 interaction further stimulates the transcriptional ability of TR2[36].

An interaction of both TR2 and TR4 with the comodulator RIP140 was reported both *in vitro* and *in vivo*[21,23]. In contrast, TR4 was shown to be unable to interact with SMRT[21]. TR2 interacts with RIP140 through its C-terminal end containing the putative activating function AF-2. RIP140 acts as a corepressor on TR2-mediated transcription[23]. Another study suggests that TR4 inhibits PPAR-mediated transactivation by competition for the interaction with RIP140, which acts as a coactivator of liganded PPARα[21].

Ligands

No ligand has been described for TR2, TR4 or DHR78. Interestingly, DHR78 the *Drosophila* homologue of TR2 and TR4 was proposed to be a ligand-activated receptor[17]. This proposal was based on the observation that, in contrast to other *Drosophila* orphan receptors, such as TLL, SVP, FTZ-F1 or DHR3, ectopic expression of DHR78 in wild-type flies has no effect on viability or development. This suggests that the activity of DHR78 is controlled post-translationally and that a signal is necessary to induce its activity. If this putative signal is absent, the receptor is silent and its ectopic expression has no effect. The nature of the signal is still elusive and there is no decisive argument that it could really be a ligand.

Three-dimensional structure

No 3D structure is available for TR2 or TR4.

Expression

The first studies on rat TR2 gene expression detected a transcript of 2.5 kb in prostate, liver, testis, seminal vesicle and kidney, and suggest that the transcript observed in prostate was negatively regulated by androgens[1]. A prominent testis expression was described in mouse, although the expression was also observed in numerous other tissues[18]. The very high expression in testis was confirmed by subsequent studies that also demonstrate that TR2 expression starts at E8.5 days of development, reaches a maximal level at E12.5 and declines at E14.5[8]. The corresponding protein was located in the seminiferous tubules, where advanced germ cells reside[8]. *In situ* hybridization located TR2 expression in many mouse tissues, including kidney and intestine during late embryogenesis[13]. In axolotl, in addition to an expression in the central nervous system during late embryogenesis, TR2 expression was found in early embryos during the cleavage stage[11]. The expression decreased during the mid-blastula transition. This suggests that TR2 is a maternal mRNA and this is confirmed by the observation of a high expression of this gene during oogenesis. The mRNA level increases again during gastrulation and reaches a peak during neurulation[10,11].

It was shown that TR2 expression is down-regulated by ionizing irradiation and that this effect is mediated by p53, which behaves as a down-regulator of TR2 expression[26]. A regulation of TR2 expression by retinoic acid was suggested, since mice fed with a vitamin-A-depleted diet exhibit no germ cells and no TR2 expression in testis[8]. This view has been substantiated by the identification of an IR0 element, which behaves as a retinoic acid response element in the mouse TR2 promoter[34,38]. In contrast, the expression of the TR2 homologue in the axolotl is apparently not regulated by retinoic acid[11]. Interestingly, overexpression of TR2 in P19 cells was shown to induce apoptosis. Since TR2 is a target gene of retinoic acid receptors, this suggests that it may be a downstream effector of RA-induced apoptosis of these cells[38].

The mouse TR4 is expressed as two mRNAs of 8 kb and 3.5 kb. The 8 kb form appears ubiquitously expressed, whereas the 3.5 kb is restricted to some tissues, such as brain, or at much lower levels in the lungs[5]. A small 2.8 kb TR4 transcript was observed in testis, mainly in spermatocytes[3]. Another study has detected a specific 4.5 kb transcript in rat spleen[4]. The relationship between these various small transcripts (4.5, 3.5 and 2.8 kb) are still unclear. The large transcript exhibits a predominant expression in rat central nervous system (see reference 24 for a review), with a prominent expression in hippocampus, cerebellum and hypothalamic area, as well as in adrenal gland, spleen, testis and prostate[4,28,35]. The importance of the expression of TR4 in adult brain as well as in neuronal precursors during embryogenesis has been confirmed by in situ hybridization[29]. It has been suggested that TR4 expression in brain is located in areas known to be involved in the stress response and that TR4 expression is up-regulated by NMDA[28].

In Drosophila, DHR78 is expressed ubiquitously in all organs throughout development[17]. Interestingly, its expression peaks in third instar larvae as well as in pre-pupae, and these peaks correlate with known ecdysteroid pulses[6,7]. It was shown in the same studies that DHR78 is an ecdysteroid-regulated gene. In accordance with these data, the null mutation of DHR78 leads to lethality during the third larval instar with an inability to pupariate in response to the late larval pulse of ecdysone[17].

Activity and target genes

Both TR2 and TR4 were shown to be repressors of the SV40 major late promoter through binding to a specific DR2 response element[15,25]. This suggests that these orphan receptors may have an intrinsic repressor activity. This is confirmed by the observation that RIP140 interacts with them and acts as a corepressor on transient transfection assays[21,23]. Other genes, such as CRABP-I or erythropoietin, are directly repressed by either TR2 or TR4[27,32].

TR2 and TR4 have been shown to activate transcription in specific cases. For example, TR4 is able to induce the intronic enhancer of the human ciliary neurotrophic factor receptor α (CNTFRα) gene through a DR1 element[29]. The reasons why in certain cases TR4 is a transcriptional repressor, whereas in other contexts it behaves as an activator, are not understood. It was shown that TR4 may act as a specific activator through the DR4 response element[20], whereas the element through which it activates the CNTFRα gene is a DR1. TR4 activates the DR4 elements located inside the rat α-myosin heavy chain and spot14 genes[20] as well as a DR0 element found upstream of the human luteinizing hormone (LH) gene[37]. TR2 activates the human aldolase-A gene, also through a DR1 element[30]. Interestingly, TR2 was also shown to be an activator of the RARβ-2 promoter through a DR5 element[36]. Thus, the nature of the response element is apparently not the determinant of the activator or repressor status of TR2 or TR4 in a given promoter. In accordance with the positive regulatory effect of TR2, two activation domains AF-1 and AF-2, respectively, in the A/B and E domains were mapped in this protein[36].

Since TR4 binds to the same response elements as the RAR–RXR or the TR–RXR heterodimers, its role in regulating their activity has been tested in numerous studies[8,14]. It has been shown that increasing amounts of TR4 inhibit a retinoic acid-mediated transactivation on direct repeat elements but not on palindromic ones. This is consistent with the known DNA binding preferences of TR4. In accordance with these reports, the DNA-binding activity of TR4 is required for its inhibitory action[14]. This suggests that the inhibition is due to a competition between TR4 homodimers and RXR–RAR heterodimers for the binding to the direct repeats. The same activity has been found for transactivation mediated by estrogen receptors, thyroid hormones, RXR–PPAR heterodimers or RXR homodimers[12,14,21]. TR2 has also been shown to repress RXR–RAR-mediated transactivation[13]. Interestingly, DHR78, the *Drosophila* homologue of TR2 and TR4, inhibits the induction by ecdysone of reporter genes, suggesting that the cross-talk existing between TR2/TR4 and liganded nuclear receptors is conserved during evolution[7]. Since DHR78 protein binds to many ecdysteroid-inducible puffs visible in polytene chromosome, it is likely that this cross-talk is of significant importance *in vivo*[17].

Knockout

No knockout data are available for TR2 and TR4.

Associated disease

The mouse TR2 gene is located on chromosome 10[18] close to the dominant cataract mutation cat3 but it was excluded as a candidate gene[22]. A strong polymorphism was detected in this gene, especially in the exon 7 coding part of the LBD. This polymorphism appears to be related to mouse strain differences[22].

Gene structure, promoters and isoforms

The mouse TR2 gene spans more than 50 kb and is organized into 14 exons with the first exon containing the 5′UTR[18,38]. The position of intron 3 inside the loop of the first zinc finger at a position strictly similar to the one for RXR fits with its phylogenetical placement in subfamily II. The position of intron 4, 9 amino acids after the last cysteine of the second zinc finger is identical to most members of the superfamily. In the same study, the promoter of the mouse gene was cloned and shown to be a TATA-less promoter containing several putative response elements for TCF transcription factors. It has independently been shown that the 5′UTR of the TR2 gene contains a strongly conserved IR7 element that may be important for the regulation of its expression[33]. Another conserved element containing two inverted AGGTCA motifs (IR0) was located in the mouse TR2 promoter[34]. Interestingly, this element confers retinoic-acid response to the TR2 promoter[34,38]. The human promoter was also cloned and is conserved when compared to its mouse homologue[38,39].

The human TR2 gene may express at least four different isoforms. The TR2-11 isoform is the full-size protein, whereas the TR2-5, TR2-9 and TR2-7 isoforms represent truncated proteins at different levels in the LBD[2]. The shorter of these, TR2-7 contains only the A/B and C domains. The significance of these isoforms is unknown. In mouse an isoform of TR2 with a deleted LBD exhibits an expression pattern in testis as well as functional characteristics opposite to the full-size protein[31].

The human and rat TR4 genes exhibit two isoforms called TR4α1 and TR4α2, which differ by the presence of 19 AA in the A/B domain, which correspond to a separate exon. Both isoforms are expressed as a large transcript in the 9 kb range in numerous tissues[19]. Another isoform called TR4-NS differs from the two others by the presence of an intron inside the LBD[28]. This results in a protein truncated in the end of this domain. This isoform is different from the ones known to arise inside the TR2 LBD.

Chromosomal location

The human TR4 gene is located on chromosome 3p25, a region deleted in some forms of cancer[16,19]. The same region contains RARβ, TRβ and PPARγ. The mouse TR2 gene is located on the distal region of chromosome 10 (18).

DHR78 is located to the chromosomal region 78D6-7 on chromosome 3L[6,7].

Amino acid sequence for human TR2 (NR2C1)

Accession number: M29960 (isoform TR2-11)

```
  1 MATIEEIAHQIIEQQMGEIVTEQQTGQKIQIVTALDHNTQGKQFILTNHDGSTPSKVILARQDSTPGKVF
 71 LTTPDAAGVNQLFFTTPDLSAQHLQLLTDNSPDQGPNKVFDLCVVCGDKASGRHYGAVTCEGCKGFFKRS
141 IRKNLVYSCRGSKDCIINKHHRNRCQYCRLQRCIAFGMKQDSVQCERKPIEVSREKSSNCAASTEKIYIR
211 KDLRSPLTATPTFVTDSESTRSTGLLDSGMFMNIHPSGVKTESAVLMTSDKAESCQGDLSTLANVVTSLA
281 NLGKTKDLSQNSNEMSMIESLSNDDTSLCEFQEMQTNGDVSRAFDTLAKALNPGESTACQSSVAGMEGSV
351 HLITGDSSINYTEKEGPLLSDSHVAFRLTMPSPMPEYLNVHYIGESASRLLFLSMHWALSIPSFQALGQE
421 NSISLVKAYWNELFTLGLAQCWQVMNVATILATFVNCLHNSLQQDKMSTERRKLLMEHIFKLQEFCNSMV
491 KLCIDGYEYAYLKAIVLFSPDHPSLENMELIEKFQEKAYVEFQDYITKTYPDDTYRLSRLLLRLPALRLM
561 NATITEELFFKGLIGNIRIDSVIPHILKMEPADYNSQIIGHSI 603
```

Amino acid sequence for human TR4 (NR2C2)

Accession number: L27586

```
  1 MTSPSPRIQIISTDSAVASPQRIQGSEPASGPLSVFTSLNKEKIVTDQQTGQKIQIVTAVDASGSPKQQF
 71 ILTSPDGAGTGKVILASPETSSAKQLIFTTSDNLVPGRIQIVTDSASVERLLGKTDVQRPQVVEYCVVCG
141 DKASGRHYGAVSCEGCKGFFKRSVRKNLTYSCRSSQDCIINKHHRNRCQFCRLKKCLEMGMKMESVQSER
211 KPFDVQREKPSNCAASTEKIYIRKDLRSPLIATPTFVADKDGARQTGLLDPGMLVNIQQPLIREDGTVLL
281 ATDSKAETSQGALGTLANVVTSLANLSESLNNGDTSEIQPEDQSASEITRAFDTLAKALNTTDSSSSPSL
351 ADGIDTSGGGSIHVISRDQSTPIIEVEGPLLSDTHVTFKLTMPSPMPEYLNVHYICESASRLLFLSMHRA
421 RSIPAFQGLGQDCNTSLVRACWNELFTLGLAQCAQVMSLSTILAAIVNHLQNSIQEDKLSGDRIKQVMEH
491 IWKLQEFCNSMANWDIDGYEYAYLKAIVLFSPDHPGLTSTSQIEKFQEKAQMELQDYVQKTYSEDTYRLA
561 RILVRLPALRLMSSNITEELFFTGLIGNVSIDSIIPYILKMETAEYNGQITGVSL  615
```

Amino acid sequence for drosophila DHR78 (NR2D1)

Accession number: U36791

```
  1 MDGVKVETFIKSEENRAMPLIGGGSASGGTPLPGGGVGMGAGASATLSVELCLVCGDRASGRHYGAISCE
 71 GCKGFFKRSIRKQLGYQCRGAMNCEVTKHHRNRCQFCRLQKCLASGMRSDSVQHERKPIVDRKEGIIAAA
141 GGSSTSGGGNGSSTYLSGKSGYQQGRGKGHSVKAESAATPPVHSAPATAFNLNENIFPMGLNFAELTQTL
211 MFATQQQQQQQQQHQQSGSYSPDIPKADPEDDEDDSMDNSSTLCLQLLANSASNNNSQHLNFNAGEAPTA
281 LPTTSTMGLIQSSLDMRVIHKGLQILQPIQNQLERNGNLSVKPECDSEAEDSGTEDAVDAELEHMELDFE
351 CGGNRSGGSDFAINEAVFEQDLLTDVQCAFHVQPPTLVHSYLNIHYVCETGSRIIFLTIHTLRKVPVFEQ
421 LEAHTQVKLLRGVWPALMAIALAQCQGQLSVPTIIGQFIQSTRQLADIDKIEPLKISKMANLTRTLHDFV
491 QELQSLDVTDMEFGLLRLILLFNPTLLQQRKERSLRGYVRRVQLYALSSLRRQGGIGGGEERFNVLVARL
561 LPLSSLDAEAMEELFFANLVGQMQMDALIPFILMTSNTSGL  601
```

References

1. Chang, C. and Kokontis, J. (1988) Biochem. Biophys. Res. Commun. 155, 971–977.
2. Chang, C. et al. (1989) Biochem. Biophys. Res. Commun. 165, 735–741.
3. Hirose, T. et al. (1994) Mol. Endocrinol. 8, 1667–1680.
4. Chang, C. et al. (1994) Proc. Natl Acad. Sci. USA 91, 6040–6044.
5. Law, S. et al. (1994) Gene Expression 4, 77–84.
6. Fisk, G.J. and Thummel, C.S. (1995) Proc. Natl Acad. Sci. USA 92, 10604–10608.
7. Zelhof, A.C. et al. (1995) Proc. Natl Acad. Sci. USA 92, 10477–10481.
8. Lee, C.H. et al. (1996) Mol. Reprod. Dev. 44, 305–314.
9. Kontrogianni-Konstantopoulos, A. et al. (1996) Dev. Biol. 177, 371–382.
10. Huard, V. and Séguin, C. (1997) DNA Seq. 8, 1–8.
11. Wirtanen, L. et al. (1997) Differentiation 62, 159–170.
12. Harada, H. et al. (1998) Endocrinology 139, 204–212.
13. Lin, T.M. et al. (1995) J. Biol. Chem. 270, 30121–30128.
14. Hirose, T. et al. (1995) Biochem. Biophys. Res. Commun. 211, 83–91.
15. Lee, H.J. et al. (1995) J. Biol. Chem. 270, 30129–30133.
16. Lee, Y.F. et al. (1999) Proc. Natl Acad. Sci. USA 96, 14724–14729.
17. Fisk, G.J. and Thummel, C.S. (1998) Cell 93, 543–555.
18. Lee, C.H. et al. (1995) Genomics 30, 46–52.
19. Yoshikawa, T. et al. (1996) Genomics 35, 361–366.
20. Lee, Y.F. et al. (1997) J. Biol. Chem. 272, 12215–12220.

21 Yan, Z.H. et al., (1998) J. Biol. Chem. 273, 10948–10957.
22 Immervoll, T. et al. (1998) Biol. Chem. 379, 83–85.
23 Lee, C.H. et al. (1998) Mol. Cell. Biol. 18, 6745–6755.
24 Lopes da Silva, S. and Burbach, J.P.H. (1995) Trends Neurosci. 18, 542–548.
25 Lee, H.J. and Chang, C. (1995) J. Biol. Chem. 270, 5434–5440.
26 Lin, D.L. and Chang, C. (1996) J. Biol. Chem. 271, 14649–14652.
27 Lee, H.J. et al. (1996) J. Biol. Chem. 271, 10405–10412.
28 Yoshikawa, T. et al. (1996) Endocrinology 137, 1562–1571.
29 Young, W.J. et al. (1997) J. Biol. Chem. 272, 3109–3116.
30 Chang, C. et al. (1997) Biochem. Biophys. Res. Commun. 235, 205–211.
31 Lee, C.H. et al. (1997) J. Endocrinol. 152, 245–255.
32 Chimpaisal, C. et al. (1997) Biochemistry 36, 14088–14095.
33 Le Jossic, C. and Michel, D. (1998) Biochem. Biophys. Res. Commun. 245, 64–69.
34 Lee, C.H. and Wei, L.N. (1999) Biochemistry 38, 8820–8825.
35 Van Schaik, H.S. et al. (2000) Brain Res. Mol. Brain Res. 77, 104–110.
36 Wei, L.N. et al. (2000) J. Biol. Chem. 275, 11907–11914.
37 Zhang, Y. and Dufau, M.L. (2000) J. Biol. Chem. 275, 2763–2770.
38 Lee, C.H. and Wei, L.N. (2000) Biochem. Pharmacol. 60, 127–136.
39 Lin, D.L. and Chang, C. (2000) Endocrine 12, 89–97.
40 Mouillet, J.F. et al. (1999) Eur. J. Biochem. 265, 972–981.
41 Nuclear Receptor Nomenclature Committee (1999) Cell 97, 161–163.

Names

The *Drosophila* tailless mutation produces embryos where the structures derived from both anterior and posterior portions of body are missing. The gene was demonstrated to play a key role in the establishment of non-metameric units of the embryo as well as in nervous system development. The gene cloned by Judith Lengyel's group was named following the name of the *Drosophila* mutant[1]. The vertebrate homologue was identified by Kazuhiko Umesono's and Gunther Schutz's laboratories and named TLX or MTLL, since it is also important for the development of the forebrain[3,4]. More recently, two other members of this group have been found: in *Drosophila* the dissatisfaction gene (DSF), which is essential for normal sexual behavior and neural development[5], and in vertebrates a gene called PNR (photoreceptor-specific nuclear receptor) or RNR (retina-specific nuclear receptor), which is specifically expressed in photoreceptors of the eyes[6,7]. Nematode homologues are also known[24]. Altogether, these genes form group E subfamily 2 in the nuclear receptors superfamily.

Species	Other names	Accession number	References
Official name: NR2E1			
Human	TLX	Y13276	11
Mouse	TLX, MTLL	S77482	3, 4
Chicken	TLX	S72373	3
Xenopus	XTLL	U67886	8
Medaka	TLL, tailless	AJ131390	Unpub.
Official name: NR2E2			
Drosophila	TLL, Tailless	AF019362	1
D. virilis	TLL, Tailless	AF019361	17
Musca	TLL, Tailless	not available	2
Tribolium	TLL, Tailless	AF219117	27
Official name: NR2E3			
Human	PNR, RNR	AF121129	6
Mouse	RNR	AF204053	7, 25
Official name: NR2E4			
Drosophila	DSF, dissatisfaction	AF106677	5
Official name: NR2E5			
Caenorhabditis	FAX-1	AF176087	24

Isolation

The TLL gene was cloned by a classical chromosome walk strategy in order to find genes able to rescue the Tailless phenotype[1]. The resulting gene was shown to be a new member of the nuclear receptor superfamily that is expressed at both sides of the embryo, as expected given the Tailless phenotype[9]. In order to see if the developmental gene hierarchy controlling early insect development is conserved, the Tailless gene was cloned in *Drosophila virilis* and in the domestic fly *Musca domestica* and shown to be strongly conserved, both structurally and functionally[2,17]. The *Drosophila* TLL proteins contain a PEST sequence, consistent with their rapid turnover *in vivo*[17].

The mouse TLX gene was cloned by low-stringency screening of a chick embryonic cDNA library with a mouse RXRβ cDNA fragment in a search for new orphan receptors[3] as well as by screening of a mouse embryonic cDNA library with a Drosophila TLL probe encompassing the DBD[4]. The TLX gene harbors 81% and 47% identity with its Drosophila homologue in the DBD and LBD, respectively[3,4]. The mouse and chicken sequences are extremely well conserved (97% in overall) as is the Xenopus sequence (91% with the mouse sequence in overall)[3,8].

Recently, two groups reported the sequence of a new vertebrate member of the TLL group, called RNR or PNR. This gene was cloned by a degenerate PCR approach[6] or by a low-stringency screening procedure of an adult retina cDNA library with a mix of nuclear receptor cDNA probes[7]. The gene harbors 69% and 46% amino-acid identities with the human TLL amino acid sequence in the DBD and LBD, respectively. The mouse and human PNR sequences are extremely well conserved (100% in the DBD and 95% in the LBD)[7]. In the DBD, PNR exhibits 74% sequence identity with TLX, 68% with TLL, 70% with DSF and 81% with FAX-1. For the LBD, the amino-acid identity is 57% for TLX, 34% for TLL, 56% for DSF and 20% for FAX-1.

In nematode, a gene called FAX-1 has recently been reported[24]. From a phylogenetical tree, this gene appears to be a homologue of PNR rather than TLX. These data, together with the existence of an unpublished sequence from nematode called NHR-67, suggest that TLX and PNR are not two paralogous genes but result from a more ancient duplication. The gene was identified first by the analysis of a mutant worm exhibiting neurons of the ventral nerve cord with an abnormal pathfinding of axons. It was cloned by a standard transformation rescue approach and shown to be related to PNR. It exhibits 81% and 20% amino-acid identity with the DBD and LBD of PNR. For TLX, these values are 70% and 13%, and for TLL 70% and 18%.

The Drosophila mutation called dissatisfaction was identified in 1997 as a mutant harboring abnormal sexual behavior and sex-specific neural development[10]. The gene was cloned by a strategy of phenotype rescue using several ordered cosmids and was shown to encode a gene related to TLL expressed in both sexes in an extremely limited set of neurons in regions of the brain that are implicated in sexual behavior[5]. In contrast to other members of the TLL group, DSF exhibits a long hinge region. The sequence harbors 74% and 63% sequence identities with the human TLL amino acid sequence in the DBD and LBD, respectively. The LBD of the Drosophila TLL harbors only 38% sequence identity with DSF, whereas the identity value with the LBD of PNR is 57%. In the DBD, DSF exhibits 74% sequence identity with TLL, 70% with PNR and 63% with FAX-1.

All the known TLL group members have unique structural features in common: (1) a very short A/B region (12–46 amino acids, except for FAX-1, which harbors an A/B region with 101 amino acids); (2) an unusual P-box sequence in the DBD (CDGCSG or CDGCAG or CNGCSG); and (3) a D-box containing six (PNR, FAX-1), seven (TLL, TLX) or nine (DSF) amino acids compared to five for most other nuclear receptors. In addition, the T/A box implicated in dimerization or sequence recognition is strongly conserved among TLL group members. The TLL group is related to the COUP-TF and TR2/TR4 groups of subfamily II[12].

DNA binding

In *Drosophila*, TLL binds to regulatory regions of target genes, such as Krüppel and Knirps, in which a common hexameric sequence (AAGTCA) was found. The fact that this sequence does not follow the classical consensus AGGTCA known for other nuclear receptors is in accordance with the unusual sequence of the P-box and D-box of TLL.

Since TLX harbors the same unusual sequence in its DNA-binding region, this has prompted a study of its DNA-binding properties[3]. It was shown that TLX is able to bind to the TLL-binding site found in the Knirps promoter. Of note, this sequence is A/T rich in 5' (AATTAAG). Interestingly, the binding site was a unique half-site sequence, but TLX was apparently able to recognize it as a monomer and as a homodimer[3]. In contrast, the RAR–RXR heterodimer was unable to bind to this sequence. The second position of the response element is a G for most nuclear receptors but an A in the TLL response element and this difference accounts for the specificity for TLL or TLX binding. Since the P-box of TLL and TLX does not contain a conserved lysine, as found in most other nuclear receptors, but instead an alanine in TLL and a serine in TLX, it has been proposed, but not yet proved, that this substitution accounts for the specific recognition of AAGTCA rather than AGGTCA elements. The same features were found for PNR but this protein requires two half-site motifs of the AAGTCA sequence to bind efficiently to DNA[6].

The DNA-binding specificity of the DSF protein is not known. Since it contains the same unusual P-box (CDGCSG) as other members of the group, but an even longer D-box (nine amino acids) and completely different T/A boxes, it is suspected that its DNA-binding properties will be different from those of other members of the group[5].

Partners

There are no known partners for TLL group members. Nevertheless, it was shown that PNR can bind to the cellular retinaldehyde binding protein (CRALBP) only in the presence of RAR and RXR. Whether this indicates a direct interaction with these factors remains to be investigated[7].

Ligands

There are no known ligands for any of the TLL group members.

Three-dimensional structure

No 3D structure data is available to date.

Expression

In *Drosophila*, a 2 kb transcript is detected, which is dynamically expressed during embryogenesis. Expression of TLL in a transient mirror-image symmetrical pattern extending on both sides of the embryo at nuclear cycle 11 was first detected by *in situ* hybridization[1]. Then, TLL exhibits two phases for zygotic transcripts localization. The TLL mRNAs are initially detected at both embryonic termini with a posterior cap and an anterior dorsal stripe of expression[1]. After cellularization, TLL transcripts become limited to the developing brain including the optic lobes[1]. A transient expression is also observed in peripheral nervous system. The early expression pattern is conserved in *Musca*[2]. In *Drosophila*, it has been clearly demonstrated that bicoid and the torso pathway activates tailless expression via the raf-1 kinase[16,17].

It has recently been shown that in *Drosophila* TLL is also important for the control of cell fate in the embryonic visual system[21]. A strong expression of TLL in the optic lobe primordium was observed, and gain and loss of function mutants of TLL reveal an important role for this gene for normal eye formation. This is a striking result given that, in vertebrates, TLX has been shown to be important for eye formation[4,8].

TLX is expressed in chicken as a major 3.2 kb band present only in the head[3]. By *in situ* hybridization during chicken development it was shown that expression starts at stage 8 in the head ectoderm, at stage 11 in the primary optic vesicles, at stage 16 in the forebrain and eye, and at stage 21 in neuroretina, optic stalk and dorsal midbrain, as well as in telecephalon and in localized regions of the diencephalon[3]. In mouse, the pattern is similar to the one observed in chicken[3,4]. The expression starts at day 7.5 in the head ectoderm and is specifically localized in the developing telencephalon and in dorsal midbrain. It also harbors a dynamic expression pattern in two structures whose development requires inductive signals from the forebrain: the eye and the nose[4]. This expression in the central nervous system persists until birth.

In *Xenopus*, TLX was shown to be expressed during early eye development[8]. The expression is first detected in the area corresponding to the eye anlagen within the open neural plate and, during the evagination of the optic vesicle, the expression was prominent in the optic stalk as well as in the dorsal tip of the forming vesicle. Later, the expression is particularly strong in the ciliary margin of the optic cup. The importance of TLL for eye formation in *Xenopus* was demonstrated by experiments in which TLL function was inhibited by the use of a constitutive repressor mutant of TLL. These experiments have demonstrated that TLL was required for the evagination of the eye vesicle[8].

The importance of the eye for the TLL group members has been extended by the finding that PNR exhibits a retina-specific expression pattern[6,7,23,25]. This gene is expressed in human as 7.5 kb, 3 kb and 2.3 kb transcripts, whereas in mouse a unique 2.3 kb species was detected[6,7]. These transcripts are exclusively expressed in the retina and *in situ* hybridization experiments reveal that the expression was restricted to the outer nuclear layer, which contains the nuclei of cone and rod photoreceptor cells[6,7]. These data suggest that one conserved function of TLL group members is eye development. This has been substantiated by knockout experiments as well as by the implication of PNR in a disorder of retinal fate in human (see below).

DSF is expressed as a single 3.7 kb mRNA. The gene is expressed in a small subset of neurons that are implicated in sexual behavior[5]. The expression is detected in larvae, pupae and adults. In adult, the DSF-expressing cells were several groups of cells in the anterior regions of the proto-cerebrum, but not in other regions of the brain or ventral nerve cord or in other tissues. In larvae, the salivary gland was also expressing faint levels of DSF transcripts[5].

Activity and target genes

It has been shown that ectopic expression of TLL induces activation of hunchback expression and repression of Krüppel and Knirps[15]. TLL response elements were found in the regulatory regions of Knirps and Krüppel, and TLL was effectively shown to repress their expression[13,14].

Like its *Drosophila* homologue, it was suggested that TLX is both able to repress and activate transcription and a number of potential target genes, such as OTX1, NURR1 or CHX10, have recently been proposed[26]. Furthermore, the same study shows that TLX directly regulates the PAX-2 gene, which is known to be involved in retinal development in mouse and human[26]. TLX represses PAX-2 gene expression through two conserved TLX sites that are present in the PAX-2 promoter. This has been confirmed *in vivo* by ectopic expression of TLX in eye, which effectively repress endogenous PAX-2 expression.

PNR has been shown to be a strong inhibitor of transcriptional activity using the GAL-4 reporter system[7]. No target genes are known up to now.

Knockout

Drosophila mutant in the TLL gene exhibit a very specific phenotype[1,9]. Cuticles of mutant embryos lack the abdominal segment 8 and the telson in the posterior part of the embryo, as well as the dorsal bridge and dorsal arms of the cephalopharyngeal skeleton in the anterior part of the embryo. This is consistent with the early expression pattern of TLL. In addition, the mutant embryos lack most of the brain, including the optic lobe in accordance with the high expression of TLL in embryonic brain. In addition, defects in the posterior midgut and Malpighian tubules have been observed[1,19]. Ectopic expression of TLL results in differentiation of ectopic terminal structures and leads to the activation of the hunchback gene that is required for these structures, and the repression of segmentation genes such as Krüppel and Knirps (NR0A1)[15]. A function for TLL in eye formation was also demonstrated using gain and loss of function mutants[21].

Since TLX and TLL have similar structure, pattern of expression and target gene specificity, the ability of TLX to regulate the TLL genetic cascade in *Drosophila* has been tested[3]. Indeed, it has been shown that transgenic flies expressing TLX ectopically under the control of a heat-shock protein promoter exhibit the same defects as flies expressing ectopic TLL[3,15]. The

Knirps and Engrailed genes, which are down-regulated by TLL, were also inhibited by TLX, demonstrating that TLX can mimic TLL action *in vivo*[3]. This is also confirmed by the fact that expression of a dominant repressive version of the *Xenopus* TLX gene perturb eye formation, in accordance with the recently described role of the *Drosophila* gene in this organ[8,21].

TLX knockout mice are viable at birth, indicating that TLX is not required for prenatal survival[22]. Nevertheless, these animals exhibit a marked forebrain phenotype with a reduction in the size of rhinencephalic and limbic structures. In addition, both males and females are more aggressive than usual and the females lack normal maternal instincts. All these data further stress the important role of TLX for correct brain development. Recently, it was shown that TLX is essential for vision, since the knockout mice exhibit a progressive retinal and optic nerve degeneration with associated blindness[26].

The mouse PNR gene has been found to be responsible of the rd7 phenotype in the mouse[25]. These mice are models for hereditary retinal degeneration and it was shown that they carry a mutation in the PNR gene, reinforcing the importance of this gene for eye development. The mutation found is an internal deletion of 380 nucleotides in the coding region that creates a frame shift and a premature stop codon in the D domain. This is fully consistent with the data regarding human mutations in the PNR gene[23].

DSF mutant flies have a striking behavioral phenotype[5]. DSF males are bisexual and mate poorly, while mutant females resist male courtship and fail to lay eggs. Males and females have sex-specific neural abnormalities[5,10]. The close analysis of these animals allows us to place DSF in the genetic cascade that controls sexual behavior in *Drosophila*[5,20].

Associated disease

There is no known associated disease for TLX. In contrast, PNR was found mutated in a disorder of retinal cell fate, S-cone syndrome[23]. This disorder is a retinal degenerative disease that manifests a gain in function of photoreceptors with patients exhibiting an increased sensitivity to blue light. In these patients, visual loss and retinal degeneration also occur. The PNR gene was found expressed in the outer nuclear layer of human retina and a number of mutations either splice site mutation, short deletion or missense mutations in both the DBD and LBD of the protein were detected. Since the TLX gene was also found in mice to be essential for vision, this highlights the importance of this group of genes for normal eye development and function[23,26].

Gene structure, promoter and isoforms

The genomic organization of TLL has been described in *D. melanogaster*[17]. A unique short intron was found in the A/B domain. The comparison of *D. melanogaster* and *D. virilis* regulatory regions upstream of TLL transcriptional start sites allow elements in this promoter to be found that are the targets of the regulation by the raf-1 kinase[17]. This element, called torRE (for torso response element), is a 11 bp sequence on which a GAGA factor and the NTF-1 factor binds and represses TLL transcription.

Interestingly, the NTF-1 factor is the target of the phosphorylation by the torso/raf-1 pathway. This phosphorylation by an MAP kinase led to the relief of the GAGA/NTF-1 repression of TLL transcription[18].

The genomic organization of TLX and PNR have been published[6,11]. The PNR gene harbors eight exons within approximately 60 kbp of genomic DNA[6]. An intron was found in the A/B region as for TLL and TLX. Furthermore, the position of the intron lying at the end of the DBD is conserved with TLX and other nuclear receptors. The intron between the two zinc fingers has an unusual location very close to the D-box of the second zinc finger.

No complete characterization of TLX or PNR promoter was performed. Nevertheless, it was shown that the 5′ part of the PNR gene contains a region with promoter activity that contains a TATA-box[6].

The genomic organization of DSF has not been described. The examination of the DSF regulatory region suggests it may be a target to the TRA gene, which is implicated in sex determination in drosophila, since it contains a single copy of the sequence on which TRA binds[5]. The generation of possible sex-specific isoforms from the DSF primary transcript was studied but nothing was found (see reference 20 for a review).

Chromosomal location

TLL is localized on *Drosophila* chromosome 3R at position 100B1. TLX is located on human chromosome 6q21, a region in which chromosomal abnormalities are frequent in hematological malignancies[11]. PNR is located on mouse chromosome 9[25] and on human chromosome 15q24, a location that was known to be associated with retinal diseases[6,23]. DSF is at position 26A6–8 on chromosome 2L.

Amino acid sequence for human TLX (NR2E1)

Accession number: Y13276

```
  1  MSKPAGSTSRILDIPCKVCGDRSSGKHYGVYACDGCSGFFKRSIRRNRTYVCKSGNQGGCPVDKTHRNQC
 71  RACRLKKCLEVNMNKDAVQHERGPRTSTIRKQVALYFRGHKEENGAAAHFPSAALPAPAFFTAVTQLEPH
141  GLELAAVSTTPERQTLVSLAQPTPKYPHEVNGTPMYLYEVATESVCESAARLLFMSIKWAKSVPAFSTLS
211  LQDQLMLLEDAWRELFVLGIAQWAIPVDANTLLAVSGMNGDNTDSQKLNKIISEIQALQEVVARFRQLRL
281  DATEFACLKCIVTFKAVPTHSGSELRSFRNAAAIAALQDEAQLTLNSYIHTRYPTQPCRFGKLLLLLPAL
351  RSISPSTIEEVFFKKTIGNVPITRLLSDMYKSSDI 385
```

Amino acid sequence for *Drosophila* TLL (NR2E2)

Accession number: AF019362

```
  1  MQSSEGSPDMMDQKYNSVRLSPAASSRILYHVPCKVCRDHSSGKHYGIYACDGCAGFFKRSIRRSRQYVC
 71  KSQKQGLCVVDKTHRNQCRACRLRKCFEVGMNKDAVQHERGPRNSTLRRHMAMYKDAMMGAGEMPQIPAE
141  ILMNTAALTGFPGVPMPMPGLPQRAGHHPAHMAAFQPPPSAAAVLDLSVPRVPHHPVHQGHHGFFSPTAA
211  YMNALATRALPPTPPLMAAEHIKETAAEHLFKNVNWIKSVRAFTELPMPDQLLLLEESWKEFFILAMAQY
281  LMPMNFAQLLFVYESENANREIMGMVTREVHAFQEVLNQLCHLNIDSTEYECLRAISLFRKSPPSASSTE
351  DLANSSILTGSGSPNSSASAESRGLLESGKVAAMHNDARSALHNYIQRTHPSQPMRFQTLLGVVQLMHKV
421  SSFTIEELFFRKTIGIDITIVRLISDMYSQRKI 452
```

Amino acid sequence for human PNR (NR2E3)

Accession number: AF121129

```
  1  METRPTALMSSTVAAAAPAAGAASRKESPGRWGLGEDPTGVSPSLQCRVCGDSSSGKHYGIYACNGCSGF
 71  FKRSVRRRLIYRCQVGAGMCPVDKAHRNQCQACRLKKCLQAGMNQDAVQNERQPRSTAQVHLDSMESNTE
141  SRPESLVAPPAPAGRSPRGPTPMSAARALGHHFMASLITAETCAKLEPEDADENIDVTSNDPEFPSSPYS
211  SSSPCGLDSIHETSARLLFMAVKWAKNLPVFSSLPFRDQVILLEEAWSELFLLGAIQWSLPLDSCPLLAP
281  PEASAAGGAQGRLTLASMETRVLQETISRFRALAVDPTEFACMKALVLFKPETRGLKDPEHVEALQDQSQ
351  VMLSQHSKAHHPSQPVRFGKLLLLLPSLRFITAERIELLFFRKTIGNTPMEKLLCDMFKN 410
```

Amino acid sequence for *Drosophila* DSF (NR2E4)

Accession number: AF106677

```
  1  MGTAGDRLLDIPCKVCGDRSSGKHYGIYSCDGCSGFFKRSIHRNRIYTCKATGDLKGRCPVDKTHRNQCR
 71  ACRLAKCFQSAMNKDAVQHERGPRKPKLHPQLHHHHHAAAAAAAAHHAAAAHHHHHHHHHAHAAAAHHA
141  AVAAAAASGLHHHHHAMPVSLVTNVSASFNYTQHISTHPPAPAAPPSGFHLTASGAQQGPAPPAGHLHHG
211  GAGHQHATAFHHPGHGHALPAPHGGVISNPGGNSSAISGSGPGSTLPFPSHLLHHNLIAEEAASKLPGITA
281  TAVAAVVSSTSTPYASAAQASSPSSNNHNYSSPSPSNSIQSISSIGSRSGGGEEGLSLGSESPRVNVETE
351  TPSPSNSPPLSAGSISPAPTLTTSSGSPQHRQMSRHSLSEATTPPSHASLMICASNNNNNNNNNNNNNNG
421  EHKQSSYTSGSPTPTTPTPPPPRSGVGSTCNTASSSSGFLELLLSPDKCQELIQYQVQHNTLLFPQQLLD
491  SRLLSWEMLQETTARLLFMAVRWVKCLMPFQTLSKNDQHLLLQESWKELFLLNLAQWTIPLDLTPILESP
561  LIRERVLQDEATQTEMKTIQEILCRFRQITPDGSEVGCMKAIALFAPETAGLCDVQPVEMLQDQAQCILS
631  DHVRLRYPRQATRFGRLLLLLPSLRTIRAATIEALFFKETIGNVPIARLLRDMYTMEPAQVDK 693
```

References

1   Pignoni, F. et al. (1990) Cell 62, 151–163.
2   Sommer, R. and Tautz, D. (1991) Development 113, 419–430.
3   Yu, R.T. et al. (1994) Nature 370, 375–379.
4   Monaghan, A.P. et al. (1995) Development 121, 839–853.
5   Finley, K.D. et al. (1998) Neuron 21, 1363–1374.
6   Kobayashi, M. et al. (1999) Proc. Natl Acad. Sci. USA 96, 4814–4819.
7   Chen, F. et al. (1999) Proc. Natl Acad. Sci. USA 96, 15149–15154.
8   Holleman, T. et al. (1998) Development 125, 2425–2432.
9   Strecker, T.R. et al. (1988) Development 102, 721–734.
10 Finley, K.D. et al. (1997) Proc. Natl Acad. Sci. USA 94, 913–918.
11 Jackson, A. et al. (1998) Genomics 50, 34–43.
12 Nuclear Receptor Nomenclature Committee (1999) Cell 97, 161–163.
13 Pankratz, M.J. et al. (1992) Science 255, 986–999.
14 Hoch, M. et al. (1992) Science 256, 94–97.
15 Steingrimsson et al. (1991) Science 254, 418–421.
16 Pignoni, F. et al. (1992) Development 115, 239–251.
17 Liaw, G.J. et al. (1993) Proc. Natl Acad. Sci. USA 90, 858–862.
18 Liaw, G.J. et al. (1995) Genes Dev. 9, 3163–3176.
19 Diaz, R.J. et al. (1996) Mech. Dev. 54, 119–130.
20 O'Kane, C. and Aszlalos, Z. (1999) Curr. Biol. 9, R289–R292.
21 Daniel, A. et al. (1999) Development 126, 2945–2954.
22 Monaghan, A.P. et al. (1997) Nature 390, 515–517.
23 Haider, N.B. et al. (2000) Nature Genet. 24, 127–131.
24 Much, J.W. et al. (2000) Development 127, 703–712.

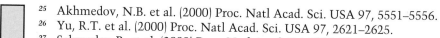

25 Akhmedov, N.B. et al. (2000) Proc. Natl Acad. Sci. USA 97, 5551–5556.

26 Yu, R.T. et al. (2000) Proc. Natl Acad. Sci. USA 97, 2621–2625.

27 Schroeder, R. et al. (2000) Proc. Natl Acad. Sci. USA 97, 6591–6596.

COUP-TF

Names

COUP-TF genes form with EAR2 a strongly conserved group of nuclear orphan receptors inside subfamily II. COUP-TFI was first isolated as a v-erbA-related gene under the name EAR3 (erbA related), together with EAR2[1]. It was independently characterized as a chicken ovalbumine gene upstream promoter transcription factor and was thus called COUP-TF and shown to be identical to EAR3[14]. COUP-TFII was found as a regulator of apolipoprotein-A1 regulatory protein (ARP-1). The close relationship with COUP-TFI explains why it was rapidly called COUP-TFII. The *Drosophila* homologue was characterized as a mutant implicated in the fate of the retina photoreceptor R7 and was thus called seven-up or SVP[10]. The zebrafish homologues of the COUP-TF genes, using the *Drosophila* name, were called SVP44 and SVP40[5,9]. A third gene exists in zebrafish called SVP46, which is also found under the name COUP-TFIII[5]. Its relationships with the third *Xenopus* gene, COUPA, is not clear[4,8]. For clarity we called the vertebrate genes COUP-TFI, COUP-TFII and COUP-TFIII. The arthropod genes are called SVP, whereas the unique homologue of the three COUP-TF genes found in echinoderms (sea urchin) and chordates (amphioxus) are called COUP-TF[4,11].

Species	Other names	Accession number	References
Official name: NR2F1			
Human	COUP-TFI, EAR3, COUP-TFA	X12795	1, 14
Mouse	COUP-TFI, COUP-TFA	U07625	2
Rat	COUP-TFI, COUP-TFA	U10995	3
Cow	COUP-TFI, COUP-TFA	AJ249440	Unpub.
Xenopus	COUP-TFI COUP-TFA	AF157558	4
Zebrafish	COUP-TFI, SVP44	X70299	5
Official name: NR2F2			
Human	COUP-TFII, ARP-1, COUP-TFB	M64497	6
Mouse	COUP-TFII, COUP-TFB	U07635	2
Rat	COUP-TFII, COUP-TFB	AF003944	Unpub.
Cow	COUP-TFII, COUP-TFB	AJ249441	Unpub.
Chicken	COUP-TFII, COUP-TFB	U00697	7
Xenopus	COUP-TFII, COUP-TF-B	Not available	4, 13
Zebrafish	COUP-TFII, SVP40	S80986	9
Official name: NR2F3			
Drosophila	SVP, seven-up	M28863	10
Shistocerca	SVP, seven-up	U36622	15
Sea urchin	COUP-TF, SpCOUP-TF	L01104, L01105	11
Amphioxus	COUP-TF, AmphiCOUP-TF	Not available	4
Official name: NR2F4			
Xenopus	COUP-TFIII, COUP-TFIV, COUP-TF-A	X63092	8
Official name: NR2F5			
Zebrafish	COUP-TFIII, SVP46, COUP-TFIV	X70300	5
Official name: NR2F6			
Human	EAR2	X12794	1
Mouse	EAR2	L25674	55
Rat	EAR2	AF003926	Unpub.

Isolation

The two first members of the COUP-TF group, called EAR2 and EAR3, were cloned in 1988 by Tadashi Yamamoto's laboratory by the screening of a human-embryo fibroblast cDNA library with a v-erbA probe[1]. EAR2 is considered as a member of the COUP group distantly related to the regular COUP-TF genes. Later, it was revealed that the EAR3 sequence was identical to the COUP-TF gene that was cloned independently in 1989 in Bert O'Malley's laboratory[14]. COUP-TF was first defined as a factor required for the transcription of the chicken ovalbumine gene in addition to another non-DNA binding factor that was called S300-II[16,17]. The factor was biochemically purified and also shown to bind to an upstream promoter element of the rat insulin II gene, suggesting that it was a factor generally implicated in gene regulation[18,19]. Characterization of antiserum raised against the purified protein and screening of a HeLa cell cDNA expression library led to the isolation of a human COUP-TF cDNA that was identical to EAR3[14].

The other member known in mammals, ARP-1 or COUP-TFII, was identified as a regulator of apolipoprotein A1 gene expression by Sotirios Karathanasis's laboratory in 1991[6]. Human placenta cDNA expression libraries were screened with an oligonucleotide corresponding to a regulatory region of the ApoA1 gene in which a specific protein was shown to bind. The resulting clone was closely related but different to COUP-TFI. COUP-TFII was also found as a COUP-TFI-related gene by Ming-Jer Tsai's laboratory[6,9].

When other teams isolated COUP-TFs homologues in *Xenopus*[4,8,13] and zebrafish[5,9], three different sequences were found. It was clear that there were orthologues of COUP-TFI and COUP-TFII, but the identity of the third clones, SVP46 in zebrafish and COUP-TFIII in *Xenopus*, and their relationships with other members of the COUP-TF group were unclear. Some authors suggested that these two genes represent two different members of the COUP-TF group, which, in this case will contain four genes, two of which are missing in mammals. Another possibility is that both of them are orthologues and represent a third gene that was not found, or does not exist in mammals. A discussion on these possibilities can be found in reference 4. Since unique homologues were found in echinoderms and chordates, it is clear that the two COUP-TF existing in mammals are paralogous genes resulting from a vertebrate specific gene duplication[4,11].

The *Drosophila* homologue was identified as a lethal insertion in an 'enhancer trap' screen for genes expressed in photoreceptors during eye development[10]. The identification of a transcription unit close to the P-element insertion point and the cloning of the corresponding gene, which was called seven-up or SVP, reveals that it was a *Drosophila* homologue of vertebrates COUP-TFs. The gene was shown to control photoreceptor cell fates[10] (see reference 22 for a review). Short COUP-TF signatures were identified in a number of other metazoans, including the coelenterate *Hydra*, showing that it is present in all of the animal kingdom[21].

The COUP-TF genes are the most conserved nuclear receptors[20,23]. In the DBD COUP-TFI and COUP-TFII share 98.5% amino acids identity and these two proteins are 92% and 94% identical to the *Drosophila* SVP gene, respectively. The COUP-TFIII genes are a little more distant when compared to COUP-TFI and COUP-TFII (91–93%) and even more with SVP (88%). EAR2 is 80–88% identical to the regular COUP-TFs in the DBD. In the LBD, COUP-TFI and COUP-TFII are 96% identical, the value with SVP is 90% and with the two COUP-TFIII 87–88%. The COUP-TFIII and SVP are 85% identical in the LBD. EAR2 is clearly different, with 68% identical amino acids in the LBD with other COUP-TF group members. These data clearly show that, for COUP-TFs, the LBD is as much conserved as the DBD in contrast to other nuclear receptors. An *Hydra* COUP-TF cDNA was shown to be very well conserved when compared to mammals or *Drosophila* sequences, further stressing the strong conservation of COUP-TFs (B. Galliot and V. Laudet., unpublished). For a general review on the COUP-TF group, see references 12 and 20.

DNA binding

Since both COUP-TFI and COUP-TFII were cloned by virtue of their DNA-binding activity, their interaction with DNA rapidly became known[6,14]. The DNA-binding site in which COUP-TFI was found to interact in the chicken ovalbumine promoter was a DR1 element containing some mutations in the 5′ half-site (GTGTCAaAGGTCA)[16]. The rat insulin 2 promoter element recognized by the purified COUP-TFI protein contains a clear half-site motif (GGGTCA) and is in fact an imperfect DR6 element[19,20]. COUP-TFII was shown to bind to the A site of the human apolipoprotein A1 gene, which has a DR2 structure (GGGTCAagGGTTCA)[6]. Nevertheless, the same study revealed that it was able to bind to a number of other gene promoters, such as site C of ApoA1, ApoCIII or ApoB, which have a clear DR1 structure. In addition, it was shown that COUP-TF binds to the palindromic TRE as well as poorly to an ERE[6].

These data were rapidly confirmed with COUP-TFs from various species as well as on different DNA elements[4,11,20,25–29]. It is now believed that COUP-TFs recognize a large number of DNA elements based on the classical half-site sequence A/GGGTCA. Systematic comparison of the relative affinities of COUP-TFs for the various elements reveal that DR1 is the preferred element, and then DR6, DR4, DR8, DR0 and DR11[26]. Palindromic and inverted repeats are also recognized, but with a lower affinity. In contrast, monomeric elements are not efficiently recognized by COUP-TFs (see reference 4 for an example).

EAR2 was also shown to bind to DR1 element[29]. It was recently shown that it seems to be much more specific to this element than other members of the COUP-TF group, since it does not interact with other direct repeats[31]. Another study revealed that EAR2 may also interact with the thyroid hormone response element (TRE) found in the chicken lysozyme gene[39].

SVP was shown to bind with a strong affinity to DR1 elements and with a lower strength to DR2, DR3, DR4, DR5[30]. It binds weakly to the EcRE of the hsp27 gene as well as to the EcRE of the Eip 28/29 gene. It does not bind to half-sites, suggesting that it binds to DNA as a homodimer, like its

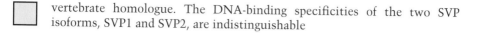

vertebrate homologue. The DNA-binding specificities of the two SVP isoforms, SVP1 and SVP2, are indistinguishable

Partners

It was immediately clear that COUP-TFs form homodimers when binding to direct repeat elements and are present as a dimer in solution[6,9,24]. A mutant of COUP-TFII containing only the A/B and C domains can bind as both a monomer and dimer to DNA[6]. Further studies of COUP-TFII mutants reveal that it binds to DNA cooperatively as an homodimer and that the C-terminal part of the protein inhibits the monomeric DNA binding[6].

COUP-TFI and COUP-TFII can form homodimers in solution as well as on DNA, but they are also able to heterodimerize[26]. The DNA-binding properties of the heterodimer are apparently indistinguishable to those of each of the homodimers[26]. More recently, a heterodimer between COUP-TFII and EAR2 was also shown to exist in solution as well as on direct repeat elements[31]. Interestingly, the specificity of the heterodimer is subtly different than that of the COUP-TFII and clearly different from that of EAR2 homodimers[31].

The fact that COUP-TF was able to heterodimerize or not with RXR has given rise to conflicting results. It was shown by gel-shift experiments that in vitro translated COUP-TFI and RXR proteins form heterodimers[24,32,36] but this finding was not reproduced by other authors[34,35]. Two-hybrid experiments has shown that COUP-TFII homodimers are formed in preference to COUP-TFII–RXR heterodimers[33]. No interaction was found between SVP and the Drosophila RXR homologue, USP[30]. Thus, it is unlikely that COUP-TFII may inhibit RXR transactivation via the formation of inactive heterodimers. Other mechanisms, such as competition of the promiscuous COUP-TF homodimer for the DNA elements recognized by RXR and its partners, appear more likely.

COUP-TFs were shown to interact with a number of other nuclear receptors and, as for RXR, the data were in some cases confusing because several groups found opposite results. COUP-TFI was shown to interact with TR and RAR and thus disrupt their functions[34,37,38]. Again, results from the two-hybrid system suggest that the efficiency of this heterodimer formation is much weaker than the strength of homodimer formation[33]. Nevertheless, it was recently shown in a two-hybrid screen performed with TRβ1 (NR1A2) as a bait that EAR2 interacts efficiently with this receptor[39]. This interaction was confirmed by GST pull-down experiments. EAR2 interacts with the C-terminal region of TRβ1. It was also shown that EAR2 decreased the binding of TRβ1 to its DNA response elements and inhibits TRβ1-mediated transactivation. An interaction between EAR2, ER (NR3A1) and GR (NR3C1) was also found in the same study[39]. Other nuclear receptors for which an interaction with COUP-TF was found include the estrogen receptor, ERα although the role and in vivo significance of this interaction remains speculative[41,42], PPARγ (NR1C3)[52] and the orphan receptors nur77 (NR4A1)[43] and HNF4 (NR2A1)[87]. In the case of nur77, the interaction exists in solution and does not require the C-terminal region of nur77. In addition, an interaction exists between SVP and the ecdysone receptor (NR1H1)[30]. As in mammals, a negative regulation of EcR–USP heterodimer by SVP exists both *in vitro* and *in vivo*.

COUP-TF also interacts with other transcription factors. One very interesting case concerns the interaction between COUP-TFI and transcription factors that regulate the LTR of HIV. It has been shown that COUP-TF is an activator of HIV gene expression via stimulation of LTR activity (see below). A direct physical interaction exists both *in vitro* and *in vivo* between COUP-TF and the Sp1 transcription factor, leading to functional synergism between the two factors[44]. It is the DNA-binding domain of COUP-TFI that is implicated in the interaction. In contrast, the interaction between COUP-TFI and NF-IL6, a member of the C/EBP family, which also requires the DBD of COUP-TFI, leads to mutual transcriptional repression[45]. Finally, COUP-TF is able to physically interact and cooperate with the viral transactivator Tat[46]. The combination of COUP-TF and Tat results in Sp1-independent enhanced transcription. Once again it is the DBD of COUP-TF and more precisely the first zinc finger that is implicated in the interaction. All these results highlight the crucial role of COUP-TF in HIV gene transcription. Other examples of transcription factors that interact with COUP-TF include the aryl hydrocarbon receptor[47], two C2H2 zinc finger proteins, CTIP1 and CTIP2, which may be implicated in the transcriptional repression mediated by COUP-TF in the brain[48], and the Oct1 factor that plays a role in the activation of the vHNF gene promoter activation by COUP-TF[83]. The distantly related COUP-TF group member, EAR2, was recently shown to interact with the transcription factor CBFA2, also called PEBP2αB, and to inhibit its activity[49]. Since CBFA2 plays a pivotal role in the differentiation of myeloid progenitors, the effect of EAR2 in this pathway was studied. It was shown that overexpression of EAR2 can prevent G-CSF-induced myeloid differentiation, suggesting that EAR2 may be a key regulator of granulocytic differentiation[49].

COUP-TFI has been shown to interact with the general transcription factor TFIIB[54,76]. In fact, S300-II, the partner found associated with COUP-TF during the biochemical purification that precedes its cloning was identified as TFIIB[17,54].

The last class of protein that interacts with COUP-TFs is corepressors and coactivators. The case of CTIP proteins that may play a role in COUP-induced repression in the nervous system[48] has already been mentioned but other cases are known. In accordance with the well-known repressing activity of COUP-TFI, it was shown that interaction with NCoR and SMRT mediates this gene silencing effect[53]. The C-terminal 35 amino acids form the repression domain of COUP-TFI, which interacts with NCoR and SMRT. Cotransfection of NCoR or SMRT effectively potentiates the silencing activity of COUP-TFI. This was confirmed by another study which also revealed a direct interaction between COUP-TF and histone deacetylase[50]. In *Drosophila* an interaction between SVP and Alien, a new corepressor that is also strongly conserved in vertebrates, is believed to play an important part in SVP repressing activity[51]. The coactivator SRC-1 is able to bind to either EAR2 and COUP-TFI, which is also able to bind the related factor GRIP1[39,80]. GRIP-1 and SRC-1 potentiate the transcriptional activity of COUP-TFI on the PEPCK gene promoter (see below). Recently, an interaction between p62 a protein that interacts with the tyrosine kinase p56[lck] and COUP-TFII has been demonstrated. It has been proposed that p62 may act as a coactivator of COUP-TFII[96].

Ligands

There are no known ligands for COUP-TF.

Three-dimensional structure

There is no 3D structure available for COUP-TF.

Expression

In humans, COUP-TFI is expressed as two transcripts of 4.6 and 4.8 kb that show a broad expression pattern. The expression pattern of COUP-TFI is well conserved in mouse[2,55] and zebrafish[5], the different species where it was studied (see reference 20 for a detailed review). In the mouse, COUP-TFI expression was first detected at 7.5 dpc, peaked at 14–15 days dpc and strongly declined before birth when organogenesis stops[2]. At 8.5 dpc COUP-TFI is expressed in specific regions of the rostral brain, in stripes in the presumptive hindbrain as well as in the anteriormost somites. Later, it exhibits a more complex expression pattern in CNS, which is clearly different from the one of COUP-TFII. Interestingly, it is confined to specific segmental neuromere compartments[2]. COUP-TFI is expressed throughout the neural tube with a small increased level in motor neurons. The expression also appears restricted in other organs and is always weaker than that of COUP-TFII[55,57]. COUP-TFI is found at high levels in the tongue, the follicles of the vibrissae, the cochlea and in the stroma of the nasal septum. In organs that require mesenchymal and epithelial interactions, COUP-TFI is expressed in the mesenchymal cells, not in the terminally differentiated epithelium. In adult, the expression is strongly reduced, and COUP-TFI was found in the rostral and caudal part of the mouse brain, in particular in the supraoptic nucleus[58]. Interestingly, the expression pattern of COUP-TFI (SVP44) in zebrafish is similar, with an expression in the anterior half of the midbrain and in the posterior part of the diencephalon at 13 hours of development[5]. As for COUP-TFI, a segment-like stripe of expression was found at sites corresponding to rhombomere primordia in the anterior region of the hindbrain. It is also expressed in the intermediate mesoderm that will give rise to the kidney and the urogenital system. Zebrafish COUP-TFI is also found in the retina, as its *Drosophila* homologue SVP.

In the mouse, the COUP-TFII gene generates a transcript of 4.5 kb found in all tissues examined[6]. COUP-TFII expression is similar to that of COUP-TFI in the sense that it exhibits a complex expression pattern in the CNS and a broad expression in other tissues (reviewed in reference 20). The beginning of the expression at 7.5 dpc is identical to the one of COUP-TFI[2]. A segmented expression in the diencephalic neuromeres was also found but the specific set of neuromeres which express COUP-TFII is different, although overlapping with COUP-TFI. In the hindbrain, the specific rhombomeres in which the genes are expressed are also distinct. In the neural tube the expression is restricted to the motor neurons. It was shown in chick that COUP-TFII expression correlates with differentiation of the motor neurons[59]. The levels of COUP-TFII expression are in general higher than those for COUP-TFI, especially in salivary gland, lung, esophagus, stomach, pancreas, kidney and prostate[55,57]. It is also found at lower levels in testes, ovary, retina, skin, inner ear or limb bud. The expression in the mesenchymal portion of places where mesenchyme–epithelial interactions occurs was also found. In the adult the expression is also higher than that of COUP-TFI in supraoptic nucleus[58]. As for COUP-TFI, this expression pattern is, overall, well conserved in zebrafish[9] as well as in *Xenopus*[13]. In *Xenopus* a complex expression pattern in the CNS, as well as the expression in the eye and mesodermal tissues was noticed[13]. In zebrafish COUP-TFII (SVP40) is also found in specific regional and segmented domains of the developing CNS as in mouse and xenopus[9]. The expression pattern suggest an involvement of COUP-TFII in neural patterning.

Interestingly, the expression of COUP-TFI and COUP-TFII was shown to be regulated by all-*trans* retinoic acid in the neural tube[55,60]. *In vivo* effects of retinoic acid were also found on COUP-TFII expression in zebrafish[9]. These data show that retinoic acid may regulate the expression of COUP-TF genes, the products of which can antagonize retinoic acid effects on transcription. In addition, a regulation of COUP-TFII by Sonic hedgehog has been found by Tsai's group (cited in reference 20).

The other vertebrate genes, SVP46 in zebrafish[5] and COUP-TFIII in *Xenopus*[13], have a very similar expression pattern, reinforcing the notion that these genes are probably orthologues (see references 4 and 20 for opposing viewpoints). Both exhibit a complex expression pattern in CNS, in the primordia of diencephalon, midbrain, hindbrain and anterior part of the spinal cord. An expression on specific rhombomeres was found in the two species. Outside CNS an expression in the eye, during formation in the somites was found[5,13].

The unique COUP-TF homologue found in other deuterostomes, such as in the sea urchin (*Strongylocentrotus purpuratus*) and in the chordate amphioxus (*Branchiostoma floridae*), in contrast exhibits different expression patterns. In amphioxus, no expression was found in the embryo, in sharp contrast to all other species[4]. Expression was only found in the nerve cord of late larvae, in a region corresponding to hindbrain and probably anterior spinal cord. Although the amphioxus nerve cord appears unsegmented at the anatomical level, COUP-TF expression was observed segmented with stripes of expressing cells that may correspond to visceral and somatic motor neurons. This confirms the importance and conservation of neuronal COUP-TF expression and suggests that the roles of COUP-TF in patterning of the nerve cord evolved prior to the appearance of vertebrates[4]. In sea urchin, the unique COUP-TF homologue was found expressed at low levels during oogenesis and embryogenesis with a small increase at the pluteus stage as well as in the adult[11]. In contrast to amphioxus, the expression was found very early on during embryogenesis and it was found that COUP-TF is a maternal mRNA that is deposited unevenly in the oocytes[62]. A complex subcellular trafficking of the COUP-

TF protein was observed in early sea-urchin embryos during the first rapid cell divisions[63]. The embryonic expression is spatially restricted in the oral ectoderm and, at later stages in the cells of the ciliated band, the neurogenic cell lineage of the sea urchin embryo. This confirms once again the conservation of neuronal COUP-TF expression.

The *Drosophila* SVP gene also exhibits an expression in neuronal cells in addition to other organs. The most studied regions of expression of SVP was of course the eye since SVP was cloned as a gene crucial for the fate of photoreceptors[10,22]. SVP was found to be expressed in four of the eight photoreceptors (R1, R3, R4 and R6, but not R7) inside each ommatidia and is critical for correct eye development, since ectopic SVP expression causes cell-fate changes during ommatidia assembly[30,63,64,68]. It has been shown that SVP requires the ras pathway for its function in ommatidia development[63,64]. SVP is also expressed in the fly CNS[10]. By an enhancer trap strategy, an expression of SVP in fat cell lineage was also noticed and it has been shown that SVP plays a role in fat-body specific expression of some fat-cell differentiation genes[65,66]. Recently, it was shown that SVP controls cell proliferation in the insect kidney[67]. In this organ, SVP was shown to be a downstream target of the EGF signaling pathway and to control cell-cycle gene expression. These interesting recent data prompt a re-evaluation of SVP function outside eye formation.

Few data are available on the expression pattern of EAR2. A 2.5 kb transcript was found in human fetal liver (as well as in placenta, heart, muscle and pancreas), weak expression was noticed in kidney, and no expression was detected in lung and brain or kidney[1,39]. Other studies in mouse, including *in situ* hybridization, also found a broad expression pattern of EAR2. In contrast to COUP-TFI and II, the expression pattern of this gene is not regulated by retinoic acid[31,55]. Consistent with its interaction with the CBFA2 transcription factor, which plays an important role in granulocytic differentiation, EAR2 was found expressed in myeloid progenitor cells and down-regulated during differentiation induced by G-CSF. Overexpression of EAR2 in these cells blocks the differentiation, suggesting that it may be a negative regulator of granulocytic differentiation[49]. An expression of EAR2 in uterine epithelial cells was also described[56].

Activity and target genes

The first functional studies immediately revealed that COUP-TFI and II behaves as potent transcriptional repressors[6,24,26,32,35]. In accordance with this notion, it has been shown that COUP-TFI interacts with the corepressors NCoR and SMRT[53]. Even more importantly, given the promiscuous DNA-binding ability of COUP-TF, it is able to repress the transactivation mediated by a number of activating nuclear receptors, either liganded or orphans (see reference 20 for a review). This effect was found on retinoic acid receptors RAR and RXR[24,26,32], TRs[24,26], VDR[24,26], PPARs[27,70], ER[72,110], HNF4[71], SF1[73], etc. This effect was observed outside vertebrates, since the COUP-TF of amphioxus is able to repress RXR, RAR–RXR, TR, ERα- and ERβ-mediated transactivation[4]. The *in vivo* importance of this cross-talk among nuclear receptors has been emphasized by the spectacular effect of an ectopic expression of COUP-TF in *Xenopus* development[72]. These results revealed that COUP-TFI is a potent regulator of RAR-induced gene expression and that misexpression of this gene causes deficiencies in anterior neural structures and head development in Xenopus embryos that are reminiscent of abnormalities in the RA-pathway[72]. Interestingly, it has been shown in another completely different system, lung cancer cell, that inhibition of RA signaling by COUP-TF may be relieved by the orphan receptor nur77 (NR4A1) via a direct interaction between COUP-TF and nur77[43].

The mechanism of this repression is still a matter of debate and it is likely that several different mechanisms may operate (reviewed in reference 20). It is clear that, in some cases, a simple model of competition for the occupancy of DNA response elements may be adequate. Through direct competition with VDR, TR and RAR for the available binding sites, COUP-TFs have been shown to repress the hormonal induction of target genes of VDR, TR and RAR, such as osteocalcin, myosin heavy chain and RARβ promoters, respectively, in transient transfection assays[24,32]. Another mechanism, the competition for RXR, appears relatively unlikely given the weak strength of the COUP-TF–RXR interaction in the two-hybrid system[33]. The fact that COUP-TF interacts directly with the corepressor NCoR and SMRT suggests that, in addition to competition for DNA binding, active repression may play an important role in the COUP-TF inhibiting effect[53]. Transrepression, i.e. formation of a complex between COUP-TF and the inhibited receptor via their respective LBD, without any role played by the DBD, was shown to play an important part in the mechanisms of repression[38]. For example, a GAL4 LBD construct of COUP-TF can inhibit TR, RAR or RXR-mediated transactivation. It is clear from several structure–function studies that the C-terminal end of COUP-TF is required for repression[24,38,89,90]. Some authors suggest that the N-terminal A/B region can also play a role in this process[90]. Since active repression (DNA dependent) and transrepression (DNA independent) were, in addition to DNA-binding competition, the two most likely mechanisms to explain the negative effect of COUP-TF on transcription, the domains responsible for these two mechanisms were mapped in the COUP-TFII protein[78]. The active repressor domain of COUP-TFII was mapped by studying the ability of several GAL4–COUP-TF chimeras to repress transcription and was shown to include all the C-terminal part of the protein from helix 3 of the LBD to helix 12[78]. The region necessary for transrepression was studied in a non-DNA-binding form of COUP-TF that was still able to repress transcription. This region contains both the DBD and the LBD from the end of helix 1 to helix 12. Thus the domains necessary for active repression and transrepression do not coincide[78].

In addition to behaving as repressors, COUP-TF was shown to be able to activate transcription in a precise situation. Indeed, it is interesting to recall that COUP-TF was first characterized as an activator of ovalbumine and rat insulin II genes expression[16,18,19]. Activation of COUP-TF by physiological concentrations of the neurotransmitter dopamine was observed in transient transfection assays, although the mechanism of this provocative finding has not been studied further[75]. In *in vitro* transcription assays, COUP-TFII was shown to behave as an activator on the site A of the apoA1 gene and to interact with TFIIB, suggesting that it possesses intrinsic transcription activation potential[76]. COUP-TFI may act as a repressor of MMTV transcription via a DR1 element but a DR5 element located downstream of the transcription start site is able to sustain a transcriptional activation[77]. Other examples of genes positively regulated by COUP-TF include the transferrin gene in hepatoma cells[81], fatty acid binding protein[82], vHNF1[83], ornithine transcarbamylase[84], NGFI-A[91], arrestin[88] and CYP7A genes[86]. COUP-TF was also shown to be an accessory factor for activation of the PEPCK gene by GR[85] and of the ERα gene by ER[40,79]. Consistent with this positive action on transcription, COUP-TFI was shown to interact with the coactivators SRC-1 and GRIP-1 via its C-terminal moiety. The site of interaction was not further mapped[80]. It has recently been shown that p62, a ligand of the tyrosine kinase p56[lck] interacts with COUP-TFII, possibly serving as a coactivator[96]. In conclusion, the positive or negative effects of COUP-TFs on transcription depend on the organization of the DNA-binding site, the promoter architecture and the cell context, i.e. the possible COUP-TF interactors that are present and available.

In *Drosophila*, SVP have been shown to inhibit the EcR–USP heterodimer[30]. As for COUP-TFs, several non-mutually exclusive mechanisms may explain this inhibition. It has been shown that SVP can bind to some, but not all, EcRE, suggesting that it may compete with the EcR–USP heterodimer for the occupancy of some DNA-binding sites. In addition, a direct physical interaction between EcR and SVP was shown, suggesting that SVP can directly inhibit the EcR–USP heterodimer by protein–protein interactions[30].

Few functional studies were done on EAR2, but it is clear that it behaves like COUP-TF as a transcriptional repressor that inhibits several other signaling pathways. This is the case for thyroid hormone regulation of gene expression as well as for HNF4[39,100].

A very large number of target genes were described for COUP-TF and EAR2 in mammals. In accordance with their numerous interactions with nuclear receptors, COUP-TF also regulates some nuclear receptor genes, such as ERα[40,79] or RARβ[92], suggesting feedback mechanisms in the cross-talk between these factors. Interestingly, in both cases, COUP-TF participates in the activation of gene expression. In contrast, the RXRγ2 promoter that is regulated by retinoids is repressed by COUP-TF via a DR1 element implicated in the RA response[93]. COUP-TF also regulates other classes of transcription factors, such as vHNF1, which is activated by both COUP-TFI and COUP-TFII via an interaction with Oct1 protein[83]. Other Oct factors such as Oct4 are inhibited by COUP-TFs, once again stressing the interplay existing between target genes of COUP-TFs and COUP-TF interactors[89,94]. It has been shown that COUP-TFII affects the expression of the MyoD1 and Myogenin genes by inhibiting their thyroid hormone up-regulation[95]. COUP-TFII is expressed in proliferating myoblasts and its expression is reduced in differentiating cells. All these data thus suggest a role for COUP-TF in myogenic differentiation.

COUP-TFII was cloned as a regulator of ApoA1 expression[6] and it was rapidly clear that COUP-TFs are important regulators of other apolipoprotein expression, such ApoB[29], ApoCIII[29,97], ApoAII[29,98] or ApoCII[99,100], although precise *in vivo* data are still missing to ascertain the physiological meaning of this regulation. In most cases a repression of promoter activity was observed when COUP-TFs or EAR2 were used in transient transfection assays on ApoA1 promoters[6,29]. It has been shown in a number of situations that COUP-TF represses the induction of apolipoprotein genes by other nuclear receptors. This is particularly the case for HNF4 (NR2A1), which has been shown to be a positive regulator of apolipoprotein genes both in intestine and liver, where both HNF4 and COUP-TFII are expressed[29,97]. However, in other cases, COUP-TFII can have a stimulatory effect on ApoCII promoter via an interaction with HNF4[99,100]. Of note, COUP-TFII alone represses the ApoCII promoter[99]. In addition, COUP-TFII was shown to activate transcription *in vitro* via the site A of ApoA1, which was used for its discovery[6,76].

COUP-TFI was shown to bind to a negative regulatory region in the HIV LTR[101]. This striking finding was followed by several studies that have narrowed the contribution of COUP-TF to HIV genes expression. It has been shown that the same region of the HIV LTR, called the nuclear receptor responsive element (NRRE), binds and can be activated by other nuclear receptors, such as RAR–RXR, RXR–PPAR or HNF4[102]. Whereas in T cells COUP-TF apparently represses the activity of the HIV LTR in human oligodendroglioma cells but not in astrocytoma cells, it dramatically activates transcription[103]. COUP-TFI was found expressed in these cells as well as in neuroblastoma cells[105]. Interestingly, it has been shown that dopamine in association with COUP-TFI stimulates HIV-1 gene transcription in both oligodendroglioma and lymphocytes by a mechanism that does not require the NRRE sequence[103,104]. The CRB factor was also shown to play a role in the action of dopamine via COUP-TF in lymphocytes[104]. The precise study of the mechanisms by which COUP-TF stimulates HIV LTR activity in the nervous system has led to the discovery that COUP-TF interacts with Sp1[44], HIV Tat[46] and C/EBP[45] (see above). Of note COUP-TF is also implicated in gene regulation in other viruses, such as the MMTV[77], SV40[106] or HBV[107].

A number of genes implicated in metabolism are regulated by COUP-TFs. It is the case of the PEPCK gene, which is stimulated by glucocorticoids, thereby increasing the rate of gluconeogenesis. Glucocorticoids bind to two GRE in the PEPCK promoter but the full activation of the promoter by this hormone necessitates at least three accessory factor binding sites among which two are recognized by COUP-TF[85,108]. Mutations that reduce COUP-TF binding to these elements also reduce the glucocorticoid response, suggesting that in this case COUP-TF contributes to transcriptional activation[108]. An association between COUP-TFI and the coactivators SRC-1 and GRIP-1 was found by studying the positive action of COUP-TF on glucocorticoid induction of the PEPCK gene[80]. Other genes encoding metabolic enzymes that are regulated by COUP-TF include the hydratase–deshydrogenase implicated in fatty acid β-oxydation on which COUP-TF antagonizes PPAR-mediated induction[27], malic enzyme, which is modulated by 9-*cis* retinoic acid and repressed by COUP-TF[70], lipoprotein lipase in which once again COUP-TF down-regulates PPAR induction[52], ornithine-transcarbamylase, an HNF4 target inhibited by COUP-TF[84] or L-type pyruvate kinase[109]. The fact that COUP-TFI and II mutations are lethal early on has impaired the systematic study of the physiological relevance of these regulations that should nevertheless be carefully verified.

Another COUP-TF target gene interesting to cite is the ovalbumin gene that was used to clone COUP-TF. It was shown recently that the COUP-TF binding site is essential and has a dual role, positive or negative according to the context, in ovalbumin gene regulation[113]. Other interesting target genes are the globin genes in which COUP-TFII may play a role in embryonic/adult switching[111], the c-*mos* proto-oncogene[112], the α-fetoprotein[114] or the lactoferrin gene[115]. Many other target genes are known.

Knockout

Both COUP-TFI and COUP-TFII genes were inactivated in the mouse and each of them gave a lethal phenotype[116,117]. COUP-TFI animals die at birth from starvation and dehydration[116]. In fact, these animals exhibit defects in morphogenesis of the ninth cranial ganglion and nerve, resulting from an excess cell death in the ganglionic precursor cells. In addition, axonal guidance and arborization defects were noticed in several regions. This results in a fewer number of cells in the ninth ganglia and abnormal nerve projections toward the hindbrain. Since the ninth ganglia supplies sensory and motor innervation to the pharynx and root of the tongue as well as to the ear and soft palate, this may explain the inability of the mutant animals to obtain exogenous nutriments and thereby cause death by starvation and dehydration[116]. These data show that COUP-TFI is required for proper fetal development and is essential for post-natal survival. This has also shown that COUP-TFI possesses vital physiological functions that are distinct from COUP-TFII despite their largely overlapping expression pattern and their indistinguishable biochemical function.

The phenotype of the COUP-TFII knockout mice further strengthens the idea of different functions for each of these two genes[117]. The function of COUP-TFII appears absolutely indispensable for normal development, since homozygous mutants die around embryonic day 10, whereas two-thirds of heterozygous animals die during the first weeks of life. Examination of the homozygous embryos show that they are growth retarded with severe hemorrhage and edema just before death. Histological analysis revealed enlarged blood vessels, lack of normal heart development and malformed cardinal veins. The vascular system exhibits a decrease in the complexity of the microvasculature in the head and spine regions, suggesting that vasculogenesis (i.e. *de novo* formation of blood vessels from mesodermal precursor cells) and vascular remodeling (conversion of the primary capillaries of the plexus into large and small vessels of the mature vasculature) are defective in COUP-TFII mutants. These defects are consistent with a need of COUP-TFII function in the mesenchymal compartments of the head, spine and heart. Molecular analysis revealed that the Angiopoietine 1 gene, which is important for the development of both the vascular system and the heart, is down-regulated in mutant animals. All these data suggest that COUP-TF is required for the signaling between the endothelial and mesenchymal compartments.

Ectopic expression of COUP-TFI in *Xenopus* dramatically affects development, since these embryos lack anterior structures at the tadpole stage[74]. The observed abnormalities include loss of the eyes, deletion of the cement gland, malformation of brain structure and truncation of head structure. Molecular analysis of this phenotype have led to the observation that numerous genes required for the correct establishment of anterior structures (Engrailed-2, Krox-20, X-twist, etc.) are down-regulated when COUP-TFI is overexpressed. A link with the retinoic acid effect was proposed by the authors of this study[74].

The phenotype of the SVP *Drosophila* mutants has shown that this gene is required to specify photoreceptor subtypes during development of the compound eye of the fly[22]. The absence of SVP causes the transformation of the R1, R3, R4 and R6 photoreceptors to the R7 cell fate[10]. Consistent with the notion that SVP prevents the differentiation of R1–R6 cells to R7, ectopic expression of SVP causes the transformation of the R7 cell to an R1–R6 cell fate[68]. In certain cases, extra R7 cells are formed by recruiting non-neuronal cone cells as photoreceptor neurons. This suggests that SVP controls decisions not only between photoreceptor subtypes but also between neuronal and non-neuronal fates[68]. Other ectopic expression experiments in which SVP was placed under the control of various different promoters suggest that each cell type within the eye appears to have a different developmental time window that is sensitive to SVP[63]. In addition, these data demonstrate that SVP requires the *ras* pathway for its function in cell fate decision within photoreceptors[63,64].

Associated disease

There is no disease associated with COUP-TF.

Gene structure, promoter and isoforms

The structure of the vertebrate COUP-TF genes are remarkably simple and conserved for COUP-TFI and COUP-TFII, and even outside vertebrates in the unique sea urchin COUP-TF gene[11,118,119]. Human COUP-TFI consists of only three exons and two introns, both zinc fingers being located in the first exon. The two genes are short, i.e. they span approximately 5 kb of genomic DNA[118,119]. No isoforms of either COUP-TFI or COUP-TFII have been described, an almost unique situation for a nuclear receptor.

The genomic organization of SVP is unknown but, in contrast to its vertebrate homologues, the gene generates two isoforms, SVP1 and SVP2, which differ in their C-terminal part[10]. SVP1 encodes a bona fide nuclear receptor with the conserved LBD structure. SVP2 became different in the middle of the E domain by an alternative splicing mechanism. Despite their different structure, functional differences were found between SVP1 and SVP2[30].

Chromosomal location

COUP-TFI was localized on human chromosome 5[1], whereas COUP-TFII was found on chromosome 15[6]. These locations have been refined to 5q14 for COUP-TFI and 15q26 for COUP-TFII[119]. No precise location has been published to date. EAR-2 was localized on chromosome 19[1].

SVP is at position 87B8–9 on the *Drosophila* chromosome 3R (see Flybase for details).

Amino acid sequence for human COUP-TFI (NR2F1)

Accession number: X12795

```
  1  MAMVVSSWRDPQDDVAGGNPGGPNPAAQAARGGGGGAGEQQQQAGSGAPHTPQTPGQPGAPATPGTAGDK
 71  GQGPPGSGQSQQHIECVVCGDKSSGKHYGQFTCEGCKSFFKRSVRRNLTYTCRANRNCPIDQHHRNQCQY
141  CRLKKCLKVGMRREAVQRGRMPPTQPNPGQYALTNGDPLNGHCYLSGYISLLLRAEPYPTSRYGSQCMQP
211  NNIMGIENICELAARLLFSAVEWARNIPFFPDLQITDQVSLLRLTWSELFVLNAAQCSMPLHVAPLLAAA
281  GLHASPMSADRVVAFMDHIRIFQEQVEKLKALHVDSAEYSCLKAIVLFTSDACGLSDAAHIESLQEKSQC
351  ALEEYVRSQYPNQPSRFGKLLLRLPSLRTVSSSVIEQLFFVRLVGKTPIETLIRDMLLSGSSFNWPYMSI
421  QCS   423
```

Amino acid sequence for human COUP-TFII (NR2F2)

Accession number: M64497

```
  1  MAMVVSTWRDPQDEVPGSQGSQASQAPPVPGPPPGAPHTPQTPGQGGPASTPAQTAAGGQGGPGGPGSDK
 71  QQQQQHIECVVCGDKSSGKHYGQFTCEGCKSFFKRSVRRNLSYTCRANRNCPIDQHHRNQCQYCRLKKCL
141  KVGMRREAVQRGRMPPTQPTHGQFALTNGDPLNCHSYLSGYISLLLRAEPYPTSRFGSQCMQPNNIMGIE
211  NICELAARMLFSAVEWARNIPFFPDLQITDQVALLRLTWSELFVLNAAQCSMPLHVAPLLAAAGLHASPM
281  SADRVVAFMDHIRIFQEQVEKLKALHVDSAEYSCLKAIVLFTSDACGLSDVAHVESLQEKSQCALEEYVR
351  SQYPNQPTRFGKLLLRLPSLRTVSSSVIEQLFFVRLVGKTPIETLIRDMLLSGSSFNWPYMAIQ   414
```

Amino acid sequence for *Drosophila* SVP (NR2F3)

Accession number: M28863

```
  1  MCASPSTAPGFFNPRPQSGAELSAFDIGLSRSMGLGVPPHSAWHEPPASLGGHLHAASAGPGTTTGSVAT
 71  GGGGTTPSSVASQQSAVIKQDLSCPSLNQAGSGHHPGIKEDLSSSLPSANGGSAGGHHSGSGSGSGSGVN
141  PGHGSDMLPLIKGHGQDMLTSIKGQPTGCGSTTPSSQANSSHSQSSNSGSQIDSKQNIECVVCGDKSSGK
211  HYGQFTCEGCKSFFKRSVRRNLTYSCRGSRNCPIDQHHRNQCQYCRLKKCLKMGMRREAVQRGRVPPTQP
281  GLAGMHGQYQIANGDPMGIAGFNGHSYLSSYISLLLRAEPYPTSRYGQCMQPNNIMGIDNICELAARLLF
351  SAVEWAKNIPFFPELQVTDQVALLRLVWSELFVLNASQCSMPLHVAPLLAAAGLHASPMAADRVVAFMDH
421  IRIFQEQVEKLKALHVDSAEYSCLKAIVLFTTDACGLSDVTHIESLQEKSQCALEEYCRTQYPNQPTRFG
491  KLLLRLPSLRTVSSQVIEQLFFVRLVGKTPIETLIRDMLLSGNSFSWPYLPSM   543
```

Amino acid sequence for *Xenopus* COUP-TFIII (NR2F4)

Accession number: X63092

```
  1 MAMVVNPWQEDIPGVPGSQMNNPPGLCNQDPGGTPQTPTTPKGGIPGQDPVHSGDKGVPNVDCLVCGDKS
 71 SGKHYGQFTCEGCKSFFKRSVRRNLTYTCRSNRDCPIDQHHRNQCQYCRLKKCLKVGMRREVQRGRMSHP
141 QTSPGQYTLNNVDPYNGHSYLTGFISLLLRAEPYPTSRYGAQCLQPNNIMGIENICELAARLLFSAIEWA
211 KNIPFFPDFQLSDQVSLLRMTWSELFVLNAAQCSMPLHVAPLLARAGLHASPMSADRVVAFMDHIRIFQE
281 QVEKLKALHVDSAEYSCLKAIALFTPDAVGLSDIGHVESIQEKSQCALEEYVRNQYPNQPTRFGRLLLRL
351 PSLRIVSAPVIEQLFFVRLVGKTPIETLIRDMLLSGSSFNWPYMPMQ  397
```

Amino acid sequence for zebrafish SVP46 (NR2F5)

Accession number: X70300

```
  1 MAMVVNQWQENISADPGSQLQMCSQEPGGTPGTPSGSTPGNDALSGDKIPNVDCMVCGDKSSGKHYGQFT
 71 CEGCKSFFKRSVRRNLSYTCRGNRDCPIDQHHRNQCQYCRLKKCLKVGMRREAVQRGRMSNSQSSPGQYL
141 SNGSDPYNGQPYLSGFISLLLRAEPYPTSRYGAQCMQSNNLMGIENICELAARLLFSAVEWAKNIPFFPD
211 LQLMDQVALLRMSWSELFVLNAAQCSMPLHVAPLLAAAGLHASPMSAERVVAFMDHIRVFQEQVEKLKAL
281 QVDTAEYSCLKSIVLFTSDAMGLSDVAHVESIQEKSQCALEEYVRNQYPNQPNRFGRLLLRLPSLRIVSS
351 PVIEQLFFVRLVGKTPIETLLRDMLLSGSSYNWPYMPVQRDRPISIHYNENGP  403
```

Amino acid sequence for human EAR2 (NR2F6)

Accession number: X12794

```
  1 MAMVTGGWGGPGGDTNGVDKAGGYPRAAEDDSASPPGAASDAEPGDEERPGLQVDCVVCGDKSSGKHYGV
 71 FTCEGCKSFFKRTIRRNLSYTCRSNRDCQIDQHHRNQCQYCRLKKCFRVGMRKEAVQRGRIPHSLPGAVA
141 ASSGSPPGSALAAVASGGDLFPGQPVSELIAQLLRAEPYPAAAGRFGAGGGAAGAVLGIDNVCELAARLL
211 FSTVEWARHGFFPELPVADQVALLRMSWSELFVLNAAQAALPLHTAPLLAAAGLHAAPMAAERAVAFMDQ
281 VRAFQEQVDKLGRLQVDSAEYGCLKAIALFTPDACGLSDPAHVESLQEKAQVALTEYVRAQYPSQPQRFG
351 RLLLRLPALRAVPASLISQLFFMRLVGKTPIETLIRDMLLSGSTFNWPYGSGQ  403
```

References
1 Miyajima, N. et al. (1988) Nucl. Acids Res. 16, 11057–11074.
2 Qiu, Y. et al. (1994) Proc. Natl Acad. Sci. USA 91, 4451–4455.
3 Connor, H. et al. (1995) J. Biol. Chem. 270, 15066–15070.
4 Langlois, M.C. et al. (2000) Dev. Genes Evol. (in press).
5 Fjose, A. et al. (1993) EMBO J. 12, 1403–1414.
6 Ladias, J.A. and Karathanasis, S.K. (1991) Science 251, 561–565.
7 Lutz, B. et al. (1994) Development 120, 25–36.
8 Matharu, P.J. and Sweeney, G.E. (1992) Biochim. Biophys. Acta 1129, 331–334.
9 Fjose, A. et al. (1995) Mech. Dev. 52, 233–246.
10 Mlodzik, M. et al. (1990) Cell 60, 211–224.
11 Chan, S.M. et al. (1992) Proc. Natl Acad. Sci. USA 89, 10568–10572.
12 Qiu, Y.H. et al. (1994) Trends Endocrinol. Metab. 5, 234–239.
13 Van der Wees, J. et al. (1996) Mech. Dev. 54, 173–184.
14 Wang, L.H. et al. (1989) Nature 340, 163–166.
15 Broadus, J. and Doe, C.Q. (1995) Development 121, 3989–3996.
16 Sagami, I. et al. (1986) Mol. Cell. Biol. 6, 4259–4267.
17 Tsai, S.Y. et al. (1987) Cell 50, 701–709.

[18] Bagchi, M.K. et al. (1987) Mol. Cell. Biol. 7, 4151–4158.
[19] Hwung, Y.P. et al. (1988) Mol. Cell. Biol. 8, 2070–2077.
[20] Tsai, S.Y. and Tsai, M.J. (1997) Endocrine Rev. 18, 229–240.
[21] Escriva, H. et al. (1997) Proc. Natl Acad. Sci. USA 94, 6803–6808.
[22] Rubin, G.M. (1991) Trends Genet. 7, 372–377.
[23] Laudet, V. (1997) J. Mol. Endocrinol. 19, 207–226.
[24] Cooney, A.J. et al. (1993) J. Biol. Chem. 268, 4152–4160.
[25] Kadowaki, Y. et al. (1992) Biochem. Biophys. Res. Commun. 183, 492–498.
[26] Cooney, A.J. et al. (1992) Mol. Cell. Biol. 12, 4153–4163.
[27] Miyata, K.S. et al. (1993) J. Biol. Chem. 268, 19169–19172.
[28] Nakshatri, H. et al. (1994) J. Biol. Chem. 269, 890–902.
[29] Ladias, J.A. et al. (1992) J. Biol. Chem. 267, 15849–15860.
[30] Zelhof, A.C. et al. (1995) Mol. Cell. Biol. 15, 6736–6745.
[31] Avram, D. et al. (1999) J. Biol. Chem. 274, 14331–14336.
[32] Kliewer, S.A. et al. (1992) Proc. Natl Acad. Sci. USA 89, 1448–1452.
[33] Butler, A.J. and Parker, M.G. (1995) Nucl. Acids Res. 23, 4143–4150.
[34] Berrodin, T.J. et al. (1992) Mol. Endocrinol. 6, 1468–1478.
[35] Tran, P. et al. (1992) Mol. Cell. Biol. 12, 4666–4676.
[36] Widom, R.L. et al. (1992) Mol. Cell. Biol. 12, 3380–3389.
[37] Casanova, J. et al. (1994) Mol. Cell. Biol. 14, 5756–5765.
[38] Leng, X. et al. (1996) Mol. Cell. Biol. 16, 2332–2340.
[39] Zhu, X.G. et al. (2000) Mol. Cell. Biol. 20, 2604–2618.
[40] Lazennec, G. et al. (1997) Mol. Cell. Biol. 17, 5053–5066.
[41] Klinge, C.M. et al. (1997) J. Biol. Chem. 272, 31465–31474.
[42] Klinge, C.M. (1999) J. Steroid Biochem. Mol. Biol. 71, 1–19.
[43] Wu, Q. et al. (1997) EMBO J. 16, 1656–1669.
[44] Rohr, O. et al. (1997) J. Biol. Chem. 272, 31149–31155.
[45] Schwartz, C. et al. (2000) J. Virol. 74, 65–73.
[46] Rohr, O. et al. (2000) J. Biol. Chem. 275, 2654–2660.
[47] Klinge, C.M. (2000) Arch. Biochem. Biophys. 373, 163–174.
[48] Avram, D. et al. (2000) J. Biol. Chem. 275, 10315–10322.
[49] Ahn, M.Y. et al. (1998) Proc. Natl Acad. Sci. USA 95, 1812–1817.
[50] Smirnov, D.A. et al. (2000) Virology 268, 319–328.
[51] Dressel, U. et al. (1999) Mol. Cell. Biol. 19, 3383–3394.
[52] Robinson, C.E. et al. (1999) Endocrinology 140, 1586–1593.
[53] Shibata, H. et al. (1997) Mol. Endocrinol. 11, 714–724.
[54] Ing, N.H. et al. (1992) J. Biol. Chem. 267, 17617–17623.
[55] Jonk, L.J. et al. (1994) Mech. Dev. 47, 81–97.
[56] Chu, K. et al. (1998) Mol. Cell. Endocrinol. 137, 145–154.
[57] Pereira, F. et al. (1995) J. Steroid Biochem. Mol. Biol. 53, 503–508.
[58] Lopes da Silva, S. et al. (1995) Endocrinology, 136, 2276–2283.
[59] Lutz, B. et al. (1994) Development 120, 25–36.
[60] Brubaker, K. et al. (1996) Dev. Brain Res. 93, 198–202.
[61] Vlahou, A. et al. (1996) Development 122, 521–526.
[62] Vlahou, A. and Flytzanis, C.N. (2000) Dev. Biol. 218, 284–298.
[63] Kramer, S. et al. (1995) Development 121, 1361–1372.
[64] Begemann, G. et al. (1995) Development 121, 225–235.
[65] Hoshizaki, D.K. et al. (1995) Genome 38, 497–506.

66 Hoshizaki, D.K. et al. (1994) Development 120, 2489–2499.
67 Kerber, B. et al. (1998) Genes Dev. 12, 1781–1786.
68 Hiromi, Y. et al. (1993) Development 118, 1123–1135.
69 Wang, L.H. et al. (1991) Gene Expr. 1, 207–216.
70 Baes, M. et al. (1995) Biochem. Biophys. Res. Commun. 215, 338–345.
71 Galson, D.L. et al. (1995) Mol. Cell. Biol. 15, 2135–2144.
72 Liu, Y. et al. (1993) Mol. Cell. Biol. 13, 1836–1846.
73 Wehrenberg, U. et al. (1994) Proc. Natl Acad. Sci. USA 91, 1440–1444.
74 Schuh, T.J. and Kimelman, D. (1995) Mech. Dev. 51, 39–49.
75 Power, R.F. et al. (1991) Science 1546–1548.
76 Malik, S. and Karathanasis, S. (1995) Nucl. Acids Res. 23, 1536–1543.
77 Kadowaki, Y. et al. (1995) Proc. Natl Acad. Sci. USA 92, 4432–4436.
78 Achatz, G. et al. (1997) Mol. Cell. Biol. 17, 4914–4932.
79 Petit, F.G. et al. (1999) Eur. J. Biochem. 259, 385–395.
80 Sugiyama, T. et al. (2000) J. Biol. Chem. 275, 3446–3454.
81 Schaeffer, E. et al. (1993) J. Biol. Chem. 268, 23399–23408.
82 Rottman, J.N. and Gordon, J.I. (1993) J. Biol. Chem. 268, 11994–12002.
83 Power, S.C. and Cereghini, S. (1996) Mol. Cell. Biol. 16, 778–791.
84 Kimura, A. et al. (1993) J. Biol. Chem. 268, 11125–11133.
85 Hall, R.K. et al. (1995) Proc. Natl Acad. Sci. USA 92, 412–416.
86 Stroup, D. et al. (1997) J. Biol. Chem. 272, 9833–9839.
87 Ktistaki, E. and Talianidis, I. (1997) Mol. Cell. Biol. 17, 2790–2797.
88 Lu, X. et al. (1994) Mol. Endocrinol. 8, 1774–1788.
89 Ben-Shushan, E. et al. (1995) Mol. Cell. Biol. 15, 1034–1048.
90 Ge, R. et al. (1994) J. Biol. Chem. 269, 13185–13192.
91 Pipaon, C. et al. (1999) Mol. Cell. Biol. 19, 2734–2745.
92 Lin, B. et al. (2000) Mol. Cell. Biol. 20, 957–970.
93 Barger, P.M. and Kelly, D.P. (1997) J. Biol. Chem. 272, 2722–2728.
94 Schoorlemmer, J. et al. (1994) Mol. Cell. Biol. 14, 1122–1136.
95 Muscat, G.E.O. et al. (1995) Nucl. Acids Res. 23, 1311–1318.
96 Marcus, S.L. et al. (1996) J. Biol. Chem. 271, 27197–27200.
97 Mietus-Snyder, M. et al. (1992) Mol. Cell. Biol. 12, 1708–1718.
98 Cardot, P. et al. (1993) Biochemistry 32, 9080–9093.
99 Kardassis, D. et al. (1998) J. Biol. Chem. 273, 17810–17816.
100 Vorgia, P. et al. (1998) J. Biol. Chem. 273, 4188–4196.
101 Cooney, A.J. et al. (1991) J. Virol. 65, 2853–2860.
102 Ladias, J.A. (1994) J. Biol. Chem. 269, 5944–5951.
103 Sawaya, B.E. et al. (1996) J. Biol. Chem. 271, 23572–23576.
104 Rohr, O. et al. (1999) J. Cell Biochem. 75, 404–413.
105 Canonne-Hergaux, F. et al. (1995) J. Virol. 69, 6634–6642.
106 Zuo, F. and Mertz, J.E. (1995) Proc. Natl Acad. Sci. USA 92, 8586–8590.
107 Yu, X. and Mertz, J.E. (1997) J. Virol. 71, 9366–9374.
108 Scott, D.K. et al. (1996) J. Biol. Chem. 271, 31909–31914.
109 Lou, D.Q. et al. (1999) J. Biol. Chem. 274, 28385–28394.
110 Jiang, J.G. et al. (1997) J. Biol. Chem. 272, 3928–3934.
111 Filipe, A. et al. (1999) EMBO J. 18, 687–697.
112 Lin, H.B. et al. (1999) J. Biol. Chem. 274, 36796–36800.
113 Park, H.M. et al. (2000) Biochemistry 39, 8537–8545.
114 Thomassin, H. et al. (1996) DNA Cell Biol. 15, 1063–1074.

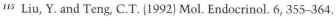

[115] Liu, Y. and Teng, C.T. (1992) Mol. Endocrinol. 6, 355–364.
[116] Qiu, Y. et al. (1997) Genes Dev. 11, 1925–1937.
[117] Pereira, F.A. et al. (1999) Genes Dev. 13, 1037–1049.
[118] Ritchie, H.H. et al. (1990) Nucl. Acids Res. 18, 6857–6862.
[119] Qiu, Y. et al. (1995) Genomics 29, 240–246.

Names

Two different ER genes called ERα and ERβ were cloned in 1986 and 1996, respectively. Since ERα was for 10 years the only estrogen receptor known, it was simply called ER. After the discovery of ERβ, it became ERα. Both receptors bind estradiol, a steroid hormone implicated in reproductive function and menstrual cycle in women. Estrogens are implicated in numerous diseases in humans, such as cancer, in particular breast cancer, and osteoporosis. For these reasons, ER functions have been studied intensively, an interest that increased further when the new estrogen receptor, ERβ was identified. Both ERs form a clear group, called NR3A, in the nuclear receptor nomenclature and are related to the orphan receptors ERR (NR3B) and to the steroid receptors GR, PR, MR and AR (NR3C).

Species	Latin name	Other names	Accession number	References
Official name: NR3A1				
Human	*Homo sapiens*	ER, ERα	X03635	1, 2
Mouse	*Mus musculus*	ER, ERα	M38651	3
Rat	*Rattus* sp.	ER, ERα	Y00102	4
Hamster	*Mesocricetus auratus*	ER, ERα	AF181077	111
Pig	*Sus scrofa*	ER, ERα	Z37167	5
Cow	*Bos taurus*	ER, ERα	Z37167	6
Sheep	*Ovis aries*	ER, ERα	Z49257	32
Horse	*Equus caballus*	ER, ERα	AF124093	Unpub.
Whiptail lizard	*Cnemidophorus uniparens*	ER, ERα	S79923	25
Green anole	*Anolis carolinensis*	ER, ERα	AF095911	Unpub.
Chicken	*Gallus gallus*	ER, ERα	X03805	7
Zebra finch	*Taeniopygia guttata*	ER, ERα	L79911	9
Xenopus	*Xenopus laevis*	ER, ERα	L20735	24
Zebrafish	*Danio rerio*	ER, ERα	AB037185	Unpub.
Atlantic salmon	*Salmo salar*	ER, ERα	X89959	10
Rainbow trout	*Oncorhynchus mykiss*	ER, ERα	AJ242740	11
Japanese medaka	*Oryzias* sp.	ER, ERα	D28954	Unpub.
Red seabream	*Chrysophrys major*	ER, ERα	AB007453	12
Gilthead seabream	*Sparus aurata*	ER, ERα	AJ006039	13
Blue tilapia	*Oreochromis aureus*	ER, ERα	X93557	26
Nile tilapia	*Oreochromis nilotica*	ER, ERα	U75604	14
Atlantic croaker	*Micropogonias undulatus*	ER, ERα	AF298183	30
Channel catfish	*Ictalurus punctatus*	ER, ERα	AF253505	31
Official name: NR3A2				
Human	*Homo sapiens*	ERβ	AB006590	15
Rhesus monkey	*Macaca mulatta*	ERβ	AF119229	113
Marmoset	*Callithrix jacchus*	ERβ	Y09372	Unpub.
Mouse	*Mus musculus*	ERβ	U81451	16
Rat	*Rattus* sp.	ERβ	U57439	17
Cow	*Bos taurus*	ERβ	Y18017	18
Sheep	*Ovis aries*	ERβ	AF177936	Unpub.
Chicken	*Gallus gallus*	ERβ	AB036415	Unpub.
Quail	*Coturnix coturnix*	ERβ	AF045149	19
Common starling	*Sturnus vulgaris*	ERβ	AF113513	8
Zebrafish	*Danio rerio*	ERβ	AJ275911	Unpub.
Rainbow trout	*Oncorhynchus mykiss*	ERβ	AJ289883	Unpub.
Goldfish	*Carassius auratus*	ERβA	AF061269	33
Goldfish	*Carassius auratus*	ERβB, ERβ2	AF177465	21

Atlantic croaker	*Micropogonias undulatus*	ERβA	AF298181	30
Atlantic croaker	*Micropogonias undulatus*	ERβB, ERγ	AF298182	30
Gilthead seabream	*Sparus aurata*	ERβ	AF136980	Unpub.
Nile tilapia	*Oreochromis nilotica*	ERβ	U75605	14
Channel catfish	*Ictalurus punctatus*	ERβ	AF185568	22
Japanese eel	*Anguilla japonica*	ERβ	AB003356	20

Isolation

The human estrogen receptor identified by Pierre Chambon and Geoffrey Greene laboratories was one of the first nuclear receptors to be cloned together with the glucocorticoid receptor[1,2]. It was cloned from a human breast cancer cell line MCF-7 cDNA library that was enriched in estrogen receptor mRNA using sucrose gradient centrifugation and *in vitro* translation[23]. The expression library was screened with a monoclonal antibody specifically directed to recognize ER or using synthetic oligonucleotides designed from peptide sequences of purified ER. The resulting cDNA was expressed in HeLa cells and shown to bind estradiol with high affinity. A high level of sequence identity between ER, GR and the v-erbA oncogene was noticed[1,2].

Soon after the cloning of the human ER cDNA, a chicken cDNA was isolated and characterized[7]. Comparison between the human and chicken ER reveals striking differences in the level of sequence identity and led to the delineation of six domains, domains A–F, which represent various levels of identity. Some of these domains, such as domain C and to a lesser extent domain E, were shown to be conserved with GR and v-erbA. This nomenclature for the domains is still used for the whole nuclear receptor superfamily. Numerous orthologues were found in other vertebrates from mouse to several fishes.

For 10 years the ER field was working with the idea that only one ER existed. The existence of a second ER was never seriously considered even after the observation that ERα knockout mice still contain high affinity estradiol binding sites[28,29]. Then surprisingly, the Jan-Ake Gustafsson's laboratory reported the existence of a second ER, called ERβ that was identified by a search for orphan receptors expressed in rat prostate[17]. The same sequence was also reported by another group in humans[27]. Ironically, a homologue was already known in fish, since the Japanese eel ER, which was described as unexpectedly distant to ERα from human or other fish, is in fact an ERβ orthologue[20]. Since then ERβ has been found in a number of species and was found duplicated in several fish species. This fish-specific duplication explains why some fish have three ERs. Some groups improperly described these fish-specific duplicates as ERγ. All available data, including the complete sequence of the human genome, nevertheless suggest that there is no ERγ in vertebrates.

ERα exhibits 94% amino-acid identity with ERβ in the DBD and 69% in the LBD. ERs exhibit 68–72% sequence identity with the ERRs in the C domain and 35–37% in the E domain. These levels of identity are

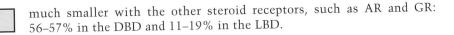
much smaller with the other steroid receptors, such as AR and GR: 56–57% in the DBD and 11–19% in the LBD.

DNA binding

It was known even before the cloning era that, as for other steroid receptors, ERs were able to bind to specific DNA sequences in the promoters of estrogen-regulated genes, such as the vitellogenin genes (see reference 54 for a review). The interaction of ER with an ERE found in the first intron of the *Xenopus* vitellogenin gene was visualized by electron microscopy by Walter Wahli's laboratory and these authors observed that estrogens increased the binding of ER to DNA, whereas tamoxifen, an ER antagonist, reduced it[35].

Analysis of the ER sequence reveals that a Cys-rich domain, called domain C, exhibits a structure reminiscent of zinc fingers[7]. To test if this region was indeed implicated in specific DNA binding, an experiment using 'finger swap' was done during which the C domain of ER was replaced by the C domain of the glucocorticoid receptor[34]. The resulting chimera was still estradiol responsive but was able to bind to a GRE and not an ERE. This was confirmed by extensive mutagenesis of ER[38]. The basis of the specific recognition of the ERE (versus the GRE) is linked to the first zinc finger[39], and within this finger to only three amino acids at the base of the first zinc finger, the region now known as the P-box[40,55,56]. The minimal DNA-binding domain of ER is composed of the 66-amino-acid C domain composed of the two zinc fingers plus 30 amino acids of the D domain, since the core of the C domain alone does not bind to DNA[44].

An ERE was defined as a palindrome of the sequence AGGTCA with a 3 bp spacing, whereas a GRE is a palindrome of the sequence AGAACA with no space between each core sequence. The modification of the core sequence of an ERE towards the sequence of the GRE is sufficient to confer corticoid inducibility to a target gene (see reference 36 as an example). The EREs are often imperfect when compared to the consensus sequence and they are often found in several copies in the promoters of target genes. A synergistic effect of the binding of ER to these elements has been observed using several types of EREs[41,42]. Cooperative binding of ER was observed on imperfect elements in contrast to perfect EREs[41,42]. ER was also shown to bind to other response elements that contain the AGGTCA core motif. For example, ERα (but not ERβ, see below) recognize, as a dimer the sequence TCA AGGTCA, called the SFRE, which contains only one core element[53].

ER binds to the ERE as a homodimer, which is enhanced by the binding of the ligand to the receptor[37]. The effect of hormone on the DNA-binding ability of the receptor has been an extremely controversial point (see below)[37,45]. The dimerization is apparently required for DNA binding, whereas this has also been controversial[44,45,49,59]. One possible explanation for the contradictory results obtained by different groups is the structure of the various EREs used. In fact it has been demonstrated that differential interaction of ER with different types of EREs (vitellogenin or pS2 EREs) brings about global changes in ER conformation[46]. This may have important repercussions for the regulation of target genes as well as the cross-talk with other transcription factors. The major dimerization domain is located in the E domain in which the dimerization interface is associated with heptad repeats of hydrophobic amino acids[43]. A minor dimerization interface also exists in the DNA-binding domain that is thus able to dimerize and interact with DNA when expressed alone[44].

To interact with DNA, a receptor should first be nuclear: several nuclear localization signals (NLS) were found in the estrogen receptor[47]. A hormone-inducible NLS is located in the LBD, whereas three basic NLS are present in the D domain. These domains cooperate in the presence of estrogen and tamoxifen but not in presence of the pure antagonist ICI 164,384[47]. Using fluorescent non-steroidal estrogens that allow visualization of ER in cultured cells, it has been shown that the N-terminal A/B domain influences the intranuclear localization of ER[48]. Recently, the use of GFP-fusion proteins has allowed to better understand the reorganization of the ER nuclear localization induced by ligand binding[118].

The DNA binding of ER is enhanced by its interaction with the HMG1 protein[50]. HMG1 is able to form a ternary complex with ER bound to the ERE. This effect is due to a decreased dissociation rate of the ER–ERE complex and is observed using the isolated ER DBD. ER binding to DNA causes the DNA to bend with an angle of approximately 34°. An increase in the degree of bending was observed when two EREs are present[51]. Interestingly, the interaction of ER with its recognition sequence can be strongly influenced by the topology of the DNA, for example by pre-bending the DNA[52].

As for ERα, ERβ binds to the ERE sequence[16,17]. ERβ is also able to dimerize and, interestingly, to form heterodimers with ERα[58,59]. It appears that these heterodimers are preferentially formed over each type of homodimer and have distinct transcriptional activation potential[59,254]. Nevertheless, in contrast to ERα, ERβ is unable to bind to the SFRE sequence (TCA AGGTCA), which is the binding site of the estrogen receptor-related ERR orphan receptors[53]. The structural basis for this discrepancy is not known.

Partners

As TR, RAR or GR, ER has been used as a paradigm in the nuclear hormone receptor superfamily and thus many proteins have been identified by virtue of their interaction with it. It has long been known that, like GR or PR, unliganded ER interacts with heat-shock proteins and in particular hsp90 (see reference 60 for a review). This hsp90-based chaperone complex is thought to repress ER's transcriptional activity while maintaining the receptor in a conformation competent for high-affinity steroid binding. The binding of ligand dissociates this complex and allows the receptor to activate transcription. The LBD of ER is necessary for the binding of hsp90 complexes. It has been shown that this hsp90 complex, which contains several proteins, such as immunophilins as well as the p23 protein, plays an important role in ER signal transduction. The interaction between hsp90 complexes and ER, which is reversible by estrogens, maintains ER in an inactive state. Unliganded ER can inhibit other DNA-binding domains and even unrelated proteins. Many fusion proteins, such as Fos-ER and myc-ER have thus been constructed with the ER LBD (see reference 63 and 64 among many others). This has allowed the development of estrogen-inducible factors that are extremely useful to decipher the *in vivo* function of a given protein. The myc-ER LBD chimera, for example, has facilitated the production of estrogen-dependent transformation of cells, allowing a better understanding of the role of *myc* in cell transformation[63]. The fusion of a mutated ER LBD with the Cre recombinase is used for generating conditional knockout in the mouse[143,144].

ERα has been shown to form homodimers as well as heterodimers with ERβ[37,58,59] (see above). In addition, ERs interact functionally and physically with numerous transcription factors. This is the case, for example, for Sp1, which interacts physically and functionally with ERα and ERβ, resulting in an increased Sp1 DNA-binding activity and functional synergy[65,121,122]. Functional interactions, either stimulatory or inhibitory between ERα with CTF/NF1, BRCA1, GATA or Islet1 have also been demonstrated[66,72,119,281]. An interaction between ERα and NFκB has been discovered by the study of the down-regulation of the IL6 gene by estrogens[208,209]. Positive or negative functional interaction between ERα and AP-1 has been described but the precise mechanism underlying these effects are poorly understood[67,68]. The vitellogenin gene promoter contains a consensus ERE that is activated by liganded ER. Cotransfection of c-*fos* and c-*jun* in MCF-7 cells inhibits this positive action of ER on the vitellogenin promoter by an unknown mechanism[68]. This may be due to a squelching for common coactivators, such as CBP/P300. In contrast, stimulating effects of ER on AP-1 has been demonstrated in several promoters. For example, one of the half-site of the ERE contained in the ovalbumin gene promoter (TGGGTCA) exhibits a high level of sequence identity with the AP-1 site (TGAGTCA) and indeed the ovalbumin gene can be induced by AP-1 through this element[67]. AP-1 and liganded ER coactivates the ovalbumin gene promoter, but direct interaction of ER with the target sequence is not required, since an ER deleted for its DNA-binding domain still coactivates c-*fos* and c-*jun* action on this promoter. This synergy requires the two activating functions of ER, AF-1 and AF-2, and the synergy takes place in the presence of either estrogen or antiestrogens[69,71]. Tamoxifen, for example, which is an antagonist of ER behaves as an estrogen in activating the AP-1 transcriptional activity[71]. This suggests that this enhancement arose by protein–protein interactions with unknown partners. In contrast to ERα, liganded ERβ does not stimulate AP-1 activity[70]. In contrast, several antiestrogens, such as tamoxifen, ICI 164,384 or raloxifene, are potent transcriptional activators with ERβ at an AP-1 site. These data were the first showing that ERα and ERβ may play different roles in gene regulation.

ERs were also shown to interact with a number of other proteins with miscellaneous function. An interaction between the SH2 domain of the tyrosine kinase Src and the phosphotyrosine 537 in the LBD of ERα has been demonstrated[74]. This ligand-dependent interaction, which has also been found in ERβ results in the activation of the tyrosine kinase activity of Src, which is important to trigger S-phase entry of prostate carcinoma cells LNCaP. Antiestrogens inhibit this interaction as well as re-entry in to the cell cycle (see also reference 155). Another link between ERs and the cell cycle is the fact that ERβ, but not ERα, interacts with the cell-cycle spindle assembly checkpoint protein, MAD2. The hinge region of ERβ is sufficient to mediate this interaction. The functional significance of this interaction remains unclear. The link between cell cycle and ERs is further supported by the observation that ERα physically interacts with cyclin D1, which is also able to bind to the coactivator SRC-1[114,115]. The resulting complex is believed to activate ERα in a ligand-independent fashion, resulting in cell-cycle entry. In addition, the coactivator P/CAF (p300/CBP-associated factor) associates with cyclin D1 and potentiates its activation by ER[270]. The relationship between ER and cyclins is not limited to cyclin D1 and physical interaction, since estrogens regulate the transcription of several cyclin genes and since cyclins play an important role in the regulation of ER activity by phosphorylation (see below and reference 105 for a review). ERα was also shown to interact with HET/SAF-B, a nuclear matrix protein that behaves as an inhibitor of ER-mediated transactivation, suggesting that ER specifically interacts with the nuclear matrix[117].

Detailed analysis of the transcriptional activation functions of ERα have led to the conclusion that each of them was interacting with various types of transcriptional mediator that are distinct from the ones of acidic activation domains such as that for VP16[79,80]. In line with these findings, ER has been found to interact with various types of molecules that are involved in mediating its transcriptional activity. This is the case of TBP as well as of the various TAFs with whom ER was found to interact[76]. In accordance with this concept, ER interacts with human $TAF_{II}30$ in a ligand-independent manner[77]. The AF-2 but not the AF-1 of ERα is required for this effect. The functional importance of this interaction is demonstrated by the observation that an antibody against $hTF_{II}30$ inhibited transcriptional stimulation by the ER AF-2. $TAF_{II}28$ stimulates the AF-2 function of ER but seems not to interact directly with it[78].

Beside the TAFs, several groups have searched for transcriptional coactivators that can mediate ER transcriptional activation. Indeed, GST pull-down experiments reveal that ER interacts with a set of proteins among which are species at 140 and 160 kDa[81,82]. This has led to the discovery of the first transcriptional coactivators of the nuclear receptor superfamily, among which are the p160 family and RIP140 (see references 83 and 84 for reviews). The recruitment of coactivators that directly associate with liganded receptors has been the basis for the search of synthetic peptides that can mimic these interactions and work as ER-specific antagonists[85]. This search has effectively given rise to peptides that are antagonists for ERα and ERβ[86].

The three members of the p160 family, SRC-1, TIF2 and p/CIP, also called AIB1, interact with ERα and ERβ in an AF-2 and ligand-dependent manner (see references 83 and 84 for references and reviews). The recruitment of p160 coactivators has been shown to be sufficient for hormone activation of target genes[120]. It has been shown using ERα that, in the coactivator, nuclear receptors recognize an LXXLL motif[87], whereas in ER the coactivator interacts with helix 12 and a region of helix 3 that forms an hydrophobic groove that accommodates an LXXLL motif[93,94]. Various isoforms that differ in their ability to potentiate transcription by ER exist, for example, for SRC-1[89]. SRC-1 can also interact with the AF-1 function of ERβ when it is phosphorylated by MAP kinase. Treatment of cells by EGF or by activated *ras*, which both stimulates MAP kinase enhances the interaction between ERβ and SRC-1[92]. The *in vivo* significance of the interaction between ER and coactivators of the p160 family has been confirmed by the demonstration that endogenously expressed ER and AIB1 interact in MCF-7 breast cancer cells[91]. This is consistent with the phenotype of mice bearing an inactivation of the AIB1 gene[90]. These animals have a pleiotropic phenotype showing delayed puberty, reduced female reproductive function and blunted mammary gland development. In addition, these mice exhibit an attenuated estrogen response. The SRC-1$^{-/-}$ mice also exhibit partial steroid hormone resistance[88].

Among the other coactivators bound by ERs, RIP140, which was one of the first to be identified, is interesting to mention[95]. If the ligand-dependent interaction of RIP140 with AF-2 of ERα is well established, the functional role of this interaction is far from being understood. Several experiments even proposed that RIP140 could in certain situations play the role of a corepressor (see reference 83 for a review). Another interesting case is the one of the cointegrators CBP and P300, which interact with ERα in a ligand-dependent manner. A strong synergism between ERα and P300 was found in chromatin templates *in vitro* and this system has allowed to understand better the mechanisms by which ER and P300 activates transcription[98]. As for GR, this interaction may explain the cases in which ERα antagonizes AP-1 action but some additional explanations of this phenomenon have also been proposed (see references 83 and 96 for a review). The link between ER, transcriptional activation and chromatin has been further reinforced by the identification of TIF1 as a ER coactivator, since it is clear that TIF1 is tightly associated with euchromatin[99,101,146,295]. Of note, the binding of ERα to DNA does not significantly affect its hormone-dependent association with TIF1[100].

In their unliganded forms, ERs are unable to interact with the corepressors NCoR or SMRT[287]. However, when they are bound by antagonists such as tamoxifen or ICI 164,384, they associate with corepressors[288,289]. Furthermore, SMRT overexpression blocked SRC-1 coactivation of tamoxifen-stimulated gene expression and preferentially inhibited tamoxifen agonist activity whether or not SRC-1 was present[289]. It was thus concluded that the relative expression of coactivators and corepressors can modulate tamoxifen regulation of ERα transcriptional activity and may contribute to the tissue-specific ability of mixed antiestrogens to activate or inhibit ER-mediated gene expression. Interestingly, these *in vitro* data were confirmed *in vivo* by the knockout of the NCoR gene[290]. In fibroblasts derived from the knockout mice, tamoxifen exhibit a full agonist activity, in contrast to its antagonist activity in fibroblasts derived from wild-type mice. These data demonstrate that NCoR was required for the antagonist effect of tamoxifen on ER activity.

It is not possible to cite all the molecules that have been shown to interact with ERs and to behave as coactivator or corepressors. The interested reader can find more information in reviews devoted to these aspects[83,84]. Among those molecules we can briefly cite those which have been discovered using the estrogen receptors as a model system. This is the case of the yeast protein SPT6[102], REA, which is a repressor of ER activity[104,116], a p68 RNA helicase[103], the POU transcription factors Brn-2a and Brn-3b[106], and the enigmatic SRA, which is an RNA with coactivator activity[107]. Other interesting cases among others are PSU1[108], NSD1[109], L7/SPA[112] or the yeast ADA complex[110].

Ligands

The natural ligand for ERs is the classical estrogenic compound 17β-estradiol[1,2,17]. The K_d for this compound is of 0.1 nM for ERα and 0.4 nM for ERβ. It has been clearly demonstrated using ERα that the LBD encompass the E region, which also contains a dimerization interface and the activating function AF-2[38,43]. It has also been shown after a long controversy that high-affinity binding of ER to DNA does not require estrogen binding to the receptor[45,49]. As discussed below (see 'Associated diseases'), the first ERα clone isolated from the breast cancer cell line MCF-7 contains a mutation within the LBD that alters its hormone-binding properties[158,200]. Several structure–function studies, using, for example, photoaffinity labeling, have tried to define more precisely the regions of the LBD that are directly implicated in ligand recognition (see reference 203 for an example). These data have now been extended by the resolution of the 3D structure of the LBD of ERα and ERβ in the apo- and holo-forms (see below).

Because estrogens are the most frequently prescribed medications in developed countries for the cure of many diseases, such as osteoporosis and breast cancer (see below), and since estrogen receptors are expressed in numerous tissues and estrogens exhibit multisystemic effects, numerous laboratories are characterizing selective estrogen-receptor modulators (SERM; previously called antiestrogens) that would produce specific effects on a given tissue or a given pathology (see references 210 and 211 for reviews). This has become particularly important since the realization that long-term hormonal replacement therapy using estrogens for osteoporosis treatment induces an increase in the risk of breast cancer[210]. Agents that can maintain the benefits of estrogens but avoid the risks are therefore needed. Some of these agents act as antagonists in human reproductive tissues, including mammary gland, but as partial agonists on the skeletal system and on serum lipoproteins and might thus be alternatives for prevention of osteoporosis. A large number of SERMs are now available and each agent has its own unique spectrum of activities with qualitative and quantitative variability in its agonist and antagonist properties at different target tissues. Two of these compounds tamoxifen and raloxifen are now in clinical use for the treatment of breast cancer and osteoporosis[210,211,298]. The first developed SERM, tamoxifen, was shown to be a tissue-specific antagonist of ERs, since it behaves as an antagonist on breast cancer cells but as an agonist in bone. Other analogues are under development. Two agents that were previously thought to be purely antiestrogenic ICI 164,384 and ICI 182,780 have some tissue-selective estrogenic activities on non-reproductive target tissues[210].

The precise description of the mechanisms of variable action of these compounds is beyond the scope of this book. Mechanisms involve differential binding to different estrogen receptor subtypes, different conformations produced with each agent when bound to the receptors, the availability of different coactivators and corepressors in different tissues as well as differential binding of these proteins to different estrogen–ER complexes (reviewed in reference 223). It is clear from 3D structure determination that the conformation of the receptor bound by different SERMs is different from the 17β-estradiol-bound receptor and that these differences could explain the partial antagonist activities of SERMs (see below and references 206, 212 and 223). Various types of SERMs by inducing variable repositioning of helix 12 are able to display a very large spectrum of agonistic or antagonistic activities[128,130,133,212]. It has been shown that the agonistic effects of tamoxifen on the full-length ERα is due to the cell-type specific and promoter context-dependent activity of AF-1, the transcriptional activation function that is located in the A/B region of the receptor[218]. The antagonist ICI 164,384 does not exhibit estrogenic activity and acts in cell transfections as a pure antagonist even though it does not inhibit the activity of isolated AF-1[218]. The antagonist-bound receptor is able to bind to the ERE as a dimer but with an altered mobility that is suggestive of a different conformation[215,217,218]. Of note, SERMs like tamoxifen can modulate the ligand-independent activation of human ERα by growth factors and cytokines such as dopamine[216]. This may further complicate the establishment of specific effects of SERMs, since the physiological context could modify SERM effects through phosphorylation of the receptors. Interestingly, the application of ICI 164,834 or tamoxifen on an dominant negative ERα mutated in helix 12 result in a potent agonistic activity of the compound[213,214]. The presence of an intact AF-1 function is required for this effect. This may explain how some breast cancer cells that contain mutations in ERα are positively selected by SERM treatment and finally became estrogen independent.

Because ERs are critical mediators of growth, molecules that bind to and activate ERs can potentially increase the risk of breast cancer. A number of natural and industrial chemicals in the environment possess estrogenic activities and may therefore pose a health risk[224]. These compounds called xenoestrogens are structurally extremely variable and can also have deleterious effects on reproductive development in wildlife. A nice illustration of the structural diversity of these compounds is the observation that the heavy metal cadmium is able to activate ERα and may thus be considered as a xenoestrogen[220]. It has recently been shown that there is a genetic variation in the susceptibility to endocrine disruption by xenoestrogens in mice[219].

The discovery of ERβ led to a detailed comparison of the abilities of ERα and ERβ to bind a wide variety of known agonists and antagonists[17,204] (reviewed in reference 205). It has clearly been shown that ERα and ERβ exhibit differences in relative ligand-binding affinity for various estrogenic compounds. The most interesting difference was found for 17α-estradiol, which has a five times higher affinity for ERα than for ERβ[204]. The isoflavone phytoestrogen genistein shows a 20-fold higher binding affinity to ERβ than to ERα and this may be used to understand the implication of ERα or β in various physiological processes, such as the vasculoprotective effects of estrogen's[207]. The antiestrogen 4-hydroxytamoxifen displays an agonistic activity on ERα when assayed on a basal promoter linked to an ERE but this effect is not observed with ERβ[16]. Such differences may contribute to the selective action of ER agonists and antagonists in different tissues, since ERα and ERβ expression profiles are also different.

Three-dimensional structure

The 3D structure of the DBD of ERα in solution has been determined by using NMR124. The structure is very similar to the one of the GR DBD with the two 'zinc-fingers', forming a unique module bearing two helices perpendicular to each other[124]. This structure is clearly distinct from the zinc fingers of other transcription factors, such as TFIIIA (reviewed in reference 132). The structure of the ERα DBD bound to DNA has also been solved by X-ray crystallography[125]. The comparison with the structure of the GR DBD bound to the GRE allowed the structural basis of the specific recognition of the ERE by ERα to be explored. This was further explored by analysis of the structure of the GR DBD mutant containing the ERα P-box bound to a GRE[126]. In this structure a specific contact that exists between the side chain of ERα's Lys 210 and DNA is prevented in the mutant, resulting in a gap filled by water molecules that result in a weaker non-specific interaction (see reference 127 for a review).

The first reported ER LBD structure was the one of ERα bound to either an agonist estradiol[128,129] or an antagonist the raloxifene[128]. The LBDs in both structures have architectures and binding modes very similar to the other reported structures but with one important difference, the position of the helix 12, which contains the AF-2–AD function. In the ERα-raloxifene LBD, the large piperidine extension of raloxifene prevents helix 12 from closing the hydrophobic pocket containing the ligand and pushes it into interactions with helices 3 and 11. This new position of helix 12 should impair coactivator binding, providing a structural basis for antagonism (reviewed in reference 131). Despite this large extension, raloxifene has interactions in the binding pocket extremely similar to those of estradiol. Interestingly, another study has reported that the binding of another antagonist, tamoxifen, to ERα also results in a displacement of helix 12[133]. Nevertheless, these data do not explain how these antagonists may exhibit agonistic activity in other tissues. More recently, the structure of the ERβ LBD bound to an agonist, the phytoestrogen genistein or the raloxifene, has confirmed this model[130]. However, despite the fact that genistein is totally buried inside the LBD hydrophobic pocket, the ERβ helix 12 does not adopt the distinctive agonist conformation but instead lies in a similar orientation to that induced by ER antagonists. This is consistent with the partial agonist character of genistein.

The dimer interface of ERα was shown to be similar to that of unliganded RXR and distinctly different from the much smaller dimer interface of the PR LBD[128,129]. The dimer interface contains residues from helices 7, 8 and 9 as well as the loop between helices 8 and 9, but is dominated by the conserved hydrophobic region at the N-terminus of helix 10/11.

Expression

The human ERα has been shown to be expressed in MCF-7 and T47D cells as a 6.2 kb transcript[23]. In fact, the ERα gene can generate several transcripts giving rise to various isoforms. In many species, such as *Xenopus*, several transcripts can be observed by northern blot (see below and reference 24). ERα, as ERβ, is expressed in a wide variety of tissues and this explains the wide spectrum of physiological effects of estrogens (see reference 230 for a review). Expression was found at variable levels in bones (in osteoblastic[232,233] and osteoclastic[237] cells), in uterus, bladder, ovary, prostate, testis, epididymys, kidney, breast, heart, vessel wall, pituitary and hypothalamus[204]. No expression was detected in gastrointestinal tract, lung and hippocampus (reviewed in references 205, 229 and 230). The expression in the CNS has been studied in detail since the effects of estrogens in CNS received increasing attention[229]. During mouse development, ERα has been shown to be expressed in pre-implantation embryos[231], as well as during early embryogenesis[238]. ERα is also expressed during pituitary development[234], very early (E9.5) during heart development as well as in kidney, gonads and, transiently (E10.5) in midgut[227]. The expression in brain starts relatively late (E16.5), after that of ERβ.

It is known that ERα is down-regulated in the presence of estradiol, such as in the breast cancer cell line MCF-7, although conflicting data have been published[161,235]. In other physiological situations, such as in the rainbow trout liver, an induction of ERα expression by estrogens was noticed[166,236]. It has been shown that the ubiquitin proteasome pathway plays a role in the autologous down-regulation of ERα (reference 73 and references therein). Interestingly, deletion of helix 12 abrogates this ligand-dependent down-regulation as do point mutations in the LBD that impair coactivator binding[73]. This suggests that protein interactions with the ERα coactivator binding surface are important for ligand-mediated receptor down-regulation. This also indicates that receptor and coactivator turnover contribute to ERα transcriptional activity.

The human ERβ gene is expressed in thymus, testis, ovary and spleen as 8 and 10 kb transcripts[27]. A small 1.3 kb transcript was also observed in testis. *In situ* hybridization experiments done in rat indicate a strong expression in prostate epithelial cells, whereas the expression in prostate smooth muscle cells and fibroblasts was low[17]. High expression was also visible in granulosa cell layer of primary, secondary and mature follicles in the ovary, whereas primordial follicles, oocytes and corpora lutea appeared completely negative[17,225]. Expression was also detected in bone[226], bladder, uterus, testis, epididymis, gastrointestinal tract (in contrast to ERα), kidney, breast, heart, vessel wall, immune system, lung (in contrast to ERα), pituitary, hippocampus and hypothalamus[204] (reviewed in reference 205). Interestingly, no ERβ expression was found in liver in which ERα is expressed. A detailed comparison of ERα and ERβ protein expression in the mammary gland reveals that the cellular distribution of both types is distinct[228]. Cell-specific expression and regulation by steroid hormones and prolactin was also observed for ERα and ERβ in rat decidua cells[294]. Cells coexpressing ERα and ERβ are rare during pregnancy, a proliferative phase, but represent up to 60% of the epithelial cells during lactation, a post-proliferative phase. This is due to an induction of ERα expression during lactation. During development, ERβ was found expressed early (E10.5) in kidney, heart mesentery and brain, whereas expression in bone appears weaker than ERα[227]. Relatively few reports described the regulation of ERβ expression (see reference 205 for a review). It has been shown that ERβ mRNA expression is down-regulated by gonadotropins in rat ovary[225].

Activity and target genes

Characterization of the functional domains of the estrogen receptor led to the conclusion that it contains at least two autonomous transactivating functions AF-1 and AF-2, located in the A/B and E domains, respectively[38,243]. These two activating functions are different and distinct from those of acidic activation domains, such as the one for VP16[244]. AF-1 and AF-2 can act in synergy and have been shown to compete for factors that mediate their positive effect on transcription[79,80,291]. This important observation was the starting point for the identification of transcriptional coactivators (see references 83 and 84 for reviews). These two activation domains have also been mapped in ERβ[16,253]. The respective contributions of each activating function of each monomer in the ERα–ERβ heterodimer are still not completely known[253,254]. The AF-1 function is active in the absence of estrogen and exhibits a marked promoter- and cell-specific activity[244]. Depending on the context, it can be much stronger than AF-2 and may also be responsible in some cases of ligand-dependent transcriptional activation through the binding of coactivators (see above). Its precise delineation has been hampered by the lack of amino-acid conservation of the A/B region. Using the yeast as a model system, Métivier et al. have recently narrowed the AF-1 function of the rainbow trout ERα AF-1 domain to a 11-amino-acid region located in the extreme N-terminal part of the receptor[239]. The AF-2 function has been shown to be ligand-dependent[241]. Despite the fact that it requires a specific region, called AF-2–AD that is located in helix 12, it is not as simple as AF-1 and corresponds to a surface created from dispersed elements in H3, H5 and H12[131,242,247]. The AF-2–AD region lies between residues 538 and 552 (corresponding to helix H12) of the mouse estrogen receptor in a region conserved in other nuclear receptors[247]. Recently, a third autonomous activation domain, AF-3, was mapped within the human ERα protein[248]. This function is located between amino acids 282–351 of the human ERα in the N-terminal part of the LBD, in the region of H1–H3.

The human ERα has been shown to be functional in yeast and this system has been widely used to characterize the activating functions of ERs[249]. As in mammalian cells, AF-1 exhibits a promoter-specific ligand-independent activity in yeast, whereas AF-2 is ligand dependent[249,250]. ERα has also been shown to function *in vitro* using cell-free transcription system[251,252]. The vitellogenin promoter has been used in such a system to demonstrate an estrogen-dependent induction using liver nuclear extracts[251].

ERα is a phosphoprotein and it has been known for a long time that estrogen treatment induces its phosphorylation, which may play a role in its ability to bind the ligand[149]. Antiestrogens, such as tamoxifen or ICI 164,384, also induce ERα phosphorylation, although to a lesser extent[150]. There are multiple phosphorylation sites in human ERα but Ser 118 in the A/B region is probably one of the most important, since its mutation to Ala causes a drastic reduction in ERα transcriptional activity[150]. This mutation does not affect the DNA-binding activity or the nuclear localization of the receptor, suggesting that Ser 118 is directly implicated in the AF-1 function of the receptor. Importantly, this residue is phosphorylated by MAP kinase *in vitro* and mediates the effects of growth factors, such as EGF or IGF-1, on ERα activity[151,152]. It has been directly shown that the phosphorylation of apo-ERα on this site by MAP kinase increases its transcriptional activity[151,152]. The activation of ERs by growth factor through phosphorylation of a specific residue can occur in the absence of ligand and has become the paradigm of ligand-independent activation of nuclear receptors. Neurotransmitters such as dopamine are also able to activate ERα in a ligand-independent manner[216]. The cross-talk between growth-factor signaling pathways and ERs is more complex than the phosphorylation of ER by growth factor activated-kinases, since it is now clear that estrogens also stimulate the MAP kinase pathway via a direct interaction between ER and Src (see reference 74 and 155 for references). Also, it has recently been demonstrated that Ser 118 can be phosphorylated in a MAP-kinase-independent manner after ligand binding. In this situation, it is the cyclin-dependent kinase cdk7, a member of the TFIIH complex that phosphorylates Ser-118[157]. This phosphorylation of Ser 118 requires an interaction between two subunits of TFIIH, XPD and p62, and the AF-2 domain of ERα. Another residue of the A/B domain, Ser 167, is directly phosphorylated by the kinase pp90$_{rsk1}$ and, as Ser 118 is important for AF-1 activity[154]. The situation seems extremely similar for ERβ, since the phosphorylation of AF-1 by MAP kinase leads to the recruitment of SRC-1 by ERβ[92]. Two residues, Ser 106 and Ser 124, of mouse ERβ are critical for the physical association with SRC-1 and the transcriptional activation. Another residue, Tyr 537, of the human ERα protein (Tyr 541 in the mouse protein), located immediately N-terminal to helix 12, which is conserved in other ERs including ERβ (Tyr 443 in rat ERβ), is phosphorylated and induces a ligand-independent constitutive activity as well as coactivators binding when mutated (see references 153 and 240, and references therein). This suggests that the phosphorylation of this Tyr can also play a role in the modulation of ERα transcriptional activity, but this time through the AF-2 function. ERα is phosphorylated at multiple other sites. In human ERα, an interesting case is Ser 236 located in the DBD at the end of the second zinc finger, which is phosphorylated by protein kinase A[156]. The mutation of this residue to Glu prevents DNA binding by inhibiting dimerization of ERα, whereas its mutation to Ala has little effect. Both the ERα homodimerization or the heterodimerization with ERβ are affected by PKA phosphorylation of this residue[156].

A vast number of target genes have been described for estrogen receptors. We will only mention some of them by discussing the main tissues or physiological processes in which ERs are active (see references 105, 141, 187, 194, 205, 229, 230 and 272 for reviews discussing ER target genes). This is the case, for example, of the functions of ERs associated with reproduction. However, it is important to note that many genes are regulated by estrogen in several organs. This is the case for the progesterone receptor that is under estrogenic control in normal reproductive tissues as well as in MCF-7 human breast cancer cells. Of note, this classic estrogen regulation is done in an ERE-independent manner. It has recently been shown that estrogenic regulation of PR promoter A involved a half-ERE and an Sp1 binding site that are both required for inducibility. Interestingly, the same synergy, including direct physical interaction, between ER and Sp1 was involved in estrogenic regulation of another nuclear receptor, RARα1[264]. This gene also provides an example of an estrogen target gene that exhibits a differential effect of ERα and ERβ[265].

One of the first estrogen target genes characterized were the vitellogenin in chicken, *Xenopus* or fishes. These proteins secreted by the liver are stored in the egg of these organisms and display a very clear estrogen induction. Several vitellogenin genes exist and are often coordinately regulated. The study of these genes allowed the estrogen response elements to be defined (see references 245 and 246 and references therein). As it is often the case for other steroid receptors complex interplay with other transcription factors has been shown to take place in these promoters. For example, the GATA6 factor has been shown to be required for full estrogen inducibility of the chicken vitellogenin II gene[255]. Another well-known target gene is the ovalbumin gene that is expressed in chicken oviduct. It has been shown that, despite its strong estrogen inducibility *in vivo*, the chicken ovalbumine gene, which was used as a model system to clone the estrogen receptor, was devoid of consensus ERE[256]. Further studies facilitated the identification of several half-site motifs (GGGTCA) that act synergistically to regulate the ovalbumin gene[257]. In fact, one of these half-sites can be recognized by AP-1, which can induce the ovalbumine gene through this element[67]. This has led to the discovery of the AP-1–ERα cross-talk: AP-1 and liganded ER coactivates the ovalbumin gene promoter in a ER DNA-binding-independent manner. This synergy requires the two activating functions of ER, AF-1 and AF-2, and the synergy takes place in the presence of either estrogen or antiestrogens[69,71]. The regulation of this promoter is complex, since many other factors have been shown to be important for its steroid hormone regulation[258]. Other estrogen-responsive genes in the reproductive tissues involved the EGF gene in uterine epithelial cells[259], negative regulation of bone morphogenetic protein BMP7 in chicken oviduct[260], repression of the HoxA10 gene in adult mouse uterus[261] or induction of the telomerase gene in human ovary epithelium cells[262].

Given the proliferative effects of estrogen on breast cancer cells, numerous laboratories have searched for estrogen target gene in these cells (reviewed in reference 194). The pS2 gene, also known as TFF1 (trefoil factor 1), the function of which remains poorly understood, has been cloned as an estradiol-regulated gene and it has been demonstrated that its induction by estradiol in MCF-7 cells is a primary transcriptional event (see references 266, 267 and 297 for references). The promoter of this gene contains an element responsive to estradiol but also to TPA, EGF, the c-Ha-ras oncogene and the AP-1 transcription factor[266]. Interestingly, this gene is also expressed in stomach mucosa in which its transcription is estrogen independent, suggesting a complex tissue-specific regulation by estrogens. The ERE that was localized in the pS2 gene promoter is an imperfect palindrome[267]. Of special interest is the intricate network of regulatory interactions that exist between the estrogen receptors and the cyclin genes that regulate the cell cycle in breast cancer cells. It has been proposed that cyclin regulation by estrogens, as well as direct cyclin D1–ERα physical interaction may explain at least in part the cell proliferation effect of estrogens (see reference 105 for a review). Many cyclin genes are induced by estrogens in breast cancer cells[268]. The promoter of the cyclin D1 gene is estrogen responsive. The sequence responsible for this effect does not bear much resemblance with EREs but can be recognized by AP-1, suggesting that estrogen regulates this gene through cross-talk with AP-1[269]. Both the AF-1 and AF-2 functions of ER are required for this effect. Other cyclin genes, such as cyclin B1 or cyclin E, are also transcriptionally regulated by estrogens[105,194]. Estrogens also regulate nuclear proto-oncogenes that are implicated in the control of cell proliferation (see reference 272 for a review). This is the case for the c-*myc* proto-oncogene[230,271] as well as of the two partners that form the AP-1 transcription factor, c-*jun* and c-*fos*[194,272]. The c-*jun* proto-oncogene is also regulated in other tissues, such as uterus[273] or avian oviduct, in which the antiestrogen tamoxifen can also induce its expression[274]. A functional antagonism has been described between Fos and ERα for the regulation of the c-*fos* gene expression[275]. This gene is regulated by estrogens via a specific divergent ERE that can also bind the AP-1 transcription factor[276]. Other genes that are regulated by estrogens and that may be directly linked to cell proliferation in breast cancer include the growth factor insulin-like growth factor, IGF-1[278], the anti-apoptotic factor Bcl-2[279], the E2F1 gene[280], the c-*src*, and c-*myb* proto-oncogenes[194], or the BRCA1, p21[wafl] and p53 genes[194]. The lysosomal proteinase cathepsin D is also induced by estrogens in the mammary gland[277].

Another important estrogen target tissue is the bone (reviewed in references 187 and 188). Among the numerous target genes characterized in this tissue, it is important to cite the TGFβ3 that is activated by estrogen metabolites and the SERM raloxifene through a polypurine sequence that cooperates with other unknown promoter elements to achieve full inducibility[284]. No classic ERE was found in this promoter nor in the IL6 promoter[286]. The IL6 gene is also down-regulated by estrogens through a complex mechanism involving an interaction with NFκB[208,209]. The TNFα gene is also down-regulated by estrogen in bone[285]. The bone matrix protein osteopontin is a classic estrogen target tissue in bone cells as well as in breast cancer. The estrogen regulation arose not through classical EREs but by the binding of ERα, but not ERβ to half-sites sequences reminiscent of the SFRE (TCA AGGTCA)[53].

Other estrogen target tissues that are interesting to mention include erythroid cells in which estrogen receptors cooperate with the TGFα receptor in regulation of cell renewal[282] or the regulation of the vascular endothelial growth factor (VEGF) by ERα and β in endothelial cells[283].

Knockout

The knockout of the ERα gene in mice (generating the so-called ERKO mice) had been performed in 1993 at a time when the ERβ gene was not known (see reference 141 for an extensive review on ERα and β knockout mice). These animals survive to adulthood, but females are infertile and males have decreased fertility[28]. Females have hypoplastic uteri and abnormal ovaries. Strikingly, these animals conserve 5% of normal estradiol binding in the uterus[28]. This has been linked to an abnormal splicing event generating smaller ER products that still bind the ligand but the existence of ERβ can also fully account for this residual binding[17,29,141]. A new mutant line with a fully disrupted ERα gene has recently been generated[139]. The mutation of ERα led to pleiotropic effects and a number of other abnormalities in gonads, mammary glands, reproductive tracts, sexual behavior, adipose and skeletal tissues have been described (see reference 141). It has, for example, been shown that the ERKO mice exhibit a modest decrease in bone density and mineral content, suggesting a direct role of ERα in bone metabolism[140,141]. This effect is more clearly visible on the males than on the females[140,141]. Nevertheless, the magnitude of these effects is far from expected given the importance of estrogen in human osteoporosis.

The phenotype of the ERβ[−/−] mice (called the BERKO mice) is different from that for the ERKO mice and both types of animals have been compared in a detailed manner[134-137,139,141]. The BERKO females have a poor reproductive capacity but are not completely sterile as the ERKO mice[134,139]. The analysis of the phenotypes of these animals in ovary reveal a functional redundancy between ERα and ERβ for folliculogenesis[134,139]. No effect in bone density was noticed in the BERKO male mice, suggesting that ERβ plays no important role in the skeleton of male mice during growth and maturation[140]. In adult females, a striking increase in bone mineral density was observed in the BERKO animals, suggesting a repressive effect of ERβ in regulation of bone growth[147]. An increase in prostate cell proliferation has been detected in BERKO mice[134] but this is not confirmed by other studies[139].

Mice lacking both ERα and ERβ (DERKO animals) have also been generated[138,139]. These animals exhibit normal reproductive tract development but are infertile. Interestingly, ovaries of adult DERKO mice exhibit follicle transdifferentiation in structures resembling testis including Sertoli-like cells and Müllerian inhibiting substance expression. This is reminiscent of a sex reversal, at a post-natal stage, and this suggests that both ERα and ERβ are required for the maintenance of female sex characteristics in the post-natal ovary[138]. The role of ERα and ERβ defined from the analysis of the knockout animals in controlling prostate cell proliferation is still controversial, since some laboratories detect a strong effect of ERβ[134], whereas others observed no alteration of prostate cell proliferation in BERKO or DERKO mice[139]. The skeleton of the DERKO mice exhibits the same range of abnormality as that of the ERKO mice, clearly stressing the importance of ERα, but not β for bone maturation in the male[140]. The male DERKO mice also show a profound alteration in sexual behavior[296].

Transgenic mice overexpressing mouse ERα under the control of the mouse metallothionein promoter exhibit an aberrant reproductive phenotype with a tendency toward reduced fertility and difficulties in parturition[142]. A transgenic rat model expressing dominant negative versions of ERα that inhibit both ERα and ERβ reveal no alteration of bone density but a lack of estrogen sensitivity[145]. A mutated ER LBD that specifically recognizes tamoxifen as a fusion protein with the Cre recombinase provided the basis of an efficient conditional gene targeting technology when expressed in transgenic mice[143,144].

Associated diseases

ER is associated, directly or indirectly, with numerous diseases and it is beyond the scope of this book to give a detailed description of its role in these pathologies. The two major pathologies in which estrogens and ERs are implicated are cancer, most notably breast cancer, and osteoporosis. Nevertheless, estrogens exert multisystemic effects, including a possible reduction in atherosclerotic disease through beneficial effects on blood lipids, and direct effects on the endothelial tissue of the arterial wall[221]. Some evidence is beginning to accumulate to suggest that estrogens exert positive effects on the central nervous system, possibly reducing the incidence and severity of Alzheimer-type diseases and maintaining normal cognitive function in healthy post-menopausal women[222]. For these reasons, many laboratories are devoted to the characterization of selective estrogen receptors modulators (SERMs) that would produce beneficial tissue-specific effects on these various pathologies[210].

A case of estrogen insensitivity due to a spontaneous mutation in the ERα gene has been described[186]. The homozygous mutation in the ERα gene was located in exon 2 and results in a premature stop codon, suggesting that it corresponds to a null mutant. This patient exhibited a severe osteoporosis as well as a significant increase in the longitudinal growth of bones and alteration of the cardiovascular system. The individual had no gender-identity disorder and normal external genitalia and sperm density. Estradiol levels were elevated and estrogen replacement therapy had no effect on this elevated level nor on the bone phenotype.

Osteoporosis is a common disease affecting the majority of older women and a significant proportion of older men. It is defined as the gradual reduction in bone strength with advancing age, such that bones fracture with minimal trauma, particularly in women after the menopause, when the titer of circulating estrogens decreases dramatically. At the cellular level, it is caused by the rupture of the dynamic equilibrium existing between the two major cell types of the bone: osteoblasts that produce bone and osteoclasts that resorb it (see reference 188 for a review). Because osteoporosis is frequent in post-menopausal women and since estrogens have a beneficial effect on osteoporosis in women, the effect of estrogen in bones and more generally on the skeleton has become a major field of interest (see reference 187 for a review). Estrogen replacement therapy is widely used as a treatment for osteoporosis and more generally for disorders linked to the menopause, such as cardiovascular diseases, for which estrogens have a protective role[190]. Nevertheless, this therapy may have major side-effects among which the incidence of breast cancer is a concern[191]. For this reason, alternatives to the use of estrogen in post-menopausal women have been developed (see reference 189 for a review). The question of the mechanism of estrogen action on bone is still open and the relatively mild bone phenotype of the mouse with inactivation of ERα, ERβ or both has even increased our uncertainty on the direct role of these receptors in bone. As discussed above, both osteoclasts and osteoblasts expressed ERs but at a relatively low level. It is thus not clear whether the ERs alone may explain the effects of estrogens in bone. Indirect effects via growth factors and cytokines such as IL6 or snowball effects via orphan receptors have been proposed as possible mechanisms to explain the strong role of estrogens despite the low effects of ER gene knockout[53,187,188].

Epidemiological studies suggest that estrogens are critical factors in breast cancer etiology. A recent clinical trial demonstrated that the antiestrogen tamoxifen significantly reduces the incidence of breast cancer in women at high risk of developing the disease[192]. A correlation has been made between estrogen-induced lobuloalveolar hyperplasia in the breast and susceptibility to estrogen-induced mammary tumors in the rat[193]. This strengthens the idea that breast cancer risk is proportional to the cumulative number of ovulatory cycles and the repetitive stimulation of mammary-cell proliferation by ovarian estrogens. Nevertheless, the molecular mechanisms through which estrogens contribute to the development of breast cancer remain largely unknown. It is well known that the proliferation and metastatic potential of breast cancer cells are markedly increased by estrogens and this explains the wide use of tamoxifen as an agent able to control proliferation of breast cancer cells[197] (reviewed in reference 194). This has been further confirmed by the use of dominant negative versions of ERα that abrogate estrogen-stimulated gene cancer and breast cancer cell proliferation, highlighting the importance of estrogen receptors for the control of cell proliferation[196]. Nevertheless, what now seems clear in the normal breast and in breast cancer cells growing *in vitro* or *in vivo* is less so in tumor samples obtained from breast cancer patients. In such samples, it is believed that ERs are present predominantly in the more differentiated cells. In addition, several reports highlight the fact that the expression of ER in human breast carcinomas is associated with a lower rate of cellular proliferation and with less aggressive tumors (see e.g. reference 198). This paradox is still unresolved (see reference 194 for a review). It is clear, however, that the ultimate effect of estrogen on breast cancer cells will depend on the network of target genes that the receptors may regulate in a given cell in response to the hormone[194]. A large number of studies have shown that, in breast cancer cells, ERs regulate a number of genes directly implicated in the control of cell growth, such as proto-oncogenes (e.g. jun, myc or fos families), growth factors (EGF, TGFα, IGFBPs), growth factor receptors (c-erbB-2, EGF receptor), cytokines (CSF-1), cell cycle regulatory proteins (cyclins, p2^{Waf1}, BRCA1, p53), etc. (see reference 194 for a review).

One problem with the use of antiestrogens such as tamoxifen or ICI 164,384 is that these compounds often have only a temporary effect and finally engender resistance in breast cancer patients as well as in animal models (see references 195, 199 and 211 for reviews). Ironically, there is a point mutation in the LBD of the ERα cDNA that was first isolated by Pierre Chambon's group from the breast cancer cell line, MCF7[158]. This mutation, a Gly to Val mutation at codon 400, decreases the affinity for estradiol and is an example of a mutation that may arise in a breast cancer[200]. Since the understanding of the mechanisms of drug resistance may lead to the development of new drug treatment therapies, a large number of authors have studied these mechanisms. Mutated, truncated or alternatively spliced versions of ERα have been detected in hormone-resistant breast cancer samples. Mutations in all the functional domains of the protein have been associated in most cases with a decreased sensitivity to the hormone and an enhanced transcriptional activity in the absence of hormone[199]. Several alternatively spliced versions of ER as well as abnormalities of methylation pattern of the ERα gene have also been detected and characterized from breast cancers[171,172] (see below). Progression of breast cancer to hormone insensitivity may be the result of a mutational event, followed by clonal selection. The search for genes involved in antiestrogen resistance was initiated by several laboratories using hormone-dependent cell lines. Transduction of such cells, like ZR75-1, with an expression vector carrying the EGF receptor, resulted in proliferation independent of estradiol[201]. These cells have been also subjected to mutagenesis, for example, using retrovirus-based insertional mutagenesis[202]. Three chromosomal loci, designated breast cancer antiestrogen resistance (BCAR) locus 1–3 have been identified using this strategy. The BCAR3 gene has been cloned and shown to contain an SH2 domain found in many proteins implicated in signal transduction. Ectopic expression of BCAR3 induces estrogen independence in ZR75-1 and MCF7, suggesting that this gene plays a dominant role in antiestrogen-arrested breast cancer cell proliferation[202].

Gene structure, promoter and isoforms

The genomic organization of the human ERα gene has been determined[158]. The gene is long (more than 140 kb) and split into eight coding exons, the positions of which are very well conserved when compared to other steroid receptors such as PR[158]. Each zinc finger is encoded by a separate exon, the boundary between the two exons being located 10 amino acids after the last Cys of the first zinc finger. The LBD is encoded by five exons. This structure is conserved in fish ERα[26,159]. The structure of the human ERβ gene is similar to that for ERα[148]. The gene is much shorter than ERα, since it spans only 40 kbp[148].

The promoter of the mouse ERα gene was the first to be characterized[3]. Multiple start sites were identified in a region of at least 62 bp. A TATA box motif exists in this promoter and coincides with the start of one cDNA clone[3]. This promoter has been used to generate transgenic mice[163]. This has revealed that the 400 proximal nucleotides are sufficient to direct widespread expression in mouse organs. Nevertheless, the first exon and intron of the mouse ERα gene are necessary to achieve sexually dimorphic expression of the transgene in neurons located at specific sites in the CNS. In humans, the situation appears more complex since two promoters were discovered: a promoter called P1 or promoter A, which is upstream of the major transcript cap site, and promoter P0 or promoter B located 3 kbp upstream, which exhibits a significant level of sequence identity with the mouse ERα promoter[1,160,169]. The precise location of the start sites within this promoter is still unclear[161,162]. In addition, an important regulatory element that would be responsible for ERα overexpression in some breast cancer is located downstream of the P1 promoter[168]. A third promoter, called promoter C, is located 21 kb upstream of promoter A[162]. Several lines of evidence suggest that the level of expression of ERα is determined by the number of promoters used rather than the selective use of specific promoters[162]. The promoter A is responsive to estradiol and contains three half EREs that are responsible for this inducibility[161]. The two other promoters are also estrogen responsive, although the location of EREs has not been done[162]. Detailed analysis of the 5′ region of the human ERα gene reveals an extreme complexity with at least six different exons containing 5′UTR sequences[173,180]. This suggests that at least six different promoters exist in the human ERα gene[180]. The three new promoters predicted to exist have not been characterized in detail. At least four promoters governing the formation of various 5′ isoforms also exist for the chicken ERα gene[177,292]. An ERα promoter is also known in *Xenopus laevis*[164]. As in mammals it contains a TATA box but no clear ERE sequence. Nevertheless, an imperfect ERE is present and active in the coding region of the gene. The promoter regulating ERα expression has also been characterized in the rainbow trout[165]. As in many other species ERα expression is activated by estrogen and the rainbow trout promoter contains an ERE located downstream of the transcription start site[166]. It was further demonstrated that other regions of this promoter are required for maximal inducibility, suggesting that cell-specific transcription factors cooperate with ERα.

The human ERβ promoter was recently cloned[293]. It contains both a TATA box and a GC-rich region containing an Inr element. This promoter was found to be active in ERβ positive cell lines but not in ERβ-negative lines.

Both ERα and ERβ can give rise to numerous isoforms. Alternative splicing events can generate isoforms of human ERα with deletion of either exon 2, 3, 4, 5 or 7, giving rise to a large variety of possible protein isoforms[170-172]. Some of these alternative splicing, such as ERαΔE3 or ERαΔE5 can reach a high concentration in cancer cells[174]. There is a large diversity of possible isoforms of ERα present in breast cancer cells; in some cases, the sequence downstream of an exon–intron junction is replaced by unrelated sequences, such as LINE-1 elements[171]. These isoforms can exhibit different expression patterns *in vivo* as exemplified by their variation during pituitary gland ontogeny[234]. Of note, ERαΔE3, which lacks the exon encoding the second zinc finger in the DBD, is able to inhibit ERα-mediated transcription in a dominant negative fashion and to dimerize with ERα[170,172]. These protein isoforms, as other single-exon skipped products, display a dramatic reduction or even suppression in DNA-binding activity[172]. ERαΔE3 can bind 17β-estradiol but no other single-exon skipped product can do so. These variants are transcriptionally inactive and in addition to ERαΔE3, ERαΔE5 can behave as a dominant negative inhibitor but can also exert a positive action on ERα-mediated transcription depending on the promoter context[172]. The possible regulation exerted by these isoforms is complex as exemplified by ERαΔE3, which is a positive regulator of gene expression on promoters containing AP-1 sites and which also interact with SRC-1. Recently, a new isoform of ERα, hERα46, was shown to exist in humans[173]. This product lacks the N-terminal 173 amino acids present in the bona fide ERα. Interestingly, this region corresponds to the A/B domain that contains the AF-1 function and indeed hERα46 contains a functional AF-2 but no AF-1 function[173]. The hERα46 product thus acts as an AF-1 competitive inhibitor of ERα. Similar N-terminally truncated isoforms exist in chicken[177] as well as in *Xenopus*[178] and rainbow trout[179]. Apparently all these shorter isoforms are generated from specific promoters (see references 173 and 177 for references). In addition, in the rat, a truncated version of ERα, called TERP, has been characterized[175]. This isoform, which is expressed in the pituitary, is transcribed from a unique start site and contains exons 5–8 of ERα. The resultant protein contains the majority of the LBD, and lacks both the DBD and the AF-1 regions and cannot bind estradiol effectively. It can act as a dominant negative inhibitor and is able to dimerize with either ERα or ERβ[176].

The ERβ gene is also expressed as numerous isoforms. An 18-amino-acid insertion within helix 6 of the LBD was found in isoform ERβ2, which displays reduced affinity for 17β-estradiol and weaker transcriptional activity[181,182]. Other similar insertions exist within helix 10[183]. Specific deletions of exon 3 or 4 have also been characterized as well as an isoform, called ERβcx, in which the C-terminal amino acids are replaced by a specific exon[184,185].

Chromosomal location

The human ERα gene is located on chromosome 6q25.1[123], whereas the human ERβ gene is located on chromosome 14q22–q24[148]. The mouse ERβ gene is located on the central region of chromosome 12[16].

Amino acid sequence for human ERα (NR3A1)

Accession number: X03635

```
  1  MTMTLHTKASGMALLHQIQGNELEPLNRPQLKIPLERPLGEVYLDSSKPAVYNYPEGAAYEFNAAAAANA
 71  QVYGQTGLPYGPGSEAAAFGSNGLGGFPPLNSVSPSPLMLLHPPPQLSPFLQPHGQQVPYYLENEPSGYT
141  VREAGPPAFYRPNSDNRRQGGRERLASTNDKGSMAMESAKETRYCAVCNDYASGYHYGVWSCEGCKAFFK
211  RSIQGHNDYMCPATNQCTIDKNRRKSCQACRLRKCYEVGMMKGGIRKDRRGGRMLKHKRQRDDGEGRGEV
281  GSAGDMRAANLWPSPLMIKRSKKNSLALSLTADQMVSALLDAEPPILYSEYDPTRPFSEASMMGLLTNLA
351  DRELVHMINWAKRVPGFVDLTLHDQVHLLECAWLEILMIGLVWRSMEHPVKLLFAPNLLLDRNQGKCVEG
421  MVEIFDMLLATSSRFRMMNLQGEEFVCLKSIILLNSGVYTFLSSTLKSLEEKDHIHRVLDKITDTLIHLM
491  AKAGLTLQQQHQRLAQLLLILSHIRHMSNKGMEHLYSMKCKNVVPLYDLLLEMLDAHRLHAPTSRGGASV
561  EETDQSHLATAGSTSSHSLQKYYITGEAEGFPATV  595
```

Amino acid sequence for human ERβ (NR3A2)

Accession number: AB006590

```
  1 MNYSIPSNVTNLEGGPGRQTTSPNVLWPTPGHLSPLVVHRQLSHLYAEPQKSPWCEARSLEHTLPVNRET
 71 LKRKVSGNRCASPVTGPGSKRDAHFCAVCSDYASGYHYGVWSCEGCKAFFKRSIQGHNDYICPATNQCTI
141 DKNRRKSCQACRLRKCYEVGMVKCGSRRERCGYRLVRRQRSADEQLHCAGKAKRSGGHAPRVRELLLDAL
211 SPEQLVLTLLEAEPPHVLISRPSAPFTEASMMMSLTKLADKELVHMISWAKKIPGFVELSLFDQVRLLES
281 CWMEVLMMGLMWRSIDHPGKLIFAPDLVLDRDEGKCVEGILEIFDMLLATTSRFRELKLQHKEYLCVKAM
351 ILLNSSMYPLVTATQDADSSRKLAHLLNAVTDALVWVIAKSGISSQQQSMRLANLLMLLSHVRHASNKGM
421 EHLLNMKCKNVVPVYDLLLEMLNAHVLRGCKSSITGSECSPAEDSKSKEGSQNPQSQ  477
```

References

1 Green, S. et al. (1986) Nature 320, 134–139.
2 Greene, G. L. et al. (1986) Science 231, 1150–1154.
3 White, R. et al. (1987) Mol. Endocrinol. 1, 735–744.
4 Koike, S. et al. (1987) Nucl. Acids Res. 15, 2499–2513.
5 Bokenkamp, D. et al. (1994) Mol. Cell. Endocrinol. 104, 163–172.
6 Rosenfeld, C.S. et al. (1999) Biol. Reprod. 60, 691–697.
7 Krust, A. et al. (1986) EMBO J. 5, 891–897.
8 Bernard, D.J. et al. (1999) Endocrinology 140, 4633–4643.
9 Jacobs, E.C. et al. (1996) J. Steroid Biochem. Mol. Biol. 59, 135–145.
10 Rogers, S.A. et al. (2000) Comp. Biochem. Physiol. B, Comp. Biochem. 125, 379–385.
11 Pakdel, F. et al. (1990) Mol. Cell. Endocrinol. 71, 195–204.
12 Touhata, K. et al. (1998) Fisheries Sci. 64, 131–135.
13 Munoz-Cueto, J.A. et al. (1999) DNA Seq. 10, 75–84.
14 Chang, X.T. et al. (1999) Zool. Sci. 16, 653–658.
15 Ogawa, S. et al. (1998) Biochem. Biophys. Res. Commun. 243, 122–126.
16 Tremblay, G.B. et al. (1997) Mol. Endocrinol. 11, 353–365.
17 Kuiper, G.G. et al. (1996) Proc. Natl Acad. Sci. USA 93, 5925–5930.
18 Walther, N. et al. (1999) Mol. Cell. Endocrinol. 152, 37–45.
19 Lakaye, B. et al. (1998) J. Neuroreport 9, 2743–2748.
20 Todo, T. et al. (1996) Mol. Cell. Endocrinol. 119, 37–45.
21 Ma, C.H. et al. (2000) Biochim. Biophys. Acta 1490, 145–152.
22 Xia, Z. et al. (2000) Gen. Comp. Endocrinol. 118, 139–149.
23 Walter, P. et al. (1985) Proc. Natl Acad. Sci. USA 82, 7889–7893.
24 Weiler, I.J. et al. (1987) Mol. Endocrinol. 1, 355–362.
25 Young, L.J. et al. (1995) J. Steroid Biochem. Mol. Biol. 55, 261–269.
26 Tan, N.S. et al. (1996) Mol. Cell. Endocrinol. 120, 177–192.
27 Mosselman, S. et al. (1996) FEBS Lett. 392, 49–53.
28 Lubahn, D.B. et al. (1993) Proc. Natl Acad. Sci. USA 90, 11162–11166.
29 Couse, J.F. et al. (1995) Mol. Endocrinol. 9, 1441–1454.
30 Hawkins, M.B. et al. (2000) Proc. Natl Acad. Sci. USA 97, 10751–10756.
31 Xia, Z. et al. (1999) Gen. Comp. Endocrinol. 113, 360–368.
32 Madigou, T. et al. (1996) Mol. Cell. Endocrinol. 121, 153–163.
33 Tchoudakova, A. et al. (1999) Gen. Comp. Endocrinol. 113, 388–400.
34 Green, S. and Chambon, P. (1987) Nature 325, 75–78.
35 ten Heggeler-Bordier, B. et al. (1987) EMBO J. 6, 1715–1720.
36 Martinez, E. et al. (1987) EMBO J. 6, 3719–3727.

37 Kumar, V. and Chambon, P. (1987) Cell 55, 145–156.

38 Kumar, V. et al. (1987) Cell 51, 941–951.

39 Green, S. et al. (1988) EMBO J. 7, 3037–3044.

40 Mader, S. et al. (1989) Nature 338, 271–274.

41 Martinez, E. and Wahli, W. (1989) EMBO J. 8, 3781–3791.

42 Ponglikitmongkol, M. et al. (1990) EMBO J. 9, 2221–2231.

43 Fawell, S.E. et al. (1990) Cell 60, 953–962.

44 Mader, S. et al. (1993) Nucl. Acids Res. 21, 1125–1132.

45 Furlow, J.D. et al. (1993) J. Biol. Chem. 268, 12519–12525.

46 Wood, J.R. et al. (1998) Mol. Cell. Biol. 18, 1927–1934.

47 Ylikomi, T. et al. (1992) EMBO J. 11, 3681–3694.

48 Miksicek, R.J. et al. (1995) Mol. Endocrinol. 9, 592–604.

49 Zhuang, Y. et al. (1995) Mol. Endocrinol. 9, 457–466.

50 Romine, L.E. et al. (1998) Mol. Endocrinol. 12, 664–674.

51 Nardulli, A.M. and Shapiro, D.J. (1992) Mol. Cell Biol. 12, 2037–2042.

52 Kim, J. et al. (1997) Mol. Cell. Biol. 17, 3173–3180.

53 Vanacker, J.M. et al. (1999) EMBO J. 18, 4270–4279.

54 Green, S. and Chambon, P. (1988) Trends Genet. 4, 309–314.

55 Umesono, K. and Evans, R.M. (1989) Cell 57, 1139–1146.

56 Danielsen, M. et al. (1989) Cell 57, 1131–1138.

57 Klein-Hitpass, L. et al. (1989) Mol. Cell. Biol. 9, 43–49.

58 Cowley, S.M. et al. (1997) J. Biol. Chem. 272, 19858–19862.

59 Pettersson, K. et al. (1997) Mol. Endocrinol. 11, 1486–1496.

60 Pratt, W.B. and Toft, D.O. (1997) Endocrinol. Rev. 18, 306–360.

61 Schlatter, L.K. et al. (1992) Mol. Endocrinol. 6, 132–140.

62 Knoblauch, R. and Garabedian, M.J. (1999) Mol. Cell. Biol. 19, 3748–3759.

63 Eilers, M. et al. (1989) Nature 340, 66–68.

64 Superti-Furga, G. et al. (1991) Proc. Natl Acad. Sci. USA 88, 5114–5118.

65 Porter, W. et al. (1997) Mol. Endocrinol. 11, 1569–1580.

66 Martinez, E. et al. (1991) Mol. Cell. Biol. 11, 2937–2945.

67 Gaub, M.P. et al. (1990) Cell 63, 1267–1276.

68 Doucas, V. et al. (1991) EMBO J. 10, 2237–2245.

69 Webb, P. et al. (1999) Mol. Endocrinol. 13, 1672–1685.

70 Paech, K. et al. (1997) Science 277, 1508–1510.

71 Webb, P. et al. (1995) Mol. Endocrinol. 9, 443–456.

72 Fan, S. et al. (1999) Science 284, 1354–1356.

73 Lonard, D.M. et al. (2000) Mol. Cell 5, 939–948.

74 Migliaccio, A. et al. (2000) EMBO J. 19, 5406–5417.

75 Poelzl, G. et al. (2000) Proc. Natl Acad. Sci. USA 97, 2836–2839.

76 Brou, C. et al. (1993) Nucl. Acids Res. 21, 5–12.

77 Jacq, X. et al. (1994) Cell 79, 107–117.

78 May, M. et al. (1996) EMBO J. 15, 3093–3104.

79 Meyer, M.E. et al. (1989) Cell 57, 433–442.

80 Tasset, D. et al. (1990) Cell 62, 1177–1187.

81 Halachmi, S. et al. (1994) Science 264, 1455–1458.

82 Cavailles, V. et al. (1994) Proc. Natl Acad. Sci. USA 91, 10009–10013.

83 McKenna, N.J. et al. (1999) Endocrine Rev. 20, 321–344.

84 Glass, C.K. and Rosenfeld, M.G. (2000) Genes Dev. 14, 121–141.

[85] Norris, J.D. et al. (1999) Science 285, 744–746.

[86] Chang, C.Y. et al. (1999) Mol. Cell. Biol. 19, 8226–8239.

[87] Heery, D.M. et al. (1997) Nature 387, 733–736.

[88] Xu, J. et al. (1998) Science 279, 1922–1925.

[89] Kalkhoven, E. et al. (1998) EMBO J. 17, 232–243.

[90] Xu, J. et al. (2000) Proc. Natl Acad. Sci. USA 97, 6379–6384.

[91] Tikkanen, M.K. et al. (2000) Proc. Natl Acad. Sci. USA 97, 12536–12540.

[92] Tremblay, A. et al. (1999) Mol. Cell 3, 513–519.

[93] Mak, H.Y. et al. (1999) Mol. Cell. Biol. 19, 3895–3903.

[94] Feng, W. et al. (1998) Science 280, 1747–1749.

[95] Cavailles, V. et al. (1995) EMBO J. 14, 3741–3751.

[96] Goodman, R.H. and Smolik, S. (2000) Genes Dev. 14, 1553–1577.

[97] Kamei, Y. et al. (1996) Cell 85, 403–414.

[98] Kraus, W.L. and Kadonaga, J.T. (1998) 12, 331–342.

[99] Le Douarin, B. et al. (1995) EMBO J. 14, 2020–2033.

[100] Thénot, S. et al. (1999) Mol. Endocrinol. 13, 2137–2150.

[101] vom Baur, E. et al. (1996) EMBO J. 15, 110–124.

[102] Baniahmad, C. et al. (1995) Mol. Endocrinol. 9, 34–43.

[103] Endoh, H. et al. (1999) Mol. Cell. Biol. 19, 5363–5372.

[104] Montano, M.M. et al. (1999) Proc. Natl Acad. Sci. USA 96, 6947–6952.

[105] Pestell, R.G. et al. (1999) Endocrine Rev. 20, 501–534.

[106] Budhram-Mahadeo, V. et al. (1998) Mol. Cell. Biol. 18, 1029–1041.

[107] Lanz, R.B. et al. (1999) Cell 97, 17–27.

[108] Gaudon, C. et al. (1999) EMBO J. 18, 2229–2240.

[109] Huang, N. et al. (1998) EMBO J. 17, 3398–3412.

[110] vom Baur, E. et al. (1998) Genes Dev. 12, 1278–1289.

[111] Bhat H.K. and Vadgama, J.V. (2000) J. Steroid Biochem. Mol. Biol. 72, 47–53.

[112] Jackson, T.A. et al. (1997) Mol. Endocrinol. 11, 693–705.

[113] Wu, W.X. et al. (2000) Am. J. Physiol. 278, C190–C198.

[114] Zwijsen, R.M.L. et al. (1997) Cell 88, 405–415.

[115] Neuman, E. et al. (1997) Mol. Cell. Biol. 17, 5338–5347.

[116] Martini, P.G.V. et al. (2000) Mol. Cell. Biol. 20, 6224–6232.

[117] Oesterreich, S. et al. (2000) Mol. Endocrinol. 14, 369–381.

[118] Stenoien, D.L. et al. (2000) Mol. Endocrinol. 14, 518–534.

[119] Gay, F. et al. (2000) Mol. Endocrinol. 14, 1627–1648.

[120] Shang, Y. et al. (2000) Cell 103, 843–852.

[121] Saville, B. et al. (2000) J. Biol. Chem. 275, 5379–5387.

[122] Batistuzzo de Medeiros, S.R. et al. (1997) J. Biol. Chem. 272, 18250–18260.

[123] Menasce, L.P. et al. (1993) Genomics 17, 263–265.

[124] Schwabe, J.W.R. et al. (1990) Nature 348, 458–461.

[125] Schwabe, J.W.R. et al. (1993) Cell 75, 567–578.

[126] Gewirth, D.T. and Sigler, P.B. (1995) Nature Struct. Biol. 2, 386–394.

[127] Arbuckle, N.D. and Luisi, B. (1995) Nature Struct. Biol. 2, 341–346.

[128] Brzozowski, A.M. et al. (1997) Nature 389, 753–758.

[129] Tanenbaum, D.M. et al. (1998) Proc. Natl Acad. Sci. USA 95, 5998–6003.

[130] Pike, A.C.W. et al. (1999) EMBO J. 18, 4608–4618.

131 Weatherman, R.V. et al. (1999) Annu. Rev. Biochem. 68, 559–581.
132 Schwabe, J.W.R. and Rhodes, D. (1991) Trends Biochem. Sci. 16, 291–296.
133 Shiau, A.K. et al. (1998) Cell 95, 927–937.
134 Krege, J.H. et al. (1998) Proc. Natl Acad. Sci. USA 95, 12887–12892.
135 Ogawa, S. et al. (1999) Proc. Natl Acad. Sci. USA 96, 12887–12892.
136 Karas, R.H. et al. (1999) Proc. Natl Acad. Sci. USA 96, 15133–15136.
137 Weihua, Z. et al. (2000) Proc. Natl Acad. Sci. USA 97, 5936–5941.
138 Couse, J.F. et al. (1999) Science 286, 2328–2331.
139 Dupont, S. et al. (2000) Development 127, 4277–4291.
140 Vidal, O. et al. (2000) Proc. Natl Acad. Sci. USA 97, 5474–5479.
141 Couse, J.F. and Korach, K.S. (1999) Endocrine Rev. 358–417.
142 Davis, V.L. et al. (1994) Endocrinology 135, 379–386.
143 Feil, R. et al. (1996) Proc. Natl Acad. Sci. USA 93, 10887–10890.
144 Li, M. et al. (2000) Nature 407, 633–636.
145 Ogawa, S. et al. (2000) J. Biol. Chem. 275, 21372–21379.
146 Remboutsika, E. et al. (1999) J. Cell Sci. 112, 1671–1683.
147 Windahl, S.H. et al. (1999) J. Clin. Invest. 104, 895–901.
148 Enmark, E. et al. (1997) J. Clin. Endocrinol. Metab. 82, 4258–4265.
149 Auricchio, F. et al. (1987) EMBO J. 6, 2923–2929.
150 Ali, S. et al. (1993) EMBO J. 12, 1153–1160.
151 Kato, S. et al. (1995) Science 270, 1491–1494.
152 Bunone, G. et al. (1996) EMBO J. 15, 2174–2183.
153 White, R. et al. (1997) EMBO J. 16, 1427–1435.
154 Joel, P.B. et al. (1998) Mol. Cell. Biol. 18, 1978–1984.
155 Migliaccio, A. et al. (1998) EMBO J. 17, 2008–2018.
156 Chen, D. et al. (1999) Mol. Cell. Biol. 19, 1002–1015.
157 Chen, D. et al. (2000) Mol. Cell. 6, 127–137.
158 Ponglikitmongkol, M. et al. (1988) EMBO J. 7, 3385–3388.
159 Le Roux, M.G. et al. (1993) Biochim. Biophys. Acta 1172, 226–230.
160 Keaveney, M. et al. (1991) J. Mol. Endocrinol. 6, 111–115.
161 Treilleux, I. et al. (1997) Mol. Endocrinol. 11, 1319–1331.
162 Donaghue, C. et al. (1999) Mol. Endocrinol. 13, 1934–1950.
163 Cicatiello, L. et al. (1995) Mol. Endocrinol. 9, 1077–1090.
164 Lee, J.H. et al. (1995) DNA Cell Biol. 14, 419–430.
165 Lazennec, G. et al. (1995) Gene 166, 243–247.
166 Le Dréan, Y. et al. (1995) J. Mol. Endocrinol. 15, 37–47.
167 Lazennec, G. et al. (1996) Mol. Endocrinol. 10, 1116–1126.
168 De Coninck, E.C. (1995) Mol. Cell. Biol. 15, 2191–2196.
169 Grandien, K.F.H. et al. (1993) J. Mol. Endocrinol. 10, 269–277.
170 Wang, Y. and Miksicek, R.J. (1991) Mol. Endocrinol. 5, 1707–1715.
171 Dotzlaw, H. et al. (1992) Mol. Endocrinol. 6, 773–785.
172 Bollig, A. and Miksicek, R.J. (2000) Mol. Endocrinol. 14, 634–649.
173 Flouriot, G. et al. (2000) EMBO J. 19, 4688–4700.
174 Erenburg, I. et al. (1997) Mol. Endocrinol. 11, 2004–2015.
175 Friend, K.E. et al. (1995) Proc. Natl Acad. Sci. USA 92, 4367–4371.
176 Resnick, E.M. et al. (2000) J. Biol. Chem. 275, 7158–7166.
177 Griffin, C. et al. (1999) Mol. Endocrinol. 13, 1571–1587.
178 Claret, F.X. et al. (1994) J. Biol. Chem. 269, 14047–14055.

[179] Pakdel, F. et al. (2000) Endocrinology 141, 571–580.

[180] Flouriot, G. et al. (1998) Mol. Endocrinol. 12, 1939–1954.

[181] Chu, S. and Fuller, P.J. (1997) Mol. Cell. Endocrinol. 132, 195–199.

[182] Hanstein, B. et al. (1998) Mol. Endocrinol. 13, 129–137.

[183] Moore, J.T. et al. (1998) Biochem. Biophys. Res. Commun. 247, 75–78.

[184] Petersen, D.N. et al. (1998) Endocrinology 139, 1082–1092.

[185] Ogawa, S. et al. (1998) Nucl. Acids Res. 26, 3505–3512.

[186] Smith, E.P. et al. (1994) N. Engl. J. Med. 331, 1056–1061.

[187] Turner, R.T. et al. (1994) Endocrine Rev. 15, 275–300.

[188] Manolagas, S.C. (2000) Endocrine Rev. 21, 115–137.

[189] Pinkerton, J.V. and Santen, R. (1999) Endocrine Rev. 20, 308–320.

[190] Barrett-Connors, E. and Bush, T.L. (1991) JAMA 265, 1861–1867.

[191] Lufkin, E.G. and Ory, S.J. (1995) Trends Endocrinol. Metab. 6, 50–54.

[192] Fisher, B. et al. (1998) J. Natl Cancer Inst. 90, 1371–1388.

[193] Harvell, D.M.E. et al. (2000) Proc. Natl Acad. Sci. USA 97, 2779–2785.

[194] Ciocca, D.R. and Fanelli, M.A. (1997) Trends Endocrinol. Metab. 8, 313–321.

[195] Clarke, R. and Brünner, N. (1996) Trends Endocrinol. Metab. 7, 291–301.

[196] Lazennec, G. et al. (1999) Mol. Endocrinol. 13, 969–980.

[197] Brünner, N. et al. (1989) Cancer Res. 49, 1515–1520.

[198] Garcia, M. et al. (1992) Proc. Natl Acad. Sci. USA 89, 11538–11542.

[199] Sluyser, M. (1994) Crit. Rev. Oncogenesis 5, 539–544.

[200] Tora, L. et al. (1989) EMBO J. 8, 1981–1986.

[201] Van Agthoven, T. et al. (1992) Cancer Res. 52, 5082–5088.

[202] Van Agthoven, T. et al. (1998) EMBO J. 17, 2799–2808.

[203] Pakdel, F. and Katzenellenbogen, B.S. (1992) J. Biol. Chem. 267, 3429–3437.

[204] Kuiper, G.G. et al. (1997) Endocrinology 138, 863–870.

[205] Nilsson, S. et al. (1998) Trends Endocrinol. Metab. 9, 387–395.

[206] Paige, L.A. et al. (1999) Proc. Natl Acad. Sci. USA 96, 3999–4004.

[207] Mäkelä, S. et al. (1999) Proc. Natl Acad. Sci. USA 96, 7077–7082.

[208] Stein, B. and Yang, M.X. (1995) Mol. Cell. Biol. 15, 4971–4979.

[209] Valentine, J.E. et al. (2000) J. Biol. Chem. 275, 25322–25329.

[210] Cosman, F. and Lindsay, R. (1999) Endocrine Rev. 20, 418–434.

[211] Jordan, V.C. and Morrow, M. (1999) Endocrine Rev. 20, 253–278.

[212] Nichols, M. et al. (1998) EMBO J. 17, 765–773.

[213] Montano, M.M. et al. (1996) Mol. Endocrinol. 10, 230–242.

[214] Mahfoudi, A. et al. (1995) Proc. Natl Acad. Sci. USA 92, 4206–4210.

[215] Metzger, D. et al. (1995) Mol. Endocrinol. 9, 579–591.

[216] Smith, C.L. et al. (1993) Proc. Natl Acad. Sci. USA 90, 6120–6124.

[217] Sabbah, M. et al. (1991) Proc. Natl Acad. Sci. USA 88, 390–394.

[218] Berry, M. et al. (1990) EMBO J. 9, 2811–2818.

[219] Spearow, J.L. et al. (1999) Science 285, 1259–1261.

[220] Stoica, A. et al. (2000) Mol. Endocrinol. 14, 545–553.

[221] Barrett-Connor, E. et al. (1999) Trends Endocrinol. Metab. 10, 320–325.

[222] Henderson, V.W. (1997) Neurology 48, S27–S35.

[223] McDonnell, D.P. (1999) Trends Endocrinol. Metab. 10, 301–311.

[224] Sonnenschein, C. and Soto, A.M. (1998) J. Steroid Biochem. Mol. Biol. 65, 143–150.

225 Byers, M. et al. (1997) Mol. Endocrinol. 11, 172–182.
226 Onoe, Y. et al. (1997) Endocrinology 138, 4509–4512.
227 Lemmen, J.G. et al. (1999) Mech. Dev. 81, 163–167.
228 Saji, S. et al. (2000) Proc. Natl Acad. Sci. USA 97, 337–342.
229 McEwen, B.C. and Alves, S.E. (1999) Endocrine Rev. 20, 279–307.
230 Ciocca, D.R. and Vargas Roig, L.M. (1995) Endocrine Rev. 16, 35–62.
231 Hou, Q. and Gorski, J. (1993) Proc. Natl Acad. Sci. USA 90, 9460–9464.
232 Komm, B.S. et al. (1988) Science 241, 81–84.
233 Eriksen, E.F. et al. (1988) Science 241, 84–86.
234 Pasqualini, C. et al. (1999) Endocrinology 140, 2781–2789.
235 Berkenstam, A. et al. (1989) Mol. Endocrinol. 3, 22–28.
236 Pakdel, F. et al. (1989) Mol. Endocrinol. 3, 44–51.
237 Oursler, M.J. et al. (1991) Proc. Natl Acad. Sci. USA 88, 6613–6617.
238 Bonnelye, E. et al. (1997) Mech. Dev. 65, 71–85.
239 Métivier, R. et al. (2000) Mol. Endocrinol. 14, 1849–1871.
240 Weis, K.E. et al. (1996) Mol. Endocrinol. 10, 1388–1398.
241 Webster, N.J.G. et al. (1988) Cell 54, 199–207.
242 Webster, N.J.G. et al. (1989) EMBO J. 8, 1441–1446.
243 Lees, J.A. et al. (1989) Nucl. Acids Res. 17, 5477–5488.
244 Tora, L. et al. (1989) Cell 59, 477–487.
245 Burch, J.B.E. and Fisher, A.H. (1990) Nucl. Acids Res. 18, 4157–4165.
246 Chang, T.C. et al. (1992) Mol. Endocrinol. 6, 346–354.
247 Daniellan, P.S. et al. (1992) EMBO J. 11, 1025–1033.
248 Norris, J.D. et al. (1997) Mol. Endocrinol. 11, 747–754.
249 Metzger, D. et al. (1988) Nature 334, 31–36.
250 Pham, T.A. et al. (1992) Mol. Endocrinol. 6, 1043–1050.
251 Corthésy, B. et al. (1988) Science 239, 1137–1139.
252 Elliston, J.F. et al. (1990) Mol. Cell. Biol. 10, 6607–6612.
253 Tremblay, G.B. et al. (1999) Mol. Cell. Biol. 19, 1919–1927.
254 Pettersson, K. et al. (2000) Oncogene 19, 4970–4978.
255 Davis, D.L. and Burch, J.B.E. (1996) Mol. Endocrinol. 10, 937–944.
256 Tora, L. et al. (1988) EMBO J. 7, 3771–3778.
257 Kato, S. et al. (1992) Cell 68, 731–742.
258 Nordstrom, L.A. et al. (1993) J. Biol. Chem. 268, 13193–13202.
259 Huet-Hudson, Y.M. et al. (1990) Mol. Endocrinol. 4, 510–523.
260 Monroe, D.G. et al. (2000) Mol. Cell. Biol. 20, 4626–4634.
261 Ma, L. et al. (1998) Dev. Biol. 197, 141–154.
262 Misiti, S. et al. (2000) Mol. Cell. Biol. 20, 3764–3771.
263 Petz, L.N. and Nardulli, A.M. (2000) Mol. Endocrinol. 14, 972–985.
264 Sun, G. et al. (1998) Mol. Endocrinol. 12, 882–890.
265 Zou, A. et al. (1999) Mol. Endocrinol. 13, 418–430.
266 Nunez, A.M. et al. (1989) EMBO J. 8, 823–829.
267 Berry, M. et al. (1989) Proc. Natl Acad. Sci. USA 86, 1218–1222.
268 Musgrove, E.A. et al. (1993) Mol. Cell. Biol. 13, 3577–3587.
269 Sabbah, M. et al. (1999) Proc. Natl Acad. Sci. USA 96, 11217–11222.
270 McMahon, C. et al. (1999) Proc. Natl Acad. Sci. USA 96, 5382–5387.
271 Dubik, D. and Shiu, R.P.C. (1992) Oncogene 7, 1587–1594.
272 Weisz, A. and Bresciani, F. (1993) Crit. Rev. Oncogenesis 4, 361–388.
273 Weisz, A. et al. (1990) Mol. Endocrinol. 4, 1041–1050.

[274] Lau, C.K. et al. (1991) Proc. Natl Acad. Sci. USA 88, 829–833.
[275] Ambrosino, C. et al. (1993) Mol. Endocrinol. 7, 1472–1483.
[276] Weisz, A. and Rosales, R. (1990) Nucl. Acids Res. 18, 5097–5106.
[277] Augereau, P. et al. (1994) Mol. Endocrinol. 8, 693–703.
[278] Lee, A.V. et al. (1999) Mol. Endocrinol. 13, 787–796.
[279] Perillo, B. et al. (2000) Mol. Cell. Biol. 20, 2890–2901.
[280] Wang, W. et al. (1999) Mol. Endocrinol. 13, 1373–1387.
[281] Blobel, G.A. et al. (1995) Mol. Cell. Biol. 15, 3147–3153.
[282] Schroeder, C. et al. (1993) EMBO J. 12, 951–960.
[283] Mueller, M. et al. (2000) Proc. Natl Acad. Sci. USA 97, 10972–10977.
[284] Yang, N.N. et al. (1996) Science 273, 1222–1225; 275, 1249.
[285] An, J. et al. (1999) Proc. Natl Acad. Sci. USA 96, 15161–15166.
[286] Ray, A. et al. (1994) J. Biol. Chem. 269, 12940–12946.
[287] Horlein, A.J. et al. (1995) Nature 377, 397–403.
[288] Lavinsky, R.M. et al. (1998) Proc. Natl Acad. Sci. USA 95, 2920–2925.
[289] Smith, C.L. et al. (1997) Mol. Endocrinol. 11, 657–666.
[290] Jepsen, K. et al. (2000) Cell 102, 753–763.
[291] Bocquel, M.T. et al. (1989) Nucl. Acids Res. 17, 2581–2595.
[292] Griffin, C. et al. (1998) Endocrinology 139, 4614–4625.
[293] Li, L.C. et al. (2000) Biochem. Biophys. Res. Commun. 275, 682–689.
[294] Tessier, C. et al. (2000) Endocrinology 141, 3842–3851.
[295] Thénot, S. et al. (1997) J. Biol. Chem. 272, 12062–12068.
[296] Ogawa, S. et al. (2000) Proc. Natl Acad. Sci. USA 97, 14737–14741.
[297] Ribieras, S. et al. (1998) Biochim. Biophys. Acta 1378, F61–F77.
[298] Bryant, H.U. et al. (1999) J. Steroid Biochem. Mol. Biol. 69, 37–44.

Names

Three closely related ERR genes are known. Since the ERR genes were first names using a numbering system and then by a Greek lettering system, the nomenclature is sometimes confusing. The following table should help the reader.

Species	Other names	Accession number	References
Official name: NR3B1			
Human	ERR1, ERRα, ESRL1	X51416	1
Mouse	ERRα1	U85259	3
Official name: NR3B2			
Human	ERR2, ERRβ	AF094517	6
Mouse	ERR2, ERRβ, Estrrb	X89594	2
Rat	ERR2	X51417	1, 6
Official name: NR3B3			
Human	ERRγ, ESRRG	AF058291	6, 7
Mouse	ERR3	AF117254	8

Isolation

ERR1 (NR3B1) and ERR2 (NR3B2) were the first orphan receptors to be isolated in 1988[1]. Both were obtained by screening human fetal kidney and adult heart cDNA libraries with a probe encompassing the DBD of the human estrogen receptor α. The study of these receptors was then relatively neglected until 1996. Of note, it was recently shown that the original human ERR2 cDNA described in reference 1 is in fact of rat origin[6]. ERR1 was also identified by three other groups as a regulator of SV40 major late promoter[9] of the human lactoferrin gene[10] and of the human aromatase gene[17].

ERR3 (NR3B3) was recently found by three groups. In humans, it was identified as a gene located in the critical region for Usher syndrome in chromosome 1, whereas it is apparently not directly implicated in the disease[7]. It was also identified by a bioinformatic search using EST and inverse PCR to isolate the full-length clone[6]. In the mouse, Michael Stallcup's laboratory identified the orthologous gene by a two-hybrid screen for proteins that interacts with the coactivator GRIP1[8].

ERR1, ERR2 and ERR3 exhibit high levels of sequence identity in their DBD (93–98%). Concerning the LBD, ERR2 and ERR3 exhibit 73% sequence identity, whereas ERR1 and ERR3 are less related (57%). ERR1 and ERR2 exhibit 63% sequence identity in the same domain. Interestingly, ERR2 and ERR3 also harbor high levels of sequence identity in the A/B domain (c. 60%), suggesting that the function of this domain is conserved in these two receptors[6–8]. This is not the case for ERR1. The three receptors are more strongly related to estrogen receptors (ERα and β, NR3A1 and 2) than to any other member of the NHR superfamily. They exhibit c. 68–73% sequence identity with the ERs in the C domain and c. 35–37% in the E domain.

DNA binding

ERR1 was shown by a number of teams to bind to synthetic or natural elements encompassing an extended half-site TCA AGGTCA[3, 9-15]. Of note, in a PCR-based site-selection assay, the consensus sequence TNA AGGTCA was found[15]. This was confirmed by an examination of the relative affinities of ERR1 for half-site motifs from a variety of cellular and viral promoters[12]. The TCA AGGTCA sequence is already known as the SF1 response element (SFRE). In line with this observation, ERR1 and SF1 harbor sequence similarities in the T and A boxes, which are responsible for the recognition of the 5′ extension of the half-site motif[3,13]. The DNA-binding specificities of SF1 and ERR1 on various mutations of the SFRE were compared and were shown to be distinct[14].

ERR1 was first described as binding as a monomer to the SFRE[12,13]. Nevertheless, it was shown by mixing experiments in gel-shift assays that, even on the extended half-site sequence, ERR1 binds to DNA as a homodimer[14]. The discrepancy between these reports is due to the fact that the ERR1 homodimer is very stable and requires cotranslation of the wild-type and mutated clones to be observed in mixing experiments performed by gel shift assays.

Interestingly, ERR2 was shown to bind to a different sequence, the ERE, which is also recognized by estrogen receptors[2]. In the same study, DNA binding on direct-repeat elements was tested and shown to be negative. It was also shown that ERR2 can form homodimers in solution. The apparent difference of DNA-binding specificity between ERR1 and ERR2 was in fact resolved by the demonstration that ERRs and ERs shared both the ERE and the SFRE as response elements[16]. This study revealed that both ERRs bind SFREs and EREs, and that ERR1 interacts with these elements as a dimer. Interestingly, ERRs recognized only a subset of EREs. For example, ERR1 recognizes EREs presenting the sequence AG AGGTCA NNN TGACCT CT, whereas ERR2 recognizes (C/A)G AGGTCA NNN TGACCT G(C/A). ERα but not ERβ is also able to bind to the SFRE element in addition to the classical EREs. These data were confirmed in a recent study which clearly showed the dimerization of ERR1 using the two-hybrid system[31].

ERR3 was shown to bind to the ERE[8] as well as to the SFRE[29].

Partners

In GST pull-down experiments, ERR1 was shown to interact with ERα[10,12]. In addition, it has been shown that ERR1 modulates the estrogen-mediated response of the human lactoferrin gene[10]. Mutation of the site recognized by ERR1 on the human lactoferrin promoter reduces estrogen activation of this promoter, whereas ERR1 alone is not able to support estrogen activation. An interaction between ERR1 and the general transcription factor TFIIB was also observed in GST pull-down assays[12]. The functional significance of this interaction is unknown.

ERR2 was shown by coimmunoprecipitation to interact with the hsp90 protein, as the estrogen receptor[2]. No functional data regarding the significance of this interaction have been published.

ERRs interact with the coactivators of the p160 family (SRC-1, TIF2 and p/CIP) in yeast and mammalian cells as well as *in vitro* (GST pull-down)[8,18,31]. Recently, ERR1 was found to interact in an AF-2-dependent manner with a newly discovered coactivator called PNRC (proline-rich nuclear receptor coregulatory protein)[32]. The interaction with this coactivator effectively enhances the transcriptional activity of ERR1.

Ligands

No ligand is known for ERRs. Some groups report the absence of effect of estrogen or estrogen-related molecules on ERR activity[1,14]. Several authors propose that they are constitutively active receptors that do not require a ligand (i.e. a true orphan)[10,18]. Of note, the constitutive transcriptional activity is transferable to a heterologous receptor, such as the glucocorticoid receptor, by the transfer of the LBD of ERRs[18,19]. This suggests that the LBD of ERRs is in an intrinsically activated state.

In contrast, other reports suggest that ERRs may have a ligand that is still to be identified. It has for example been shown that the transcriptional activity of ERR1 and ERR2 requires normal serum[14]. If the transcriptional activity of ERRs is assayed in cells cultivated in charcoal-treated serum (i.e. serum in which small hydrophobic molecules such as steroids or retinoids are depleted), the transcriptional activity is barely detectable. In this test, the ERRs behave with the same kinetic as the ERs. Nevertheless, in contrast to the ERs, the addition of estradiol or of an antagonist such as ICI 164,384 does not modify the activity of the ERRs. These data are only an indication that a ligand could exist. There are many other possible explanations to this effect, such as, for example, the phosphorylation of the receptor through the activation of a kinase cascade induced by an hydrophobic component of the serum binding to a membrane receptor.

Consistent with the ligand hypothesis, two organochloride pesticides, toxaphene and chlordane, were recently proposed to be antagonists of ERR1[20]. The assay was conducted both in yeast cells and in the SK-BR-3 breast cancer cell line, both with native receptor or with a GAL4 fusion protein. The repressive effect on ERR1 transcriptional activity was detected using a synthetic response element or, on the human aromatase promoter, an ERR1 target gene[17]. The effect of toxaphene was visible at a concentration of 0.1 μM and a maximal concentration of 10 μM was tested. Chlordane was only active at 10 μM. No direct binding was demonstrated. These findings may suggest that as CARβ (NR1I4) ERRs are active without ligand and that ligand promotes the dissociation of coactivators and a decrease of the transcriptional activity of the receptor. However, it is too early to draw any firm conclusions from these findings based on two synthetic compounds acting at high concentration.

Three-dimensional structure

The 3D structure of the DBD of ERR2 in solution was determined by NMR spectroscopy. The structure of the domain from Asn 96 to Ser 194 was shown to be extremely similar to that of other DBDs of NHRs, such as ERα or GR. Two helices were located in positions homologous to other receptors: one in the P-box from Glu 121 to Gly 133, and the other in the second zinc finger from Gln 156 to Gly 167. Interestingly, the carboxy-terminal extension that in RXRα forms a third helix even in solution is completely unstructured in ERR2. In classic monomer-binding orphan receptors that recognize extended half-sites, this region is predicted from functional studies to be important for the discrimination of the sequence of the 5'-extension of the response element. The structural basis of this effect remains thus unknown.

No structure for the LBD is available.

Expression

ERR1 is expressed in both human and rodents as 2.2–2.6 kb mRNAs expressed in a wide variety of tissues[1]. In northern blots of adult rodent tissues, a large amount was detected in cerebellum, hippocampus and kidney. Moderate levels were found in gut, heart, hypothalamus, liver and lung[1,3]. By Western blot analyses, multiple forms of ERR1 proteins were detected in adult mouse liver, kidney, brain and heart. A major 53 kDa band was observed in the reproductive tissues (uterus, vagina, cervix)[3]. Another study reported two products present in HeLa cells: a major 53 kDa species and a minor 58 kDa one[12].

Three studies described the expression of ERR1 during mouse embryogenesis using *in situ* hybridization[13,15,22]. Consistent with the complex expression pattern of this gene, these studies focused on different aspects of ERR1 expression. Vincent Giguère's group stressed the persistent high level of ERR1 expression in brown adipose tissue and intestinal mucosa[15]. The same study reported an induction of ERR1 expression during adipose differentiation, using the classical NIH 3T3-L1 pre-adipocytes and HIB cells as a model of brown fat differentiation. In addition, this report described the regulation by ERR1 of the medium-chain acyl coenzyme A deshydrogenase (MCAD) gene, which encodes an enzyme involved in the mitochondrial β-oxidation of fat. Another study also showed the parallel regulation of ERR1 and MCAD during adipocyte differentiation[26]. Bonnelye et al. pointed out an expression of ERR1 during bone development[13]. Interestingly, both for endochondral and intramembranous bone differentiation, ERR1 expression coincides with a period of highly active ossification. In bone tissue, the expression of ERR1 was detected in osteoblasts and not in osteoclasts. These authors also describe the osteopontin gene (which encodes a protein expressed by osteoblasts and release in the bone matrix), as a direct target of ERR1. These two studies are not contradictory and simply confirm that ERR1 has a function in several differentiation systems. This has been confirmed in the study reporting the complete expression pattern of ERR1 in mice[22]. The expression starts at very early stages in ES cells and at 8.5 days post-coitum in the yolk sac. Later, the gene exhibits a complex expression pattern in the nervous system, the muscles, the epidermis and several endodermal derivatives, such as the epithelium of intestine and urogenital system. In several differentiation systems, such as muscle (but also in the peripheral nervous system), a burst of ERR1 expression coincides with the post-mitotic status of the cells. The *in vivo* significance of this observation, which may suggest a connection of ERR1 function with the regulation of the cell cycle, is still unknown.

By contrast, the expression pattern of ERR2 is much more specific. In adult tissue, ERR2 expression was detected as a 4.8 kb transcript expressed at very low levels in kidney and heart[1]. A low expression in rat supraoptic nucleus was also detected by RT-PCR[23]. *In situ* hybridization studies reveal a prominent expression during a narrow developmental window in trophoblast progenitor cells between days 6.5 and 7.5 post-coitum[2,24]. These extra-embryonic cells are implicated in placental formation, consistent with the knockout phenotype[24]. After 8.5 days the expression is not detectable. This is interesting to correlate with the early expression of ERR1, since it is apparently the only situation (with ES cells and adult kidney) in which both receptors may be coexpressed.

ERR3 expression was mainly studied by Northern blot and RT-PCR[6-8,29]. This shows that, despite its close sequence similarity with ERR2, ERR3 harbors an expression pattern as complex as that for ERR1. Expression as a 5.5 kb transcript was found in a variety of human and mouse adult tissues, such as brain, kidney, testis, lung, adrenal gland, pancreas, placenta or bone marrow[6-8,29]. The different studies are not in accordance for the expression in heart but this may be linked to the differential expression of transcripts coding for specific isoforms[7,8,29]. Hong et al. reported an absence of expression in mouse embryos 7 days post coitum and a high expression at 11, 15 and 17 days[8]. This suggests that, in contrast to ERR1 and 2, ERR3 is not expressed during early embryogenesis. A recent report using *in situ* hybridization showed a complex expression pattern of ERR3 in the adult mouse brain[30].

Activity and target genes

Although the question was unclear for a while[9,15], it is now demonstrated that ERRs are potent transcriptional activators[13,14,19]. The fact, whether they are constitutively active is still a matter of discussion[14,18]. It has been recently demonstrated that all three ERRs can interact with the coactivators of the p160 family (SRC-1, TIF2 and p/CIP) in yeast and mammalian cells as well as *in vitro* (GST pull-down)[8,18]. In the coactivators, ERRs interact with the NR boxes II and III that harbor a LXXLL motif[8,18]. The interaction with the coactivators increases the transcriptional activation mediated by ERRs[8,18,20]. The interaction with the coactivator is AF-2 dependent, as is the transcriptional activation itself, suggesting that most of the activity of ERRs is carried out through AF-2-dependent mechanisms[14,18,20]. Several data suggest that the AF-1 function is not required for the activity of ERR1[14,16,18].

Numerous target genes of ERRs were described and apparently ERRs can modulate positively and negatively the expression of their targets. It has been shown that ERR2 can repress glucocorticoid receptor transcriptional activity by an unknown mechanism[25]. Such an effect may explain negative modulation of transcription in certain promoter context. The first ERR target to be described was the major late promoter (MLP) of the SV40 virus that is repressed by ERR1[9,11,12]. ERR1 was purified from HeLa cells nuclear extract as a factor binding to the MLP that represses its activity both in COS-1 cells and in *in vitro* cell-free transcription assay[12]. Another described target was the human lactoferrin gene on which ERR1 modulates an estrogen receptor-mediated response[10]. The promoter of this gene contains an imperfect ERE that is alone unable to sustain a high transactivation by ER. In order to reach a complete estrogenic response, another element called FP1, which is recognized by ERR1, is required. Thus, on this promoter, ERR1 acts synergistically with the estrogen receptor in order to reach maximal induction. Interestingly, if the imperfect ERE is transformed into a perfect ERE, the action of ERR1 became unnecessary to reach a maximal estrogen induction. These data thus suggest that ERR1 can, probably by a direct interaction with ER that was demonstrated in the same study, synergize with estrogen receptor, specifically in some promoter context[10].

From these data, it was difficult to conclude on the ability of ERR1 to regulate positively or negatively transcription. This was even more complicated by the finding that ERR1 was able to bind to an element, called NRRE-1, present in the promoter of the human medium-chain acyl coenzyme A dehydrogenase gene[15,26], but was apparently unable to activate its transcription[15]. In the same study, ERR1 was shown to repress retinoic-acid induction of the MCAD gene, which is mediated through the same NRRE-1 element[15]. Nevertheless, consistent with its action on synthetic elements, several other teams described genes positively regulated by ERR1. These genes, which all contain classic SFRE elements, are the osteopontin gene[13,28], the human thyroid hormone receptor α gene[27] or the human aromatase gene[17]. In all these cases, the expression pattern of ERR1 in osteoblasts, heart, testis, brown fat or during breast development correlate well with the identified target genes.

Knockout

The knockout of ERR2 was conducted by Vincent Giguère's laboratory and was found to result in an embryonic lethal phenotype[24]. The homozygous mutant embryos have severely impaired placental formation and die at 10.5 days post-coitum. The mutants display abnormal chorion development. This is consistent with the early expression pattern of ERR2[2,24]. This phenotype can be rescued by aggregation of ERR2 homozygous mutants with tetraploid wild-type cells, which contribute exclusively to extra-embryonic tissues. In such experiments, the embryos at 12.5 days post-coitum were phenotypically indistinguishable from their wild-type littermates. No information is available on the phenotype of these rescued embryos at birth. These results indicate that ERR2 has an important role in early placentation but the roles of ERR2 in adult, if any, remain unclear.

Associated disease

There is no associated disease for ERR. ERR3 was shown to be located in the critical region for Usher's syndrome but the gene is apparently not mutated in 70 tested patients with Usher's syndromes[7].

Gene structure, promoter and isoforms

The human ERR1 gene consists of seven exons and six exons, and occupies 20 kb of genomic DNA[5]. ERR1 is thus a much smaller gene than ERα, which spans more than 140 kb. The two zinc fingers are encoded by two adjacent exons and share splice patterns identical to those of the estrogen receptors, consistent with the high ERR1–ER relatedness. In contrast, the LBD of ERR1 contains one less intron than ERα[5].

In the same study, the promoter of the human ERR1 gene was cloned and the start sites of transcription mapped[5]. This promoter is a GC-rich promoter with multiple start sites. No TATA or CAAT boxes were present, and the only putative sites for transcription factors that were noticed in the 600 bp upstream of the major start site were Sp1 binding elements and two E boxes. Interestingly, a regulation of ERR1 expression by estrogen in mouse uterus was demonstrated[3]. On the reported fragment of the human ERR1 promoter no ERE was found, suggesting that estrogen regulation takes place in another region of the gene or uses another DNA element[5].

Although no definitive results are available yet, several observations suggest that at least two N-terminal isoforms exist for the ERR1 gene[5,10,12,15]. ERR3 also contains three N-terminally different isoforms in mouse and human because of alternative splicing events, which may be due to alternative promoter usage[6-8,29]. The long isoform contains 23 supplementary amino acids that are located upstream of the ATG used in the short isoform. The functional differences between these isoforms are unknown but it is interesting to recall that the A/B region of ERR3 and ERR2 are exceptionally well conserved for NHR members (60% identity), suggesting an important function of this domain. Specific expression patterns for the three human isoforms were observed both in adult and fetal tissues[29].

Chromosomal location

The chromosomal location of human ERR1 and ERR2 genes are respectively 11q12–13 and 14q24.3[4-6]. The 14q24.3 region is interesting since it contains numerous loci putatively implicated in Alzheimer's disease and diabetes. Human ERR3 was localized on 1q41 in the critical region for Usher's syndrome[7]. Of note, a processed pseudogene (accession number U85258) of human ERRα is present in chromosome 13q12.1[4].

Amino acid sequence for human ERR1 (NR3B1)

Accession number: X51416

```
  1  MGLEMSSKDSPGSLDGRAWEDAQKPQSAWCGGRKTRVYATSSRRAPPSEGTRRGGAARPEEAAEEGPPAA
 71  PGSLRHSGPLGPHACPTALPEPQVTSAMSSQVVGIEPLYIKAEPASPDSPKGSSETETEPPVALAPGPAP
141  TRCLPGHKEEEDGEGAGPGEQGGGKLVLSSLPKRLCLVCGDVASGYHYGVASCEACKAFFKRTIQGSIEY
211  SCPASNECEITKRRRKACQACRFTKCLRVGMLKEGVRLDRVRGGRQKYKRRPEVDPLPFPGPFPAGPLAV
281  AGGPRKTAAPVNALVSHLLVVEPEKLYAMPDPAGPDGHLPAVATLCDLFDREIVVTISWAKSIPGFSSLS
351  LSDQMSVLQSVWMEVLVLGVAQRSLPLQDELAFAEDLVLDEEGARAAGLGELGAALLQLVRRLQALRLER
421  EEYVLLKALALANSDSVHIEDEPRLWSSCEKLLHEALLEYEAGRAGPGGGAERRRAGRLLLTLPLLRQTA
491  GKVLAHFYGVKLEGKVPMHKLFLEMLEAMMD 521
```

Amino acid sequence for human ERR2 (NR3B2)

Accession number: AF094517

```
  1 MSSDDRHLGSSCGSFIKTEPSSPSSGIDALSHHSPSGSSDASGGFGLALGTHANGLDSPPMFAGAGLGGT
 71 PCRKSYEDCASGIMEDSAIKCEYMLNAIPKRLCLVCGDIASGYHYGVASCEACKAFFKRTIQGNIEYSCP
141 ATNECEITKRRRKSCQACRFMKCLKVGMLKEGVRLDRVRGGRQKYKRRLDSESSPYLSLQISPPAKKPLT
211 KIVSYLLVAEPDKLYAMPPPGMPEGDIKALTTLCDLADRELVVIIGWAKHIPGFSSLSLGDQMSLLQSAW
281 MEILILGIVYRSLPYDDKLVYAEDYIMDEEHSRLAGLLELYRAILQLVRRYKKLKVEKEEFVTLKALALA
351 NSDSMYIEDLEAVQKLQDLLHEALQDYELSQRHEEPWRTGKLLLTLPLLRQTAAKAVQHFYSVKLQGKVP
421 MHKLFLEMLEAKAWARADSLQEWRPLEQVPSPLHRATKRQHVHFLTPLPPPPSVAWVGTAQAGYHLEVFL
491 PQRAGWPRAA 500
```

Amino acid sequence for human FRR3 (NR3B3)

Accession number: AF058291

```
  1 MSNKDRHIDSSCSSFIKTEPSSPASLTDSVNHHSPGGSSDASGSYSSTMNGHQNGLDSPPLYPSAPILGG
 71 SGPVRKLYDDCSSTIVEDPQTKCEYMLNSMPKRLCLVCGDIASGYHYGVASCEACKASFKRKIQANIEYS
141 CPATNECEITKRRRKSCQACRFMKCLKVGMLKEGVRLDRVRGGRQKYKRRIDAENSPYLNPQLVQPAKKP
211 YNKIVSHLLVAEPEKIYAMPDPTVPDSDIKALTTLCDCADRELVVIIGWAKHIPGFSTLSLADQMSLLQS
281 AWMEILILGFVYRSLSFEDELVYADDYIMDEDQSKLAGLLDLNNAILQLVKKYKSMKLEKEEFVTLKAIA
351 LANSDSMHIEDVEAVQKLQDVLHEALQDYEAGQHMEDPRRAGKMLMTLPLLRQTSTKAVQHFYNIKLEGK
421 VPMHKLFLEMLEAKVC 436
```

References

1   Giguère, V. et al. (1988) Nature 331, 91–94.
2   Pettersson, K. et al. (1996) Mech. Dev. 54, 211–223.
3   Shigeta, H et al. (1997) J. Mol. Endocrinol. 19, 299–309.
4   Sladek, R et al. (1997) Genomics, 45, 320–326.
5   Shi, H. et al. (1997) Genomics, 44, 52–60.
6   Chen, F. et al. (1999) Gene, 228, 101–109.
7   Eudy, J.D. et al. (1998) Genomics, 50, 382–384.
8   Hong, H. et al. (1999) J. Biol. Chem. 274, 22618–22626.
9   Wiley, S.R. et al. (1993) Genes Dev. 7, 2206–2219.
10 Yang, N. et al. (1996) J. Biol. Chem. 271, 5795–5804.
11 Zuo, F. and Mertz, J.E. (1995) Proc. Natl Acad. Sci. USA, 92, 8586–8590.
12 Johnson, S.D. et al. (1997) Mol. Endocrinol. 11, 342–352.
13 Bonnelye, E. et al. (1997) Mol. Endocrinol. 11, 905–916.
14 Vanacker, J-M. et al. (1999) Mol. Endocrinol. 13, 764–773.
15 Sladek, R. et al. (1997) Mol. Cell. Biol. 17, 5400–5409.
16 Vanacker, J-M. et al. (1999) EMBO J. 18, 4270–4279.
17 Yang, C. et al. (1998) Cancer Res. 58, 5695–5700.
18 Xie, W. et al. (1999) Mol. Endocrinol. 13, 2151–2162.
19 Lydon, J.P. et al. (1992) Gene Expression 2, 273–283.
20 Yang, C and Chen, S. (1999) Cancer Res. 59, 4519–4524.
21 Sem, D.S. et al. (1997) J. Biol. Chem. 272, 18038–18043.
22 Bonnelye, E. et al. (1997) Mech. Dev. 65, 71–85.
23 Lopes da Silva, S. et al. (1995) Endocrinology 136, 2276–2283.
24 Luo, J. et al. (1997) Nature 388, 778–782.
25 Trapp, T. and Holsboer, F. (1996) J. Biol. Chem. 271, 9879–9882.

26 Vega, R.B. and Kelly, D.P. (1997) J. Biol. Chem. 272, 31693–31699.
27 Vanacker, J-M. et al. (1998) Oncogene 17, 2429–2435.
28 Vanacker, J-M. et al. (1998) Cell Growth Diff. 9, 1007–1014.
29 Heard, D.J. et al. (2000) Mol. Endocrinol. 14, 382–392.
30 Lorke, D.E. et al. (2000) Mol. Brain Res. 77, 277–280.
31 Zhang, Z. and Teng, C.T. (2000) J. Biol. Chem. 275, 20837–20846.
32 Zhou, D. et al. (2000) Mol. Endocrinol. 14, 986–998.

Names

GR is a steroid receptor highly related to the mineralocorticoid receptor. It binds glucocorticoids, the hormones implicated in a large number of physiological processes in fully differentiated non-dividing cells, most notably carbohydrate and lipid metabolism as well as stress responses and inflammation. GR also binds mineralocorticoids that are involved in the control of salt and water homeostasis. GRs are known from a wide variety of vertebrates but are unknown outside this phylum. The receptor is a large protein of 777 amino acids (GRα) in human.

Species	Latin name	Other names	Accession number	References
Official name: NR3C1				
Human	*Homo sapiens*	GR	X03225	1, 2
Ma's night monkey	*Aotus nancymaae*	GR	U87952	257
Cotton-top marmoset	*Saguinus oedipus*	GR	U87953	175
Squirrel monkey	*Saimiri boliviensis*	GR	U87951	257
Squirrel monkey	*Saimiri sciureus*	GR	AF041834	Unpub.
Mouse	*Mus musculus*	GR	X04435	260
Rat	*Rattus* sp.	GR	M14053	261
Guinea pig	*Cavia porcellus*	GR	L13196	3
Sheep	*Ovis aries*	GR	X70407	5
Pig	*Sus scrofa*	GR	AF141371	Unpub.
Tupaia	*Tupaia belangeri*	MR	Z75079	Unpub.
African clawed frog	*Xenopus laevis*	GR	X72211	262
Rainbow trout	*Oncorhynchus mykiss*	GR	Z54210	4
Tilapia	*Oreochromis mossambicus*	GR	D66874	6
Japanese flounder	*Paralichthys olivaceus*	GR	AB013444	Unpub.

Isolation

The glucocorticoid receptor was purified using radiolabeled high-affinity glucocorticoid analogues, such as dexamethasone[7,8]. This allowed the production of specific antibodies that facilitated the isolation of clones containing human or rat glucocorticoid cDNAs[9,10]. Further analysis of these clones led to the isolation of complete human GR cDNA[1,2]. Interestingly, two different forms, GRα and GRβ, differing in their C-terminal parts were isolated (see below). The sequence was the first complete one for a member of the nuclear receptor superfamily to be published and it reveals a strong level of identity with the product of the v-erbA oncogene[2]. It contains a Cys-rich domain that harbors 44% identical amino acid to the corresponding region of v-erbA and which was proposed as the DBD. The protein produced *in vitro* by GRα but not GRβ binds radiolabeled triamcinolone acetonide, a synthetic glucocorticoid[1]. As for the biochemically purified receptor, this binding can be competed out by cold dexamethasone as well as by cortisol. Progesterone was shown to be able to compete for the binding but not estradiol or testosterone[1].

GR cDNAs were found in a number of other species by screening cDNA libraries using the human GR cDNA or by RT-PCR using GR-specific primers (see references in the table above).

In the DBD, GR exhibits 92%, 89% and 78% sequence identity with MR, PR and AR, respectively. These values are 56%, 54% and 49% in the LBD.

DNA binding

Studies performed with biochemically purified glucocorticoid receptors have shown that GR binds to specific DNA sequences and have allowed several target genes such as the MMTV[11,12] or metallothionein[13,14] to be defined. It was shown that all these target genes share a core hexamer sequence AGAG/ACA that is often present in two palindromic copies[15-17]. This element is also recognized by other steroid receptors such as PR but not by ER[17]. This may result in cross-talk between these various steroid receptors, especially when a mutated version of the receptors exerts a dominant negative activity as in some hormone resistance syndromes[70]. The basis of the specificity of the GRE has been explored and it has been shown that substitution of 1 or 2 bp can change a ERE toward a GRE[18,19].

The analysis of the functional domains of GR led to the demonstration that the Cys-rich domain is effectively the DBD but also contains a transactivation function[20-23,179]. The DBD confers specific recognition of the GRE versus the ERE. This has been demonstrated by 'finger switch' experiments during which a chimeric ER, in which the DBD was replaced by the one of GR, was found to bind a GRE and activate transcription in an estradiol-dependent manner[24]. The cysteine residues of the DBD are critical for the DNA-binding activity of the receptor and each of the two zinc fingers effectively coordinates a Zn ion[28]. In fact, the DBDs of all nuclear receptors contain nine Cys residues. It was shown that the last one (Cys 500 of the rat GR) is not involved in zinc finger formation and is dispensable for DNA-binding activity[27]. In addition, two Cys residues, Cys 476 and Cys 482, which are located in the N-terminal part of the second finger can be replaced by His without completely abolishing GR activity[27]. The determinants for the discrimination between GRE and ERE (or between TREpal and GRE) were shown to be localized in the C-terminal part of the first zinc finger, a region that was called the P-box, which forms a recognition helix that interacts with the major groove of the DNA double helix[25,26,29,30]. In the 3D structure of the DBD, this region has been found to form an α-helix, which directly contacts DNA and establishes specific contacts within the GRE sequence[26,31,32]. GR binds to DNA as a dimer and the interface between the two monomers has been genetically mapped in the second zinc finger and more precisely in the D-box, which is located between the two first Cys residues of the second zinc finger[25,29,30,38]. A specific point mutation, A458T, has been shown to block dimer formation specifically[250]. This mutant has been used to generate specific GR knockout mice that allows the DNA-binding-dependent and -independent mechanisms in the GR *in vivo* function to be separated (see below). By measuring the *in vitro* affinity of GRE binding by homodimers and monomers it was found that homodimer formation is the rate-limiting step for high-affinity DNA binding of GR[40]. In addition, since there are often multiple GREs in the promoters of target genes, it has been shown that GR binds cooperatively to adjacent recognition sites[39] (see references 33, 34, 36 and 37 for reviews on GR DNA binding).

GR regulation of the MMTV LTR transcriptional activity has been used as a model system to study how nuclear receptors bind to DNA and how the receptors may activate transcription in target sequences organized into chromatin (see references 43, 67 and 82 and references therein). GR can recognize the MMTV GREs when organized in a positioned nucleosome *in vitro*[41,42]. The MMTV LTR contains several GREs as well as other sites for transcription factors, such as NF-1. An influence of the position of the GRE within a nucleosome was found on GR affinity for the GRE[48]. A synergism exists between NF-1 and GR molecules bound in the GRE immediately upstream of the NF-1 site[44,45]. Conventional mechanisms invoking cooperative DNA binding are insufficient to explain that synergism, since both proteins compete for binding to the naked promoter[45]. It was shown that positioned nucleosomes play an active role in induction by mediating cooperative binding and functional synergism between GR and NF-1[46]. It is known that nucleosomes have a repressive effect on gene transcription. Using the MMTV promoter system it was shown that hormone activation induces nucleosome positioning *in vivo*[47]. Using glucocorticoid regulation of MMTV as a model system, it was possible functionally to link chromatin remodeling and transcriptional initiation[253].

The location of GR within the cells in the absence of its ligand has been controversial for a number of years[52,53]. It has been shown that, in the absence of hormone, GR is complexed to heat shock proteins (Hsp) and additional components of the molecular chaperone machinery in the cytoplasm (reviewed in references 49, 50 and 258; see below). Other studies using confocal microscopy of overexpressed GR have found the receptor nuclear in the absence of hormone[53]. When hormone is present, the receptor enters the nucleus but bidirectional transport of the receptor occurs, suggesting that the localization of the receptor within the cells is dynamic and that an export signal exists within the receptor[54,56]. The same dynamic behavior exists for the progesterone receptor (see entry for PR).

Within the nucleus, the binding to chromatin also appears to be an extremely dynamic event since it was observed that the hormone-occupied GR undergoes a rapid exchange between chromatin and the nucleoplasmic compartment[281].

Two different nuclear localization signals (NLS) were mapped by mutagenesis experiments[51]. The first, NL1, maps to a 28-amino-acid region just C-terminal to the C domain and contains basic amino acids such as the NLS of the SV40 T antigen. The DNA-binding activity of the receptor is apparently important to maintain it in a stable fashion in the nucleus when the hormone is present[55]. This signal mediates the interaction of GR with importin α[56]. The second NLS, NL2, is located within the LBD but was not precisely mapped, and this NLS alone is sufficient to promote hormone-dependent nuclear localization. This NL2 signal mediates a much slower transport of GR into the nucleus and is

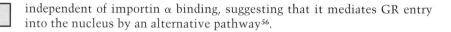

independent of importin α binding, suggesting that it mediates GR entry into the nucleus by an alternative pathway[56].

Partners

The unliganded GR forms a complex with heat-shock proteins and other chaperone molecules (see references 49 and 50 for reviews). The use of anticorticosteroids such as RU486 has shown that Hsp90 maintains the GR in a non-DNA-binding form[136]. The complex contains numerous other proteins such as Hsp56, which is an immunophilin that is able to bind immunosuppressive compounds such as rapamycine and FK506[147] or the Hsp40 protein[251]. This large heterocomplex can now be reconstituted at least partially *in vitro* and the role of each protein in complex formation can be studied (see e.g. reference 251). It has been shown that FK506 and rapamycine can potentiate nuclear translocation of GR[148]. In yeast, reduced levels of hsp90 have been shown to compromise GR action[259]. It is thought that hormone binding brings about a conformational change, which releases Hsp90, Hsp56 and other associated proteins that are found in the non-activated receptor complex and results in a transcriptionally active receptor. It has been suggested that the release of these heat-shock proteins alone is sufficient to activate the receptor[143]. However, partial agonists bind to the receptor, dissociate the non-activated complex and yet bring about transcriptional responses that are only a fraction of the response of the full agonists, suggesting that a hormone-induced conformational change separate from that required for the release of heat-shock proteins is required for full transcriptional activity.

In the nucleus, GR also interacts to calreticulin, a calcium-binding protein that is known to bind to the synthetic peptide KLGFFKR that is similar to the region immediately adjacent to the first zinc finger of GR (CKVFFKR)[97,98]. Through its binding to GR DBD, calreticulin inhibits GR DNA binding and thus glucocorticoid signaling[97]. These effects were also observed for AR and RAR[98]. GR also binds to the insulin-degrading enzyme, a metalloendoprotease that is implicated in the intracellular degradation of insulin[99]. In contrast to calreticulin, this enzyme enhances the DNA-binding activity of GR and AR.

GR binds to its sequence as an homodimer but it is also able to heterodimerize with MR, leading to synergy or inhibition of transcription depending on the promoter or the cellular context[68,69].

GR has been found to interact with a large number of transcription factors and it is beyond the scope of this book to describe these interactions exhaustively. Beside the MMTV promoter discussed above, the promoter of the tyrosine amino transferase (TAT) gene has been used to analyse synergistic action of GR with transcription factors[86]. Since these first descriptions many other cases have been described and it became clear that GR has deployed a large number of strategies to inhibit or synergize functionally with transcription factor signaling pathways, such as AP-1[71–74,107], NFκB[75–78] or Stat5[79,100,101]. It is not always easy to decipher the mechanisms by which GR interferes with other transcription factors. In some cases these functional interactions involve physical interaction but this does not always happen. Interested readers can find more information in the following reviews: references 80–85, 208. We will briefly describe below the principal findings regarding NFκB and AP-1 cross-coupling with GR.

The AP-1–GR cross-coupling was one of the first cases of mutual antagonism between a nuclear receptor and a classical transcription factor to be discovered[71-74,82,83,107]. A prominent medical usage of glucocorticoids is to inhibit inflammatory processes and cell proliferation. Examples of genes activated in these processes are the metalloprotease, such as collagenase and stromelysine, which are activated by TPA or cytokines via AP-1 signaling[83]. It has been shown that AP-1 binds and activates the collagenase-1 promoter via a specific *cis*-acting sequence that binds the Fos–Jun heterodimer[108]. This sequence is the target for repression of collagenase expression by glucocorticoids, although GR is unable to bind to it[71,72]. Similar results were obtained with reporter genes containing synthetic response elements or with other target genes and shown to be mutual[73,74,107]. Many other genes or physiological processes in which this cross-coupling mechanism occurs have been found (see references 111–114 for specific examples). Interestingly, the antagonist RU486 inhibited repression of AP-1 activity by dexamethasone[83]. Several authors found that GR interact specifically with AP-1, suggesting that the mutual antagonism may occur through a direct physical interaction[71,72,74,109]. Nevertheless, the *in vivo* significance of this interaction appears dubious (see reference 273 for a discussion). Interestingly, GR antagonizes AP-1 action without altering AP-1 site occupation *in vivo*[110]. Fos was the apparent target of GR in the Fos–Jun heterodimer[107]. The regions of the receptors that are required for the cross-coupling involved the DBD and the C-terminal LBD[72,73,107]. The DBD appears to play the major role in that process, since most mutations in that domain result in loss of repressor activity *in vivo*. In addition to the direct interaction, other mechanisms have recently been proposed to explain the AP-1–GR cross-coupling. One of these mechanisms was based on the observation that both AP-1 and GR bind to the cointegrator CBP/P300. Mutual transrepression also occurs at least in part by the sequestration of this common partner[115,116]. It has also been proposed that hormone-activated GR prevents c-Jun phosphorylation by JNK and consequently AP-1 activation[117].

The first data concerning NFκB and GR described a physical interaction between the p65 subunit of NFκB and GR that results in a functional antagonism[75,77]. This antagonism provides a molecular basis for the potent anti-inflammatory and immunosuppressive effects of glucocorticoids[85]. A large number of NFκB target genes that are implicated in inflammatory and immunological response among which IL6, IL8, IL1β, GM-CSF, ICAM-1, cyclooxygenase-2 or RANTES have been shown to be suppressed by glucocorticoids[75-78,85] (see reference 208 for a review). The repression of these genes by glucocorticoids has been shown to depend on NFκB sites within their promoters. The observed repression is mutual, since NFκB can inhibit GR-mediated transactivation on GRE-containing reporters (see reference 102 for an example). This effect was also observed for AR, PR and ER but was not mutual for PR and ER since these two receptors have no effects on p65 function[102]. The interaction between GR and p65 requires the

DBD of GR but not its DNA-binding activity, since a point mutation in the DBD that changes GR specificity does not alter GR–p65 interaction[77]. The repressive action of GR on NFκB signaling was further mapped within GR DBD and the second zinc finger was shown to be required for this activity[105]. The mutation of Arg 488 or Lys 490 of rat GR led to a GR that is unable to inhibit NFκB action. Other regions of GR are also required for this antagonism[102]. Recently, Keith Yamamoto's laboratory proposed that the interaction between NFκB and GR did not inhibit NFκB DNA binding nor the assembly of the RNA polymerase II pre-initiation complex on NFκB-regulated promoters but, rather interfered with phosphorylation of the C-terminal extension of the RNA polymerase II which is necessary for transcriptional activation of these promoters[282]. Nevertheless, an entirely different mechanism of GR inhibition of NFκB signaling based on transcriptional activation of the IκB gene by GR was also proposed[103,104]. IκB is a cytoplasmic molecule that retains NFκB in the cytoplasm and an increase of IκB levels in the cell may effectively inhibit NFκB action. Nevertheless, this model does not account for all the observed effects and IκB stimulation by GR was not found in cellular systems in which GR-inhibition of NFκB signaling occurs[85,102,106]. For example IκB induction does not explain how GR mutants that are unable to activate transcription through GREs are still able to inhibit NFκB action[77]. Since ligands of nuclear receptors that selectively induce AP-1 transrepression are known, compounds that may specifically induce inhibition of NFκB signaling while being inactive in transactivation are also under development[85]. Unfortunately, the classic GR antagonists are not active in that process: GR-mediated transrepression of NFκB was only partly induced by RU486[77,78,106], whereas the pure antagonist ZK98299 is unable to repress NFκB transactivation through GR[105]. Recently, it was proposed that GR represses NFκB-driven genes by interfering with the interactions existing between p65 and the basal transcription machinery and not by the squelching of common coactivators[254].

Besides these three classic cases, other transcription factors that functionally interact either positively or negatively with GR include Oct[87-89], CACCC box factors[90], ETS factors like PuI[91,264], GATA1[92], HNF3[93], the retinoblastoma protein[94], HIV-1 vpr[95,266] as well as the orphan nuclear receptors COUP-TF and HNF4[96] among others. The in vivo significance of the GR–Oct protein–protein interaction has been studied during early zebrafish development[169]. Recently, a direct interaction between GR and the Smad3 protein implicated in TGFβ signaling was reported. This results in a negative cross-talk between glucocorticoid and TGFβ signaling pathways[249].

GR also interacts with molecules that mediate their direct effect on transcription (see references 118 and 119 for reviews). GR interacts with the coactivators of the p160 family, such as SRC-1, TIF2 and AIB1, in a ligand-dependent fashion[120,121,146]. This interaction requires the NR box motifs LXXLL as well as another region known as the auxiliary nuclear receptor interacting domain (NIDaux)[269]. The role of coactivators of the p160 family in regulating GR activity in presence of either agonists or antagonists has been studied[146]. The accessory factors that regulate the PEPCK gene together with GR are all able to recruit SRC-1. Interestingly, the binding site of one of these accessory factors in the PEPCK gene promoter can be replaced by a GAL4 binding site if a GAL4–SRC-1 fusion protein is introduced in the cell, suggesting that the main role of these accessory factors is to increase the amount of SRC-1 molecules around the promoter[263].

Another important coactivator for GR is the CBP/p300 cointegrator and it has been suggested that this interaction plays an important role in AP-1/GR cross-coupling[115,116]. No specific interaction was observed between GR and TBP associated factors[118], but a competition between TBP and GR for binding to DNA has been linked to the GR-mediated repression of the osteocalcin gene[242]. GR interacts also with chromatin remodeling proteins, such as members of the SWI/SNF complex (see reference 124 for a review) and the SAGA complex, which exhibits histone acetyl transferase activity[255]. The human homologue of the *drosophila* gene Brahma, HBRM, potentiates the transcriptional activation by GR and this enhancement requires GR DBD[123]. GR recruits the ligand-dependent nucleosomal remodeling activity of the SWI/SNF complex in yeast[125]. Other studies have revealed that, when GR is prevented from binding to human Brahma homologues, it can activate transcription from naked DNA templates but not from organized chromatin templates[122]. GR also interacts with TIF1β to induce expression of the α1-acid glycoprotein gene[126]. The precise mechanistic consequences of this interaction are not yet understood. A fascinating example of a specific coactivator is the steroid receptor RNA activator (SRA) that was isolated in a yeast two-hybrid screen using the amino-terminal A/B domain of PR-A isoform[127]. When overexpressed in mammalian cells, SRA specifically enhances steroid receptor AF-1-mediated transactivation, including that of GR. Strikingly, SRA appears to function as an RNA transcript, not a protein. It has been suggested that SRA is present in an SRC-1 complex that is recruited by steroid receptors *in vivo*. It has been proposed that SRA functions to confer a specificity to steroid-mediated transcription, via an interaction with AF-1 function[127]. Other coactivators known for GR include the L7/SPA protein, which interacts with the hinge region[128], TLS[129], and the ASC-2 factor, which also interacts with CBP/P300 and SRC-1[130]. RIP140 also associates with GR in a ligand-dependent manner and modulates either positively or negatively, depending on the assay used, the glucocorticoid-dependent transactivation[265].

Relatively little is known about the interaction between GR and corepressors[118]. It has been proposed that antagonists of GR, such as RU486, may induce active repression by promoting the association between corepressors and steroid receptors, such as PR, ER or GR (see reference 118 for a review). The PR bound to RU486 has been studied as a model system for this interaction and GR was not studied in such great detail (see the PR chapter for references). GR and PR mutants lacking a short C-terminal portion can be activated by RU486, suggesting the existence of an intrinsic repressor function in this domain[131-133]. Indeed, it has been shown that NCoR and SMRT can bind to antagonist-occupied steroid receptors (see references 128, 134, 146 and references therein). Detailed studies of the dose–response curve for glucocorticoid receptors in the presence or absence of either coactivators or corepressors have led to a model in which the ratio of coactivator–corepressor bound to either receptor–agonist or receptor–antagonist complexes regulates the final transcriptional outcome[146]. The AF-1 region of GR has been shown to bind to both transcriptional repressors, such as TSG101, a tumor susceptibility gene that represses transcription, and to coactivators, such as DRIP150, a member of the DRIP/TRAP complex that activates transcription[189].

Ligands

Human GR binds the natural human corticoid cortisol, whereas the corticoid found in rodent is corticosterone[1,20]. The most widely used glucocorticoid is the synthetic compound dexamethasone, which is more stable than the *in vivo* ligands and can be labeled. Triamcinolone acetonide has been also used as a radioligand (see reference 1 for an example and references). It has been known for a long time that RU486 is an antagonist for GR as well as for PR[135]. RU486 stabilizes the non-DNA-binding hsp90-complexed form of the GR but can also bind to the free receptor[55,136]. An unrelated compound, ZK98299, has been also widely used as a pure antagonist (see references 105 and 145 for examples and references). The antibiotic rifampicin, which also has an immunosuppressive activity, has been shown to bind and activate GR, but these findings are controversial and it is probable that most of the effects of rifampicin are in fact elicited by PXR (NR1I2)[149,267,268].

Mutagenesis experiments have identified the C-terminal E domain as the LBD[20,21]. Several studies have used affinity labeling, which allows covalent binding to the receptor of synthetic compounds, such as dexamethasone 21-mesylate, to study the organization of the ligand-binding pocket[137,138,144]. These analyses were coupled to site-directed mutagenesis and have shown that some specific positions (Cys 640, Cys 656, Cys 661 in rat as well as Met 610, Cys 736, Tyr 770, Phe 780 in mouse and Ile 747 in human) are important for ligand binding[138–142,144]. Cys 736, Ile 747 and Phe 780 are interesting because their mutation results in receptors with altered hormone specificity[139,141,142]. It is clear that the extreme C-terminus of the receptor plays a role in agonist–antagonist discrimination, as is the case for PR (see the relevant entry)[131–133,140]. The activities of several GR ligands, such as dexamethasone, RU486 and the antagonist ZK98299, have been tested on different types of GR-responsive promoters[145]. It has been shown that unique receptor conformations induced by these ligands exhibit both promoter and tissue-specific activity.

Three-dimensional structure

The NMR structure of the GR DBD in solution was the first structure of a nuclear receptor domain to be determined[31]. This structure reveals the two helices of the DBD, one of which was proposed to interact to DNA via the P-box, whereas the second is perpendicular to the first one and do not contact the DNA directly. The two zinc fingers do not form two independent subdomains in this structure, reinforcing the difference existing between classic zinc fingers and nuclear receptor DBDs (see reference 33 for a review). These data have been confirmed by the crystallographic analysis of the interaction of GR DBD with the GRE[32]. Each GR DBD was found to interact with the DNA major groove and the two domains dimerize and interact at the level of their P-box, which forms a dimer interface. Strikingly, in the same study, the crystal of GR complexed with a GRE with an abnormal spacing of 4 bp between the two half-sites of the GRE was determined[32]. This has resulted in one subunit interacting specifically with the consensus core sequence and the other non-specifically with a non-cognate element. These observations have clearly shown that the spacing between each response element half-site is of major importance in determining the target sequence specificity (reviewed in references 34 and 37). The molecular basis of specific DNA binding has been explored by resolving the structure of the ER complexed to a GRE[35]. This study has revealed that DNA conformation as well as the presence of water molecules can influence the affinity and specificity of protein–DNA complexes. It is clear that the ratio existing within each cell between coactivators and corepressors regulates the transcriptional properties of agonist- or antagonist-bound GR[146].

The structure of the human GR LBD was recently modeled based on its high sequence identity with PR[196]. This study stressed the importance of Tyr 735 in the LBD for transcriptional activation. This residue interacts with the D-ring of dexamethasone but its mutation to Phe or Val did not reduce ligand-binding affinity, whereas it reduces maximal transactivation of an MMTV reporter[196]. In contrast, mutation to Ser caused a lower affinity for dexamethasone. These effects were even stronger using the natural ligand cortisol. This suggests that this residue plays a pivotal role in the ligand-inducible AF-2 transactivation function.

Expression

Northern blot analysis reveals that the human GR gene is expressed in fibroblastic HT1080 cells as multiple transcripts of 5.6, 6.1 and 7.1 kb, which show a 2–3-fold reduction after glucocorticoid treatment of the cells, suggesting that a negative feedback regulation has occurred[1]. The autologous regulation of GR by glucocorticoids has been demonstrated in a number of different cell types (see reference 152 for a review). Several studies have characterized the down-regulation of GR mRNA expression induced by glucocorticoid exposure[154–156]. An intragenic element in the human GR coding sequence may contribute to this auto–regulation[157]. The interference existing between AP-1 and GR has been also used to explain the negative autoregulation of GR, since AP-1 up-regulates GR gene expression[151,161]. Other mechanisms involving post-transcriptional and post-translational events have also been proposed[151,152,158,243].

It is clear that the level of cellular GR expression is closely correlated with the magnitude of the GR-mediated response[159]. This has also been shown using antisense technology[160]. Tissues of transgenic mice expressing an antisense GR transcript in which the transgene is expressed, show reduced GR levels and thus signs of glucocorticoid resistance. In the same line, GR expression and glucocorticoid responsiveness were shown to be developmentally regulated in chicken neuroretina[150]. In the undifferentiated retina at 6 days of development (E6), GR was expressed in virtually all cells, whereas in the more differentiated E10 and E12 retina, it was detected only in Müller glia cells in which GR target genes such as glutamine synthetase are well characterized[150]. It is a nice example showing that compartmentalization of GR expression during development modulates receptor activity since the responsiveness of the chicken neuroretina to corticoids strongly increases from E6 to E12.

The expression of GR varies in a tissue-specific manner, with the thymus probably expressing one of the highest number of receptors per cell in accordance with the strong and well-known effect of corticoids on T-cell apoptosis (see reference 153 for a review). Other tissues in which GR expression has been carefully studied are, among others, brain[270], bone[271] and pancreas[272].

Activity and target genes

The activation domains present within the GR were mapped by Ron Evans's and Keith Yamamoto's groups using GR mutants[23,38,179-186]. Deletion of the N- or C-terminal parts of the receptors results in molecules with constitutive activity, suggesting that the activation function was distinct from the hormone–binding domain[20,179,180]. The fact that the deletion of the LBD gives rise to a constitutively active receptor suggests that this domain normally represses receptor function[23,179]. Surprisingly, and in contrast to most other nuclear receptors, the DNA-binding domain was shown to contain all the information required to mediate DNA binding and transactivation[23,180]. The deletion of the N-terminal region was shown to reduce the ability of GR to induce transcriptional activation, suggesting that this region, later called τ1, enh-2, or AF-1, plays a role in target gene activation[23,184]. Thus, these first analyses led to the delineation of three domains important for transactivation: a transactivation domain, called enh-1, in the DBD, a modulatory domain in the N-terminal part and a repressor domain within the LBD[20,23,179,180]. An extensive mutational analysis of the DBD has allowed residues that are important for DNA binding or for transcriptional activation to be separated[38]. Analysis of specific mutants in the rat GR, P493R and S459A, allowed Keith Yamamoto's laboratory to suggest that DNA may have an allosteric effect, which allows the receptor to interact with specific regulatory targets[193]. The repressive action located inside the LBD has been linked to hsp90 binding and has been shown to be transferable to other transcription factors, such as the E1A gene product[182,183]. In addition, an inducible transcription activation function, called τ2, has also been mapped within the LBD, suggesting that several regions of this domain may have divergent functions[187]. This function has been precisely mapped between amino acids 526 and 556 of the human GR in the N-terminal part of the LBD.

The AF-1 function in the N-terminal region of GR contributes to 80% of the ligand-inducible transcriptional activity of GR. It has been mapped between amino acids 77 and 262 and a core region has been delineated to amino acids 187–227[181,197]. This has allowed the study of its possible conformation by NMR and shown that it contains three regions with α-helical conformation[194]. In accordance with its major effect, deletion of AF-1 gives rise to a mutant GR that is able to inhibit activation by wild-type receptor of some but not all target genes[195]. Experiments using cell lines on which several effects and target genes of glucocorticoids were described have helped to decipher the respective contributions of AF-1 and AF-2 in mediating the biological effects of corticoids[199].

The GR has been shown to function in yeast, i.e. to bind specifically to the GRE and enhance transcription in response to the addition of ligand[185,192]. Consistent with this observation, the expression of a high level of the N-terminal AF-1 domain in yeast induces an inhibition of endogenous yeast gene expression, suggesting that the AF-1 GR domain squelches transcriptional coactivators of yeast genes[191]. This is consistent with the observation that AF-1 interacts with two mammalian proteins, TSG101 and DRIP150, respectively, a corepressor and a coactivator that can regulate transcription in yeast[188]. Interactions between AF-1 and the members of the transcriptional SAGA complex that possess histone acetyltransferase activity has also been demonstrated[255]. Hormonal activation of a GRE-containing minimal promoter *in vitro* by purified GR was also shown using a cell-free system based on *Drosophila* embryonic nuclear extract[186]. GR is also able to function in *Drosophila* cells, whereas the *Drosophila* genome does not contain a GR homologue[190]. These data suggest that GR used a set of evolutionary conserved transcriptional mediators to stimulate transcription.

It was known even before the cloning era that GR was a phosphoprotein (see reference 205 for a review). The aporeceptor is phosphorylated with additional phosphorylation events occurring in conjunction with ligand binding[206,207]. Receptor phosphorylation occurs on serine and threonine residues in both the presence and absence of hormone, and the majority of phosphorylated residues lie within the N-terminal A/B domain[184,201]. It was shown that phosphorylation of S224 and S232 in rat GR increases in the presence of hormone, whereas residues T171 and S246 are constitutively phosphorylated[204]. The transcriptional activity of GR can be modified by phosphorylation and these phosphorylation events can be regulated. For example, GR is unable to transactivate target genes such as the TAT gene or the MMTV LTR during the G2 phase of the cell cycle in synchronized fibroblasts[198]. It appears that, during the G2 phase, agonist-bound GRs are not efficiently retained in the nucleus and this correlates with site-specific alteration of GR phosphorylation[198,203]. Glucocorticoids do not induce hyperphosphorylation during the G2 phase in contrast to S phase[203]. In line with these findings it was shown that CDK, the cyclin-dependent kinases, as well as MAP kinase phosphorylate the rat GR *in vitro* at sites that are phosphorylated *in vivo*[204]. These phosphorylation events have either positive or negative effects on the transcriptional activation through GR[202,204]. In addition, PKA up-regulates the hormone-dependent transactivation of GR, suggesting that GR could be an *in vivo* target of this enzyme[200]. The phosphorylation of GR has been shown to be required for the down-regulation of the receptor induced by glucocorticoid, since a mouse GR in which phosphorylation sites has been abrogated was no longer able to down-regulate its own synthesis[243].

Using biochemically purified GR it has been shown that several genes such as the MMTV LTR and the metallothionein gene are specifically regulated by glucocorticoids and are recognized by GR in specific regions of their promoters[11-14]. It was shown that all these target genes share a core hexamer sequence AGAG/ACA that is often present in two palindromic copies and which is also recognized by other steroid receptors but not by ER[15-17]. The MMTV LTR has been extensively used as a powerful model system to study how GR regulates transcription on chromatin templates and how this may be influenced by other transcription factors[12,43,82]. Interestingly, numerous other viruses are regulated by glucocorticoids, which are hormones implicated in stress and inflammatory responses[208,209]. This is the case of the human papilloma virus HPV-16, which contains a functional GRE in the regulatory region controlling the expression of the transforming early proteins E6 and E7[210]. Two other GREs were also found in HPV-16 and shown to be required for full corticoid induction[215]. The Epstein–Barr virus (EBV), which immortalizes B lymphocytes, is also directly regulated by GR[211,212]. Nevertheless, the effect of glucocorticoids on EBV is more complex, since indirect effects of corticoids on EBV gene expression involving AP-1 were also described[213]. The HIV LTR also contains a GRE and it has been proposed that glucocorticoid regulation of HIV-1 could be relevant to the steroid responsiveness of HIV-1[214].

A very large number of enzymes are regulated by glucocorticoids. This is the case in the liver for enzymes implicated in gluconeogenesis, such as phosphoenolpyruvate carboxykinase (PEPCK), TAT or serine deshydratase[220]. As for the MMTV, the TAT gene has been used as a model to study gene regulation by glucocorticoids[16,17,221,222]. For example, this gene has been used to study how glucocorticoids establish and maintain specific alteration in chromatin structure during gene induction[223]. The PEPCK gene is also regulated by GR through a complex response unit that contains GRE as well as binding sites for other transcription factors[226]. This gene is also regulated by cAMP through another regulatory element that is also required for full glucocorticoid induction[227,228]. Other known targets of glucocorticoids are the rat tryptophane oxygenase[224], the tyrosine hydroxylase[225], the cytosolic aspartate aminotransferase gene[229], the carbamoylphosphate synthetase gene[230], the glutamate synthetase gene[231], and the mouse α-amylase 2 gene[232]. GR also positively regulates the multidrug resistance gene, mdr[233] as well as p57[Kip2], the cell cycle progression inhibitor[234], or the insulin-like growth factor binding protein-1 (IGFBP-1)[235]. The expression of several secreted molecules is also induced by glucocorticoids. This is the case for the growth hormone gene, which contains a GRE within its first intron, the β-casein gene[237] and the thyrotropin-releasing hormone (TRH) gene[238]. Among the various transcription factors that are regulated by glucocorticoids, it is interesting to cite the c-myc gene, the transcription of which is inhibited by GR[218,219], and the pituitary-specific gene Pit-1, which is implicated in the activation of prolactine and growth hormone genes[217]. Given the famous cross-talk between GR and AP-1, it is interesting to note that liganded GR represses the c-*jun* promoter via the interference with an AP-1 site that is located in this promoter[274,275].

GR negatively regulates several genes, such as the rat pro-opiomelanocortine (POMC) gene[239] or the human glycoprotein hormone α-subunit[240] (see reference 188 for a review). GR-mediated repression can require the DNA-binding activity of the receptor or can be DNA-binding independent. In this latter case, the repressive activity has been shown to be due to negative interaction with other transcription factors, such as AP-1 and NFκB (see 'partners' for a more precise discussion on these interactions and 'Knockout' for their *in vivo* relevance)[188,208]. Examples of DNA-dependent silencing of gene expression mediated by GR involved the CRH gene. This gene participates in the regulation of corticoid synthesis by the hypothalamic–pituitary–adrenal (HPA) axis and is down-regulated by glucocorticoids by a negative feedback mechanism. Specific high-affinity binding of GR was observed in specific regions of the CRH promoter that are important for glucocorticoid repression[241]. Glucocorticoids repress several keratin genes via specific negative GRE on which GR binds as a monomer[256]. An interesting case is the osteocalcin gene that is inhibited by glucocorticoids via the binding of GR to the TATA box, resulting in the displacement of the TATA-binding protein and gene repression[242]. Other down-regulated genes for which DNA binding of GR was required are the vasoactive intestinal peptide (VIP)[246] and the proliferin[74,247,248] genes. The down-regulation of the POMC gene by glucocorticoids involved the titration by GR of a dimer of the orphan receptor NGFI-B (NR4A1), which is a strong activator of the POMC gene[244,245]. It has recently been shown that GR inhibits TGFβ signaling by direct interaction with the Smad3 protein which transduces the TGFβ signal[249]. This results in the inhibition of TGFβ responsive genes, such as the type-1 plasminogen activator[249].

Knockout

The knockout of the GR gene in mice has been performed by Gunther Schutz's laboratory[162]. The homozygous animals die within a few hours after birth owing to respiratory failure caused by incomplete lung maturation. In addition, these animals have a reduced capacity to activate genes for key gluconeogenic enzymes such as PEPCK, TAT or serine deshydratase in the liver. These results underscore the importance of glucocorticoids for the activation of perinatal gluconeogenesis. Another salient feature of these animals is their strongly elevated ACTH and corticosterone levels owing to an abnormal feedback regulation via the HPA axis. In these animals, the adrenal cortex is enlarged owing to a vast excess of corticosteroid synthesis and exhibit increased expression of key enzymes such as steroid 11β-hydroxylase, side-chain cleavage enzyme and aldosterone synthase. These animals also show impaired proliferation of erythroid progenitor cells induced by stress or hypoxia, showing that glucocorticoids are important for stress erythropoiesis[167]. This is consistent with reports describing the importance of glucocorticoids in the differentiation of erythroid progenitors[216]. Detailed study of the adrenal medulla development in mice with different inactivated GR alleles has shown that, in contrast to expectations based on a large number of *in vitro* experiments, these mice have a normal number of adrenal chromaffin cells but do not adopt a neuronal phenotype and do not express classical neuronal markers[166].

Since numerous effects of GR, such as the immunosuppressive effects of glucocorticoids, have been described linked to an AP-1 or NFκB cross-talk via protein–protein interactions and not to a direct effect on transcription via GR bound to DNA, mice carrying a dimerization-defective GR, which is unable to bind to DNA, were generated (see references 163–165 for reviews). These mice expressed a GR in which Ala 458 in the D-loop is mutated over a Thr, resulting in a protein that fails to bind to DNA and cannot transactivate GRE-dependent promoters but that is still able to repress AP-1-dependent promoters, such as the collagenase-1 promoter[250]. In contrast to the GR$^{-/-}$ mice, these mice, called GR$^{dim/dim}$, are viable and do not exhibit lung development delay, revealing the *in vivo* relevance of DNA-binding-independent activities of the GR[163]. As expected, the regulation of gluconeogenesis enzymes such as TAT whose transcription in the liver is dependent on the binding of GR to an upstream promoter is not responsive to glucocorticoids in the GR$^{dim/dim}$ mice. The analysis of the HPA axis in these animals reveals that it depends both on DNA-binding dependent and -independent functions of GR, since the ACTH level in the serum is not elevated (in contrast to GR$^{-/-}$ mice), whereas corticosterone levels are increased (as in GR$^{-/-}$ mice). Finally, T-cell apoptosis was normal in the GR$^{dim/dim}$ animals, suggesting that these events are controlled by DNA-binding-independent mechanisms (see reference 153 for a review). Analysis of the GR$^{dim/dim}$ animals show that stress erythropoiesis is abnormal in these animals as in the GR$^{-/-}$ mice[167]. In contrast, normal erythropoiesis is not impaired in the GR$^{dim/dim}$ or GR$^{-/-}$ animals[167]. The function of GR in skin is also linked to the AP-1 transrepression as revealed by the study of the GR$^{dim/dim}$ mice[277].

The *in vivo* analysis of GR function was continued by the generation of mice with tissue-specific disruption of the GR gene, using the Cre/loxP system (see references 165 and 278 for a review). For example, targeted disruption of the GR gene in the nervous system was performed using a transgenic mouse strain expressing the Cre recombinase under the control of the nestin promoter[279]. Using this strategy, it was shown that GR is essential for the regulation of the HPA axis and the stress response as well as for the control of emotional behavior. Indeed, the mice carrying a mutation of the GR gene in their nervous system exhibit reduced anxiety[279].

Before the construction of mice with inactivation of the GR gene, transgenic animals expressing an antisense GR transcript under the control of a human neurofilament gene promoter were generated[160]. These animals have an impaired GR function, have greatly increased fat deposition and can be twice the weight of normal mice. These animals have an increased ACTH and corticosterone levels reminiscent of the abnormal HPA axis found in the knockout animals. Interestingly, these characteristics are also found in depression and treatment of these animals with an antidepressant increased hypothalamic GR mRNA levels[168].

Transgenic mice with overexpression of GR in pancreas (via the use of the insulin promoter) allowed the important function of GR in pancreatic β cells to be demonstrated[170]. These animals exhibit a reduced glucose tolerance and a dramatic decrease in acute insulin response consistent with the diabetogenic action of glucocorticoids. It has recently been shown that an increased GR gene dosage, using yeast artificial chromosome containing the GR gene, alters the basal regulation of the HPA axis, resulting in reduced circulating glucocorticoid levels[280]. In addition, the thymocytes of these transgenic animals exhibit enhanced sensitivity to glucocorticoid-induced apoptosis. These animals also show enhanced resistance to stress and endotoxic shock. These results clearly highlight the importance of tight control of GR expression in target tissues and may explain differences in the susceptibility of humans to inflammatory diseases and stress[280].

Associated disease

The GR is associated to the generalized inherited glucocorticoid resistance or familial glucocorticoid resistance (FGR) (see references 172 and 178). Individuals diagnosed with signs of FGR have a mutation or a deletion of the GR gene. This abnormality interferes with glucocorticoid feedback signaling and, as a consequence, these individuals exhibit high levels of circulating corticoids. The first subjects with clinical signs of FGR were diagnosed with hypertension and exhibited GRs with reduced binding affinity for glucocorticoid, owing to a missense mutation giving rise to a single amino-acid substitution in the LBD[176]. In another case, a splice deletion in exon 6 led to a FGR case in which low GR numbers were observed in blood cells[177]. This female patient exhibits hirsutism, menstrual irregularities and acne due to ACTH-induced hypersecretion of androgens. Many other mutations were found in the GR gene, the list of which can be found in a specific database, the glucocorticoid receptor resource (http://biochem1.basic-sci.georgetown.edu/grr/grr.html).

In addition to FRT, a syndrome of acquired glucocorticoid resistance with tissue-specific resistance has been described[172]. No specific mutations were linked to the GR gene to explain this syndrome. Of note, the specific C-terminal isoform has been implicated in this partial resistance in several studies (see reference 172 for a review). Nevertheless, this has been subject to much debate and some authors think that GRβ has no specific role in the modulation of GR action[62,63].

Glucocorticoid resistance has been characterized and linked to GR in several variant cell lines, such as S49 mouse lymphoma cells, which were screened for their inability to undergo apoptosis in response to dexamethasone[173,174]. Analysis of the GR gene in these resistant cells has allowed the study of specific mutations within the GR gene that explain the resistance syndrome. More striking is the behavior of some animal species, such as the neotropical cotton-top marmoset (*Saguinus oedipus*), which exhibits an increased plasma cortisol concentration when compared to other species. The GR gene from this species has been cloned but the reason for its resistance is still unclear[175]. The same kind of analysis has been carried out on other glucocorticoid-resistant monkeys, such as the squirrel monkey (*Saimiri boliviensis*)[257]. The guinea pig has a GR with mutations, which explains its low affinity for the ligand and its glucocorticoid insensitivity[3].

Gene structure, promoter and isoforms

The genomic organization of the GR gene has been determined in human and mouse[59,60]. In both cases, the gene is large (*c.* 100 kbp). In the human gene 10 exons were found[59]. The ATG is in exon 2 and the stop codon for the GRα isoform is in exon 9α. The GRα specific C-terminal end is encoded by exon 9α, which is located in the vicinity of exon 9α. The two exons encoding the DBD are exons 3 and 4, and their boundaries are identical to that of other steroid receptors. The LBD is encoded by five exons. The mouse gene was shown to contain three alternative exons, 1A, 1B and 1C, but does not contain alternative 3' exons in accordance with the fact that the GRβ isoform was not found in mouse[60]. The structure of the gene and exon boundaries are identical to the human gene. In the rat several 5' isoforms were also described and the 5' structure of the gene was shown to contain at least 11 alternative exons, which are expressed in different tissues, suggesting tissue-specific differences in promoter activity[276].

A promoter of the human GR gene was characterized and was shown to be GC-rich without TATA or CAAT boxes[59,61,161]. It does not harbor any GRE sequence. In the mouse, three different promoters were found, each in front of one of the alternative first exons[60]. One promoter (promoter A) is cell specific, being only active in T lymphocytes; the other two promoters are active in other tissues, such as liver, fibroblasts or brain. To date, no GRE has been found in these promoters[59,61,161]. The transcription factor AP-2 has been shown to be important for human promoter activity[161] (see reference 152 for a review).

Two C-terminally different isoforms of GR, GRα and GRβ, are known[1]. These isoforms are identical up to amino acid 727, but diverge beyond this position, with GRα having 50 additional amino acids found conserved in the mouse GR, whereas GRβ contains only 15 divergent amino acids. In contrast to GRα, GRβ does not bind the ligand and is unable to transactivate glucocorticoid-responsive genes but is still able to bind to DNA[1,64]. GRβ was first neglected and even considered as an artifact before it was recognized by John Cidlowski's and George Chrousos's laboratories that it is expressed *in vivo* and may modulate GRα signaling[64-66]. Interestingly, an heterodimer can exist between GRα and GRβ, and GRβ can inhibit GRα-mediated repression of NFκB-responsive promoters[252]. Nevertheless, the biological function of GRβ has remained a subject of debate and some authors still consider that it does not have a significant role in glucocorticoid signaling (see two conflicting views in references 62 and 63).

Chromosomal location

As soon as the human GR cDNA had been cloned, it was localized on human chromosome 5[1]. This location has been subsequently refined to 5q31–5q32[57,58].

Amino acid sequence for human GR (NR3C1)

Accession number: X03225

```
  1  MDSKESLTPGREENPSSVLAQERGDVMDFYKTLRGGATVKVSASSPSLAVASQSDSKQRRLLVDFPKGSV
 71  SNAQQPDLSKAVSLSMGLYMGETETKVMGNDLGFPQQGQISLSSGETDLKLLEESIANLNRSTSVPENPK
141  SSASTAVSAAPTEKEFPKTHSDVSSEQQHLKGQTGTNGGNVKLYTTDQSTFDILQDLEFSSGSPGKETNE
211  SPWRSDLLIDENCLLSPLAGEDDSFLLEGNSNEDCKPLILPDTKPKIKDNGDLVLSSPSNVTLPQVKTEK
281  EDFIELCTPGVIKQEKLGTVVYCQASFPGANIIGNKMSAISVHGVSTSGGQMYHYDMNTASLSQQQDQKPI
351  FNVIPPIPVGSENWNRCQGSGDDNLTSLGTLNFPGRTVFSNGYSSPSMRPDVSSPPSSSSTATTGPPPKL
421  CLVCSDEASGCHYGVLTCGSCKVFFKRAVEGQHNYLCAGRNDCIIDKIRRKNCPACRYRKCLQAGMNLEA
491  RKTKKKIKGIQQATTGVSQETSENPGNKTIVPATLPQLTPTLVSLLEVIEPEVLYAGYDSSVPDSTWRIM
561  TTLNMLGGRQVIAAVKWAKAIPGFRNLHLDDQMTLLQYSWMFLMAFALGWRSYRQSSANLLCFAPDLIIN
631  EQRMTLPCMYDQCKHMLYVSSELHRLQVSYEEYLCMKTLLLLSSVPKDGLKSQELFDEIRMTYIKELGKA
701  IVKREGNSSQNWQRFYQLTKLLDSMHEVVENLLNYCFQTFLDKTMSIEFPEMLAEIITNQIPKYSNGNIK
771  KLLFHQK 777
```

References

1. Hollenberg, S.M. et al. (1985) Nature 318, 635–641.
2. Weinberger, C. et al. (1985) Nature 318, 670–672.
3. Keightley, M.C. and Fuller, P.J. (1994) Mol. Endocrinol. 8, 431–439.
4. Ducouret, B. et al. (1995) Endocrinology 136, 3774–3783.
5. Yang, K. et al. (1992) J. Mol. Endocrinol. 8, 173–180.
6. Tagawa, M. et al. (1997) Gen. Comp. Endocrinol. 108, 132–140.
7. Simons, S.S. and Thompson, E.B. (1981) Proc. Natl Acad. Sci. USA 78, 3541–3545.
8. Gehring, U. and Hotz, A. (1983) Biochemistry 22, 4013–4018.
9. Miesfeld, R. et al. (1984) Nature 312, 779–781.
10. Weinberger, C. et al. (1985) Science 228, 740–742.
11. Ringold, G.M. et al. (1975) Cell 6, 299–305.
12. Scheidereit, C. et al. (1983) Nature 304, 749–752.
13. Hager, L.J. and Palmiter, R.D. (1981) Nature 291, 340–342.
14. Karin, M. et al. (1980) Nature 286, 295–297.
15. Karin, M. et al. (1984) Nature 308, 513–519.
16. Jantzen, H.M. et al. (1987) Cell 49, 29–38.
17. Strähle, U. et al. (1987) Proc. Natl Acad. Sci. USA 84, 7871–7875.
18. Klock, G. et al. (1987) Nature 329, 734–736.
19. Nordeen, S.K. et al. (1990) Mol. Endocrinol. 4, 1866–1873.
20. Giguère, V. et al. (1986) Cell 46, 645–652.
21. Rusconi, S. and Yamamoto, K.R. (1987) EMBO J. 6, 1309–1315.
22. Carlstedt-Duke, J. et al. (1987) Proc. Natl Acad. Sci. USA 84, 4437–4440.
23. Hollenberg, S.M. et al. (1987) Cell 49, 39–46.
24. Green, S. and Chambon, P. (1987) Nature 325, 75–78.
25. Danielsen, M. et al. (1989) Cell 57, 1131–1138.
26. Alroy, I. and Freedman, L.P. (1992) Nucl. Acids Res. 20, 1045–1052.
27. Saverne, Y. et al. (1988) EMBO J. 7, 2503–2508.
28. Freedman, L.P. et al. (1988) Nature 334, 543–546.
29. Umesono, K. and Evans, R.M. (1989) Cell 57, 1139–1146.
30. Mader, S. et al. (1989) Nature 338, 271–274.
31. Härd, T. et al. (1990) Science 249, 157–160.

32 Luisi, B.F. et al. (1991) Nature 352, 497–505.
33 Schwabe, J.W.R. and Rhodes, D. (1991) TIBS 16, 291–296.
34 Freedman, L.P. and Luisi, B.F. (1993) J. Cell. Biochem. 51, 140–150.
35 Gewirth, D.T. and Sigler, P.B. (1995) Nature Struct. Biol. 2, 386–394.
36 Arbuckle, N.D. and Luisi, B.F. (1995) Nature Struct. Biol. 2, 341–346.
37 Zilliacus, J. et al. (1995) Mol. Endocrinol. 9, 389–400.
38 Schena, M. et al. (1989) Genes Dev. 3, 1590–1601.
39 Schmid, W. et al. (1989) EMBO J. 8, 2257–2263.
40 Drouin, J. et al. (1992) Mol. Endocrinol. 6, 1299–1309.
41 Perlmann, T. and Wrange, Ö. (1988) EMBO J. 7, 3073–3079.
42 Pina, B. et al. (1990) Cell 60, 719–731.
43 Beato, M. et al. (1995) Cell 83, 851–857.
44 Miksicek, R. et al. (1987) EMBO J. 6, 1355–1360.
45 Brüggemeier, U. et al. (1990) EMBO J. 9, 2233–2239.
46 Chavez, S. and Beato, M. (1997) Proc. Natl Acad. Sci. USA 94, 2885–2890.
47 Belikov, S. et al. (2000) EMBO J. 19, 1023–1033.
48 Li, Q. and Wrange, Ö. (1993) Genes Dev. 7, 2471–2482.
49 Pratt, W.B. and Toft, D.O. (1997) Endocrinol. Rev. 18, 306–360.
50 Cheug, J. and Smith, D.F. (2000) Mol. Endocrinol. 14, 939–946.
51 Picard, D. and Yamamoto, K.R. (1987) EMBO J. 6, 3333–3340.
52 Wikstrom, A.C. et al. (1987) Endocrinology 120, 1232–1242.
53 Martins, V.R. et al. (1991) Mol. Endocrinol. 5, 217–225.
54 Madan, A.P. and De Franco, D.B. (1993) Proc. Natl Acad. Sci. USA 90, 3588–3592.
55 Sackey, F.N.A. et al. (1996) Mol. Endocrinol. 10, 1191–1205.
56 Savory, J.G.A. et al. (1999) Mol. Cell. Biol. 19, 1025–1037.
57 Francke, U. and Foellmer, B.E. (1989) Genomics 4, 610–612.
58 Thériault, A. et al. (1989) Hum. Genet. 83, 289–291.
59 Encio, I.J. et al. (1991) J. Biol. Chem. 266, 7182–7188.
60 Strähle, U. et al. (1992) Proc. Natl Acad. Sci. USA 89, 6731–6735.
61 Zong, J. et al. (1990) Mol. Cell. Biol. 10, 5580–5585.
62 Vottero, A. and Chrousos, G.P. (1999) Trends Endocrinol. Metab. 10, 333–338.
63 Carlstedt-Duke, J. (1999) Trends Endocrinol. Metab. 10, 339–342.
64 Bamberger, C.M. et al. (1995) J. Clin. Invest. 95, 2435–2441.
65 De Castro, M. et al. (1996) Mol. Med. 2, 597–607.
66 Oakley, R.H. et al. (1996) J. Biol. Chem. 271, 9550–9559.
67 Archer, T.K. et al. (1998) Trends Endocrinol. Metab. 8, 384–390.
68 Trapp, T. et al. (1994) Neuron 13, 1457–1462.
69 Liu, W. et al. (1995) Proc. Natl Acad. Sci. USA 92, 12480–12484.
70 Yen, P.M. et al. (1997) Mol. Endocrinol. 11, 162–171.
71 Jonat, C. et al. (1990) Cell 62, 1189–1204.
72 Yang-Yen, H.F. et al. (1990) Cell 62, 1205–1215.
73 Schüle, R. et al. (1990) Cell 62, 1217–1226.
74 Diamond, M.I. et al. (1990) Science 249, 1266–1272.
75 Ray, A. and Prefontaine, K.E. (1994) Proc. Natl Acad. Sci. USA 91, 752–756.
76 Mukaida, N. et al. (1994) J. Biol. Chem. 269, 13289–13295.

[77] Caldenhoven, E. et al. (1995) Mol. Endocrinol. 9, 401–412.
[78] Scheinman, R.I. et al. (1995) Mol. Cell. Biol. 15, 943–953.
[79] Stöcklin, E. et al. (1996) Nature 383, 726–728.
[80] Landers, J.P. and Spelsberg, T.C. (1992) Crit. Rev. Euk. Gene Expr. 2, 19–63.
[81] Schüle, R. and Evans, R.M. (1991) Trends Genet. 7, 377–381.
[82] Truss, M. and Beato, M. (1993) Endocrine Rev. 14, 459–479.
[83] Pfahl, M. (1993) Endocrine Rev. 14, 651–658.
[84] Miner, J.N. and Yamamoto, K.R. (1991) TIBS 16, 423–426.
[85] Van der Burg, B. et al. (1997) Trends Endocrinol. Metab. 8, 152–156.
[86] Strähle, U. et al. (1988) EMBO J. 7, 3389–3395.
[87] Wieland, S. et al. (1991) EMBO J. 10, 2513–2521.
[88] Kutoh, E. et al. (1992) Mol. Cell. Biol. 12, 4960–4969.
[89] Préfontaine, G.G. et al. (1998) Mol. Cell. Biol. 18, 3416–3430.
[90] Muller, M. et al. (1991) Mol. Endocrinol. 5, 1498–1503.
[91] Gauthier, J-M. et al. (1993) EMBO J. 12, 5089–5096.
[92] Chang, T.J. et al. (1993) Mol. Endocrinol. 7, 528–542.
[93] Wang, J.C. et al. (1996) Mol. Endocrinol. 10, 794–800.
[94] Singh, P. et al. (1995) Nature 374, 562–565.
[95] Refaeli, Y. et al. (1995) Proc. Natl Acad. Sci. USA 92, 3621–3625.
[96] Hall, R.K. et al. (1995) Proc. Natl Acad. Sci. USA 92, 412–416.
[97] Burns, K. et al. (1994) Nature 367, 476–480.
[98] Dedhar, S. et al. (1994) Nature 367 480–483.
[99] Kupfer, S.R. et al. (1994) J. Biol. Chem. 269, 20622–20628.
[100] Cella, N. et al. (1998) Mol. Cell. Biol. 18, 1783–1792.
[101] Wyszomierski, S.L. et al. (1999) Mol. Endocrinol. 13, 330–343.
[102] McKay, L.I.L. and Cidlowski, J.A. (1998) Mol. Endocrinol. 12, 45–56.
[103] Scheilman, R.I. et al. (1995) Science 270, 283–286.
[104] Auphan, N. et al. (1995) Science 270, 286–290.
[105] Liden, J. et al. (1997) J. Biol. Chem. 272, 21467–21472.
[106] Wissink, S. et al. (1998) Mol. Endocrinol. 12, 355–363.
[107] Lucibello, F.C. et al. (1990) EMBO J. 9, 2827–2834.
[108] Angel, P. et al. (1987) Cell 49, 729–739.
[109] Touray, M. et al. (1991) Oncogene 6, 1227–1234.
[110] König, H. et al. (1992) EMBO J. 11, 2241–2246.
[111] Berko-Flink, Y. et al. (1994) EMBO J. 13, 646–654.
[112] Helmberg, A. et al. (1995) EMBO J. 14, 452–460.
[113] Zhou, F. and Thompson, E.B. (1996) Mol. Endocrinol. 10, 306–316.
[114] Oren, A. et al. (1999) Mol. Cell. Biol. 19, 1742–1750.
[115] Kamei, Y. et al. (1996) Cell 85, 403–414.
[116] Chakravarti, D. et al. (1996) Nature 383, 99–103.
[117] Caelles, C. et al. (1997) Genes Dev. 11, 3351–3364.
[118] McKenna, N.J. et al. (1999) Endocrine Rev. 20, 321–344.
[119] Glass, C.K. and Rosenfeld, M.G. (2000) Genes Dev. 14, 121–141.
[120] Onate, S.A. et al. (1995) Science 270, 1354–1357.
[121] Chen, H. et al. (1997) Cell 90, 569–580.
[122] Fryer, C.J. and Archer, T.K. (1998) Nature 393, 88–91.
[123] Murchardt, C. and Yaniv, M. (1993) EMBO J. 12, 4279–4290.
[124] Kingston, R.E. and Narlikar, G.J. (1999) Genes Dev. 13, 2339–2352.

[125] Ostlund Farrants, A.K. et al. (1997) Mol. Cell. Biol. 17, 895–905.
[126] Chang, C.J. et al. (1998) Mol. Cell. Biol. 18, 5880–5887.
[127] Lanz, R.B. et al. (1999) Cell 97, 17–27.
[128] Jackson, T.A. et al. (1997) Mol. Endocrinol. 11, 693–705.
[129] Powers, C.A. et al. (1998) Mol. Endocrinol. 12, 4–18.
[130] Lee, S.K. et al. (1999) J. Biol. Chem. 274, 34283–34293.
[131] Xu, J. et al. (1996) Proc. Natl Acad. Sci. USA 93, 12195–12199.
[132] Vegeto, E. et al. (1992) Cell 69, 703–713.
[133] Lanz, R.B. and Rusconi, S. (1994) Endocrinology 135, 2183–2195.
[134] Wagner, B.L. et al. (1998) Mol. Cell. Biol. 18, 1369–1378.
[135] Baulieu, E.E. (1989) Science 245, 1351–1357.
[136] Groyer, A. et al. (1987) Nature 328, 624–626.
[137] Simons, S.S. et al. (1981) Proc. Natl Acad. Sci. USA 78, 3541–3545.
[138] Chakraborti, P.K. et al. (1992) J. Biol. Chem. 267, 11366–11373.
[139] Chen, D. et al. (1994) Mol. Endocrinol. 8, 422–430.
[140] Zhang, S. et al. (1996) Mol. Endocrinol. 10, 24–34.
[141] Roux, S. et al. (1996) Mol. Endocrinol. 10, 1214–1226.
[142] Lind, U. et al. (1996) Mol. Endocrinol. 10, 1358–1370.
[143] Danielsen, M. et al. (1987) Mol. Endocrinol. 1, 816–822.
[144] Carlstedt-Duke, J. et al. (1988) J. Biol. Chem. 15, 6842–6846.
[145] Guido, E.C. et al. (1996) Mol. Endocrinol. 10, 1178–1190.
[146] Szapary, D. et al. (1999) Mol. Endocrinol. 13, 2108–2121.
[147] Tai, P.K. et al. (1992) Science 256, 1315–1318.
[148] Ning, Y.M. et al. (1993) J. Biol. Chem. 268, 6073–6076.
[149] Calleja, C. et al. (1998) Nature Med. 4, 92–96.
[150] Gorovits, R. et al. (1994) Proc. Natl Acad. Sci. USA 91, 4786–4790.
[151] Vig, E. et al. (1994) Mol. Endocrinol. 8, 1336–1346.
[152] Bamberger, C.M. et al. (1996) Endocrine Rev. 17, 245–261.
[153] Thompson, E.B. (1999) Trends Endocrinol. Metab. 10, 353–358.
[154] Okret, S. et al. (1986) Proc. Natl Acad. Sci. USA 83, 5899–5903.
[155] Dong, Y. et al. (1988) Mol. Endocrinol. 2, 1256–1264.
[156] Rosewicz, S. et al. (1988) J. Biol. Chem. 263, 2581–2584.
[157] Burnstein, K.L. et al. (1994) Mol. Endocrinol. 8, 1764–1773.
[158] Hoeck, W. et al. (1989) J. Biol. Chem. 264, 14396–14402.
[159] Vanderbilt, J.N. et al. (1987) Mol. Endocrinol. 1, 68–74.
[160] Pepin, M.C. et al. (1992) Nature 355, 725–728.
[161] Nobukuni, Y. et al. (1995) Biochemistry 34, 8207–8214.
[162] Cole, T.J. et al. (1995) Genes Dev. 9, 1608–1621.
[163] Reichardt, H.M. et al. (1998) Cell 93, 531–541.
[164] Karin, M. (1998) Cell 93, 487–490.
[165] Tronche, F. et al. (1998) Curr. Opin. Genet. Dev. 8, 532–538.
[166] Finotto, S. et al. (1999) Development 126, 2935–2944.
[167] Bauer, A. et al. (1999) Genes Dev. 13, 2996–3002.
[168] Pepin, M.C. et al. (1992) Mol. Pharmacol. 42, 991–995.
[169] Wang, J.M. et al. (1999) Mol. Cell. Biol. 19, 7106–7122.
[170] Delaunay, F. et al. (1997) J. Clin. Invest. 100, 2094–2098.
[171] Heck, S. et al. (1997) EMBO J. 13, 4087–4095.
[172] De Kloet, E.R. et al. (1997) Trends Endocrinol. Metab. 8, 26–33.
[173] Ashraf, J. and Thompson, E.B. (1993) Mol. Endocrinol. 7, 631–642.

[174] Dieken, E.S. et al. (1990) Mol. Cell. Biol. 10, 4574–4581.

[175] Brandon, D.D. et al. (1991) J. Mol. Endocrinol. 7, 89–96.

[176] Hurley, D.M. et al. (1991) J. Clin. Invest. 87, 680–686.

[177] Karl, M. et al. (1993) J. Clin. Endocrinol. Metab. 76, 683–689.

[178] Brönnegard, M. and Carlstetdt-Duke, J. (1995) Trends Endocrinol. Metab. 6, 160–164.

[179] Godowski, P.J. et al. (1987) Nature 325, 365–368.

[180] Miesfeld, R. et al. (1987) Science 236, 423–427.

[181] Hollenberg, S.M. et al. (1988) Cell 55, 899–906.

[182] Picard, D. et al. (1988) Cell 54, 1073–1080.

[183] Oro, A.E. et al. (1988) Cell 55, 1109–1114.

[184] Godowski, P.J. et al. (1988) Science 241, 812–816.

[185] Schnena, M. and Yamamoto, K.R. (1988) Science 241, 965–967.

[186] Freedman, L.P. et al. (1989) Science, 245, 298–301.

[187] Webster, N.J.G. et al. (1988) Cell 54, 199–207.

[188] Webster, J.C. and Cidlowski, J.A. (1999) Trends Endocrinol. Metab. 10, 396–402.

[189] Hittelman, A.B. et al. (1999) EMBO J. 18, 5380–5388.

[190] Yoshinaga, S.K. and Yamamoto, K.R. (1991) Mol. Endocrinol. 5, 844–853.

[191] Wright, A.P.H. et al. (1991) Mol. Endocrinol. 5, 1366–1372.

[192] Wright, A.P.H. and Gustafsson, J.A. (1991) J. Biol. Chem. 267, 11191–11195.

[193] Lefstin, J.A. et al. (1994) Genes Dev. 8, 2842–2856.

[194] Dahlman-Wright, K. et al. (1995) Proc. Natl Acad. Sci. USA 92, 1699–1703.

[195] Delaunay, F. et al. (1996) Eur. J. Biochem. 242, 839–845.

[196] Ray, D.W. et al. (1999) Mol. Endocrinol. 13, 1855–1863.

[197] Dahlman-Wright, K. et al. (1994) Proc. Natl Acad. Sci. USA 91, 1619–1623.

[198] Hsu, S.C. et al. (1992) EMBO J. 11, 3457–3468.

[199] Rogatsky, I. et al. (1999) Mol. Cell. Biol. 19, 5036–5049.

[200] Rangarajan, P.N. et al. (1992) Mol. Endocrinol. 6, 1451–1457.

[201] Bodwell, J. et al. (1991) J. Biol. Chem. 266, 7549–7555.

[202] Mason, S.A. and Housley, P.R. (1993) J. Biol. Chem. 268, 21501–21504.

[203] Hu, J.M. et al. (1997) Mol. Endocrinol. 11, 305–311.

[204] Krstic, M.D. et al. (1997) Mol. Cell. Biol. 17, 3947–3954.

[205] Orti, E. et al. (1992) Endocrine Rev. 13, 105–128.

[206] Hoeck, W. and Groner, B. (1990) J. Biol. Chem. 265, 5403–5408.

[207] Orti, E. et al. (1989) J. Biol. Chem. 264, 9728–9731.

[208] Mc Kay, L.I. and Cidlowski, J.A. (1999) Endocrine Rev. 20, 435–459.

[209] Sapolsky, R.M. et al. (2000) Endocrine Rev. 21, 55–89.

[210] Chan, W.K. et al. (1989) J. Virol. 63, 3261–3269.

[211] Kupfer, S.R. and Summers, W.C. (1990) J. Virol. 64, 1984–1990.

[212] Schuster, C. et al. (1991) Mol. Endocrinol. 5, 267–272.

[213] Sinclair, A.J. et al. (1992) J. Virol. 66, 70–77.

[214] Ghosh, D. (1992) J. Virol. 66, 586–590.

[215] Mittal, R. et al. (1993) J. Virol. 67, 5656–5659.

[216] Wessely, O. et al. (1997) EMBO J. 16, 267–280.

217 Jong, M.T.C. et al. (1994) Mol. Endocrinol. 8, 1320–1327.

218 Rories, C. et al. (1989) Mol. Endocrinol. 3, 391–1001.

219 Ma, T. et al. (1992) Mol. Endocrinol. 6, 960–968.

220 Simons, S.S. et al. (1992) Mol. Endocrinol. 6, 995–1002.

221 Becker, P.B. et al. (1986) Nature 324, 686–688.

222 Strähle, U. et al. (1989) Nature 339, 629–632.

223 Reik, A. et al. (1991) EMBO J. 10, 2569–2576.

224 Schüle, R. et al. (1988) Nature 332, 87–90.

225 Lewis, E.J. et al. (1987) Proc. Natl Acad. Sci. USA 84, 3550–3554.

226 Imai, E. et al. (1990) Mol. Cell. Biol. 10, 4712–4719.

227 Imai, E. et al. (1993) J. Biol. Chem. 268, 5353–5356.

228 Scott, D.K. et al. (1998) Mol. Endocrinol. 12, 482–491.

229 Garlatti, M. et al. (1994) Mol. Cell. Biol. 14, 8007–8017.

230 Christoffels, V.M. et al. (1998) Mol. Cell. Biol. 18, 6305–6315.

231 Richardson, J. et al. (1999) Mol. Endocrinol. 13, 546–554.

232 Slater, E.P. et al. (1993) Mol. Endocrinol. 7, 907–914.

233 Altuvia, S. et al. (1993) J. Biol. Chem. 268, 27127–27132.

234 Samuelsson, M.K.R. et al. (1999) Mol. Endocrinol. 13, 1811–1822.

235 Suh, D.S. et al. (1996) Mol. Endocrinol. 10, 1227–1237.

236 Slater, E.P. et al. (1985) Mol. Cell. Biol. 5, 2984–2992.

237 Doppler, W. et al. (1989) Proc. Natl Acad. Sci. USA 86, 104–108.

238 Tavianini, M.A. et al. (1989) Mol. Endocrinol. 3, 605–610.

239 Drouin, J. et al. (1989) Genome 31, 510–519.

240 Akerblom, I.E. et al. (1988) Science 241, 350–354.

241 Malkoski, S.P. et al. (1997) Mol. Cell. Endocrinol. 127, 189–199.

242 Meyer, T. et al. (1997) J. Biol. Chem. 272, 30709–30714.

243 Webster, J.C. et al. (1997) J. Biol. Chem. 272, 9287–9293.

244 Philips, A. et al. (1997) Mol. Cell. Biol. 17, 5952–5959.

245 Drouin, J. et al. (1998) J. Steroid Biochem. Mol. Biol. 65, 59–63.

246 Pei, L. (1996) J. Biol. Chem. 271, 20879–20884.

247 Mordacq, J.C. and Linzer, D.I. (1989) Genes Dev. 3, 760–769.

248 Starr, D.B. et al. (1996) Genes Dev. 10, 1271–1283.

249 Song, C.Z. et al. (1999) Proc. Natl Acad. Sci. USA 96, 11776–11781.

250 Heck, S. et al. (1994) EMBO J. 13, 4087–4095.

251 Dittmar, K.D. et al. (1998) J. Biol. Chem. 273, 7358–7366.

252 Oakley, R.H. et al. (1999) J. Biol. Chem. 274, 27857–27866.

253 Sheldon, L.A. et al. (1999) Mol. Cell. Biol. 19, 8146–8157.

254 De Bosscher K. et al. (2000) Proc. Natl Acad. Sci. USA 97, 3919–3924.

255 Wallberg, A.E. et al. (1999) Mol. Cell. Biol. 19, 5952–5959.

256 Radoja, N. et al. (2000) Mol. Cell. Biol. 20, 4328–4339.

257 Reynolds, P.D. et al. (1997) J. Clin. Endocrinol. Metab. 82, 465–472.

258 Pratt, W.B. and Dittmar, K.D. (1998) Trends Endocrinol. Metab. 9, 244–252.

259 Picaed, D. et al. (1990) Nature 348, 166–168.

260 Danielsen, M. et al. (1986) EMBO J. 5, 2513–2522.

261 Mielfeld, R. et al. (1986) Cell 46, 389–399.

262 Gao, X. et al. (1994) Biochim. Biophys. Acta 1218, 194–198.

263 Stafford, J.M. et al. (2000) J. Biol. Chem. (in press).

264 Aittomäki, S. et al. (2000) J. Immunol. 164, 5689–5697.

265 Windahl, S.H. et al. (1999) J. Steroid Biochem. Mol. Biol. 15, 93–102.
266 Sherman, M.P. et al. (2000) J. Virol. 74, 8159–8165.
267 Herr, A.S. et al. (2000) Mol. Pharmacol. 57, 732–737.
268 Huss, J.M. and Kasper, C.B. (2000) Mol. Pharmacol. 58, 48–57.
269 Hong, H. et al. (1999) J. Biol. Chem. 274, 3496–3502.
270 Sousa, R.J. et al. (1989) Mol. Endocrinol. 3, 481–494.
271 Abu, E.O. et al. (2000) J. Clin. Endocrinol. Metab. 85, 883–889.
272 Matthes, H. et al. (1994) Endocrinology 135, 476–479.
273 Shemshedini, L. et al. (1991) EMBO J. 10, 3839–3849.
274 Vig, E. et al. (1994) Mol. Endocrinol. 8, 1336–1346.
275 Wei, P. et al. (1998) Mol. Endocrinol. 12, 1322–1333.
276 McCormick, J.A. et al. (2000) Mol. Endocrinol. 14, 506–517.
277 Tuckerman, J.P. et al. (1999) J. Cell. Biology. 147, 1365–1370.
278 Reichardt, H.M. et al. (2000) Biol. Chem. 381, 961–964.
279 Tronche, F. et al. (1999) Nature Genet. 23, 99–103.
280 Reichardt, H.M. et al. (2000) Mol. Cell. Biol. 20, 9009–9017.
281 Mc Nally, J.G. et al. (2000) Science 287, 1262–1265.
282 Nissen, R.M. and Yamamoto, K.R. (2000) Genes Dev. 14, 2314–2329.

Names

MR is a steroid receptor closely related to the glucocorticoid receptor. It binds mineralocorticoids, the hormones involved in the control of salt and water homeostasis, by regulating sodium, potassium and hydrogen transport across tight epithelia. MR are known from a wide variety of vertebrates but are unknown outside this phylum. The receptor is a large protein of 984 amino acids in humans.

Species	Latin name	Other names	Accession number	References
Official name: NR3C2				
Human	*Homo sapiens*	MR	M16801	1
Squirrel monkey	*Saimiri sciureus*	MR	AF245224	Unpub.
Rat	*Rattus* sp.	MR	M36074	43
Tupaia	*Tupaia belangeri*	MR	Z75077	Unpub.
African clawed frog	*Xenopus laevis*	MR	U15133	48
Rainbow trout	*Oncorhynchus mykiss*	MR	AF209873	49

Isolation

The MR cDNA was identified by Ron Evans's laboratory by low-stringency screening of a genomic library with a human glucocorticoid receptor probe[1]. The resulting cDNA was expressed in COS cells and shown to bind aldosterone with high affinity and to activate transcription in response to aldosterone. It was also demonstrated in the same report that this molecule shows high affinity for glucocorticoids and can regulate transcription in response to this hormone, suggesting a complex interplay between the two hormones. The MR gene has been cloned recently in *Xenopus* as well as in fish[48,49]. In the DBD, MR exhibits 91%, 90% and 77% sequence identity with GR, PR and AR, respectively. These values are 56%, 55% and 51% in the LBD.

DNA binding

When it was discovered, the MR cDNA clone was shown to code a protein able to bind specifically to the LTR of the MMTV and to regulate transcription of an MMTV-CAT reporter plasmid in response to aldosterone[1]. This is consistent with the sequence of the P-box of this receptor, which is identical to that of GR, PR and AR (CGSCKV). As for these receptors, MR binds to the palindromic element GGTACAnnnTGTTCT. No specific target sequence for MR has ever been found in contrast to the situation for other steroid receptors such as AR.

MR binds to its sequence as an homodimer but recent data suggest that MR is also able to heterodimerize with GR, leading to synergy or inhibition of transcription depending on the promoter or the cellular context[3,4].

The subcellular distribution of MR has been a matter of discussion. In studies using antibodies and confocal microscopy it has been shown that, without the ligand, the MR is entirely cytoplasmic and that 30 minutes after addition of the ligand, the receptor was perinuclear and after 60 minutes predominantly nuclear[5]. These data have been refined using GFP fusion proteins[6]. It has been shown using this method that, without the ligand, MR is both cytoplasmic and nuclear, and that ligand addition in 10 minutes promotes the reorganization of the fusion protein in the nucleus where the labeling is present as thousands of small spots that may be clustering of the protein on defined chromatin regions.

Partners

The MR without ligand is bound to heat-shock proteins, such as hsp70, hsp90 or the immunophilin FKBP-52[7-9]. The binding of the ligand dissociates this complex and promotes the entrance in the nucleus of the receptor[7,24]. As for other steroid receptors, these proteins may function as chaperones in order to stabilize the MR structure. Other possible roles of the interaction between steroid receptors (including MR) and heat-shock proteins is still a matter of controversy (see references 10 and 47 for a review).

As discussed above, MR and GR can cooperatively form heterodimers on classic GRE. The transcriptional response pattern of these heterodimers is distinct from that obtained by MR or GR alone, suggesting that it plays a hitherto unrecognized role[3,4] (see reference 11 for a review).

MR has been shown to interact with the TIF2 coactivator[50].

Ligands

It has been demonstrated that aldosterone binds to MR with a high affinity $(K_d$ 1.3 nM)[1]. Nevertheless, it was immediately realized that glucocorticoids also bind equally well to MR[1,31]. In fact, MR was found to be approximately 10-fold more sensitive to cortisol than GR. At the transcriptional levels, it has been shown that the level of transcriptional activation obtained with glucocorticoids markedly differs between MR and GR. The maximal activation by MR is only 5–15% of the levels obtained with GR[12]. This suggests that each of these receptors has its own dynamic of response to glucocorticoids and forms a binary response system for these hormones[13].

In these conditions, how can a specific regulation by aldosterone be achieved in target tissues? It has been demonstrated that this specificity is mediated by the metabolism of the hormone in the target tissue rather than at the level of the receptors[14]. Glucocorticoids, which circulate at approximately 1000-fold higher levels than aldosterone, are excluded from mineralocorticoids target tissues by the action of the enzyme 11β-hydroxysteroid deshydrogenase type 2, which converts cortisol to its inactive metabolite cortisone[14]. Aldosterone, in contrast, is not a substrate for this enzyme and is thus protected from degradation. Additional specificity-conferring mechanisms include intrinsic MR specificity for aldosterone[15], specific cellular factors modulating ligand sensitivity[16], intracellular hormone availability[17,18] and differential interference with other transcription factors such as AP-1[19]. This latter report has shown that, in contrast to GR, MR does not interfere with AP-1 activity.

Several antagonists of MR, such as the synthetic steroids spirolactones or RU26752, are used in the treatment of sodium-retaining states and as hypertensive agents[20]. These compounds bind the receptor with an affinity identical to that of aldosterone and induce a receptor conformation that is transcriptionally silent, despite their smaller size compared with agonists[21]. Strikingly, progesterone also acts as an antagonist on the MR[23]. The structural basis of the action of agonists and antagonists on MR has been studied by constructing a homology model of the human MR LBD as well as by mutation analysis of the LBD[22,24].

Three-dimensional structure

No available 3D structures are available for either the LBD or the DBD of MR. Nevertheless, a homology model of the 3D structure of the LBD of human MR was built taking as a basis the structure of the LBD of RARγ[22]. The 3D model reveals the existence of two polar sites located at the extremities of the elongated hydrophobic ligand-binding pocket and provide a structural basis for the activity of antagonist ligands. This work suggests that on antagonist-bound receptors, the AF-2 core region is destabilized by the loss of contact between the antagonist and the H12 helix. The structural basis of the aldosterone-binding selectivity of MR has also been explored by another study mixing structure–function analysis and homology model building[46].

Expression

When it was discovered, MR was shown to be expressed as a *c.* 6 kb mRNA in kidney, gut, brain, pituitary, hypothalamus and heart[1]. Classic aldosterone target tissues like sodium-transporting epithelia, such as, in the kidney, the distal part of the nephron, the distal colon, the salivary and sweat glands express MR. Data are accumulating that support the expression of MR and the action of aldosterone in many other tissues, such as the central nervous system (hippocampus), the cardiovascular system, blood cells and epidermis among others (reviewed in references 20, 25, 32).

Activity and target genes

MR is a transcriptional activator less potent than its close relative, GR[12,27]. In addition, in contrast to GR, MR does not show cooperative activity on promoters containing multiples GREs[27]. Most importantly, it has been demonstrated that the N-terminus of MR does not provide the strong AF-1 transactivation function present in the equivalent GR domain[27].

Physiological evidence suggests a cross-talk between cAMP and aldosterone signaling. This has been demonstrated directly by showing that protein kinase A modulates the transcriptional activity of human MR[26]. A synergistic activation was achieved when cells were treated with both aldosterone and cAMP. Of note, the binding of MR to DNA was enhanced by PKA and the N-terminal of MR was required for this effect. Nevertheless, PKA is apparently unable to phosphorylate MR directly and it was concluded that PKA acts indirectly on MR, probably by relieving the effects of a repressor[26].

Known target genes for MR are linked to the physiological action of mineralocorticoids. Of note, given the lack of specificity of MR action versus GR *in vitro* or in transient transfection systems, the *in vivo* significance of target genes is even more important to establish than for other nuclear receptors. The most prominent action of aldosterone is the control of Na^+ and K^+ homeostasis, which is maintained mainly by the regulation of Na^+ reabsorption in the kidney. The Na^+/K^+-ATPase in the collecting ducts of the kidney was thus scrutinized as a target gene for aldosterone on which a synergistic action of cAMP and aldosterone was observed[26]. The same situation holds for the 11β-hydroxysteroid deshydrogenase, which is, logically, a target gene of MR (see reference 26 for references). The gene encoding amiloride-sensitive Na^+ channels (ENaC) in the kidney is also a target of aldosterone, although the precise mechanism of action of the hormone on this gene is still unclear. The cyclooxygenase-2 (COX-2) gene has been shown to be a target gene to GR and MR in the rat renal cortex[44]. Interestingly, results obtained on knockout mice suggest that the expression of Na^+/K^+-ATPase and ENaC is not directly controlled by MR[28].

Knockout

To gain insight into the function of MR and to separate it from that of GR knockout of the MR gene was performed by Gunther Schutz's laboratory[28]. Mice with a null mutation in the MR gene die c. 10 days after birth and exhibit impaired amiloride-sensitive Na^+ reabsorption, leading to changes similar to type 1 pseudohypoaldosteronism. The impaired Na^+ reabsorption results in severe Na^+ and water loss causing mortality in neonates because they are unable to compensate these losses by increasing Na^+ and water uptake. Strikingly, amiloride-sensitive Na^+ reabsorption is reduced in these mice but the abundance of the transcripts encoding the ENaC and Na^+/K^+-ATPase is unchanged in knockout mice. This leads to the conclusion that expression of these genes is not MR dependent. This suggests that the MR-mediated control of Na^+ reabsorption is not achieved by transcriptional regulation of ENaC and Na^+/K^+-ATPase but by the transcriptional control of other yet unidentified genes[28]. It was of interest to study the expression of the components of the renin–angiotensin system that regulated blood pressure via aldosterone in these mice, since they present a condition of extreme sodium depletion that is never observed in wild-type animals[29]. It was shown that renin was the most stimulated component in these mice in kidney and adrenal. Angiotensinogen and angiotensin II receptor AT_1 were also increased in the liver only. Other components of the renin–angiotensin system were not affected[29].

Associated disease

Pseudohypoaldosteronism is a rare syndrome of mineralocorticoid resistance presenting in the newborn with urinary salt loss and dehydratation. Studies of patients with pseudohypoaldosteronism have shown that no major rearrangement was present in the human MR gene and that no abnormality was found in the coding region, 5'UTR and regulatory regions[33-35]. This syndrome appears to be linked to defects in the subunits of ENaC (see references 15 and 25 for references).

It was recently shown that the transcript of an isoform of MR, MRβ (see below), is greatly reduced in the sweat gland ducts of patients in the situation of positive sodium balance found in hyperaldosteronism (Conn's syndrome) or when the epithelium sodium channels are constitutively activated (Liddle's syndrome)[36].

A mutation located in the LBD of MR has been found associated with early-onset hypertension that is markedly exacerbated by pregnancy[51]. This mutation, S810L, located in helix H5 disrupts an interaction between helix H5 and H3 that is highly conserved among nuclear receptors. This result in a constitutively active receptor with an altered specificity versus progesterone and other steroids lacking a 21-hydroxyl group.

Gene structure, promoter and isoforms

The genomic organization of the human MR gene has been determined[38]. The gene contained 10 exons, including two alternative ones (1α and 1β) that encode two different 5' UTR. The location of the intron inside the conserved regions of the coding sequence is identical to that of other steroid receptors.

The two alternative first exons generate different transcripts that encode the same protein, since the ATG is in exon 2, but are controlled by two different promoters. These two promoters are TATA-less and GC-rich[38,41]. Interestingly, the MRβ (and not α) isoform expression is decreased in human patients with a positive sodium balance[36]. The distinct tissue-specific utilization of these two alternative promoters was studied by an elegant study in which the oncogenic large T antigen of the SV40 virus was placed under the control of each of them[45]. These results indicate that P1 is active in all MR-expressing tissues, directing strong expression in testis, salivary glands, moderate expression in lung, brain, uterus, liver, and heart, and low expression, in contrast to its mouse homologue, in colon and kidney. P2 was 10 times less active than P1 with no activity in brain and colon[45].

In rat, at least three different isoforms exist. Interestingly, the specific part of rat and human MRβ shares 85% of sequence identity, suggesting that it contains conserved and important regulatory elements. Again these three different isoforms start on three different promoters[39,40,43].

In rat, a variant of MR with a 12 bp insertion in the exons coding the DBD, resulting in the insertion of four amino acids in this domain[42]. This isoform is expressed in brain, liver, kidney and heart. Interestingly, an homologous isoform was detected in human[42]. Functional and physiological relevance of this isoform has not been studied (see reference 25 for an excellent review on MR isoforms).

Chromosomal location

The human MR gene was located to chromosome 4 when it was discovered[1]. This localization has been refined to 4q31.1[2,37].

Amino acid sequence for human MR (NR3C2)

Accession number: M16801

```
  1 METKGYHSLPEGLDMERRWGQVSQAVERSSLGPTERTDENNYMEIVNVSCVSGAIPNNSTQGSSKEKQEL
 71 LPCLQQDNNRPGILTSDIKTELESKELSATVAESMGLYMDSVRDADYSYEQQNQQGSMSPAKIYQNVEQL
141 VKFYKGNGHRPSTLSCVNTPLRSFMSDSGSSVNGGVMRAIVKSPIMCHEKSPSVCSPLNMTSSVCSPAGI
211 NSVSSTTASFGSFPVHSPITQGTPLTCSPNAENRGSRSHSPAHASNVGSPLSSPLSSMKSSISSPPSHCS
281 VKSPVSSPNNVTLRSSVSSPANINNSRCSVSSPSNTNNRSTLSSPAASTVGSICSPVNNAFSYTASGTSA
351 GSSTLRDVVPSPDTQEKGAQEVPFPKTEEVESAISNGVTGQLNIVQYIKPEPDGAFSSSCLGGNSKINSD
421 SSFSVPIKQESTKHSCSGTSFKGNPTVNPFPFMDGSYFSFMDDKDYYSLSGILGPPVPGFDGNCEGSGFP
491 VGIKQEPDDGSYYPEASIPSSAIVGVNSGGQSFHYRIGAQGTISLSRSARDQSFQHLSSFPPVNTLVESW
561 KSHGDLSSRRSDGYPVLEYIPENVSSSTLRSVSTGSSRPSKICLVCGDEASGCHYGVVTCGSCKVFFKRA
631 VEGQHNYLCAGRNDCIIDKIRRKNCPACRLQKCLQAGMNLGARKSKKLGKLKGIHEEQPQQQQPPPPPPP
701 PQSPEEGTTYIAPAKEPSVNTALVPQLSTISRALTPSPVMVLENIEPEIVYAGYDSSKPDTAENLLSTLN
771 RLAGKQMIQVVKWAKVLPGFKNLPLEDQITLIQYSWMCLSSFALSWRSYKHTNSQFLYFAPDLVFNEEKM
841 HQSAMYELCQGMHQISLQFVRLQLTFEEYTIMKVLLLLSTIPKDGLKSQAAFEEMRTNYIKELRKMVTKC
911 PNNSGQSWQRFYQLTKLLDSMHDLVSDLLEFCFYTFRESHALKVEFPAMLVEIISDQLPKVESGNAKPLY
981 FHRK   984
```

References

1 Arriza, J.L. et al. (1987) Science 237, 268–275.
2 Morrison, N. et al. (1990) Hum. Genet. 85, 130–132.
3 Trapp, T. et al. (1994) Neuron 13, 1457–1462.
4 Liu, W. et al. (1995) Proc. Natl Acad. Sci. USA 92, 12480–12484.
5 Robertson, N.M. et al. (1993) Mol. Endocrinol. 7, 1226–1239.
6 Fejes-Toth, G. et al. (1998) Proc. Natl Acad. Sci. USA 95, 2973–2978.
7 Couette, B. et al. (1996) Biochem. J. 315, 421–427.
8 Bruner, K.L. et al. (1997) Recept. Signal Transduc. 7, 85–98.
9 Bamberger, C.M. et al. (1997) Mol. Cell. Endocrinol. 131, 233–240.
10 Smith, D.F. and Toft, D.O. (1993) Mol. Endocrinol. 7, 4–11.
11 Trapp, T. et al. (1996) Trends Pharmacol. Sci., 17, 145–149.
12 Arriza, J.L. et al. (1988) Neuron 1, 887–900.
13 Evans, R.M. and Arriza, J.L. (1989) Neuron 2, 1105–1112.
14 Funder, J.W. et al. (1988) Science, 242, 583–586.
15 Lombès, M. et al. (1994) Endocrinology 135, 834–840.
16 Lim-Tio, S.S. et al. (1997) Endocrinology 139, 2537–2543.
17 Ueda, K. et al. (1992) J. Biol. Chem. 267, 24248, 24252.
18 Kralli, A. et al. (1995) Proc. Natl Acad. Sci. USA 92, 4701–4705.
19 Pearce, D. and Yamamoto, K.R. (1993) Science 259, 1161–1165.
20 Funder, J.W. (1997) Annu. Rev. Med. 48, 231–240.
21 Couette, B. et al. (1996) Biochem. J. 315, 421–427.
22 Fagart, J. et al. (1998) EMBO J. 17, 3317–3325.
23 Souque, A. et al. (1995) Endocrinology, 136, 5651–5658.
24 Couette, B. et al. (1998) Mol. Endocrinol. 12, 855–863.
25 Zennaro, M-C. and Lombès, M. (1998) Curr. Opin. Endocrinol. Diabetes 5, 183–188.
26 Massaad, C. et al. (1999) Mol. Endocrinol. 13, 57–65.
27 Rupprecht, R. et al. (1993) Mol. Endocrinol. 7, 597–603.
28 Berger, S. et al. (1998) Mol. Endocrinol. 95, 9424–9429.

[29] Hubert, C. et al. (1999) Mol. Endocrinol. 13, 297–306.

[30] Staub, O. et al. (1996) EMBO J. 15, 2371–2380.

[31] Funder, J.W. (1992) J. Steroid Biochem. Mol. Biol. 43, 389–394.

[32] Joels, M. and Kloet, E.R. (1992) Trends Neurosci. 15, 25–30.

[33] Zennaro, M–C. et al. (1994) J. Clin. Endocrinol. Metab. 79, 32–38.

[34] Komesaroff, P. et al. (1994) J. Clin. Endocrinol. Metab. 79, 27–31.

[35] Arai, K. et al. (1995) J. Clin. Endocrinol. Metab. 80, 814–817.

[36] Zennaro, M-C. et al. (1997) J. Clin. Invest. Metab. 82, 1345–1352.

[37] Fan, Y-S. et al. (1989) Cytogenet. Cell Genet. 52, 83–84.

[38] Zennaro, M-C. et al. (1995) J. Biol. Chem. 270, 21016–21020.

[39] Castren, M. and Damm, K. (1993) J. Neuroendocrinol. 5, 461–466.

[40] Kwak, S.P. et al. (1993) Endocrinology 133, 2344–2350.

[41] Listwak, S.J. et al. (1996) J. Steroid Biochem. Mol. Biol. 58, 495–506.

[42] Bloem, L.J. et al. (1995) J. Steroid Biochem. Mol. Biol. 55, 159–162.

[43] Patel, P.D. et al. (1989) Mol. Endocrinol. 3, 1877–1885.

[44] Zhang, M.J. et al. (1999) Proc. Natl Acad. Sci. USA 96, 15280–15285.

[45] Le Menuet, D. et al. (2000) J. Biol. Chem. 275, 7878–7886.

[46] Rogerson, F.M. et al. (1999) J. Biol. Chem. 274, 36305–36311.

[47] Cheung, J. and Smith, D.F. et al. (2000) Mol. Endocrinol. 14, 939–946.

[48] Csikos, T. et al. (1995) Recent Prog. Horm. Res. 50, 393–396.

[49] Colombe, L. et al. (2000) Steroids 65, 319–328.

[50] Hong, H. et al. (1997) Mol. Cell. Biol. 17, 2735–2744.

[51] Geller, D.S. et al. (2000) Science 289, 119–123.

PR

Names

Progesterone receptor is a steroid receptor related to GR, AR and MR. It binds progesterone, a hormone involved in female reproductive physiology, especially during pregnancy, in mammals as well as a well-known antagonist used for contragestion, RU486. PR is known from a wide variety of vertebrates but is unknown outside this phylum. The receptor is a large protein of 933 amino acids in human.

Species	Other names	Accession number	References
Official name: NR3C3			
Human	PR	M15716	1
Mouse	PR	M68915	2
Rat	PR	L16922	3
Rabbit	PR	M14547	4
Sheep	PR	Z66555	Unpub.
Chicken	PR	Y00092	5
Crocodile	PR	AF030321	Unpub.
Lizard	PR	S79937	6
Eel	PR	AB032075	7

Isolation

The progesterone receptor was cloned by the laboratories of Pierre Chambon, Edwin Milgrom and Bert O'Malley in 1986 using essentially the same strategies[4,11,14]. It has been shown that PR exists in various forms, called A and B, that are both able to bind to DNA and to the hormone[8,9]. Using affinity-labeling techniques, two forms of the PR were purified from chicken oviduct and antibodies that were able to recognize them were produced[10]. These antibodies were used to isolate cDNA clones using cDNA expression libraries[4,11,14]. The resulting sequence was shown to contain a Cys-rich domain, as in GR, ER and v-erbA, and thus to be a member of the superfamily. This was confirmed by the cloning of the full-sized cDNA that harbors a strong identity level with GR and ER and has the same structural domains[4,5,11,14]. It was realized that the two forms of PR are products of the same gene, the A form corresponding to an N-terminally truncated form B.

In the DBD, the progesterone receptor exhibits 90%, 88% and 80% amino-acid identity with MR, GR and AR, respectively. For the LBD, these values are 57%, 54% and 54%, respectively.

DNA binding

Deletion analysis of the chicken PR clone demonstrates that the C domain was responsible for the specific DNA binding and was not implicated in transactivation[5]. It was already known from experiments using purified PR that it was able to bind to DNA in a sequence-specific manner[15]. The direct binding of PR to DNA was observed directly by electron microscopy[16]. Gel-shift experiments demonstrate that PR binds to DNA as a dimer, with each monomer recognizing in a cooperative fashion a half-site sequence of the hormone response element of the TAT gene that contains the imperfect palindromic sequence GGTACAnnnTGTTCT[17,18]. This sequence is also bound by other steroid receptors, AR, GR and MR. No specific target sequences have been described for this receptor, which binds DNA in a manner indistinguishable from that of GR[21]. As with other steroid receptors, PR can dimerize *in vitro* in the absence of hormone or DNA[19]. The hormone was nevertheless found to induce dimerization. Interestingly, it was shown that the PR-A and PR-B isoforms can form homodimers as well as heterodimers[138]. Heterodimerization with the other steroid receptors has not been demonstrated. In contrast, the binding sites of ER and PR are different and the molecular basis of these differences has been studied in the context of the MMTV promoter. This has confirmed the important role of the P-box and the D-box for the specific recognition of the HRE[20].

The binding of PR to DNA in the context of reconstituted chromatin has been extensively studied by Miguel Beato group (see references 12 and 26, and references therein). In contrast to its behavior as naked DNA, the MMTV promoter assembled in minichromosomes can be activated synergistically by PR and the transcription factor NF-1 in an ATP-dependent manner[26]. It has been proposed that the receptor triggers a chromatin remodeling event that facilitates the access of NF-1, which in turn stabilizes an open nucleosomal conformation required for efficient binding and full transactivation.

The PR is able to bend its DNA target as TR, ER or RXR[22]. The two isoforms PR-A and PR-B induced different distortions to the DNA molecule but the ligands, either agonists or antagonists, do not induce any change in the bending produced by the receptor. In addition, the HMG-1 protein, which acts as an accessory factor for DNA binding by PR, had minimal influence on the DNA bending.

The intracellular trafficking of the progesterone receptor has been studied extensively. The rabbit PR was always found to be nuclear when transiently expressed into different cell types. Mutation analysis reveals an NLS harboring identity with the SV40 T antigen NLS, in the D domain[23]. In addition, two other NLS signals were localized, one close at the very end of the DBD, adjacent to the first defined signal and the other in the LBD[27]. It has been shown that these three signals cooperate for efficient nuclear localization of the protein. It has been shown that a mutant PR, in which the classical NLS was deleted, can enter in the nucleus after hormone treatment because it interacts with wild-type receptors that contain NLS. In addition, further experiments show that, in living cells, the nuclear residency of PR reflects a dynamic situation. The receptor diffuses into the cytoplasm, but it is constantly and actively transported into the nucleus[24]. Recently, it has been shown using GFP-fusion proteins that the PR-A and PR-B isoforms have distinct intracellular distributions into living cells, PR-A being more nuclear than PR-B, especially in cells untreated by progesterone[25].

Partners

In the absence of hormone or other activating signals, steroid receptors, such as PR, exist in heteromeric complexes with heat-shock proteins (Hsp) and additional components of the molecular chaperone machinery (reviewed in references 28 and 29). It has been shown that the interaction between steroid receptors and Hsp allows the complete folding of the receptor, and that ligand binding induces the dissociation of the complex and the nuclear entry and DNA binding of the receptor[28,29]. The assembly pathway for steroid receptor–chaperone complexes can involve at least 10 chaperone components, five of which are obligatory *in vitro* using reticulocytes lysates as an assay system for the assembly of the PR–chaperone complex[30,33]. Three of the obligatory factors are constitutively expressed forms of Hsp70, DnaJ/Hsp40 and Hsp90[28,29,30]. Two cochaperone proteins, the Hsp90 binding protein, p23 and the Hsp70/Hsp90 organizing protein, Hop, which brings Hsp70 and Hsp90 together in a common complex, are also required[29,32]. Other non-essential factors include two Hsp70-binding proteins, Hip (Hsp70 interacting protein) and BAG-1. Four Hsp90-binding proteins are also found in these complexes. These are the FK506-binding immunophilins, FKBP51 and FKBP52, the cyclosporin-A-binding immunophilin cyclophilin 40 (Cyp40) and the protein phosphatase PP5. Each of these proteins compete for Hop for binding and, like Hop, contain a tetratricopeptide repeat (TPR) domain that mediates their competitive binding to Hsp90[28,29,31]. As in many chaperone processes, ATP is required for the assembly of the PR–chaperone complexes. In addition to the chaperone function of this complex, it has recently been proposed that the immunophilin-binding FKBP51 protein can have a physiologically relevant modulatory effect on the receptor function that goes beyond the folding and stabilization of a non-native receptor substrate[29]. It has also been proposed that chaperone–PR interaction not only stabilizes a hormone-binding conformation of PR but also inhibits its ability to bind DNA, dimerize or interact with coactivators[29,33,34]. In that sense, these complexes play an important repressive role, which ensures that the receptor will not activate transcription inappropriately when activating signals are not present. In addition, it has been proposed that Hsp complexes may play a role after the binding of the ligand to its receptor, for example, in the recycling of the receptor. More experiments are nevertheless needed to demonstrate this hypothesis experimentally[29].

It has been shown that PR also interacts with the high mobility group protein HMG-1 that enhances its binding to DNA[35,36]. HMG-1 is active on both PR-A and PR-B isoforms and does not modify the bending of the DNA molecule induced by PR binding[22,35]. An interaction with the related HMG-2 protein was also noticed and it was also shown that this interaction occurred with other steroid receptors, such as GR, but not with RAR, RXR or VDR[37]. It is the HMG domain B of HMG-1 and -2 that is apparently able to enhance PR DNA binding. A ternary complex between DNA, PR and HMG protein was observed as well as a direct HMG–PR complex in the absence of DNA. Of note, this interaction results in an enhancement of the PR transcriptional activity[37].

Interactions between PR and classical transcription factors were also observed. This is the case for the Stat5 factor that transduces the signal induced by cell-surface growth factors and cytokines[38]. It has been shown that the PR–Stat5 interaction results in an increased Stat5 nuclear accumulation, and that Stat5 and PR synergize on some promoters, such as p21^{WAF} or c-*fos*. In addition, it has been shown that Stat5 gene expression is up-regulated by progesterone, suggesting that a convergence and synergy of the two signaling pathways may occur in some organs, such as the mammary gland[38]. An interaction was also demonstrated between PR and another transcription factor, p65, the subunit of the NFκB complex expressed from the RelA gene[64,139]. This interaction results in a mutual transrepression effect that was also observed for the glucocorticoid receptor (see the entry for GR).

An interaction between PR and the general transcription factor TAF$_{II}$110, a subunit of the TFIID complex has been described[39]. Surprisingly, the AF-1 and AF-2 functions of PR are not required for this interaction, which depends on the DBD. It has been shown in transient transfection assays that the DBD of PR exhibits some transactivation potential[39]. Thus it is believed that the PR–TAF$_{II}$110 interaction plays a role in the transcriptional activation by the liganded PR. The role of the hormone in this interaction has not been studied. Interestingly, PR may also interact with TFIID via another subunit of this complex. It has been shown in yeast that PR interacts with RSP5, a protein that belongs to the same activation pathway as the SPT3 factor, which directly contacts yeast TBP[40]. SPT3 enhances the positive effects of RSP5 on PR transactivation both in yeast and mammalian cells. This activating pathway is conserved in mammals, since PR interacts with RPF1, the human homologue of RSP5, and since SPT3 is the yeast homologue of TAF$_{II}$18, a component of the TFIID complex. This reinforces the importance of the functional link between PR and the TFIID complex for efficient transcriptional activation.

A number of studies have recently discussed the interaction existing between PR and coactivators and corepressors. Concerning corepressors, it has been shown that antagonist-bound PR has a conformation different from agonist-bound receptor and interacts with NCoR and SMRT[41]. Functional studies of the relevance of this interaction demonstrate that NCoR and SMRT expression may play a pivotal role in PR pharmacology (see below).

Hormone-dependent interactions between PR and the coactivators SRC-1 and CBP were demonstrated[42,45–47]. These interactions stimulate PR transcriptional activity in cell-free transcription assays in a ligand-dependent manner even from a naked DNA template, suggesting that the chromatin acetylation activity of these coactivators may not explain all their effects on PR activity[42]. The same study also demonstrates that the interaction between PR and SRC-1 and CBP is disrupted by the Adenovirus E1A protein that repressed PR-mediated transactivation. An interaction between PR and the Jun activation domain binding protein JAB1, which also interacts with SRC-1, resulting in an increased PR transcriptional activity, was recently demonstrated[44]. Interestingly, differential binding of SMRT and SRC-1 to the two isoforms of PR, PR-A and PR-B was shown to explain the opposite transcriptional activities of these two isoforms (see below)[43].

The PR LBD occupied by RU486 was used in a two-hybrid screen and, in addition to NCoR, L7/SPA, a 27-kDa protein containing a basic region

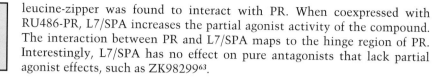

leucine-zipper was found to interact with PR. When coexpressed with RU486-PR, L7/SPA increases the partial agonist activity of the compound. The interaction between PR and L7/SPA maps to the hinge region of PR. Interestingly, L7/SPA has no effect on pure antagonists that lack partial agonist effects, such as ZK98299[63].

Ligands

The cloned PR expressed in HeLa cells binds to progestins in a manner indistinguishable from the receptor found in living cells[5]. The purified E domain of PR fused to the β-gal enzyme exhibit a K_d of 1.0–1.5 nM to progesterone and can be specifically cross-linked to the synthetic progestin R5020[48].

A number of different antagonists have been developed against PR. Among those, the most studied is the famous abortifacient RU486 compound, which has generated an extensive literature[50]. This compound inhibits the action of progestins in human, but not in chicken or hamster. This differential effect has been shown to be linked to a unique amino acid, Cys 575, in the LBD of the chicken PR, which is substituted by a Gly in the homologous position (722) of the human receptor[53]. This molecule binds to both PR and GR, and it has been shown that it exerts its antagonistic effects on these two receptors by different mechanisms[50,56]. Concerning PR, it has been shown that the RU486-bound PR can bind to DNA, suggesting that the antagonistic action of this compound is mainly at a post-DNA-binding step[51,52,56]. Interestingly, it has been shown that the dimerization interface of PR/RU486 and PR/R5020 are non-compatible, since no heterodimer between each complex can be observed[52]. Nevertheless, *in vivo*, using chromatin and not naked DNA, it has been suggested that, in contrast to the abovementioned model, antiprogestins interfere with receptor function by preventing its DNA binding[57]. It has been shown that the RU486/PR has an inactive AF-2 function, whereas the AF-1 function of the PR-B/RU486 complex remains active. In accordance with this notion, the shorter isoform PR-A is completely inactive when bound to RU486. Analysis of mutant PR in yeast and mammalian cells emphasized the importance of the C-terminal region of the receptor for the differential recognition of agonist and antagonists ligands, such as R5020 and RU486, respectively[54]. The C-terminal part of the receptor is required to bind progestins while antagonists bind to a site located closer to the N-terminal. This region has been shown to contain a 12-amino-acid domain that behaves as a transcriptional repressor domain inhibiting the transcriptional activity of the RU486-bound PR[61]. This model has been confirmed by the demonstration that agonists and antagonists induce different conformations of the receptor[49,55]. The inability of the antagonist-occupied receptor to activate transcription is hypothesized to be a consequence of its inability to associate with coactivators[45,46] and possibly its enhanced ability to recruit a corepressor[54,61]. This interaction with corepressors, consistent with the existence of a repressor domain[61] was demonstrated by the isolation of NCoR in a two-hybrid screen using the LBD of PR bound by RU486[63]. This was also confirmed by the observation that overexpression of NCoR or SMRT repress the partial agonist activity of RU486-bound PR[63].

In addition to RU486, other antiprogestins, such as ZK299, ZK98299, etc. were used and characterized for their effects on PR. These compounds also induce novel conformation of the receptor and some of them exhibit interesting mixed agonist activities[58-60,62]. It was demonstrated recently that the different transcriptional activities induced by the various types of PR ligands (agonists, antagonists and mixed agonists) may result from differential interactions with NCoR and SMRT as a direct result of the unique receptor conformational changes that these ligands induce on binding[41,60].

Three-dimensional structure

No 3D structure of the PR DBD is available. In contrast, the structure of the PR bound to progesterone has been determined by Paul Sigler's laboratory[65,66]. The overall fold of the receptor is similar to that found in related receptors, such as RAR, TR or ER. Nevertheless, the dimerization interface of PR is quite different to the one of other receptors[65,66]. The position of H12 and of the C-terminal extension prevents the formation of the classic dimer interface. Instead, a smaller and less stable interface is formed with residues of H11 and H12[65]. This structure reveals a ligand-induced stabilization of the C-terminal secondary structure that also suggests how RU486 may work. This compound is likely to displace H12 and the C-terminal extension from its normal position, clearly suggesting that indeed antagonist-bound PR exhibits a different conformation from that of progesterone-bound receptor[65]. Based on the 3D structure of this structure, a detailed mutagenesis analysis of PR and GR have explored the basis of the specificities of these receptors for their ligands[67].

Expression

The first studies of PR expression revealed that the chicken PR gene is expressed in oviduct as a series of three transcripts of 4.5, 4.0 and 3.9 kb, the amount of which increased after estradiol treatment[5,14]. No PR expression was found in brain, lung, liver and kidney in these preliminary analysis. In rabbit, a unique 5.9 kb transcript was observed in uterus and vagina and absent from liver, spleen, kidney or intestine[4]. In human, at least nine different transcripts from 2.5 to 11 kb were characterized (see reference 128 and references therein). PR expression was found in classical progesterone target tissues, such as uterus, ovary, vagina, fallopian tubes and breast, as well as some central nervous system regions, such as hypothalamus (see reference 50 for references). Further experiments reveal an expression of PR in the ventral medial nucleus of the hypothalamus as well as in pre-optic area and is required for a normal sexual behavior (see reference 113 and references therein).

In the ovary, studies performed in rhesus monkeys or women during the menstrual cycle consistently detected PR expression in the theca and surrounding stroma of growing follicles, but not in the granulosa cells (see references 69 and 70 for reviews). Expression of PR in luteinizing granulosa cells was detected after LH treatment. No expression in the oocyte was detected[70].

PR is expressed in the mammary gland (see reference 116 for a review). A colocalization between the Wnt4 protooncogene and PR was observed in the luminal compartment of the ductal epithelium[117]. The PR expression pattern in the mammary epithelium changes from puberty to pregnancy, starting from a uniform to a scattered expression pattern as branching and budding is initiated[119]. The effects of progesterone on breast tissue have been studied in a detailed manner but remain controversial. Depending on the experimental model system, the cell context and the duration of the treatment, progesterone can elicit either proliferative or antiproliferative effects on breast cancer epithelial cells (reviewed in reference 68).

Other sites of expression of the progesterone include, among others, thymus[71]. It has been shown that PR in thymic epithelial cells is required for pregnancy-related thymic involution, which is implicated in the lack of rejection of the fetus by the mother's immune system.

It has been known for a long time that in mammals PR expression is up-regulated by estrogen in reproductive tissues, whereas it is down-regulated by progestins (see references 50, 68, 70 for reviews, and see below).

Activity and target genes

The progesterone receptor is a transcriptional activator in the presence of its ligand and this effect is dependent on both the N-terminal A/B region and the LBD, which were later called AF-1 and AF-2 (see references 5 and 52 among others). It rapidly became obvious that the PR-A and PR-B isoforms contain different AF-1-activating functions, which account for their different transactivating potential[73]. The AF-1 function of human PR has been narrowed down to a 91-amino-acid region located in the C-terminal part of the A/B region, which is common between the A and B isoforms[78]. A third transactivation function, called AF-3, was mapped in the unique N-terminal segment of the PR-B isoform[80]. Depending on the promoter and cells used, AF-3 can activate transcription autonomously or can functionally synergize with AF-1 or AF-2. The AF-2 region was mapped in the C-terminal part of the LBD[141]. Interestingly, direct physical interactions exist between the N-terminal and C-terminal parts of the receptors that account for the transcriptional synergism between AF-1 and AF-2[83]. These interactions require a hormone agonist-induced conformational change of the LBD and are thus not produced by antagonists[83].

Using in vitro transcription assays and the MMTV promoter as a target gene, it was possible to observe a progesterone-dependent transcriptional activation in cell-free systems[75,76]. This allowed to demonstrate the fact that ligand-activated PR acts by facilitating the formation of a stable preinitiation complex at the target promoter[76].

Two lines of evidence led to the suggestion that the activating functions are regions of the receptor that interact with specific coactivators. First, the demonstration of a cell- and promoter-specific activity of these regions suggests that these coactivators are not present in the same amount in each cell type[74]. Second, a transcriptional interference or squelching was observed between ER and PR, suggesting that both receptors can interact with common coactivators present in small amounts in the cells[72].

A negative regulatory region was mapped in the C-terminal part of the progesterone receptor. When this region is mutated there is an increase of the transcriptional activity of the receptor[79]. This region was further studied and a transcriptional repressor domain of 12 amino acids was characterized that confers a repressive activity to an heterologous protein such as GAL4[61,82]. Interestingly, the C-terminal region of PR is implicated in the differential binding of agonists and antagonists such as RU486[54]. It was proposed that this region silences transcription of the receptor in the absence of the agonist that relieves its effect. Antagonists are not able to interfere with this repression domain[54]. In addition, it was shown that NCoR can interact with the LBD of PR bound by RU486, suggesting that antagonists may modulate the interaction between PR and corepressors[63]. This has been also shown for mixed antagonists[41]. Of note, a repressor domain was also mapped in the first 140 amino acids of the PR-A isoform[134]. Its functional relationship with the C-terminal repressor domain remains unclear.

Whereas steroid receptors such as GR or ER are known to exhibit a strong mutual transrepression with AP-1 factors, no such effect was shown for PR. In contrast, it was shown that apo-PR strongly increased AP-1 transcriptional activity and that the addition of the ligand reversed this stimulating effect[81]. In contrast, the mutual transrepression that exists between NFκB and GR was also observed between PR and NFκB. One possible mechanism for this effect is a direct physical interaction between the p65 subunit of NKκB and PR[64,139].

One of the striking activities of PR is its ability to be a ligand-dependent repressor of estrogen receptor activity[135]. The magnitude of the effect is dependent on the PR isoform and of the ligand type (agonist or antagonist). This effect is believed to be due to the ability of PR to interfere with the contacts that exist between ER and the transcriptional machinery, a process known as quenching.

As for ER, a ligand-independent activation of the progesterone receptor can occur. Dopamine was able to activate PR in transient transfection assays and induces chicken PR translocation from the cytoplasm to the nucleus[85]. Activation by dopamine and progesterone were dissociable and a unique residue, Ser 628, was shown to be required for dopamine activation but not for progesterone activation[85]. The fact that this residue may be phosphorylated in certain conditions *in vivo* suggests that the dopamine effect may be linked to a phosphorylation of the receptor.

Indeed, PR is phosphorylated and treatment with progesterone *in vivo* stimulates this phosphorylation[84]. A number of studies have attempted to map phosphorylation sites, to study the role of the hormone as well as of DNA binding and to link these phosphorylation events to specific pathways[86–92,94,95]. The strong transcriptional effect of 8-bromo-cAMP, an activator of protein kinase A (PKA), was extensively studied. It was first believed that it induces PR phosphorylation but it was recently shown that the net phosphorylation of PR does not change in response to 8-bromo-cAMP treatment[87,95]. It is now believed that this compound decreases the ability of the receptor to interact with corepressors, leading to activation[41]. Human PR is phosphorylated on multiple Ser residues in a manner that involves distinct groups of sites coordinately regulated by hormone and different kinases. For example, Ser 530, is phosphorylated in a hormone-dependent manner, an event which facilitates hormone-dependent activation of chicken PR[90]. Three phosphorylation sites were identified in the A/B region of human PR[91]. Two of these sites, Ser 294 and Ser 345, are common to PR-A and PR-B whereas the third, Ser 102, is unique to PR-B. In contrast to other sites, such as Ser 530, these sites are phosphorylated in response to hormonal treatment with a relatively slow kinetic[91]. In accordance with this low kinetic, Ser 294 is phosphorylated by MAP kinase and this event signals the degradation of PR by proteasome[92]. Other kinases implicated in PR phosphorylation are casein kinase II, which phosphorylates Ser 81 of human PR-B and cyclin-dependent kinases on Ser 162, 190 and 400 (see reference 93 for references). It is clear that PR-A and PR-B isoforms are differentially phosphorylated but the precise role of these phosphorylations on the activity of the two isoforms is far from being understood[93].

A large number of target genes of PR have been characterized. Some, such as the tyrosine amino transferase (TAT) gene and the MMTV LTR were used extensively as models for functional analysis of PR[96–98]. In addition, the PR gene itself was studied as a progesterone responsive gene (see below). PR target genes can be grouped in two main overlapping classes: (1) those which are regulated by progesterone in reproductive tissues such as ovary and uterus; and (2) those which are regulated in the mammary gland and breast cancer. Among the first group it is interesting to cite the vitellogenin II gene on which a synergism with estrogen exists[99], the chicken lysozyme gene that is regulated by several PREs located in the promoter of the gene and in a far upstream enhancer[100], the β2 adrenergic receptor gene in pregnant rat myometrium[101], matrix metalloproteinases such as collagenase and gelatinases A and B in human endometrium[102] as well as cathepsin L and the proteinase ADAMTS-1 during ovulation[105]. Interestingly, the homeotic gene Hoxa-10 is also regulated by progesterone during the implantation of the embryo in the mouse[104]. The anti-apoptotic role of progesterone in both mammary gland and reproductive tissue is currently being studied extensively (see reference 69 for a review). In mammary gland, the role of progesterone in controlling cell proliferation is still a matter of discussion (see reference 68 for a review). It has been shown that progestin has a biphasic action on the cell cycle, since it first stimulates and then inhibits the breast cancer cell cycle. Of note, progesterone increases the expression of TGFα, EGF receptor, c-fos and c-myc genes, while it inhibits the activity of CDK–cyclin complexes[106,108]. This may be due at least in part to an induction of CDK inhibitors such as p27[Kip1] or p18[Ink4c], but other mechanisms such as decreased G1 CDK activity are also possible[136]. The stimulatory effect of progesterone on the cell cycle in mammary epithelial cells has been explained by an induction of cyclin D1 expression by progestins[140]. In contrast, in mouse uterine epithelial cells, other data have suggested that progesterone inhibits estrogen-induced cyclin D1 and cdk4 nuclear translocation, resulting in a decreased cell proliferation[137]. These data highlight the complex and profound effects that progestins may have on cell proliferation both in breast as well as in reproductive tissues. The molecular basis of the effects of progesterone on cell cycle genes is not yet understood. Members of the jun family, such as c-jun and jun-B, are also regulated by progesterone, c-jun being up-regulated, whereas jun-B expression decreased[107]. Finally, it is interesting to note that some genes such as the α-fetoprotein gene are repressed by progesterone in the presence of either agonists or antagonists[109].

Knockout

The knockout of the PR gene has been carried out in the mouse[113]. Both male and female embryos homozygous for the PR gene mutation developed normally to adulthood. However, the adult PR-null females displayed defects in all reproductive tissues, such as an inability to ovulate, uterus hyperplasia and inflammation, limited mammary gland development and an inability to exhibit sexual behavior. All these results clearly suggest that progesterone plays the role of a pleiotropic coordinator of the reproductive function. Further study of the PR$^{-/-}$ mice led to the observation that PR is implicated in the immunological mechanism by which the fetus is not rejected by the mother's immune system[71]. It was shown that progesterone induces thymus involution during pregnancy, which induces a block of T-cell lymphopoiesis during pregnancy.

The PR$^{-/-}$ mice exhibit abnormal mammary gland development. The ductal growth accompanying puberty is not compromised in these mice but lateral ductal branching and lobulo-alveolar growth characteristic of development accompanying pregnancy do not occur[113]. Nevertheless, since these animals do not exhibit estrous cycles and fail to become pregnant, the full extent of the mammary gland defect cannot be studied. For that reason, PR$^{-/-}$ breasts were transplanted into wild-type mice[114]. In this situation the development of the mammary gland is arrested at the stage of the simple ductal system found in the virgin mouse. It was demonstrated that the primary target for progesterone is the mammary epithelium and that the mammary stroma is not required for the progesterone response[114]. The relationships between C/EBPβ and PR were also studied in the context of the PR$^{-/-}$ mice[119]. It was suggested that PR acts downstream of C/EBPβ.

Recently, an interesting connection between progesterone and Wnt signalling was observed[116,117]. The defect in mammary gland development observed in PR$^{-/-}$ animals can be overcome by ectopic expression of the proto-oncogene Wnt1[117]. In addition, PR and Wnt4, another member of the Wnt family, are coexpressed in the luminal compartment of the ductal epithelium. In accordance with this coexpression, progesterone regulates Wnt4 expression in mammary epithelial cells. It remains to be established whether this regulation is direct or not. Interestingly, it was shown that Wnt4 has an essential role in side-branching of the mammary ducts early in pregnancy. This new model suggests that progesterone induces Wnt4 production, which in turn induces the branching of mammary ducts during puberty and pregnancy (see reference 116 for a review).

Transgenic mice overexpressing the PR-A isoform exhibit abnormal mammary gland development with excessive lateral ductal branching[118]. This confirms that alterations in PR signaling may have important consequences to mammary development. Interestingly, mice exhibiting overexpression of the PR-B isoform in the mammary gland also display an abnormal mammary development characterized by inappropriate alveolar growth[115]. In addition, these mice undergo a premature arrest in ductal growth that was not observed in PR-A transgenic mice. These studies thus highlight the importance of PR signaling for appropriate cell-fate decision (ductal versus lobulo-alveolar growth) during normal mammary gland development[115,118].

Associated disease

There are no known mutations of PR directly known to be associated with a human disease. Nevertheless, PRs are key markers of steroid hormone dependence and indicators of disease prognosis in breast cancer. The loss of PR signals development of an aggressive tumor phenotype associated with acquisition of enhanced sensitivity to growth factors (see reference 68 for a review).

Gene structure, promoter and isoforms

The chicken PR gene was the first member of the superfamily whose genomic organization was deciphered by Bert O'Malley's group in 1987[13]. The gene was shown to span 38 kbp; each zinc finger being encoded by a separate exon. Multiple RNAs were shown to be generated from this gene either by alternative polyadenylation or by the use of several promoters[125].

The PR gene generates in all species examined except rabbit, two major isoforms called PR-B and PR-A that differ in their N-terminal A/B region[5,11]. In human, the A isoform is 164 amino acids shorter than the B isoform (128 amino acids in chicken). The two proteins are found in approximately equal amounts in human tissues. Each isoform corresponds to a separate transcript, suggesting that the isoforms are not generated from alternative translation but by the use of alternative promoters[125-127]. The extensive studies of the 5' region of the PR gene that were performed in different species allowed two questions to be answered: how these isoforms are generated, and how estrogen regulation of the PR gene expression is achieved.

Promoters governing the expression of the PR gene were characterized in different species including chicken[13,77,125], rabbit[120], rat[121] or human[126]. These studies reveal that two promoters govern the expression of the PR gene. A distal promoter is located in front of the start codon of the B isoform and a proximal promoter is located close to the start codon of the A isoform[121,126]. Results concerning estrogen inducibility were conflicting: in humans, both promoters were found to be estrogen inducible but no clear ERE can be noticed in their sequences[126], whereas in rat only the proximal promoter was responsive to estrogens[121]. Further studies locate multiple EREs in the rat proximal promoter that governs the synthesis of isoform A[122]. In humans, an ERE was located in the proximal promoter, in a position apparently similar to some of the rat EREs[123]. This promoter was also down-regulated by progestins, consistent with the physiological observations performed in mammals[123]. A down-regulation of the human proximal promoter by retinoic acid and AP-1 was also found. Strikingly, estrogen stimulation and the repression by progestins, retinoic acid and AP-1 were performed by the same ERE element that is recognized by ER but not by PR, RARs or AP-1[123,124]. In chicken the situation is different, since PR is up-regulated by progesterone and estrogens in chick oviduct[77]. The study of the chicken promoter reveals the existence of progestins- but not estrogens-responsive elements, in contrast with rats or humans[77].

More recently, a third isoform, called PR-C, was characterized in humans[128]. This isoform is an N-terminally truncated receptor that starts at Met 595 in exon 2. This isoform contains only the second zinc finger, the hinge and the LBD, and was found in vivo. Several lines of evidence suggest that PR-C is generated from the internal Met 595 by alternative translation[129]. The PR-C protein binds progesterone and, despite the lack of the first zinc finger of the DBD, it can enhance the transcriptional activity of PR-A or PR-B[129]. Another unusual isoform generated from the chicken gene by alternative splicing and coding only the A/B region and the first zinc finger was detected. Its function, if any, remains obscure[13,125].

The respective functions of PR-A and PR-B and the interplay of these two proteins were extensively studied. The two isoforms were compared for virtually all functions of the progesterone receptor: DNA binding, ligand binding, antagonist effects, cofactor binding, etc. (see reference 130 for a review). Both exhibit similar DNA-binding and ligand-binding abilities, except that the bending angle induced by PR on DNA is different for the two isoforms[22]. In contrast, it rapidly became clear that, both in humans and chickens, PR-B is a stronger transactivator than PR-A and that both isoforms exhibit different promoter specificities[52,73,74,126,135]. This is consistent with the existence of a specific activation function, called AF-3, in the unique N-terminal segment of the PR-B isoform[80]. One of the most fascinating observations concerning these two isoforms is that the short PR-A product is a cell- and promoter-specific repressor of PR-B[131,132]. In cells in which PR-A is transcriptionally inactive or when antagonists were present, an inhibition of PR-B-mediated activation was observed when the two receptors were cotransfected[131,132]. PR-A also inhibits GR, AR or MR, but not VDR. This inhibitory effect was also observed using a PR-A mutated in its DBD, suggesting that DNA binding is not required for PR-A inhibitory activity[131]. This inhibitory effect of PR-A is linked to the repressor domain that is present in the C-terminal part of PR[43,133]. This domain is present in both isoforms but is functionally active only in the context of PR-A. A repressor domain was also found in the first 140 amino acids of the PR-A isoform but it was shown that this domain cannot act autonomously[135]. This suggests that this domain may need the C-terminal repressor domain to produce a strong repressive activity. These observations may be linked to the differential ability of the AF-1 and AF-2 domains of PR-A and PR-B to interact together[83]. Indeed, it was shown that PR-A and PR-B display different cofactor interactions, with PR-A displaying a stronger interaction with the corepressor SMRT than PR-B[43]. In addition, PR-A is unable to recruit coactivators such as SRC-1, in contrast to PR-B. These two mechanisms (enhanced corepressor interaction and decreased coactivator binding) may explain why PR-A is transcriptionally inactive and represses PR-B activity[43].

Chromosomal location

The human PR gene is located to human chromosome 11 at position q22[110,111]. Another close location (11q13) was also published[112].

Amino acid sequence for human PR (NR3C3)

Accession number: M15716

```
  1  MTELKAKGPRAPHVAGGPPSPEVGSPLLCRPAAGPFPGSQTSDTLPEVSAIPISLDGLLFPRPCQGQDPS
 71  DEKTQDQQSLSDVEGAYSRAEATRGAGGSSSSPPEKDSGLLDSVLDTLLAPSGPGQSQPSPPACEVTSSW
141  CLFGPELPEDPPAAPATQRVLSPLMSRSGCKVGDSSGTAAAHKVLPRGLSPARQLLLPASESPHWSGAPV
211  KPSPQAAAVEVEEEDGSESEESAGPLLKGKPRALGGAAAGGGAAAVPPGAAAGGVALVPKEDSRFSAPRV
281  ALVEQDAPMAPGRSPLATTVMDFIHVPILPLNHALLAARTRQLLEDESYDGGAGAASAFAPPRSSPCASS
351  TPVAVGDFPDCAYPPDAEPKDDAYPLYSDFQPPALKIKEEEEGAEASARSPRSYLVAGANPAAFPDFPLG
421  PPPPLPPRATPSRPGEAAVTAAPASASVSSASSSGSTLECILYKAEGAPPQQGPFAPPPCKAPGASGCLL
491  PRDGLPSTSASAAAAGAAPALYPALGLNGLPQLGYQAAVLKEGLPQVYPPYLNYLRPDSEASQSPQYSFE
561  SLPQKICLICGDEASGCHYGVLTCGSCKVFFKRAMEGQHNYLCAGRNDCIVDKIRRKNCPACRLRKCCQA
631  GMVLGGRKFKKFNKVRVVRALDAVALPQPVGVPNESQALSQRFTFSPGQDIQLIPPLINLLMSIEPDVIY
701  AGHDNTKPDTSSSLLTSLNQLGERQLLSVVKWSKSLPGFRNLHIDDQITLIQYSWMSLMVFGLGWRSYKH
771  VSGQMLYFAPDLILNEQRMKESSFYSLCLTMWQIPQEFVKLQVSQEEFLCMKVLLLLNTIPLEGLRSQTQ
841  FEEMRSSYIRELIKAIGLRQKGVVSSSQRFYQLTKLLDNLHDLVKQLHLYCLNTFIQSRALSVEFPEMMS
911  EVIAAQLPKILAGMVKPLLFHKK   933
```

References

[1] Misrahi, M. et al. (1987) Biochem. Biophys. Res. Commun. 143, 740–748.

[2] Schott, D.R. et al. (1991) Biochemistry 30, 7014–7020.

[3] Park-Sage, O.K. and Mayo, K.E. (1994) Endocrinology 134, 709–718.

[4] Loosfelt, H. et al. (1986) Proc. Natl Acad. Sci. USA 83, 9045–9049.

[5] Gronemeyer, H. et al. (1987) EMBO J. 6, 3985–3994.

[6] Young, L.J. et al. (1995) J. Steroid Biochem. Mol. Biol. 55, 261–269.

[7] Todo, T. et al. (2000) FEBS Lett. 465, 12–17.

[8] Gronemeyer, H. and Govindan, M.V. (1986) Mol. Cell. Endocrinol. 46, 1–19.

[9] Horwitz, K.B. et al. (1985) Recent Prog. Horm. Res. 41, 249–316.

[10] Gronemeyer, H. et al. (1985) J. Biol. Chem. 260, 6916–6925.

[11] Jeltsch, J.M. et al. (1986) Proc. Natl Acad. Sci. USA 83, 5424–5428.

[12] Truss, M. et al. (1995) EMBO J. 14, 1737–1751.

[13] Huckaby, C.S. et al. (1987) Proc. Natl Acad. Sci. USA 84, 8380–8384.

[14] Conneely, O.M. et al. (1986) Science 233, 767–770.

[15] Mulvihill, E.R. et al. (1982) Cell 24, 621–632.
[16] Théveny, B. et al. (1987) Nature 329, 79–81.
[17] Tsai, S.Y. et al. (1988) Cell 55, 361–369.
[18] Tsai, S.Y. et al. (1989) Cell 57, 443–448.
[19] Rodriguez, R. et al. (1990) Mol. Endocrinol. 4, 1782–1790.
[20] Truss, M. et al. (1991) Mol. Cell. Biol. 11, 3247–3258.
[21] Lieberman, B.A. et al. (1993) Mol. Endocrinol. 7, 515–527.
[22] Prendergast, P. et al. (1996) Mol. Endocrinol. 10, 393–407.
[23] Guiochon-Mantel, A. et al. (1989) Cell 57, 1147–1154.
[24] Guiochon-Mantel, A. et al. (1991) EMBO J. 10, 3851–3859.
[25] Lim, C.S. et al. (1999) Mol. Endocrinol. 13, 366–375.
[26] Di Croce, L. et al. (1999) Molec. Cell 4, 45–54.
[27] Ylikomi, T. et al. (1992) EMBO J. 11, 3681–3694.
[28] Pratt, W.B. and Toft, D.O. (1997) Endocrinol. Rev. 18, 306–360
[29] Cheug, J. and Smith, D.F. (2000) Mol. Endocrinol. 14, 939–946.
[30] Kosano, H. et al. (1998) J. Biol. Chem. 273, 32973–32979.
[31] Barents, R.L. et al. (1998) Mol. Endocrinol. 12, 342–354.
[32] Johnson, J.L. et al. (1994) Mol. Cell. Biol. 14, 1956–1963.
[33] Smith, D.F. et al. (1995) Mol. Cell. Biol. 15, 6804–6812.
[34] Smith, D.F. et al. (1993) Mol. Endocrinol. 7, 1418–1429.
[35] Prendergast, P. et al. (1994) J. Steroid Biochem. Mol. Biol. 48, 1–13.
[36] Onate, S.A. et al. (1994) Mol. Cell. Biol. 14, 3376–3391.
[37] Boonyaratanakornkit, V. et al. (1998) Mol. Cell. Biol. 18, 4471–4487.
[38] Richer, J.K. et al. (1998) J. Biol. Chem. 273, 31317–31326.
[39] Schwerk, C. et al. (1995) J. Biol. Chem. 270, 21331–21338.
[40] Imhof, M.O. and McDonnell, D.P. (1996) Mol. Cell. Biol. 16, 2594–2605.
[41] Wagner, B.L. et al. (1998) Mol. Cell. Biol. 18, 1369–1378.
[42] Xu, Y. et al. (2000) Mol. Cell. Biol. 20, 2138–2146.
[43] Giangrande, P.H. et al. (2000) Mol. Cell. Biol. 20, 3102–3115.
[44] Chauchereau, A. et al. (2000) J. Biol. Chem. 275, 8540–8548.
[45] Onate, S.A. et al. (1995) Science 270, 1354–1357.
[46] Voegel, J.J. et al. (1996) EMBO J. 15, 3667–3675.
[47] Zhang, X. et al. (1996) J. Biol. Chem. 271, 14825–14833.
[48] Eul, J. et al. (1989) EMBO J. 8, 83–90.
[49] Allan, G.F. et al. (1992) Proc. Natl Acad. Sci. USA 89, 11750–11754.
[50] Baulieu, E.E. (1989) Science 245, 1351–1357.
[51] Guiochon-Mantel, A. et al. (1988) Nature 336, 695–698.
[52] Meyer, M.E. et al. (1990) EMBO J. 9, 3923–3932.
[53] Benhamou, B. et al. (1992) Science 255, 206–209.
[54] Vegeto, E. et al. (1992) Cell 69, 703–713.
[55] Weigel, N.L. et al. (1992) Mol. Endocrinol. 6, 1585–1597.
[56] Beck C.A. et al. (1993) Endocrinology 133, 728–740.
[57] Truss, M. et al. (1994) Proc. Natl Acad. Sci. USA 91, 11333–11337.
[58] Allan, G.F. et al. (1996) Mol. Endocrinol. 10, 1206–1213.
[59] Klein-Hitpass, L. et al. (1991) Nucl. Acids Res. 19, 1227–1234.
[60] Wagner, B.L. et al. (1996) Proc. Natl Acad. Sci. USA 93, 8739–8744.
[61] Xu, J. et al. (1996) Proc. Natl Acad. Sci. USA 93, 12195–12199.
[62] Garcia, T. et al. (1992) Mol. Endocrinol. 6, 2071–2078.
[63] Jackson, T.A. et al. (1997) Mol. Endocrinol. 11, 693–705.

64 Kalkhoven, E. et al. (1996) J. Biol. Chem. 271, 6217–6224.
65 Williams, S.P. and Sigler, P.B. (1998) Nature 393, 392–396.
66 Tanenbaum, D.M. et al. (1998) Proc. Natl Acad. Sci. USA 95, 5998–6003.
67 Robin-Jagerschmidt, C. et al. (2000) Mol. Endocrinol. 14, 1028–1037.
68 Lange, C.A. et al. (1999) Mol. Endocrinol. 13, 829–836.
69 Peluso, J.J. et al. (1997) Trends Endocrinol. Metab. 8, 267–271.
70 Zelinsky-Wooten, M.B. and Stouffer, R.L. (1996) Trends Endocrinol. Metab. 7, 177–183.
71 Tibbetts, T.A. et al. (1999) Proc. Natl Acad. Sci. 96, 12021–12026.
72 Meyer, M.E. et al. (1989) Cell 57, 433–442.
73 Tora, L. et al. (1988) Nature 333, 185–188.
74 Bocquel, M.T. et al. (1989) Nucl. Acids Res. 17, 2581–2595.
75 Kalff, M. et al. (1990) Nature 344, 360–362.
76 Klein-Hitpass, L. et al. (1990) Cell 60, 247–257.
77 Turcotte, B. et al. (1991) J. Biol. Chem. 266, 2582–2589.
78 Meyer, M.E. et al. (1992) J. Biol. Chem. 267, 10882–10887.
79 McDonnell, D. et al. (1992) Proc. Natl Acad. Sci. USA 89, 10563–10567.
80 Sartorius, C.A. et al. (1994) Mol. Endocrinol. 8, 1347–1360.
81 Bamberger, A.M. et al. (1996) Proc. Natl Acad. Sci. USA 93, 6169–6174.
82 Klotzbücher, M. et al. (1997) Mol. Endocrinol. 11, 768–778.
83 Tetel, M.J. et al. (1999) Mol. Endocrinol. 13, 910–924.
84 Denner, L.A. et al. (1990) Science 250, 1740–1743.
85 Power, R.F. et al. (1991) Science 254, 1636–1639.
86 Chauchereau, A. et al. (1991) J. Biol. Chem. 266, 18280–18286.
87 Beck, C.A. et al. (1992) Mol. Endocrinol. 6, 607–620.
88 Bagchi, M.K. et al. (1992) Proc. Natl Acad. Sci. USA 89, 2664–2668.
89 Takimoto, G.S. et al. (1992) Proc. Natl Acad. Sci. USA 89, 3050–3054.
90 Bai, W. et al. (1994) Mol. Endocrinol. 8, 1465–1473.
91 Zhang, Y. et al. (1995) Mol. Endocrinol. 9, 1029–1040.
92 Lange, C.A. et al. (2000) Proc. Natl Acad. Sci. USA 97, 1032–1037.
93 Clemm, D.L. et al. (2000) Mol. Endocrinol. 14, 52–65.
94 Denner, L.A. et al. (1990) J. Biol. Chem. 265, 16548–16555
95 Sartorius, C.A. et al. (1993) J. Biol. Chem. 268, 9262–9266.
96 Beato, M. (1995) Cell 83, 851–857.
97 Godowski, P.J. (1993) Sem. Virol. 4, 43–51.
98 Strähle, U. et al. (1989) Nature 339, 629–632.
99 Cato, A.C.B. et al. (1988) Mol. Cell. Biol. 8, 5323–5330.
100 Hecht, A. et al. (1988) EMBO J. 7, 2063–2073.
101 Vivat, V. et al. (1992) J. Biol. Chem. 267, 7975–7978.
102 Marbaix, E. et al. (1992) Proc. Natl Acad. Sci. USA 89, 11789–11793.
103 Vaßen, L. et al. (1999) Mol. Endocrinol. 13, 485–494.
104 Lim, H. et al. (1999) Mol. Endocrinol. 13, 1005–1017.
105 Robker, R.L. et al. (2000) Proc. Natl Acad. Sci. USA 97, 4689–4694.
106 Musgrove, E.A. et al. (1991) Mol. Cell. Biol. 11, 5032–5043.
107 Alkhalaf, M. and Murphy, L.C. (1992) Mol. Endocrinol. 6, 1625–1633.
108 Musgrove, E.A. et al. (1998) Mol. Cell. Biol. 18, 1812–1825.
109 Turcotte, B. et al. (1990) Mol. Cell. Biol. 10, 5002–5006.
110 Mattei, M.G. et al. (1988) Hum. Genet. 78, 96–97.

111 Rousseau-Merck, M.F. et al. (1987) Hum. Genet. 77, 280–282.
112 Law, M.L. et al. (1987) Proc. Natl Acad. Sci. USA 84, 2877–2881.
113 Lydon, J.P. et al. (1995) Genes Dev. 9, 2266–2278.
114 Brisken, C. et al. (1998) Proc. Natl Acad. Sci. USA 95, 5076–5081.
115 Shyamala, G. et al. (2000) Proc. Natl Acad. Sci. USA 97, 3044–3049.
116 Robinson, G.W. et al. (2000) Genes Dev. 14, 889–894.
117 Brisken, C. et al. (2000) Genes Dev. 14, 650–654.
118 Shyamala, G. et al. (1998) Proc. Natl Acad. Sci. USA 95, 696–701.
119 Seagroves, T.N. et al. (2000) Mol. Endocrinol. 14, 359–368.
120 Misrahi, M. et al. (1988) Nucl. Acids Res. 16, 5459–5472.
121 Kraus, W.L. et al. (1993) Mol. Endocrinol. 7, 1603–1616.
122 Kraus, W.L. et al. (1994) Mol. Endocrinol. 8, 952–969.
123 Savouret, J.F. et al. (1991) EMBO J. 10, 1875–1883.
124 Savouret, J.F. et al. (1994) J. Biol. Chem. 269, 28955–28962.
125 Jeltsch, J.M. et al. (1990) J. Biol. Chem. 265, 3967–3974.
126 Kastner, P. et al. (1990) EMBO J. 9, 1603–1614.
127 Kastner, P. et al. (1990) J. Biol. Chem. 265, 12163–12167.
128 Wei, L.L. et al. (1990) Mol. Endocrinol. 4, 1833–1840.
129 Wei, L.L. et al. (1996) Mol. Endocrinol. 10, 1379–1387.
130 Schrader, W.T. et al. (1993) Mol. Endocrinol. 7, 1241–1242.
131 Vegeto, E. et al. (1993) Mol. Endocrinol. 7, 1244–1255.
132 Tung, L. et al. (1993) Mol. Endocrinol. 7, 1256–1265.
133 Wen, D.X. et al. (1994) Mol. Cell. Biol. 14, 8356–8364.
134 Kraus, W.L. et al. (1995) Mol. Cell. Biol. 15, 1847–1857.
135 Giangrande, P.H. et al. (1997) J. Biol. Chem. 272, 32889–32900.
136 Swarbrick, A. et al. (2000) Mol. Cell. Biol. 20, 2581–2591.
137 Tong, W. and Pollard, J.W. (1999) Mol. Cell. Biol. 19, 2251–2264.
138 Leonhardt, S.A. et al. (1998) Mol. Endocrinol. 12, 1914–1930.
139 Wissink, S. et al. (1998) Mol. Endocrinol. 12, 355–363.
140 Said, T.K. et al. (1997) Endocrinology 138, 3933–3939.
141 Danielan, P.S. (1992) EMBO J. 11, 1025–1033.

Names

On the four steroid receptors of the NR3C group (GR, MR, PR, AR), the androgen receptor was the fourth to be cloned. It is now known in a wide variety of vertebrates but is still unknown in invertebrates. It is a large 900–920-amino-acid molecule that transduces the message of the steroid male hormone testosterone and which is implicated in a number of human diseases.

Species	Latin name	Other names	Accession number	References
Official name: NR3C4				
Human	*Homo sapiens*	AR	M20132	85
Chimpanzee	*Pan troglodytes*	AR	U94177	3
Baboon	*Papio hamadryas*	AR	U94176	3
Macaque	*Macaca fascicularis*	AR	U94179	3
Lemur	*Eulemus fulvus collaris*	AR	U94178	3
Mouse	*Mus musculus*	AR	X53779	4
Rat	*Rattus* sp.	AR	M20133	5
Rabbit	*Oryctolagus cuniculus*	AR	U16366	8
Sheep	*Ovis aries*	AR	AF105713	Unpub.
Dog	*Canis familiaris*	AR	AF197950	Unpub.
Canary	*Serinus canarius*	AR	L25901	86
Green anole	*Anolis carolinensis*	AR	AF223224	Unpub.
Xenopus	*Xenopus laevis*	AR	U67129	87
Rainbow trout	*Oncorhynchus mykiss*	ARα	AB012095	88
Rainbow trout	*Oncorhynchus mykiss*	ARβ	AB012096	88
Japanese eel	*Anguilla japonica*	ARα	AB023960	89
Japanese eel	*Anguilla japonica*	ARβ	AB025361	90
Red seabream	*Chrysophrys major*	AR	AB017158	91

Isolation

The first androgen receptor cDNAs were cloned independently by Elizabeth Wilson's and Shutsung Liao's laboratories. One of this group used the fact that the androgen-insensivity syndrome was linked to the X chromosome and that no other steroid receptor is located on this chromosome[1]. X chromosome libraries were thus screened with a consensus oligonucleotide from the DBD of PR, GR, ER and TR. One of the isolated clones corresponded to a new member of the superfamily. When this clone was expressed in COS cells, it yielded a high-affinity androgen-binding protein with steroid-binding specificity corresponding to that of native AR[1]. The other team screened a human testis cDNA library with an oligonucleotide probe homologous to the DBDs of GR, ER, PR, MR and v-erbA[2]. The resulting clone produced a protein that binds androgens and that was recognized by human autoimmune antibodies to the androgen receptor. Other groups identified the same sequence independently using similar strategies[9-11].

DNA binding

In presence of its ligand, AR binds as a homodimer to sequence called ARE (androgen response element), which harbors the classical consensus for steroid receptors (GGTACAnnnTGTTCT) first defined in the MMTV regulatory region[22,23,56]. This sequence also binds to the other steroid receptors, such as GR, MR or PR[2,56], and has been found by a random selection procedure[92]. This is consistent with the fact that the P-box of all these receptors is identical (CGSCKV). A sequence more specific to AR (GGTTCTtggAGTACT) was identified in the probasin gene promoter[25]. The specificity of this sequence came from the 5′ repeat (GGTTCT), which excludes other receptors. One important feature of this sequence is its direct repeat structure[72]. A similar sequence was found in two other AR target genes[73]. In the receptor it is the second zinc finger as well as the C-terminal extension that is responsible for this selective DNA-binding activity[71]. Four amino acids in these regions were shown to be critical for the binding to the GGTTCT[72]. Other authors have found a completely different sequence, called IDR17, by a different random selection procedure from that originally used[92]. This sequence is organized as a complex element containing both a direct repeat and an inverted repeat (GGAACGGAACATGTTCT)[26]. It is the 5′ region, called the DR1 element (GGAACGGAACA), which apparently exhibits high-affinity specific AR binding. The way in which response elements specific to AR, GR or PR may have been selected during evolution has recently been explored[74]. It has been shown that some nucleotides found in specific response elements were selected to discourage binding from inappropriate receptors, even if their presence reduces the binding affinity of the appropriate receptor. The importance of spacer and flanking regions of the response elements in these specific response elements was also emphasized in this study[74].

AR has been shown to heterodimerize with GR on its classic palindromic response element. In the heterodimer, the receptors inhibit each other's transcriptional activity[27].

AR is known to be able to repress gene transcription (see below). As for GR, it has proved difficult to isolate the sequences on which AR binds to repress transcription. One such element (called nARE) is present in the maspin gene and is different from the classical ARE (GTACTCTGATCTCC)[28]. The way AR down-regulates the activity of this gene is not clear since it is not known to bind any corepressors. It has been proposed that, on the maspin gene, as for other down-regulated genes, AR negatively regulates transcription via an interaction with transcription factors of the Ets family[23,28].

After some controversy on its subcellular localization, it has clearly been shown using GFP fusion proteins that, in the absence of ligand, AR is located in the cytoplasm and that the presence of the ligand induces its accumulation in the nucleus within an hour[38,39,56]. Interestingly, the nuclear transport does not occur at the same speed for various ligands. The non-metabolizable androgen analogue, R1881, induces a more rapid transport than dihydrotestosterone (DHT) or testosterone itself. Mutations in the LBD that cause androgen insensitivity reduce or abolish the nuclear transport of AR[38] (see reference 39 for a review).

Partners

Like other nuclear receptors, AR interacts with heat-shock proteins through its LBD. This interaction might be necessary for optimal ligand-binding activity of the receptor (see reference 35 for a review). It is believed that these proteins act as chaperones and that their interaction with AR may play a role in a disease in which AR is implicated, spinal and bulbar muscular atrophy[75].

AR has been shown to be able to form heterodimers with GR[27]. This situation is reminiscent to the MR–GR heterodimer. More recently, AR has been shown to interact with the orphan receptor TR4[76] (see entry for TR2 and TR4).

AR has numerous partners, which explains how it can regulate transcription[23,46]. A direct binding of the N-terminal domain of AR to the basal transcription factors TFIIF and TBP has been described[47]. An interaction between the A/B region of AR and cdk-activating kinase (CAK), which is a part of the TFIIH complex has also been described[77]. CAK was shown to induce AR-mediated transactivation.

Furthermore, a number of coactivators for AR were identified, such as TIF2, CBP or ARA70[48-50]. None of these coactivators appears specific to AR. Interestingly, TIF2 appears to act on AR in a different manner from that for other nuclear receptors. It was shown recently that interaction between the LBD (AF-2 function) and N-terminal domain (AF-1 function) is a prerequisite for the interaction between AR and TIF2[52]. An interaction between AR and the related factor SRC-1 in which both the AF-1 and AF-2 functions of AR were implicated was also recently demonstrated[80].

The case of ARA70, which was described as a specific coactivator for AR, is worth discussing[50,51]. Recent data suggest that ARA70 (also called ELE1α) and the closely related factor ELE1β, effectively interact with AR in a ligand-binding-dependent manner, via the helix 3 of the LBD, but that this interaction is not AR specific[51]. An interaction with GR and ER was also observed. Furthermore, ARA70 has only a minor effect on the transcriptional activity of AR, casting some doubts on its role as a coactivator[51,78].

Recently, the identification of FHL2, as a novel tissue-specific coactivator of AR was reported[53]. This gene appears to be coexpressed with AR in heart and, more importantly, in the epithelial cells of the prostate. It binds specifically to AR and increases its transcriptional activity in an AF-2- and ligand-dependent manner. This positive effect on transcription was observed on a prostate-specific target gene of AR, the probasin gene[53]. The in vivo role played by FHL2 remains nevertheless to be established, given that the transcriptional role played by the AF-2 function in AR is much reduced when compared to AF-1 (see below). Another interesting coactivator is PIAS1, a human protein that plays an inhibitory role on the STAT-1 transcription factor[79]. This protein interacts in a ligand-dependent manner with AR and GR, is highly expressed in testis at the onset of spermatogenesis and is coexpressed with AR in Sertoli and Leydig cells[79].

AR can interact with other transcription factors and negatively regulate transcription by interfering with their action (see reference 23 for a review). This has been shown for c-jun, Ets family members and it is also suspected for NFκB (see below)[42,45].

Ligands

Although a variety of androgens have been identified, testosterone, the major hormone produced by the testes, was thought to be the most important androgen given its potency and concentration in blood. However, it is a metabolite of testosterone, 5α-dihydrotestosterone produced by the enzyme 5α-reductase that is the compound that can bind to the AR with the greatest affinity ($K_d = 2.7 \times 10^{-10}$ M; reviewed in reference 46)[1,2,9-11]. A non-metabolizable analogue methyltrienolone (R1881) is an even more potent ligand[1,61]. A number of molecules with androgenic agonist or antagonists activity have been characterized and their binding and transcriptional effect on AR studied[62,63]. Among them, it is interesting to note that, 1,1-dichloro-2,2-bis(p-chlorophenyl)ethylene (p,p'-DDE), a persistent metabolite of the insecticide DDT with estrogenic activity, is able to inhibit androgen binding to the receptor, suggesting that abnormalities of male sex development induced by xenosteroids may also be mediated at the level of AR[62]. Since the full transcriptional activity of AR is known to require the physical interaction between the A/B domain and the LBD, several agonist and antagonist molecules were compared for their ability to promote this interaction[63]. It was shown that inhibition of this interaction does not necessarily reflect the activity of an antagonist and that some weak agonists, such as medroxyprogesterone acetate, activate AR via a mechanism that does not involve this interaction. This suggests that the control of AR activity may imply several independent mechanisms.

Three-dimensional structure

The 3D structure of the DBD or LBD of AR has not been reported. A model based on the LBD of PR was recently proposed. Together with a structure–function analysis, it allows the specific recognition of androgens by the receptor to be discussed[93].

Expression

AR is expressed as a predominantly large mRNA of c. 10 kb, which is present in most tissues, for example, human foreskin fibroblasts, rat testis, epididymis, kidney and prostate[1]. Of note, AR transcript is also present in osteoblasts that are known to be physiological targets of DHT as well as in osteoclasts[30,31,37]. Consistent with its role in males, AR is expressed in rat penile tissue during sexual maturation[32]. In liver, AR displays an age-regulated expression: it is almost undetectable before puberty, rises rapidly in post-pubertal life and then gradually declines during aging[34].

Recently, a survey of AR expression during mouse embryogenesis was carried out by *in situ* hybridization[33]. AR mRNA was detected from E12.5 shortly before embryonic hormone production starts at E13. AR transcripts were detected in the genital tubercules in both sexes. It was also detected in mesonephric mesenchyme of both sexes at E12.5 and then shifted in the male to ductal epithelia at E15.5. In females, AR expression was restricted to mesenchymal tissue around the Müllerian ducts. An expression was also observed during mammary development starting in both males and females at E12.5 and then declining in the male from E14.5. Further sites of AR expression were the glandular part of the pituitary, the adrenal gland and one muscle of the external genital. Minor sites of expression were found in the testis, kidney, hypothalamus, dorsal root ganglia and larynx[33].

Activity and target genes

AR is a transcriptional activator in transient transfection assays but also *in vitro*[54,55]. A striking difference between AR and other nuclear receptors is that its AF-1 function is much stronger than the AF-2 function in the LBD. In fact, the existence of an intrinsic AF-2 activity has remained elusive despite its strong sequence identity with other nuclear receptors[52,54]. The AF-1 function has been characterized by several groups (see reference 52 for references). It is now clear that the activity of the AF-1 function in the A/B region is somehow regulated by the LBD and that the two domains interact. The region of the A/B domain important for AF-1 function encompasses large polyglutamine repeats that are the target of mutations in several diseases (see below). Interestingly, an inverse correlation has been made between AR transcriptional activity and the length of the glutamine repeats[59].

The complex relationships existing between AF-1 and AF-2 functions in AR may play a part in the specificity of its action, which has always presented a problem in understanding the physiological role of androgens[23,46]. The deciphering of AR action on several promoters has led to the observation that AREs are often part of complex regulatory units which, by multiple interactions between transcription factors, provide a specificity for androgen regulation[60]. Examples of target genes are too numerous to be cited. Interested readers should refer to references 23 and 46 for examples.

AR can be a potent transcriptional repressor. It has never been found associated with corepressors such as NCoR or SMRT, and the manner by which it represses transcription is not well understood. This probably occurred via the interference with other transcription factors[23]. As discussed above, only one negative ARE (nARE) was clearly identified[28]. Apparently, all the other cases of negative regulation by AR involve transrepression mechanisms through interaction with other transcription factors. For example, AR interacts with c-*jun* to inhibit its DNA-binding activity and thereby down-regulates its activity[42]. On the human collagenase-1 gene, the target for the down-regulation is ERM, a member of the Ets family of transcription factors[45]. This interaction with Ets family members apparently occurs for many other cases[23,28]. The N-terminal part of AR has been identified as essential for this effect, but specific protein interaction surfaces were also identified in the C-terminal part and the relationship between these findings is still unclear[23]. A negative interference between AR and NFκB signaling was also reported on the IL6 promoter[43]. The precise mechanisms of this interference is not clear. It was shown that AR does not up-regulate the transcription of IκBα, the negative regulator of NFκB. It is believed that this negative regulatory action of AR on NFκB follows the same principle of protein–protein interaction described for other steroid receptors such as GR[44].

A third way of action of androgens involving growth factor enhancement of transactivation by the receptor is often discussed[23]. These effects probably implicate phosphorylation events of either the receptor itself or of a coactivator[57]. AR is known to be phosphorylated on serine residues in the A/B and hinge region. The phosphorylation in the hinge region is important for full transcriptional activity of the receptor[58].

Knockout

No knockout of the AR gene was reported, probably because there are human patients that exhibit mutations of both alleles, which totally inactivates the receptor (see e.g. reference 41).

Overexpression of AR using transgenic mice has been reported. The transgene was overexpressed under the control of a liver-specific promoter[34]. It was known that the liver is an important target of androgens and that rodent liver displays marked sex-dependent changes in androgen sensitivity, owing to the sexually dimorphic and temporally programmed expression of AR. The permanent overexpression of AR leads to a perturbation of the expression of androgen-controlled steroidogenic enzymes, which are important for androgen metabolism[34].

A transgenic mice overexpressing a mutated AR containing an expended Gln repeat found in spinal/bulbar muscular atrophy under the control of various promoters was generated[70]. The lack of phenotypic effect was linked to the stability of the poly-Gln tract, which is opposite to the variability found in humans (see below).

Associated disease

AR is associated with three types of diseases: X-linked androgen insensivity, spinal/bulbar muscular atrophy and cancer[46]. Since up to 200 different naturally occurring mutations of the AR gene have been described, a database has been set up (available at www.mcgill.ca/androgendb/)[64]. Mutations found in androgen insensitivity syndrome are mainly located in the DBDs and LBDs of the receptor. Some of these mutations alter DNA binding, whereas others impair ligand-binding activity and cause complete or partial androgen insensitivity (see the web sites and reference 46 for a review; references 81 and 82 for recent examples). Mutations in AR have also been associated with a male breast cancer[46].

Both prostate cancer and splinal/bulbar muscular atrophy have been linked to variation in the number of Gln residues found in the polyglutamine repeat, which normally contains 16–39 residues. Expansion of the repeat to 40–65 residues leads to spinal/bulbar muscular atrophy, whereas reduction in the number of repeats confers a higher risk for prostate cancer. Since the length of this repeat is inversely correlated to the transcriptional activity of the receptor, it is speculated that these variations in number confer a higher (prostate cancer) or lower (spinal/bulbar muscular atrophy) transcriptional activity[46,59,65,66]. The protective role of chaperone proteins, such as hsp70 and hsp40, in spinal/bulbar muscular atrophy has been linked to their ability to suppress aggregate formation and apoptosis in neuronal cells expressing an AR protein with an expanded Gln tract[75]. It has furthermore been demonstrated that an amplification of the androgen receptor gene is implicated in the androgen resistance of prostate tumors, although other mechanisms, among which is the suppression of caveolin expression, have also been suggested[67,68]. Recently, it was shown, in a patient with a prostate cancer resistant to androgen treatment, that two mutations in the LBD of AR allow this gene product to function as a high-affinity glucocorticoid receptor and reduce its ability to bind to androgens[82,83].

Gene structure, promoter and isoforms

The human AR gene comprises more than 90 kbp of DNA and contains eight exons[12,15]. The location of the introns in the conserved regions are similar to those found in the genes for other steroid receptors, including ER.

A TATA-less promoter lying in a GC-rich region with several Sp1 sites is used for transcription and to produce the large transcript of c. 10 kbp. This promoter was cloned in human, mouse and rat, as well as in a number of other species, and exhibits high levels of sequence identity (80% between mouse and human)[13,14,21]. In both species the start site is located c. 1 kbp upstream of the start codon. This long 5'UTR may have a regulatory function, since it has been shown to contain a strong suppressor element that reduces the transcription of the AR gene[18]. The same promoter is used for transcription in fibroblasts, prostate, or breast cancer cell lines. It is known that AR expression is down-regulated by androgens in a variety of tissues but may be up-regulated in other situations or tissues such as in prostate[16,94–97]. Using the mouse promoter region, this observation was reproduced, showing that the regulation takes place at the transcriptional level. In addition, the same study reported an up-regulation of AR gene expression by cAMP via a CREB binding site[17]. The positive autoregulation of AR by androgens was shown to be dependent on four AREs and a *myc* binding site that is located in exonic sequences of the AR gene[84,96]. In addition, the AR gene has been shown to be regulated by a number of molecules, such as prolactin, growth hormone or FSH (see reference 20 for references). Recently, it has been shown that members of the transcription factor family NF1 down-regulate the activity of the rat AR promoter[36].

During the characterization of the suppressor element present in the 5'UTR of the mouse AR gene, a second promoter located 3' to the previous one was isolated[20,98]. This promoter which does not contain an obvious TATA box is differentially regulated when compared to the first one and is controlled by androgens in the kidney[98].

Two isoforms of AR were identified in human genital skin fibroblasts[19]. These isoforms are encoded via alternative translation initiation on two different ATG codons from the same transcript. The large isoform of 110 kDa (called AR-B) corresponds to the bona fide receptor, whereas the small one of 87 kDa (AR-A) initiate on an internal initiation codon. A parallel can be made between these isoforms and the A and B isoforms of the human progesterone receptor, but in the case of AR, the functional differences between the two isoforms appear to be subtle and their specific function, if any, remains unclear[40]. Two AR isoforms differing in the A/B region and that exhibit different expression patterns are also present in *Xenopus laevis*[29].

Chromosomal location

The gene has been located to the human chromosome X between the centromere and q13 and its particular chromosomal location was used to clone the gene[1]. Later, the location was refined and determined to be on Xq11–12[6]. This location was extended to other mammals and shown to be on a highly conserved region of the X chromosome in eutherian mammals as well as in marsupials and monotremes[7].

Amino acid sequence for human NR3C4

Accession number: M20132

```
  1 MEVQLGLGRVYPRPPSKTYRGAFQNLFQSVREVIQNPGPRHPEAASAAPPGASLLLLQQQQQQQQQQQQQQ
 71 QQQQQQQQETSPRQQQQQQQGEDGSPQAHRRGPTGYLVLDEEQQPSQPQSALECHPERGCVPEPGAAVAAS
141 KGLPQQLPAPPDEDDSAAPSTLSLLGPTFPGLSSCSADLKDILSEASTMQLLQQQQQQEAVSEGSSSGRAR
211 EASGAPTSSKDNYLGGTSTISDNAKELCKAVSVSMGLGVEALEHLSPGEQLRGDCMYAPLLGVPPAVRPT
281 PCAPLAECKGSLLDDSAGKSTEDTAEYSPFKGGYTKGLEGESLGCSGSAAAGSSGTLELPSTLSLYKSGA
351 LDEAAAYQSRDYYNFPLALAGPPPPPPPPPHPHARIKLENPLDYGSAWAAAAAQCRYGDLASLHGAGAAGP
421 GSGSPSAAASSSWHTLFTAEEGQLYGPCGGGGGGGGGGGGGGGGGGGGGGGEAGAVAPYGYTRPPQGLA
491 GQESDFTAPDVWYPGGMVSRVPYPSPTCVKSEMGPWMDSYSGPYGDMRLETARDHVLPIDYYFPPQKTCL
561 ICGDEASGCHYGALTCGSCKVFFKRAAEGKQKYLCASRNDCTIDKFRRKNCPSCRLRKCYEAGMTLGARK
631 LKKLGNLKLQEEGEASSTTSPTEETTQKLTVSHIEGYECQPIFLNVLEAIEPGVVCAGHDNNQPDSFAAL
701 LSSLNELGERQLVHVVKWAKALPGFRNLHVDDQMAVIQYSWMGLMVFAMGWRSFTNVNSRMLYFAPDLVF
771 NEYRMHKSRMYSQCVRMRHLSQEFGWLQITPQEFLCMKALLLFSIIPVDGLKNQKFFDELRMNYIKELDR
841 IIACKRKNPTSCSRRFYQLTKLLDSVQPIARELHQFTFDLLIKSHMVSVDFPEMMAEIISVQVPKILSGK
911 VKPIYFHTQ 919
```

References
1 Lubahn, D.B. et al. (1988) Science 240, 327–330.
2 Chang, C. et al. (1988) Science 240, 324–326.
3 Choong, C.S. et al. (1998) J. Mol. Evol. 47, 334–342.
4 He, W.W. et al. (1990) Biochem. Biophys. Res. Commun. 171, 697–704.
5 Tan, J.A. et al. (1988) Mol. Endocrinol. 2, 1276–1285.
6 Brown, C.J. et al. (1989) Am. J. Hum. Genet. 44, 264–269.
7 Spencer, J.A. et al. (1991) J. Heredity 82, 134–139.
8 Krongrad, A. et al. (1995) J. Andrology 16, 209–212.
9 Chang, C. et al. (1988) Proc. Natl Acad. Sci. USA 85, 7211–7215.
10 Tilley, W.D. et al. (1989) Proc. Natl Acad. Sci. USA 86, 327–331.
11 Trapman, J. et al. (1988) Biochem. Biophys. Res. Commun. 153, 241–248.
12 Lubahn, D.B. et al. (1989) Proc. Natl Acad. Sci. USA 86, 9534–9538.
13 Tilley, W.D. et al. (1990) Proc. Natl Acad. Sci. USA 265, 13776–13781.
14 Baarends, W.M. et al. (1990) Mol. Cell. Endocrinol. 74, 75–84.

[15] Kuiper, G.G.J.M. et al. (1989) J. Mol. Endocrinol. 2, R1–R4.
[16] Shan, L.X. et al. (1990) Mol. Endocrinol. 4, 1636–1646.
[17] Lindzey, J. et al. (1993) Mol. Endocrinol. 7, 1530–1540.
[18] Grossman, M.E. et al. (1994) Mol. Endocrinol. 8, 448–455.
[19] Wilson, C.M. and McPhaul, M.J. (1994) Proc. Natl Acad. Sci. USA 91 1234–1238.
[20] Kumar, M.V. et al. (1994) Nucl. Acids Res. 22, 3693–3698.
[21] Faber, P.W. et al. (1991) Biochem. J. 278, 269–278.
[22] Beato, M. (1989) Cell 56, 335–344.
[23] Cato, A.C.B. and Peterziel, H. (1998) Trends Endocrinol. Metab. 9, 150–154.
[24] Cato, A.C.B. et al. (1987) EMBO J. 6, 363–368.
[25] Claessens, F. et al. (1996) J. Biol. Chem. 271 19013–19016.
[26] Zhou, Z. et al. (1997) J. Biol. Chem. 272, 8227–8235.
[27] Chen, S-Y. et al. (1997) J. Biol. Chem. 272, 14087–14092.
[28] Zhang, M. et al. (1997) Proc. Natl Acad. Sci. USA 94, 5673–5678.
[29] Fisher, L. et al. (1993) Proc. Natl Acad. Sci. USA 90, 8254–8258.
[30] Colvard, D.S. et al. (1989) Proc. Natl Acad. Sci. USA 86, 854–857.
[31] Orwoll, E.S. (1996) Trends Endocrinol. Metab. 7, 77–84.
[32] Gonzalez-Cadavid, N.F. (1991) Endocrinol. 129, 1671–1678.
[33] Crocoll, A. et al. (1998) Mech. Dev. 72, 175–178.
[34] Chatterjee, B. et al. (1996) Proc. Natl Acad. Sci. USA 93, 728–733.
[35] Smith, D.F. and Toft, D.O. (1993) Mol. Endocrinol. 7, 4–11.
[36] Song, C.S. et al. (1999) Mol. Endocrinol. 13, 1487–1496.
[37] Pederson, L. et al. (1999) Proc. Natl Acad. Sci. 96, 505–510.
[38] Georget, V. et al. (1997) Mol. Cell. Endocrinol. 129, 17–26.
[39] Guiochon-Mantel, A. et al. (1996) J. Steroid Biochem. Mol. Biol. 56, 1–6.
[40] Gao, T. and McPhaul, M.J. (1998) Mol. Endocrinol. 12, 654–663.
[41] Quigley, C.A. et al. (1992) Mol. Endocrinol. 6, 1103–1112.
[42] Kallio, P.J. et al. (1995) Mol. Endocrinol. 9, 1017–1028.
[43] Keller, E.T. et al. (1996) J. Biol. Chem. 271, 26267–26275.
[44] Heck, S. et al. (1997) EMBO J. 16, 4698-4707.
[45] Schneikert, J. et al. (1996) J. Biol. Chem. 117, 149–156.
[46] Hiipakka, R.A. and Liao, S. (1998) Trends Endocrinol. Metab. 9, 317–324.
[47] McEwan, I.J. and Gustafsson, J-A. (1997) Proc. Natl Acad. Sci. USA 94, 8485–8490.
[48] Voegel, J.J. et al. (1996) EMBO J. 15, 3667–3675.
[49] Arnisalo, P. et al. (1998) Proc. Natl Acad. Sci. USA 95, 2122–2127.
[50] Yeh, S. and Chang, C. (1996) Proc. Natl Acad. Sci. USA 93, 5517–5521.
[51] Alen, P. et al. (1999) Mol. Endocrinol. 13, 117–128.
[52] Alen, P. et al. (1999) Mol. Cell. Biol. 19, 6085–6097.
[53] Müller, J.M. et al. (2000) EMBO J. 19, 359–369.
[54] Jenster, G. et al. (1991) Mol. Endocrinol. 5, 1396–1404.
[55] De Vos, P. et al. (1994) Nucl. Acids. Res. 22, 1161–1166.
[56] Wong, C.I. et al. (1993) J. Biol. Chem. 268, 19004–19012.
[57] Yeh, S. et al. (1999) Proc. Natl Acad. Sci. USA 96, 5458–5463.
[58] Zhou, Z-X. et al. (1995) Mol. Endocrinol. 9, 605–615.
[59] Chamberlain, N.L. et al. (1994) Nucl. Acids Res. 22, 3181–3186.

60 Adler, A.J. et al. (1993) Mol. Cell. Biol. 13, 6326–6335.
61 Zhou, Z-X. et al. (1995) Mol. Endocrinol. 9, 208–218.
62 Kelce, W.R. et al. (1995) Nature 375, 581–585.
63 Kemppainen, J.A. et al. (1999) Mol. Endocrinol. 13, 440–454.
64 Gottlieb, B. et al. (1996) Nucl. Acids Res. 24, 151–154.
65 Mahtre, A.N. et al. (1993) Nature Genet. 5, 184–187.
66 La Spada, A.R. et al. (1991) Nature 352, 77–79.
67 Visakorpi, T. et al. (1995) Nature Genet. 9, 401–406.
68 Nasu, Y. et al. (1998) Nature Medicine 4, 1062-1064.
69 Lobaccaro, J-M. et al. (1993) Nature Genet. 5, 109–110.
70 Bingham, P.M. et al. (1995) Nature Genet. 9, 191–196.
71 Schoenmakers, E. et al. (1999) Biochem J. 341, 515–521.
72 Schoenmakers, E. et al. (2000) J. Biol. Chem. 275, 12290–12297.
73 Verrijdt, G. et al. (2000) J. Biol. Chem. 275, 12298–12305.
74 Nelson, C.C. et al. (1999) Mol. Endocrinol. 13, 2090–2107.
75 Kobayashi, Y. et al. (2000) J. Biol. Chem. 275, 8772–8778.
76 Lee, Y.F. et al. (1999) Proc. Natl Acad. Sci. USA 96, 14724–14729.
77 Lee, D.K. et al. (2000) J. Biol. Chem. 275, 9308–9313.
78 Gao, T. et al. (1999) Mol. Endocrinol. 13, 1645–1656.
79 Tan, J.A. et al. (2000) Mol. Endocrinol. 14, 14–26.
80 Bevan, C.L. et al. (2000) Mol. Cell. Biol. 19, 8383–8392.
81 Nazareth, L.V. et al. (1999) Mol. Endocrinol. 13, 2065–2075.
82 Zhao, X.Y. et al. (2000) Nature Med. 6, 703-706.
83 Brinkmann, A.O. and Trapman, J. (2000) Nature Med. 6, 628–629.
84 Grad, J.M. et al. (1999) Mol. Endocrinol.13, 1896–1911.
85 Lubahn, D.B. et al. (1988) Mol. Endocrinol. 2, 1265–1275.
86 Nastiuk, K.L. and Clayton, D.F. (1994) Endocrinology 134, 640–649.
87 Fischer, L. et al. (1993) Proc. Natl Acad. Sci. USA 90, 8254–8258.
88 Takeo, J. and Yamashita, S. (1999) J. Biol. Chem. 274, 5674–5680.
89 Todo, T. et al. (1999) Biochem. Biophys. Res. Commun. 254, 378–383.
90 Ikeuchi, T. et al. (1999) J. Biol. Chem. 274, 25205–25209.
91 Touhata, K. et al. (1999) Biochim. Biophys. Acta 1450, 481–485.
92 Roche, P.J. et al. (1992) Mol. Endocrinol. 6, 2229–2235.
93 Poujol, N. et al. (2000) J. Biol. Chem. 275, 24022–24031.
94 Mora, G.R. et al. (1999) Steroids 64, 587–591.
95 Yeap, B.B. et al. (1999) Endocrinology 140, 3282–3289.
96 Dai, J.L. and Burnstein, K.L. (1996) Mol. Endocrinol. 10, 1582–1594.
97 Quarmby, V.E. (1990) Mol. Endocrinol. 4, 22–28.
98 Grossmann, M.E. et al. (1994) Biochemistry 33, 14594–14600.

Names

The NGFIB group contains three orphan receptors in vertebrates and their unique homologue in arthropods and nematodes. The nomenclature of this group is extremely confusing because different names were given to orthologous versions of the same gene or even within the same species. These names are still used by different authors, leading to great difficulties in knowing which gene is studied in a given paper. We thus decided to use the official nomenclature exclusively for this group of receptors in order to gain clarity. Altogether, these four genes form the subfamily IV of the nuclear receptor superfamily.

Species	Other names	Accession number	References
Official name: NR4A1			
Human	NAK1, ST-59, TR3, NGFI-Bα	L13740	4, 5, 10
Mouse	nur/77, nur77, N10, NGFI-Bα	J04113	2, 3
Rat	NGFI-B, NGFI-Bα, TIS1	U17254	1, 9
Dog	NGFI-B, NGFI-Bα	X97226	17
Xenopus	NGFI-B, NGFI-Bα	X70700	6
Official name: NR4A2			
Human	NOT, NGFI-Bβ, TINUR, NURR1	X75918	11, 13
Mouse	NURR1, NGFI-Bβ	S53744	7
Rat	RNR-1 ??, NGFI-Bβ, NURR1, HZF-3	L08595	8
Official name: NR4A3			
Human	TEC, MINOR, CHN, NOR-1, NGFI-Bγ	D78579	12, 16, 20
Rat	NOR-1 ??, NGFI-Bγ	D38530	19
Official name: NR4A4			
Drosophila	DHR38	X89246	14, 15
Aedes	AHR38	AF165528	18
Bombyx	BHR38	X89247	15
Caenorhabditis	CNR8	U13076	21

Isolation

NR4A1 was isolated as a gene rapidly and transiently induced by nerve growth factor in the rat pheochromocytoma cell line, PC12[1]. The same gene was found as an immediate-early gene induced by serum in fibroblasts[2,3]. It was also found by low-stringency screening with probes encompassing the DBD coding region of several nuclear receptors[4,10]. The *Xenopus* and dog homologues were isolated by screening with their rat homologue[6,17]. The dog NR4A1 gene is also an early regulated gene during cAMP response of thyrocytes in primary cultures.

Another member of this group in vertebrates, NR4A2, was cloned both by its properties of immediate-early gene and by low-stringency screening. It was found by screening of a neonatal mouse brain cDNA library screened with the DBD of COUP-TF[7] and other members of the superfamily[13]. It was also isolated using differential screening analyses of cDNA libraries derived from regenerating liver[8] as well as from activated T-cells[11].

The third member was found in human as implicated in the t(9;22) (q22;q12) chromosomal translocations found in extraskeletal myxoid chondrosarcoma, fused with the EWS protein[12,20]. It has also been isolated as a mitogen-inducible gene in human Jurkat cells[16].

NR4A4, the insect homologue, was isolated during a screen for new members of the superfamily that are expressed during the onset of metamorphosis in *Drosophila* and *Bombyx*[12,20]. The mosquito homologue was also cloned by RT-PCR with primers derived from *Drosophila* and *Bombyx* sequences[18].

Identity levels between the three NR4A genes are very high: 91–97% in the DBD, the two most closely related being NR4A2 and NR4A3, and 62–71% in the LBD, the two closer genes being NR4A1 and NR4A2. The *Drosophila* NR4A4 gene is identical at 85–89% and 55–61% with the three human paralogues in the DBD and LBD, respectively.

DNA binding

The DNA binding site of NR4A1 was identified by an *in vivo* genetic selection in yeast[22]. NR4A1 was the first member of the superfamily shown to bind as a monomer to a half-site motif containing a single AGGTCA element with an A-rich sequence located in 5'[22-24]. The NBRE (NGFI-B response element) was shown to harbor the sequence AA AGGTCA, which is distinct from the one of other monomeric orphan receptors, such as SF-1 (TCA AGGTCA). In fact, it was shown that a domain distinct from the C domain, and called the A-box, was required for the recognition of the two A at the 5'-end of the NBRE[23]. The precise mechanisms that allow the binding specificity of NR4A1 and SF1 (NR5A1) to the NBRE and SFRE, respectively, were deciphered in an elegant study showing that an amino group in the minor groove of the SFRE and an Asn residue in the A-box of SF1 allow the precise recognition of SFRE by SF-1[24]. The validity of these findings was verified by the isolation of NR4A1 mutants in a yeast assay and more recently by the determination of the 3D structure of the isolated NR4A1 DBD[26,63]. It was shown that other members of the NR4A group binds to the same response elements both in vertebrates and insects[25].

Later, it was realized that the DNA-binding repertoire of NR4A group members is much larger than anticipated. NR4A1 and NR4A2 heterodimerize with RXR and this complex can in fact recognize two types of sequence: the NBRE and the DR5 elements[27,28]. When the RXR–NR4A1 or 2 complexes bind to the NBRE element, the RXR moiety does not bind to DNA, whereas when it binds to the DR5 (such as the DR5 element found in the RARβ promoter), the two partners bind to DNA. No binding was detected on direct repeats with different spacing or on palindromic elements[28]. This binding exhibits a strong polarity, since the DR5 should contain an NBRE site (i.e. an AGGTCA motif preceded by two A) in 3'. This confirmed by the observation that RXR binds to the 5' half-site, whereas NR4A1 or 2 bind to the 3' one[28]. The carboxy-terminal parts of both NR4A1 and NR4A2 are required for the heterodimerization with RXR. Interestingly (see below), the RXR–NR4A heterodimer exhibits specific transactivation properties[27,28]. Of note, NR4A3 is unable to heterodimerize with RXR, suggesting that a certain specificity of action of each of the three paralogues exists[29]. The interaction between NR4A group members and

RXR is evolutionary conserved, since the insect homologue NR4A4, both from *Drosophila*, *Aedes* and *Bombyx*, heterodimerizes efficiently with the RXR homologue, USP[15,18]. This interaction was observed using a yeast two-hybrid system but also in solution with GST pull-down experiments. The USP–NR4A4 heterodimer is able to bind to monomeric NBRE sites[14,15]. Interestingly, NR4A4 is able to disrupt the EcR–USP heterodimer bound to palindromic EcRE from the hsp27 gene but is unable to directly bind to this element[15].

A third type of DNA-binding activity for NR4A group members has been recently demonstrated by Jacques Drouin's laboratory[30-32]. In the regulatory region of the POMC gene, this group has localized a palindromic element containing two NBREs that confer a high response to NR4A group members[30]. This element, called NurRE, which has the structure of an ER10 (two AGGTCA elements as everted repeats spaced by 10 bp, the two immediate 5' bases of each element being As) binds NR4A homodimers but also heterodimers[30,32]. This NurRE element, and thus NR4A1, is responsible for the activation of the POMC gene by CRH (corticotropin-releasing hormone) in the pituitary. Apparently, each NurRE may bind various types of heterodimers, suggesting that a combinatorial code relying on specific NurRE might be responsible for the activation of subsets of target genes[32].

Partners

RXR has been shown to be a partner for NR4A1 and NR4A2 but not NR4A3[27-29]. The insect homologue NR4A4 is also able to bind to USP[15,18]. The interaction is visible both in solution or on the cognate DNA target sequence.

It has been shown that the positive action of NR4A1, relaying the signal of the CRH on the POMC promoter, may be antagonized by glucocorticoids[31]. This repression by glucocorticoids is a complex phenomenon that probably arises through several mechanisms. Among these, it has been suggested that the glucocorticoid receptor (NR3C1) may interact with NR4A group members and inhibit their DNA-binding ability[31]. Increasing amounts of GR effectively decrease NR4A1 binding to the NurRE, whereas increasing amounts of NR4A1 reduced GR binding to the GRE. This interaction requires the same domains of GR as that implicated in the GR transrepression of AP-1 activity and may rely on the same mechanisms.

It has also been demonstrated that NR4A1 interacts with COUP-TFs (NR2F) and that this interaction modulates retinoic acid sensitivity in human lung cancer cells.

The recent characterization of the AF-2 function of NR4A2 has led to the conclusion that this receptor does not interact with classical coactivators, such as SRC-1, but that the adenovirus E1A protein may increase the NR4A2 transcriptional activity[62]. The basis of this effect is not known.

Ligands

NR4A members are orphan receptors that are transcriptionally active in transient transfection assays even in the absence of serum[34,35]. This may be an argument (although a weak one) suggesting that NR4A group members are real orphans.

Interestingly, the RXR–NR4A1 (or 2) heterodimer exhibits a striking responsiveness to retinoids on DR5 elements as well as on NBREs[27–29]. Although RXR is a silent partner that is unable to respond to 9-*cis* retinoic acid (or other RXR specific ligands, such as synthetic rexinoids) in the RXR–TR or RXR–RAR complexes, it is transcriptionally active when complexed to either NR4A1 or NR4A2. The demonstration that rexinoids could activate NR4A–RXR heterodimers suggests that rexinoids could enhance the response to growth factors initiated by the rapid induction of NR4A1 and NR4A2. The NR4A–RXR heterodimer may thus be considered as a unique ligand-responsive transcriptional unit that may be a target for specific ligands.

Three-dimensional structure

The X-ray crystal structure of the DBD of NR4A1 complexed to DNA has been determined[63]. The structure of the core DBD and its interaction with DNA in the major groove are similar to those for other nuclear receptors. This study has fully confirmed the extensive mutational studies and has shown that the C-terminal extension (T- and A-boxes) form a separate structure, which is not helical and which interacts strongly with the three specific A-T base pairs located 5′ of the NBRE site[63].

Expression

NR4A1 is expressed as a 2.5 kb mRNA, which is inducible by growth factors and serum[1-3]. The NR4A2 transcript is a little longer (3.5 kb) and also exhibits a strong inducibility, for example, by membrane depolarization in PC12 cells[7,11,13]. NR4A3 is expressed as a complex pattern of transcripts (6.5 kb, 5.0 kb, 2.0 kb) in skeletal muscles, fetal heart as well as at lower levels in other tissues[12].

The three vertebrate genes are expressed in a complex pattern in the nervous system, where they are induced as part of the immediate early response to stimuli, such as growth factors, membrane depolarization and seizure[2,3,7,16,29,36–38,41,47]. Their expression pattern outside the nervous system is quite large. NR4A1 and NR4A3 have relatively similar expression profiles with NR4A1 having a late expression during development. NR4A2 is expressed more specifically in brain[1,4,8,11,12,19,29,39,40]. Major sites of expression of NR4A1 outside the brain involve pituitary, adrenal and thyroid, as well as liver, testis, ovary, thymus, muscle, lung and prostate. NR4A2 is expressed in the adult liver as well as in pituitary, thymus and osteoblasts, whereas NR4A3 is expressed in the pituitary, adrenal, heart, muscle, thymus and kidney[25].

The regulation of NR4A1 expression by growth factors has been studied in detail by the cloning and characterization of its promoter[2,3,41]. Several elements able to respond to serum have been located in the NR4A1 promoter and its response to members of the AP-1 complex, such as c-*fos* or junD has been studied[46–49].

In addition to being regulated at the transcriptional level by growth factors and other physiological events, it has been shown that the NR4A1 protein is phosphorylated in response to growth factors[42,43]. The nuclear localization, DNA binding and transcriptional activity are regulated by the phosphorylation events[35,42–45]. Major phosphorylation sites are on serine residues located just upstream of the zinc fingers of the DBD[44,45]. The *in vivo* significance of these phosphorylation events as well as the kinases implicated in these regulations are the subject of intensive studies[64,65].

By contrast, the expression of NR4A4 in insects has been much less studied. In mosquito, it has been implicated in vitellogenesis and is expressed in the fat body[18]. In *Drosophila* a function of NR4A4 in adult cuticule formation has been demonstrated[50].

Activity and target genes

All NR4A family members are transcriptional activators that apparently do not require a ligand to be active (see above)[15,18,34,35]. It has been demonstrated for NR4A1 and NR4A2 that the C-terminal part of the protein contains an AF-2 activation domain[35,62]. In addition, an activation domain has been mapped inside the A/B region in an 18-amino-acid stretch, rich in serine and threonine residues[35]. Thus, NR4A1 (and probably also the three other members of the group) contains two activation functions: an AF-1 in the A/B region and an AF-2 in the LBD.

Three main functions have been attributed to vertebrate NR4A group members. First, it has been demonstrated that NR4A1 and its close relatives NR4A2 and NR4A3 play an important role in the regulation of the hypothalamus–pituitary–adrenal (HPA) axis[30,31]. NR4A1 is expressed in the paraventricular nucleus of the hypothalamus, which also produces the CRH peptide as well as the adrenal cortex[46,47]. NR4A1 regulates both the POMC gene in the pituitary and the genes encoding steroid 21-hydroxylase and steroid 17-hydroxylase in the adrenal[30–32,51,52]. Taken together, these data imply that NR4A group members are important players in the regulation of the HPA axis, whereas to date there is no direct *in vivo* proof in favor of this role. Second, a role in dopaminergic neuron function has been proposed for NR4A2 by knockout experiments (see below). Interestingly, it has been recently demonstrated that tyrosine hydroxylase is a target gene for NR4A2 in neural progenitor cells[56]. The GRIK5 gene, which encodes the kainate-preferring glutamate receptor subunit KA2, is also a target for NR4A2[69]. Indeed, the overexpression of NR4A2, together with the presence of factors derived from astrocytes, is able to induce a midbrain dopaminergic phenotype in neural stem cells[66]. Third, a role for apoptosis of T-cells has been suggested[13,49,53–55,67]. It has been demonstrated that thymocyte apoptosis is impaired by antisens or dominant-negative NR4A1[53,54]. These effects are blocked by the immunosuppressive compound cyclosporin A through a mechanism that is not understood[55]. Recently, it has been shown that, in response to apoptotic stimuli, NR4A1 translocates from the nucleus to the mitochondria to induces cytochrome c release and apoptosis[75]. The pro-apoptotic effect of NR4A1 is not dependent on its DNA binding or

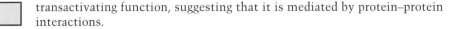

transactivating function, suggesting that it is mediated by protein–protein interactions.

Knockout

The knockout of NR4A1 has been performed by Jeff Milbrandt's laboratory[57]. Strikingly, these mice have no clear phenotype, suggesting that there is much redundancy between NR4A1 and its two relatives. For example, although NR4A1 has been shown to be important for T-cell apoptosis in cell culture[53,54], thymic and peripheral T-cell death is unimpaired in NR4A1 null mice[57]. The redundant function of NR4A1, NR4A2 and NR4A3 has been confirmed by the observation that NR4A2 expression is strongly increased in the NR4A1 mutant mice[60]. The same study has revealed that NR4A3 may replace NR4A1 in the induction of T-cell apoptosis.

In contrast, NR4A2 knockout mice have a strong phenotype that confirms its important role in the nervous sytem[58,59]. These mice died soon after birth and had no mesencephalic dopaminergic system. It has been shown that several dopaminergic cell markers, such as tyrosine hydroxylase, the retinoic acid-converting enzyme ADH2, and c-*ret*, a membrane receptor with tyrosine kinase activity, are absent in these mice[58]. This is consistent with the regulation of the tyrosine hydroxylase promoter by NR4A2[56]. Since the loss of dopaminergic neurons in midbrains associated with Parkinson's disease, it is anticipated that the study of these mice can give clues to better understand the cause of this disease.

No knockout of the NR4A3 gene has yet been published.

Mutants of the NR4A4 gene have been described in *Drosophila*. These mutants alleles cause localized fragility and rupturing of the adult cuticle, reinforcing the notion that NR4A4 is important for the formation of the adult cuticle[50].

Associated disease

The phenotype of NR4A2 knockout mice is reminiscent of Parkinson's disease[58,59] and it has been suggested but not yet tested, that rexinoids that regulate the activity of the NR4A2–RXR heterodimer could provide a novel therapeutic avenue for the treatment of this disease[25].

The entire NR4A3 coding region has been found in the t(9;22) (q22;q12) translocation fused with the N-terminal transactivation domain of the EWS oncogene in several cases of extraskeletal myxoid chondrosarcomas[12,20]. The functional analysis of the fusion protein shows that it harbors a potent transcriptional activity[68].

Gene structure, promoter and isoforms

Consistent with its function as an immediate early gene, NR4A1 is a small gene only 8 kbp long with seven exons and six introns[3,41]. In the C domain intron lies just after the P-box (KGFFK/RTV) and separates the two zinc finger, whereas another is located inside the A-box in a position conserved for most nuclear receptors. It has been shown that the NR4A1 promoter is a GC-rich TATA-less promoter that responds to growth factors with both an immediate-early and delayed-early response and contains several sites critical for this induction[2,3,38,41,46–49].

The mouse NR4A2 gene is also short (7 kbp) and contains 8 exons[70]. Its structure in the coding region is identical to the one of NR4A1. Its promoter has been characterized and contains several GRE half-site motifs[70]. The human gene spans 8.3 kb and has an identical structure[72].

The human NR4A3 gene is much longer, since it spans at least 40 kbp and contains eight exons[61,73]. Its promoter does not contain a TATA box but harbors several CREB sites[73].

An isoform of NR4A2 exists, which is truncated at the C-terminal part. This isoform is expressed in several brain locations[71]. A C-terminally truncated isoform of NR4A3 has been described[73,74]. A variant of the 5'UTR has also been described[74]. These isoforms have not been characterized functionally.

In contrast to NR4A1 and NR4A2, the *Drosophila* NR4A4 transcriptional unit spans more than 40 kb, contains five exons and produces at least four isoforms differentially expressed in development[50]. There is no intron in the DBD coding region. One intron is located in the region coding for the LBD and the position of this intron is conserved in NR4A1 and NR4A3. Two of these isoforms are greatly enriched in the pupal stage. The isoforms result both from alternative splicing and alternative promoter usage.

Chromosomal location

The mouse NR4A1 is located on chromosome 15[3].

The human NR4A2 gene was localized on 2q22–q23[11,70,72], a loci found deleted or translocated in several types of lymphoma. The implication of NR4A2 in these diseases has not been reported.

Human NR4A3 gene is located at 9q22, consistent with its implication in the t(9;22) (q22;q12) translocation found in extraskeletal myxoid chondrosarcomas[12,20,68].

DHR38 is on chromosome 2L at position 38E1–3.

Amino acid sequence for human NR4A1

Accession number: L13740

```
  1  MPCIQAQYGTPAPSPGPRDHLASDPLTPEFIKPTMDLASPEAAPAAPTALPSFSTFMDGYTGEFDTFLYQ
 71  LPGTVQPCSSASSSASSTSSSSATSPASASFKFEDFQVYGCYPGPLSGPVDEALSSSGSDYYGSPCSAPS
141  PSTPSFQPPQLSPWDGSFGHFSPSQTYEGLRAWTEQLPKASGPPQPPAFFSFSPPTGPSPSLAQSPLKLF
211  PSQATHQLGEGESYSMPTAFPGLAPTSPHLEGSGILDTPVTSTKARSGAPGPSEGRCAVCGDNASCQHYG
281  VRTCEGCKGFFKRTVQKNAKYICLANKDCPVDKRRRNRCQFCRFQKCLAVGMVKEVVRTDSLKGRRGRLP
351  SKPKQPPDASPANLLTSLVLAHLDSGPSTAKLDYSKFQELVLPHFGKEDAGDVQQFYDLLSGSLEVIRKW
421  AEKIPGFAELSPADQDLLLESAFLELFILRLAYRSKPGEGKLIFCSGLVLHRLQCARGFGDWIDSILAFS
491  RSLHSLLVDVPAFACLSALVLITDRHGLQEPRRVEELQNRIASCLKEHVAAVAGEPQPASCLSRLLGKLP
561  ELRTLCTQGLQRIFYLKLEDLVPPPPIIDKIFMDTLPF  598
```

Amino acid sequence for human NR4A2

Accession number: X75918

```
  1 MPCVQAQYGSSPQGASPASQSYSYHSSGEYSSDFLTPEFVKFSMDLTNTEITATTSLPSFSTFMDNYSTG
 71 YDVKPPCLYQMPLSGQQSSIKVEDIQMHNYQQHSHLPPQSEEMMPHSGSVYYKPSSPPTPTTPGFQVQHS
141 PMWDDPGSLHNFHQNYVATTHMIEQRKTPVSRLSLFSFKQSPPGTPVSSCQMRFDGPLHVPMNPEPAGSH
211 HVVDGQTFAVPNPIRKPASMGFPGLQIGHASQLLDTQVPSPPSRGSPSNEGLCAVCGDNAACQHYGVRTC
281 EGCKGFFKRTVQKNAKYVCLANKNCPVDKRRRNRCQYCRFQKCLAVGMVKEVVRTDSLKGRRGRLPSKPK
351 SPQEPSPPSPPVSLISALVRAHVDSNPAMTSLDYSRFQANPDYQMSGDDTQHIQQFYDLLTGSMEIIRGW
421 AEKIPGFADLPKADQDLLFESAFLELFVLRLAYRSNPVEGKLIFCNGVVLHRLQCVRGFGEWIDSIVEFS
491 SNLQNMNIDISAFSCIAALAMVTERHGLKEPKRVEELQNKIVNCLKDHVTFNNGGLNRPNYLSKLLGKLP
561 ELRTLCTQGLQRIFYLKLEDLVPPPAIIDKLFLDTLPF 598
```

Amino acid sequence for human NR4A3

Accession number: D78579

```
  1 MPCVQAQYSPSPPGSSYAAQTYSSEYTTEIMNPDYTKLTMDLGSTEITATATTSLPSISTFVEGYSSNYE
 71 LKPSCVYQMQRPLIKVEEGRAPSYHHHHHHHHHHHHHQQQHQQPSIPPASSPEDEVLPSTSMYFKQSPP
141 STPTTPAFPPQAGALWDEALPSAPGCIAPGPLLDPPMKAVPTVAGARFPLFHFKPSPPHPPAPSPAGGHH
211 LGYDPTAAAALSLPLGAAAAAGSQAAALEGHPYGLPLAKRAAPLAFPPLGLTPSPTASSLLGESPSLPSP
281 PSRSSSSGEGTCAVCGDNAACQHYGVRTCEGCKGFFKRTVQKNAKYVCLANKNCPVDKRRRNRCQYCRFQ
351 KCLSVGMVKEVVRTDSLKGRRGRLPSKPKSPLQQEPSQPSPPSPPICMMNALVRALTDSTPRDLDYSRYC
421 PTDQAAAGTDAEHVQQFYNLLTASIDVSRSWAERIPGFTDLPKEDQTLLIESAFLELFVLRLSIRSNTAE
491 DKFVFCNGLVLHRLQCLRGFGEWLDSIKDFSLNLQSLNLDIQALACLSALSMITERHGLKEPKRVEELCN
561 KITSSLKDHQSKGQALEPNESKVLVALVELRKICTLGLQRIFYLKLEDLVSPPSIIDKLFLDTLPF 626
```

Amino acid sequence for Drosophila NR4A4

Accession number: X89246

```
  1 MDEDCFPPLSGGWSASPPAPSQLQQLHTLQSQAQMSHPNSSNNSSNNAGNSHNNSGGYNYHGHFNAINAS
 71 ANLSPSSSASSLYEYNGVSAADNFYGQQQQQQQQSYQQHNYNSHNGERYSLPTFPTISELAAATAAVEAA
141 AAATVGGPPPVRRASLPVQRTVLPAGSTAQSPKLAKITLNQRHSHAHAHALQLNSAPNSAASSPASADLQ
211 AGRLLQAPSQLCAVCGDNAACQHYGVRTCEGCKGFFKRTVQKGSKYVCLADKNCPVDKRRRNRCQFCRFQ
281 KCLVVGMVKEVVRTDSLKGRRGRLPSKPKSPQESPPSPPISLITALVRSHVDTTPDPSCLDYSHYEEQSM
351 SEADKVQQFYQLLTSSVDVIKQFAEKIPGYFDLLPEDQELLFQSASLELFVLRLAYRARIDDTKLIFCNG
421 TVLHRTQCLRSFGEWLNDIMEFSRSLHNLEIDISAFACLCALTLITERHGLREPKKVEQLQMKIIGSLRD
491 HVTYNAEAQKKQHYFSRLLGKLPELRSLSVQGLQRIFYLKLEDLVPAPALIENMFVTTLPF   551
```

References

1 Milbrandt, J. (1988) Neuron 1, 183–188.
2 Hazel, T.G. et al. (1988) Proc. Natl Acad. Sci. USA 85, 8444–8448.
3 Ryseck, R.P. et al. (1989) EMBO J. 8, 3327–3335.
4 Nakai, A. et al. (1990) Mol. Endocrinol. 4, 1438–1443.
5 Bondy, G.P. et al. (1991) Cell Growth Diff. 2, 203–208.
6 Smith, T.S. et al. (1993) Biochim. Biophys. Acta 1173, 239–242.
7 Law, S.W. et al. (1992) Mol. Endocrinol. 6, 2129–2135.
8 Scearce, L.M. et al. (1993) J. Biol. Chem. 268, 8855–8861.
9 Altin, J.G. et al. (1991) J. Biol. Chem. 266, 5401–5406.
10 Chang, C. et al. (1989) J. Steroid Biochem. Mol. Biol. 34, 391–395.

[11] Mages, H.W. et al. (1994) Mol. Endocrinol. 8, 1583–1591.

[12] Labelle, Y. et al. (1995) Hum. Molec. Genet. 4, 2219–2226.

[13] Okabe, T. et al. (1995) J. Immunol. 154, 3871–3879.

[14] Fisk, G.J. and Thummel, C.S. (1995) Proc. Natl Acad. Sci. USA 92, 10604–10608.

[15] Sutherland, J.D. et al. (1995) Proc. Natl Acad. Sci. USA 92, 7966–7970.

[16] Hedvat, C.V. and Irving, S.G. (1995) Mol. Endocrinol. 9, 1692–1700.

[17] Pichon, B. et al. (1996) Endocrinology 137, 4691–4698.

[18] Zhu, J. et al. (2000) EMBO J. 19, 253–262.

[19] Ohkura, N. et al. (1994) Biochem. Biophys. Res. Commun. 205, 1959–1965.

[20] Clark, J. et al. (1996) Oncogene 12, 229–235.

[21] Kostrouch, Z. et al. (1995) Proc. Natl Acad. Sci. USA 92, 156–159.

[22] Wilson, T.E. et al. (1991) Science 252, 1296–1300.

[23] Wilson, T.E. et al. (1992) Science 256, 107–110.

[24] Wilson, T.E. et al. (1993) Mol. Cell. Biol. 13, 5794–5804.

[25] Giguère, V. (1999) Endocrine Rev. 20, 689–725.

[26] Wilson, T.E. et al. (1993) Proc. Natl Acad. Sci. USA 90, 9186–9190.

[27] Forman, B.M. et al. (1995) Cell 81, 541–550.

[28] Perlmann, T. and Jansson, L. (1995) Genes Dev. 9, 769–782.

[29] Zetterstrom, R.H. et al. (1996) Mol. Endocrinol. 10, 1656–1666.

[30] Philips, A. et al. (1997) Mol. Cell. Biol. 17, 5946–5951.

[31] Philips, A. et al. (1997) Mol. Cell. Biol. 17, 5952–5959.

[32] Maira, M. et al. (1999) Mol. Cell. Biol. 19, 7549–7557.

[33] Wu, Q. et al. (1997) EMBO J. 16, 1656–1669.

[34] Davis, I.J. et al. (1991) Mol. Endocrinol. 5, 854–859.

[35] Paulsen, R.E. et al. (1992) J. Biol. Chem. 267, 16491–16496.

[36] Watson, M.A. and Milbrandt, J. (1990) Development 110, 173–183.

[37] Wisden, W. et al. (1990) Neuron 4, 603–614.

[38] Williams, G.T. and Lau, L.F. (1993) Mol. Cell. Biol. 13, 6124–6136.

[39] Bandoh, S. et al. (1997) J. Neuroendocrinol. 9, 3–8.

[40] Hayashi, K. et al. (1996) Mol. Cell. Endocrinol. 123, 205–209.

[41] Watson, M.A. and Milbrandt, J. (1989) Mol. Cell. Biol. 9, 4213–4219.

[42] Fahrner, T.J. et al. (1990) Mol. Cell. Biol. 10, 6454–6459.

[43] Hazel, T.G. et al. (1991) Mol. Cell. Biol. 11, 3239–3246.

[44] Davis, I.J. et al. (1993) Mol. Endocrinol. 7, 953–964.

[45] Hirata, Y. et al. (1993) J. Biol. Chem. 268, 24808–24812.

[46] Parkes, D. et al. (1993) Mol. Endocrinol. 7, 1357–1367.

[47] Davis, I.J. and Lau, L.F. (1994) Mol. Cell. Biol. 14, 3469–3483.

[48] Yoon, J.K. and Lau, L.F. (1994) Mol. Cell. Biol. 14, 7731–7743.

[49] Woronicz, J.D. et al. (1995) Mol. Cell. Biol. 15, 6364–6376.

[50] Kozlova, T. et al. (1998) Genetics 149, 1465–1475.

[51] Wilson, T.E. et al. (1993) Mol. Cell. Biol. 13, 861–868.

[52] Zhang, P. and Mellon, S.H. (1997) Mol. Endocrinol. 11, 891–904.

[53] Woronicz, J.D. et al. (1994) Nature 367, 277–281.

[54] Liu, Z.G. et al. (1994) Nature 367, 281–284.

[55] Yazdanbakhsh, K. et al. (1995) Proc. Natl Acad. Sci. USA 92, 437–441.

[56] Sakurada, K. et al. (1999) Development 126, 4017–4026.

[57] Lee, S.L. et al. (1995) Science 269, 532–535.

[58] Zetterstrom, R.H. et al. (1997) Science 276, 248–250.
[59] Saucedo-Cardenas, O. et al. (1998) Proc. Natl Acad. Sci. USA 95, 4013–4018.
[60] Cheng, L.E.C. et al. (1997) EMBO J. 16, 1865–1875.
[61] Ohkura, N. et al. (1996) Biochim. Biophys. Acta 1308, 205–214.
[62] Castro, D.S. et al. (1999) J. Biol. Chem. 274, 37483–37490.
[63] Meinke, G. and Sigler, P.B. (1999) Nature Struct. Biol. 6, 471–477.
[64] Blaesert, F. et al. (2000) J. Biol. Chem. 275, 197–209.
[65] Swanson, K.D. et al. (1999) J. Biol. Chem. 274, 3385–3395.
[66] Wagner, J. et al. (1999) Nature Biotech. 17, 653–659.
[67] Amsen, D. et al. (1999) J. Biol. Chem. 96, 622–627.
[68] Labelle, Y. et al. (1999) Oncogene 18, 3303–3308.
[69] Chew, L.J. et al. (1999) J. Biol. Chem. 274, 29366–29375.
[70] Castillo, S.O. et al. (1997) Genomics 41, 250–257.
[71] Ohkura, N. et al. (1999) Biochim. Biophys. Acta 1444, 69–79.
[72] Torii, T. et al. (1999) Gene 230, 225–232.
[73] Maltais, A. and Labelle, Y. (2000) DNA Cell Biol. 19, 121–130.
[74] Ohkura, N. et al. (1998) Gene 211, 79–85.
[75] Li, H. et al. (2000) Science 289, 1159–1163.

Names

This group contains two genes conserved in all vertebrates (NR5A1 and NR5A2) as well as a gene found to date only in fish (NR5A4) and two genes known in arthropods (NR5A3 and NR5B1). These genes have been cloned by several groups using different strategies and have thus many different names. NR5A1 was identified as an embryonic protein binding to the LTR of the Moloney murine leukemia virus and was thus called ELP (for embryonal LTR-binding protein)[22]. It was also found as a regulator of steroidogenic enzyme gene expression and called SF-1 (for steroidogenic factor-1)[4]. Finally, it was isolated as a gene regulating steroidogenic P450 gene via an element of its promoter called Ad4 and was thus called Ad4BP (AD4-binding protein)[2]. Since, it was immediately clear that this gene was closely related to FTZ-F1 (NR5A3), a factor implicated in the regulation of the *Drosophila* fushi-tarazu gene, NR5A1 was also called FTZ-F1[22]. Because the original ELP clone is in fact a C-terminally truncated isoform of SF-1, this name is now widely used to designate the isoform and not the regular full-sized product of NR5A1, which is called SF-1[5].

In the same way, NR5A2 was found as a liver receptor homologous protein (LRH-1) by Stephen Green's laboratory but remained unpublished. Its *Xenopus* homologue was identified and called FTZ-F1 related orphan (xFF1r)[11] and the rat gene was identified as a regulator of α-fetoprotein gene (FTF, α-fetoprotein transcription factor)[12]. The human cDNA was also cloned later by two other groups that used the names CPF and hB1F. As it is also related to FTZ-F1, it is often called FTZ-F1β. However, we will use the first name, LRH-1.

The *Drosophila* FTZ-F1 gene encodes two N-terminal isoforms called αFTZ-F1 and βFTZ-F1[23]. The gene is often found in the literature as βFTZ-F1 because this isoform of the gene plays an important role in metamorphosis. Unfortunately, NR5B1, the other gene known in *Drosophila*, was called FTZ-F1β, and thus much confusion arose[20]. This gene has another name, DHR39, which we will use in the text of this chapter[24].

All together these genes form the fifth subfamily of nuclear hormone receptors. These genes have no particular relationship with other members, although in some analyses the GCNF receptors (NR6) appear distantly related to NR5 subfamily members. The precise position of two members of this subfamily, NR5A4 and NR5B1, is still not well understood. These two genes may represent genes that are the result of late duplications in fish and insects, respectively, and that have diverged rapidly afterwards.

Species	Other names	Accession number	References
Official name: NR5A1			
Human	SF-1, FTZ-F1	U76388	1
Mouse	SF-1, FTZ-F1, ELP, AD4BP	S65878	4, 5
Rat	SF-1	D42151–D42156	6
Cow	AD4BP	D13569	2
Pig	pSF-1, SF-1	U84399	3
Chicken	SF-1, AD4BP	AB002404	7
Trachemys (turtle)	FTZ-F1, SF-1	AF033833	28
Rana (frog)	SF-1, AD4BP	AB017352	8

Official name: NR5A2

Human	LRH1, FTF, CPF, hB1F, FTZ-F1β	U93553	9
Mouse	LRH-1, FTF	M81385	Unpub.
Rat	FTZ-F1β, LRH-1, FTF	U47280	12
Chicken	FTF, LRH-1	AB002403	7
Rana (frog)	FTZ-F1β	AB035499	10
Xenopus	xFF1rA	U05001	11
Xenopus	xFF1rB	U05003	11
Zebrafish	zFF1, FTZ-F1	AF014926	13
Medaka	FTZ-F1	AB016834	Unpub.
Rainbow trout	fFZR1	AB006153	15

Official name: NR5A3

Drosophila	FTZ-F1, βFTZ-F1	M98397	17
Aedes	AaFTZ-F1	AF274870	25
Bombyx	BmFTZ-F1	D10953	14
Metapenaeus	FTZ-F1	AF159132	18
Caenorhabditis	NHR-25, FTZ-F1	AF179214	19

Official name: NR5A4

Zebrafish	FF1b, FTZ-F1	AF198086	16

Official name: NR5B1

Drosophila	FTZ-F1β, DHR39	L06423	20
Bombyx	DHR39	AB005660	21

Isolation

The first member of the NR5 subfamily to be cloned was the *Drosophila* FTZ-F1 gene, which was found as a regulator of the fushi-tarazu (*ftz*) gene[17]. The *ftz* gene is expressed at the cellular blastoderm stage in a pattern of seven stripes and a specific element in its promoter, called the zebra element, governs this striped expression. FTZ-F1 was biochemically purified as a factor binding to a specific site harboring the sequence TCA AGGTCG of the zebra element[26]. Interestingly, these authors found that two different but related forms of FTZ-F1 exist, one expressed during early embryogenesis and implicated in the control of *ftz* expression and one expressed during post-embryonic development. These were shown later to correspond to two different N-terminal isoforms, αFTZ-F1 and βFTZ-F1[23]. The cDNA was isolated by screening an expression library with the DNA-binding site recognized by FTZ-F1 and shown to be a new member of the nuclear receptor superfamily[17]. Interestingly, it was shown in the silkworm *Bombyx mori* that a factor binding to the same target sequence was found in embryonic and silk-gland extracts[27]. The factor was cloned from PCR and silk-gland cDNA screening and shown to be the *Bombyx* homologue of FTZ-F1[14].

The other *Drosophila* member of the NR5 subfamily, DHR39 (NR5B1), was isolated by cross-hybridization with an FTZ-F1 probe[20]. It was also discovered independently as a factor implicated in the regulation of alcohol dehydrogenase (*Adh*) gene. It was cloned by screening an expression library with an element of the distal promoter of the *Adh* gene that was reminiscent of the FTZ-F1 binding site[24]. No vertebrate homologue of this gene has been found to date and it can be proposed that this gene results from a specific duplication in the *Drosophila* lineage. The fact that a

Bombyx homologue exists suggests that this duplication arose early on during insect evolution[21].

In vertebrates, two independent research strategies led to the cloning of SF-1. (1) Keith Parker's laboratory studied the regulation of steroidogenic enzyme expression[29]. This led to the definition of a shared promoter element, containing a half-site core sequence AGGTCA, present in the promoters of the genes encoding three steroidogenic enzymes: cholesterol side-chain cleavage enzyme (SCC); steroid 21-hydroxylase (21-OHase); and steroid 11β-hydroxylase (11β-OHase)[30]. These researchers used this common sequence to partially purify a protein from bovine adrenal extracts. Given the recognition sequence, they hypothesized that this factor was a nuclear receptor and they screened a cDNA library prepared with RNA from a mouse adrenocortical cell line[4]. One of the resulting cDNAs was able to bind on the recognition sequence and had an expression pattern consistent with its physiological role in steroidogenic enzyme regulation and was thus called SF-1[4]. In bovine, Ad4BP was cloned with a similar strategy using its ability to regulate the CYP11B gene that encodes a steroidogenic enzyme[2]. (2) Ohtsura Niwa's laboratory in Japan studied the mechanism of suppression of the activity of the LTR of the Moloney leukemia virus in mouse embryonal carcinoma (EC) cells[31]. It was shown that a factor called ELP was able to bind to discrete sequences of this LTR and to shut off its activity[31,32]. These authors then realized that the recognition sequence of ELP was the same as the recognition sequence of the FTZ-F1 gene of *Drosophila* and showed that FTZ-F1 binds to the ELP recognition site[22]. Thus, they cloned the mouse homologue of FTZ-F1 from a EC cells cDNA library[22]. After the cloning of SF-1 and ELP, it became clear that the two sequences were alternative isoforms of the same gene. This was formally demonstrated when the genomic organization of the mouse FTZ-F1 gene was shown to generate two isoforms: SF-1, a bona fide member of the superfamily, and ELP, which is a C-terminal alternative version[5].

Lastly, the LRH-1 gene was first cloned from a mouse liver cDNA library by Stephen Green's laboratory but remains unpublished. The *Xenopus* homologue was identified by Christine Dreyer's team in a systematic search for orphan receptors implicated in *Xenopus* development and was shown to generate two C-terminal isoforms as the mouse FTZ-F1 gene[11]. Later, the rat cDNA was found expressed in liver and implicated in the regulation of the α-fetoprotein gene via an element reminiscent of the SF-1 binding site[12].

The NR5 subfamily is variable. The vertebrate sequences harbor 89–92% sequence identity in the DBD and 53–61% in the LBD. FTZ-F1 is 85% and only 28–32% identical in the same two domains with the vertebrate members. DHR39 is even more divergent: 60% identity in the DBD with other members and only 22–28% in the LBD.

DNA binding

Since many members of this subfamily were identified by virtue of their DNA-binding abilities, it was immediately clear that they all recognize a common sequence that contains the TCA AGGTCA motif and which was called SFRE (SF-1 response element)[2,4,17,20,22,24,33]. The 5' part of the element, containing the TCA sequence, is critical for the DNA binding of NR5 subfamily members. Interestingly, in the C-terminal of the conserved C domain, all members of this subfamily harbor a conserved sequence called the FTZ-F1-box, which is now known as the T and A boxes, that are required for the DNA-binding activity of both *Drosophila* FTZ-F1 and mouse SF-1[34]. It also became apparent that these members of the superfamily are unusual in that they bind DNA as monomers like the NGFIB (NR4A1) orphan receptor. The detailed comparison of the DNA-binding abilities of NGFIB and SF-1 led to the discovery of the precise mechanisms that allow the binding specificity of NGFIB and SF-1. NGFIB binds to a core sequence AA AGGTCA, whereas SF-1 recognizes TCA AGGTCA. In an elegant study, Jeff Millbrandt's team demonstrated that an amino group in the minor groove of the SFRE and an Asn residue in the A-box of SF-1 allow the precise recognition of SFRE by SF-1[35]. This model was confirmed in the case of NGFI-B by the determination of the 3D structure of its DBD (see the relevant entry). More recently, it was shown that ERR1 (NR3B1) another orphan receptor binds to the SFRE sequence, and that SF-1 and ERR-1 have different sequence requirements for SFRE binding[36].

The fact that all members of the subfamily bind to the SFRE element has interesting functional consequences. It was, for example, shown in *Drosophila* that FTZ-F1 and DHR39 compete for the binding to the elements that are present in the *ftz* and *Adh* genes[37]. In addition, the two alternative products of the mouse SF-1 gene, SF-1 itself and ELP, compete for the binding to DNA, as well as the isoforms generated from the *Xenopus* LRH-1 gene[40].

A bipartite NLS is located in the T- and A-boxes of SF-1. This region is rich in basic amino acids that are critical for nuclear localization[41].

Despite its specific binding to the SFRE half-site, it has been shown that SF-1 may compete for DNA binding with other nuclear receptors, such as COUP-TF (NR2F)[42-45]. In fact, COUP-TFs can recognize a wide variety of response elements. For example, the bovine CYP17 gene promoter contains two half-site motifs spaced by 6 bp on which both SF-1 and COUP-TF bind[42]. It has been demonstrated that a competition exists between these two factors for the occupancy of this DNA sequence and that SF-1 activates transcription via this element, whereas COUP-TF represses it[42]. Other examples include the murine DAX-1 promoter[45], the rat P450c17 gene[44] or the human aromatase P450 gene promoter II[43]. This cross-inhibition of SF-1 and COUP-TFs is likely to be significant *in vivo*, since it has been demonstrated that COUP-TF is expressed in adrenal cells[46].

Partners

SF-1 and its *Drosophila* homologue FTZ-F1 have been shown to interact with a number of proteins that can be classified into three main groups: (1) transcription factors; (2) factors of the basic transcription machinery; and (3) coactivators.

Interestingly, many transcription factors that are partners of SF-1 are homeodomain proteins. This is the case, for example, for Ftz in *Drosophila* but also for Ptx1 in mammals. *Drosophila* FTZ-F1, which was discovered as a regulator of *ftz* gene expression, has been shown to interact directly with the homeodomain protein ftz and to increase ftz transcriptional activity[49-51]. Both factors regulate the engrailed gene synergistically and the DNA-binding affinity of *ftz* is increased by its interaction with FTZ-F1[50,51]. The interaction was observed *in vivo* and this was confirmed by the phenotype of *Drosophila* mutants in the FTZ-F1 gene that exhibits cuticular defects reminiscent of those of ftz mutant flies[50].

The mammalian SF-1 protein was shown to interact with AP1 via c-Jun[41,82]. The N-terminal part of the D domain, including the FTZ-F1 box and a proline-rich cluster, was shown to be critical for this interaction. Indeed it was demonstrated that this interaction increases SF-1 activity[41]. SF-1 and Sp1 function cooperatively in the transcription of cholesterol side-chain cleavage (CYP11A1) promoter[52]. An interaction with the factor GATA-4 has been shown to be important for the regulation by SF-1 of the Müllerian inhibiting substance (MIS). GATA-4 and SF-1 bind to adjacent sites on this promoter and synergistically activate it[53].

SF-1 is an obligatory gene for mammalian gonadogenesis and has been shown to be critical for the development of the hypothalamic–pituitary–gonadal axis. It has been shown to be required for male sex determination (see below). In accordance with these notions, SF-1 exhibits protein–protein interactions with factors implicated in sex determination, such as DAX-1 (NR0B1) and SOX9[53-57]. SF-1 was shown to interact physically with DAX-1, an unusual orphan receptor implicated in X-linked adrenal hypoplasia congenita (AHC) as well as in hypogonadotropic hypogonadism (HHG). DAX-1 inhibits SF-1-mediated transactivation and interacts with it via a repressive domain located in the C-terminus of SF-1[55,57]. The interaction between DAX-1 and SF-1 also antagonizes the synergy existing between SF-1 and WT1, the Wilm's tumor gene that is also implicated in sex determination[54]. It was thus proposed that WT1 and DAX-1 functionally oppose each other in testis development by modulating SF-1-mediated transactivation. The repressive effect of DAX-1 on SF-1 was demonstrated to be linked to the recruitment of the corepressor NCoR to SF-1[54]. SF-1 alone is unable to interact with NCoR, but DAX-1 brings NCoR to SF-1 and this results in the inhibition of SF-1 transcriptional activity. Interestingly, mutations of DAX-1 found in AHC patients allow the SF-1–DAX-1 interaction but markedly diminish NCoR recruitment. An interaction between the HMG box containing factor SOX9 and SF-1 is also critical for the regulation of the MIS gene by SF-1. It is the C-terminal part of SF-1 that is implicated in this interaction, suggesting that this region may bind to multiple factors[56]. The relative importance of the interaction with SOX9, WT-1, GATA-4 and DAX-1 for the regulation of the MIS by SF-1 has still to be compared (see reference 67 for a review).

The DAX-1–SF-1 interaction is reminiscent of the interaction recently demonstrated between LRH-1 and SHP. This interaction decreased the

LRH-1 ability to regulate transcription, leading to the repression of the CYP7A1 gene by SHP (see reference 133 for a review)[134,135].

A similar multiprotein complex, resulting in synergism between the various factors, has been shown to occur in the pituitary between SF-1, Ptx1 and Egr-1[58–60]. A synergism between Ptx1 and SF-1 was observed in gonadotrope cells of the pituitary on the βLH promoter[58]. A transcriptional synergism exists between Ptx1 and SF-1 on two SF-1 target genes, βLH and, again, MIS[59]. A direct physical interaction between the two factors occurs and implicates the N-terminal half of SF-1, resulting in transcriptional enhancement of SF-1 activity. It is an interesting case in which a protein–protein interaction mimics the activation by a ligand[59]. Finally, it has been shown that the transcription factor Egr-1, which is a target of GnRH signaling, interacts with both Ptx1 and SF-1 leading to enhancement of Ptx1- and SF-1-mediated βLH transcription[60]. Thus the role of SF-1 as a partner for many other transcription factors, leading to cell-specific transcriptional activation in a combinatorial fashion, is obviously of extreme importance.

Concerning the basal transcription machinery, SF-1 has been shown to interact with TFIIB via the FTZ-F1 box and the proline-rich region located in the D domain[41].

SF-1 has been shown to interact with the coactivator SRC-1 in an AF-2-dependent fashion[61,62]. Other coactivator partners were also found, such as PNRC (proline-rich nuclear receptor coregulatory protein), which also interacts with the SFRE-binding ERR-1 orphan receptor[63]. The interaction also depends on the integrity of the AF-2 domain, in accordance with the fact that PNRC increases SF-1 transcriptional activity.

The case of the MBF factors is of special interest[64–66]. These factors, MBF1 and MBF2, were isolated by their ability to interact with the *Bombyx* FTZ-F1 protein[64]. MBF1 and 2 are unable to bind to DNA, form a heterodimer and mediate activation of *in vitro* transcription of the ftz promoter by BmFTZ-F1. These factors interact with BmFTZ-F1 and stabilize its interaction with DNA. Of note, MBF1 interacts with the TATA-box binding protein, TBP, suggesting that indeed MBF factors may mediate FTZ-F1 transcriptional activity[64,65]. Interestingly, MBF1 is conserved in mammals and the two human isoforms of MBF1 are able to interact with SF-1 and to increase its activity[66]. This result demonstrates that the interaction between MBF1 factors and NR5 subfamily members is conserved in evolution and is important to bridge SF-1 with the basal transcription machinery.

Ligands

It has been proposed that the SF-1 transcriptional activity may be up-regulated by oxysterols, such as 25-, 26- and 27-hydroxycholesterol at concentrations that are in the micromolar range[47]. This activation has been shown to be AF-2 dependent and SF-1 specific. Nevertheless, these exciting findings have not been confirmed. Other studies have failed to reproduce any transcriptional effects of oxysterols on SF-1 transactivation[41,48]. Thus, SF-1 is still considered as an orphan receptor.

Three-dimensional structure

No 3D structure is yet available for either the DBD or LBD of SF-1 or FTZ-F1.

Expression

The mouse SF-1 gene generates two isoforms, ELP and SF-1, which have different expression patterns. ELP was shown to be expressed in EC cells and to be down-regulated during their differentiation[22]. In mouse, ELP was not found from E8 to adult, suggesting it is only expressed in early embryos[5]. In contrast, SF-1 was expressed from E9 in the urogenital ridge, the source of steroidogenic cells of the adrenals and the gonads[5,69]. Then SF-1 expression localize in the cortex of the adrenals before the expression of the steroidogenic enzymes. Interestingly a sexually dimorphic expression of SF-1 was detected[68,69]. When testicular cords are forming in males, a strong and diffuse SF-1 expression can be observed. SF-1 is expressed in the Sertoli cells. In contrast, in the ovary, SF-1 transcripts disappear between E13.5 and E16.5, and then reappear during late development (E18)[69]. These data have led to the suggestion that, in addition to a role in steroidogenesis, SF-1 was playing a role in gonadal differentiation. In the nervous system an expression of SF-1 in the pituitary as well as in the ventromedial hypothalamic nucleus has been demonstrated[70,87]. In the pituitary, the expression is specific to the gonadotrope cells[87]. A colocalization of DAX-1 and SF-1 expression was demonstrated in adrenal cortex, testis, ovary, hypothalamus and anterior pituitary[71]. In humans, transcripts at 3.5–4 kb were observed in the adrenals and gonads as well as transcripts at 4.4 kb and 8 kb[72]. An expression of human SF-1 in Sertoli cells as well as in steroidogenic Leydig cells was observed which was very similar to the mouse expression pattern[74]. The importance of SF-1 expression in steroidogenic cells was demonstrated by the ectopic expression of SF-1 in undifferentiated ES cells. In this case, the ES cells adopt a steroidogenic phenotype, clearly suggesting that SF-1 is a key factor for this differentiation[73]. In the pituitary, GnRH regulates SF-1 expression by an unknown mechanism[128].

Rat LRH1 was described as expressed mainly in adult and newborn liver, but not in brain or kidney as a 5.2 kb mRNA[12]. An mRNA of the same size was observed in mouse. *In situ* hybridization experiments revealed that mouse LRH-1 expression initiates at E8 in branchial arch, neural crest cells and yolk sac endoderm and continues at E9 in foregut endoderm during liver and pancreatic embryogenesis. In adults, expression was found in liver, pancreas and intestine. In *Xenopus*, a complex pattern of transcripts was observed. The number of these transcripts that are ubiquitously expressed increases between gastrula and tail bud stage and later decreases[11].

As for SF-1, the *Drosophila* FTZ-F1 gene generates two major isoforms differing in the A/B domain, with one, αFTZ-F1, being expressed in early embryos, whereas the other, βFTZ-F1, is expressed late and corresponds to the ecdysone-induced puff 75CD during *Drosophila* metamorphosis[17,23]. The βFTZ-F1 gene expression and its role during metamorphosis have been intensively studied. This gene is expressed as a brief pulse during mid-prepupae, a period of low ecdysone titer between the high ecdysone titer of the late third instar larvae and the high ecdysone titer of the late pre-pupae[23,76,78]. It was shown that βFTZ-F1 represses its own expression and is repressed by ecdysone, explaining its brief expression when ecdysone titer is low in mid-prepupae[76]. Further experiments have suggested that βFTZ-F1 acts as a competence factor for stage-specific response during metamorphosis[77]. A similar expression pattern was found in *Bombyx*, and in *Aedes*[14,25]. In nematodes, the FTZ-F1 gene is expressed during embryonic and larval development in the epidermis, the developing somatic gonad and some epithelial cells[19].

The zebrafish NR5A4 gene was shown to be expressed first in the diencephalon and later in the anterior part of the pancreas. In later embryos the expression disappears[16]. This expression pattern is more reminiscent of LRH-1 expression than of SF-1.

The *Drosophila* DHR39 gene is expressed at all stages of the life cycle. In early embryos, it is expressed as a maternal mRNA and then throughout the blastoderm layer. Later the expression is localized in brain, ventral chord structures and hindgut[20,24]. It is expressed during metamorphosis as an early–late transcript starting in the third larval instar, increasing during the early and mid-prepupal stages and decreasing in late pre-pupae and pupae[78,79]. It was shown that DHR39 expression is induced by ecdysone[79].

Activity and target genes

The NR5 subfamily are transcriptional activators dependent on the AF-2 domain and having no clear AF-1 function in the A/B domain[41,80,82]. Nevertheless, an activation function, called AF-1 or AF-3, has been mapped in the hinge region of SF-1 both in mammals and zebrafish[80,85].

It has been shown that the activity of SF-1 may be modulated by phosphorylation[83]. It is known for a long time that many peptide hormones such as GnRH regulate expression of their target genes via SF-1 (see reference 91 for an example). It has been suggested that this effect requires post-translational modification of SF-1. It has been suggested that SF-1 may be phosphorylated by protein kinase A and that this may explain its effect as a transcriptional regulator of cAMP-induced genes[92,93]. A mutation of the AF-2 domain has been shown to suppress PKA-dependent transcription of the bovine CYP17 gene, suggesting that indeed PKA may regulate SF-1 activity[101]. Nevertheless, the sites at which SF-1 is phosphorylated and the functional consequences of this phosphorylation are not yet understood. More recently, it has been shown by Holly Ingraham's laboratory that SF-1 may be phosphorylated on Ser 203 in the hinge, coincident with the activation function[85]. This phosphorylation is likely to be mediated by the MAP kinase signaling pathway and results in the activation of SF-1. This could be a way by which peptide hormones activate their target genes and this further stresses the importance of SF-1 in the hypothalamic–pituitary–gonadal axis[85].

Apart from the LTR of the MoMLV virus, no specific target genes are known for the ELP product of the SF-1 gene. Nevertheless, high-affinity targets of ELP were isolated using a GST–ELP fusion protein that was incubated with genomic DNA fragments[104]. The identity of the isolated DNA fragments that contain several SFRE sites is not known.

For the SF-1 isoform, such large numbers of target genes have been found that it is not possible to discuss them extensively. Some are reviewed in references 33, 67, 99 and 100. The target genes of SF-1 can be classified in three main groups: (1) steroidogenic enzymes; (2) hormones and growth factors; and (3) other genes. The steroidogenic enzymes were the first to be found as SF-1 target genes and were the basis of its discovery[4,33]. Among those we can cite are the bovine steroid hydroxylase, CYP17[42,101], the cholesterol side-chain cleavage enzyme CYP11A[10,52], the type II 3β-hydroxysteroid deshydrogenase[103], the steroidogenic acute regulatory protein (StAR) gene[91,113,114] or the aromatase gene[6,43,92]. All these genes are regulated by SF-1 in the adrenal cortex and/or in the gonads, testis (Leydig and Sertoli cells) and ovary (theca, granulosa and corpus luteum)[100]. In the gonadotrope cells of the pituitary, a number of genes encoding peptide hormones, such as LHβ[60,107–109], gonadotropin IIβ[13,115], or the α-subunit of glycoprotein hormones[58], are SF-1 target genes. The GnRH receptor[111,112] is also regulated by SF-1 as well as the prolactin receptor gene[105,106]. The GnRH[110] gene itself is regulated by SF-1. The MIS, in the Sertoli cells of the testis is one of the SF-1 target genes that has been extensively studied[53,54,56,59,116,117]. Interestingly, the MIS type II receptor in Leydig cells is also an SF-1 target gene. Finally, other diverse genes are SF-1 targets. Among others we can cite the ACTH receptor gene in the adrenal cortex[119] or the Leydig insulin-like gene in the Leydig cells of the testis[120]. It has also been proposed that the transcription of the SF-1 gene is regulated by its own product via an SFRE located in the first intron of the gene[121], although this result remains controversial[127]. SF-1 is also a positive regulator of two unusual orphan receptor genes that contain only the LBD and no C domain. These are the DAX-1 gene[45,123], the product of which also physically interacts with SF-1 and the SHP (NR0B2)[122], which is regulated by both SF-1 in adrenals and LRH-1 in the liver.

Few targets are known for LRH-1 to date. The first is, of course, the α-fetoprotein gene that was used to isolate LRH-1[12]. Other target genes described are the sterol 12α-hydroxylase, the enzyme responsible for cholic acid synthesis in the liver[98]. Interestingly, cholic acids are bile acids that were shown to be the ligand for the FXR (NR1H4) nuclear receptor, suggesting a possible interesting cross-talk between LRH1 and FXR activities in the liver. The existence of such a cross talk has recently been substantiated since it was shown that FXR bound to bile acids activates SHP transcription, which in turn physically interacts with LRH-1. This interaction led to the repression of the cholesterol 7α-hydroxylase CYP7A1 gene, which is normally up-regulated by LXRs and LRH-1 (see reference 133 for a review)[134,135]. It has also been shown that LRH-1 may regulate an enhancer of the hepatitis B virus, which is specifically active in the liver[39,97]. Cholesterol 7α-hydroxylase is also regulated by LRH-1[38].

The target genes of αFTZ-F1 and βFTZ-F1 are different in accordance with the different expression pattern of these two isoforms[23]. For αFTZ-F1 the two main targets known are the *ftz* gene by the study of which FTZ-F1 was cloned and the engrailed gene[50,51]. DHR39 also regulates *ftz* gene expression[37]. FTZ-F1 has a complex and important role in *ftz* expression. It was for example shown that it can mediate the opposite regulatory effects of runt and hairy on *ftz* expression[94]. Several other transcription factors interact with the zebra element that contains the SFRE site to achieve the complex pattern of *ftz* expression[95]. Concerning βFTZ-F1 it has been shown that being expressed in the critical mid-prepupae period during which ecdysone titer is low it is responsible for the establishment of a stage-specific response to ecdysone[76,77]. This means that βFTZ-F1 role may explain why the response to ecdysone is not identical between late third instar larvae and late pre-pupae. The binding of βFTZ-F1 has been detected to *c.* 150 chromosomal sites on the salivary polytene chromosome including FTZ-F1 itself[23]. Indeed, βFTZ-F1 down-regulates its own expression; this may partly explain why it is expressed as a short pulse[76]. Ectopic βFTZ-F1 expression leads to enhanced levels of ecdysone-induced BR-C, E74 and E75 early transcripts and premature induction of the late response gene E93. This shows that βFTZ-F1 plays an essential role in the regulatory hierarchy in response to ecdysone pulses[76]. A direct target of βFTZ-F1 has been identified in the cuticle as the EDG84A gene that is expressed slightly following βFTZ-F1 gene expression during the pre-pupal period[96].

Knockout

The knockout of SF-1 has been described extensively (reviewed in references 67 and 90). The mutant mice died early after birth with a severe deficiency in corticosterone, owing to the adrenal gland being absent[86]. The heterozygous mice also display adrenal insufficiency resulting from profound defects in adrenal development and organization[132]. The fact that the death was caused by a lack of corticosteroids was demonstrated by the prolonged survival of animals treated with exogenous glucocorticoids[89]. The absence of intrauterine death was explained by a normal level of corticosteroids in the embryos[88]. Also, both male and female null mice had female internal genitalia and lacked gonads, suggesting that in addition to a major role in the formation of steroidogenic tissues, SF-1 also plays a role in sexual differentiation[86,88]. In addition to this peripheric phenotype, these animals display a selective loss of gonadotrop markers (LHβ, FSHβ and GnRH receptor) in the pituitary, further stressing the important role played by SF-1 for gonadotropic cells. Even more spectacularly, in the hypothalamus the knockout animals exhibit a grossly abnormal ventromedial nucleus, a region important for reproductive behavior in which SF-1 is expressed[70]. Taken together, these results show that SF-1 plays a critical role in all structures implicated in the hypothalamic–pituitary–gonadal axis. Since the SF-1 gene generates two isoforms, it was of interest to discover which of the two played a major part in the phenotype, since the first knockout strategy has disrupted the two isoforms. Interestingly, a mouse in which only the SF-1 isoform, but not ELP, was disrupted had exactly the same phenotype as the first null mutant[89]. This clearly showed that ELP do not play an important role in sex determination or steroidogenesis, in accordance to its expression pattern restricted to early embryogenesis. No specific ELP knockout has been published yet.

The knockout of LRH-1 has still not been described.

The analysis of βFTZ-F1 mutants clearly demonstrates that βFTZ-F1 plays the role of a competence factor for stage-specific response to ecdysone pulses during mid-prepupae in *Drosophila* metamorphosis[77]. The mutant flies pupariate normally in response to the late larval peak of ecdysone but exhibit defects in stage-specific response to the following ecdysone peak that takes place in late pre-pupae[77]. Consistent with the known network of genes regulated by βFTZ-F1, mutant flies exhibit reduced expression levels of E74A and E75A in pre-pupae, whereas the levels of BR-C are only marginally affected. These data clearly show that βFTZ-F1 is crucial to achieve a stage-specific response to the same hormone, ecdysone.

No DHR39 mutants have been described extensively.

Associated disease

SF-1 and LRH-1 are not clearly associated with a specific disease. Nevertheless, recently a mutation in the P-box of the DBD of SF-1 (G35E) of human SF-1 has been found associated to a case of complete XY sex reversal and adrenal insufficiency, consistent with the known function of SF-1[81]. This mutation effectively reduces the DNA-binding activity of SF-1 in certain but not all response elements and its ability to regulate target gene expression[81].

Gene structure, promoter and isoforms

The genomic organization of the mouse and human SF-1 genes has been studied[1,5]. The mouse and human genes span *c.* 10 kbp. Each zinc finger of the DBD is encoded by a different exon, the intron lying just after the last cysteine residue of the first finger. The FTZ-F1 box is encoded by a third exon and the intron lying between the C domain and the FTZ-F1 box is located at a position conserved in most nuclear receptors. This genomic structure has been useful to understand the alternative splicing event that generates the ELP and FTZ-F1 isoforms that were previously discovered[4,22]. Interestingly, ELP and SF-1 are different both in the C-terminal and N-terminal, since in exon 1 they do not use the same promoter in accordance with their completely different expression pattern[1,5]. SF-1 is transcribed from the more 5′ promoter and its first exon contains only untranslated sequences, which is then spliced to the exon encoding the first zinc finger. The ELP promoter was not precisely mapped but is located within SF-1 intron 1. The physiological role of ELP remains to be established, since the specific knockout of the SF-1 gives a phenotype indistinguishable from that for the knockout of both isoforms[89].

The promoter governing the SF-1 isoform expression contains a CAAT box and Sp1 sites[127,131]. The region necessary for the adrenal expression is restricted to a 90 bp proximal fragment, which contains the CAAT box, an Sp1 site and an E-box. It has been shown that the E-box is critical for adrenal expression and that this element is most likely to be the steroidogenic cell-specific element. Conflicting results were obtained concerning the regulation of this promoter by its own product[121,127]. To date it is difficult to conclude on the contribution of SF-1 to its own transcription. The E-box motif that controls steroidogenic expression of SF-1 binds USF (upstream stimulatory factor), a member of the helix–loop–helix family of transcription factors[129]. The expression of SF-1 in the Leydig and Sertoli cells of the testis is dependent on two sites, the previously described E-box as well as the CAAT box[130].

No isoform was detected for the mammals LRH-1 gene and its genomic organization has not been described. Nevertheless, in *Xenopus laevis*, a short isoform reminiscent of the ELP isoform of SF-1 has been described[40]. The product encoded by this alternative transcript differs from the regular product in C-terminal at a position very close to the ELP/SF-1 diverging point. This isoform binds to DNA with a lower efficiency and does not activate transcription. It can also reduce the transcriptional activation mediated by the full-sized product. In contrast to ELP, this small isoform is coexpressed with the large one[40]. In zebrafish, the gene is also spliced with two isoforms[13]. The genomic organization of the zebrafish gene exhibits two promoters governing the formation of two different N-terminal isoforms[126]. The gene is spliced in eight exons among which four are common to all isoforms. The DNA-binding domain is encoded by three exons: one for each zinc finger and the last containing the FTZ-F1 box. This suggests that four different proteins products differing in N- and C-terminal parts may be generated from the zebrafish gene[13,126]. These isoforms exhibit different but overlapping expression patterns[126].

The genomic organization of FTZ-F1 has not been described. The *Drosophila* FTZ-F1 gene encodes two N-terminally different isoforms, αFTZ-F1 and βFTZ-F1, which have different expression patterns and physiological roles[23]. The αFTZ-F1 isoform plays a role in early embryogenesis as a regulator of pair-rule genes such as *ftz* whereas βFTZ-F1 is crucial for the progression of the *Drosophila* response to ecdysone during the mid-prepupal stage. The βFTZ-F1 gene is down-regulated by its own product[76]. The promoter governing the expression of βFTZ-F1 is regulated by the orphan receptor DHR3 (NR1F3), which binds DNA as a monomer on half-sites sequences called RORE (see the relevant entry)[124,125]. Three DHR3 binding sites, with the typical structure of ROREs, were identified downstream from the start sites of an isoform of FTZ-F1, called βFTZ-F1, the expression of which is activated by DHR3. Interestingly, the non-DNA-binding isoform E75B of the E75 orphan receptor (NR1D3), which is a homologue of Rev-erbs, is able to inhibit the induction of βFTZ-F1 by forming a complex with DHR3 on the βFTZ-F1 promoter[124,125].

The nematode FTZ-F1 gene, also called NHR-25, also governs the expression of two isoforms that exhibit different N-terminal parts. Interestingly, one of these isoforms, NHR-25β lacks the A/B and DNA-binding domain, suggesting that it will be unable to bind to DNA[19]. The promoter that controls the NHR-25α isoform expression has been cloned and is active in hypodermal cell lineage. The importance of NHR-25 for the development of these cells has been confirmed by RNA interference experiments that result in the death of the animals as embryos or young larva. These animals exhibit impaired hypodermal cell differentiation[19].

The DHR39 coding region is interrupted by two introns, which are located in the LBD[20]. The gene is apparently able to generate a C-terminally truncated isoform, whereas the region where this isoform differs from the bona fide

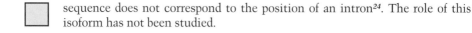

sequence does not correspond to the position of an intron[24]. The role of this isoform has not been studied.

Chromosomal location

The human SF-1 gene is located on chromosome 9q33 and to mouse chromosome 2[1]. The human LRH-1 gene is located on chromosome 1q32[9]. *Drosophila* FTZ-F1 gene maps on the 75CD puff on the chromosome 3L and indeed this puff corresponds to the expression of βFTZ-F1 during metamorphosis[23]. DHR39 is located at position 39C on chromosome 2L[20].

Amino acid sequence for human SF-1 (NR5A1)

Accession number: U76388

```
  1 MDYSYDEDLDELCPVCGDKVSGYHYGLLTCESCKGFFKRTVQNNKHYTCTESQSCKIDKTQRNRCPFCRF
 71 QKCLTVGMRLEAVRADRMRGGRNKFGPMYKRDRALKQQKKAQIRANGFKLETGPPMGVPPPPPPAPDYVL
141 PPSLHGPEPKGLAAGPPAGPLGDFGAPALPMAVPGAHGPLAGYLYPAFPGRAIKSEYPEPYASPPQPGLP
211 YGYPEPFSGGPNVPELILQLLQLEPDEDQVRARILGCLQEPTKSRPDQPAAFGLLCRMADQTFISIVDWA
281 RRCMVFKELEVADQMTLLQNCWSELLVFDHIYRQVQHGKEGSILLVTGQEVELTTVATQAGSLLHSLVLR
351 AQELVLQLLALQLDRQEFVCLKFIILFSLDLKFLNNHILVKDAQEKANAALLDYTLCHYPHCGDKFQQLL
421 LCLVEVRALSMQAKEYLYHKHLGNEMPRNNLLIEMLQAKQT 461
```

Amino acid sequence for human LRH1 (NR5A2)

Accession number: U93553

```
  1 MLPKVETEALGLARSHGEQGQMPENMQVSQFKMVNYSYDEDLEELCPVCGDKVSGYHYGLLTCESCKGFF
 71 KRTVQNNKRYTCIENQNCQIDKTQRKRCPYCRFQKCLSVGMKLEAVRADRMRGGRNKFGPMYKRDRALKQ
141 QKKALIRANGLKLEAMSQVIQAMPSDLTISSAIQNIHSASKGLPLNHAALPPTDYDRSPFVTSPISMTML
211 HGSLQGYQTYGHFPSRAIKSEYPDPYTSSPESIMGYSYMDSYQTSSPASIPHLILELLKCEPDEPQVQAK
281 IMAYLQQEQANRSKHEKLSTFGLMCKMADQTVFSIVEWARSSIFFRELKVDDQMKLLQNCWSELLILDHI
351 YRQVVHGKEGSIFLVTGQQVDYSIIASQAGATLNNLMSHAQELVAKLRSLQFDQREFVCLKFLVLFSLDV
421 KNLENFQLVEGVQEQVNAALLDYTMCNYPQQTEKFGQLLLRLPEIRAISMQAEEYLYYKHLNGDVPYNNL
491 LIEMLHAKRA 500
```

Amino acid sequence for *Drosophila* FTZ-F1 (NR5A3)

Accession number: M98397

```
  1 MLLEMDQQQATVQFISSLNISPFSMQLEQQQQPSSPALAAGGNSSNNAASGSNNNSASGNNTSSSSNNNN
 71 NNNDNDAHVLTKFEHEYNAYTLQLAGGGGSGSGNQQHHSNHSNHGNHHQQQQQQQQQQQHQQQQQEHYQ
141 QQQQQNIANNANQFNSSSYSYIYNFDSQYIFPTGYQDTTSSHSQQSGGGGGGGGGGGNLLNGSSGGSSAGGG
211 YMLLPQAASSSGNNGNPNAGHMSSGSVGNGSSGGAGNGGAGGNSGPGNPMGGTSATPGHGGEVIDFKHLFE
281 ELCPVCGDKVSGYHYGLLTCESCKGFFKRTVQNNKVYTCVAERSCHIDKTQRKRCPYCRFQKCLEVGMKL
351 EAVRADRMRGGRNKFGPMYKRDRARKLQVMRQRQLALQALRNSMGPDIKPTPISPGYQQAYPNMNIKQEI
421 QIPQVSSLTQSPDSSPSPIAIALGQVNASTGGVIATPMNAGTGGSGGGGLNGPSSVGNGNSSNGSSNGNN
491 NSSTGNGTSGGGGGNNAGGGGGGTNSNDGLHRNGGNDSSSCHEAGIGSLQNTADSKLCFDSGTHPSSTAD
561 ALIEPLRVSPMIREFVQSIDDREWQTQLFALLQKQTYNQVEVDLFELLMCKVLDQNLFSQVDWARNTVFF
631 KDLKVDDQMKLLQHSWSDMLVLDHLHHRIHNGLPDETQLNNGQVFNLMSLGLLGVPQPGDYFNELQNKLQ
701 DLKFDMGDYVCMKFLILLNPSVRGIVNRKTVSEGHDNVQAALLDYTLTCYPSVNDKFRGLVNILPEIHAM
771 AVRGEDHLITCTPSTVPAVRPPKRCSWRCCTPSARDRGRENVTRNT 816
```

Amino acid sequence for *Drosophila* DHR39 (NR5B1)

Accession number: L06423

```
  1  MPNMSSIKAEQQSGPLGGSSGYQVPVNMCTTTVANTTTTLGSSAGGATGSRHNVSVTNIKCELDELPSPN
 71  GNMVPVIANYVHGSLRIPLSGHSNHRESDSEEELASIENLKVRRRTAADKNGPRPMSWEGELSDTEVNGG
141  EELMEMEPTIKSEVVPAVAPPQPVCALQPIKTELENIAGEMQIQGKCYPQSNTQHHAATKLKLAPTQSDP
211  INLKFEPPLGDNSPLLAARSKSSSGGHLPLPTNPSPDSAIHSVYTHSSPSQSPLTSRHAPYTPSLSRNNS
281  DASHSSCYSYSSEFSPTHSPIQARHAPPAGTLYGNHHGIYRQMKVEASSTVPSSGQEAQNLSMDSASSNL
351  DTVGLGSSHPASPAGISRQQLINSPCPICGDKISGFHYGIFSCESCKGFFKRTVQNRKNYVCVRGGPCQV
421  SISTRKKCPACRFEKCLQKGMKLEAIREDRTRGGRSTYQCSYTLPNSMLSPLLSPDQAAAAAAAAAVASQ
491  QQPHQRLHQLNGFGGVPIPCSTSLPASPSLAGTSVKSEEMAETGKQSLRTGSVPPLLQEIMDVEHLWQYT
561  DAELARINQPLSAFASGSSSSSSSSSGTSSGAHAQLTNPLLASAGLSSNGENANPDLIAHLCNVADHRLYK
631  IVKWCKSLPLFKNISIDDQICLLINSWCELLLFSCCFRSIDTPGEIKMSQGRKITLSQAKSNGLQTCIER
701  MLNLTDHLRRLRVDRYEYVAMKVIVLLQSDTTELQEAVKVRECQEKALQSLQAYTLAHYPDTPSKFGELL
771  LRIPDLQRTCQLGKEMLTIKTRDGADFNLLMELLRGEH
```

References

1 Wong, M. et al. (1996) J. Mol. Endocrinol. 17, 139–147.
2 Honda, S. et al. (1993) J. Biol. Chem. 268, 7494–7502.
3 Pilon, N. et al. (1998) Endocrinology 139, 3803–3812.
4 Lala, D.S. et al. (1992) Mol. Endocrinol. 6, 1249–1258.
5 Ikeda, Y. et al. (1993) Mol. Endocrinol. 7, 852–860.
6 Lynch, J.P. et al. (1993) Mol. Endocrinol. 7, 776–786.
7 Kudo, T. and Sutou, S. (1997) Gene 197, 261–268.
8 Kawano, K. et al. (1998) Gene 222, 169–176.
9 Galarneau, L. et al. (1998) Cytogenet. Cell Genet. 82, 269–270.
10 Nakajima, T. et al. (2000) Gene 248, 203–212.
11 Ellinger-Ziegelbauer, H. et al. (1994) Mol. Cell. Biol. 14, 2786–2797.
12 Galarneau, L. et al. (1996) Mol. Cell. Biol. 16, 3853–3865.
13 Liu, D. et al. (1997) Mol. Endocrinol. 11, 877–890.
14 Sun, G-C. et al. (1994) Dev. Biol. 162, 426–437.
15 Ito, M. et al. (1998) Biochim. Biophys. Acta 1395, 271–274.
16 Chai, C. and Chan, W. (2000) Mech. Dev. 91, 421–426.
17 Lavorgna, G. et al. (1991) Science 252, 848–851.
18 Chan, S.M. and Chan, K.M. (1999) FEBS Lett. 454, 109–114.
19 Gissendanner, C.R. and Sluder, A.E. (2000) Dev. Biol. 221, 259–272.
20 Ohno, C.K. and Petkovitch, M. (1992) Mech. Dev. 40, 13–24.
21 Niimi, T. et al. (1997) Dev. Genes Evol. 207, 410–412.
22 Tsukiyama, T. et al. (1992) Mol. Cell. Biol. 12, 1286–1291.
23 Lavorgna, G. et al. (1993) Proc. Natl Acad. Sci. USA 90, 3004–3008.
24 Ayer, S. et al. (1993) Nucl. Acids Res. 21, 1619–1627.
25 Li, C. et al. (2000) Dev. Biol. 224, 96–110.
26 Ueda, H. et al. (1990) Genes Dev. 4, 624–635.
27 Ueda, H. and Hirose, S. (1990) Nucl. Acids Res. 18, 7229–7234.
28 Wibbels, T. et al. (1998) J. Exp. Zool. 262, 454–457.
29 Chaplin, D.D. et al. (1986) Proc. Natl Acad. Sci. USA 83, 9601–9605.
30 Rice, D.A. et al. (1991), Mol. Endocrinol. 5, 1552–1561.
31 Tsukiyama, T. et al. (1989) Mol. Cell. Biol. 9, 4670–4676.
32 Tsukiyama, T. et al. (1991) J. Virol., 65, 2979–2986.

[33] Parker, K.L. and Schimmer, B.P. (1993) Trends Endocrinol. Metab. 4, 46–50.
[34] Ueda, H. et al. (1992) Mol. Cell. Biol. 12, 5667–5672.
[35] Wilson, T.E. et al. (1993) Mol. Cell Biol. 13, 5794–5804.
[36] Vanacker, J.M. et al. (1999) Mol. Endocrinol. 13, 764–773.
[37] Ohno, C.K. et al. (1994) Mol. Cell. Biol. 14, 3166–3175.
[38] Nitta, M. et al. (1999) Proc. Natl Acad. Sci. USA 96, 6660–6665.
[39] Li, M. et al. (1998) J. Biol. Chem. 273, 29022–29031.
[40] Ellinger-Ziegelbauer, H. et al. (1995) Mol. Endocrinol. 9, 872–886.
[41] Li, L.A. et al. (1999) Mol. Endocrinol. 13, 1588–1598.
[42] Bakke, M. and Lunde, J. (1995) Mol. Endocrinol. 9, 327–339.
[43] Zeitoun, K. et al. (1999) Mol. Endocrinol. 13, 239–253.
[44] Zhang, P. and Mellon, S.H. (1997) Mol. Endocrinol. 11, 891–904.
[45] Yu, R.N. et al. (1998) Mol. Endocrinol. 12, 1010–1022.
[46] Shibata, H. et al. (1998) J. Clin. Endocrinol. Metab. 83, 4520–4523.
[47] Lala, D.S. et al. (1997) Proc. Natl Acad. Sci. USA 94, 4895–4900.
[48] Mellon, S.H. and Bair, S.R. (1998) Endocrinology 139, 3026–3029.
[49] Guichet, A. et al. (1997) Nature 385, 548–552.
[50] Yu, Y. et al. (1997) Nature 385, 552–555.
[51] Florence, B. et al. (1997) Development 124, 839–847.
[52] Liu, Z. and Simpson, E.R. (1997) Mol. Endocrinol. 11, 127–137.
[53] Tremblay, J.J. and Viger, R.S. (1999) Mol. Endocrinol. 13, 1388–1401.
[54] Nachtigal, M.W. et al. (1998) Cell 93, 445–454.
[55] Crawford, P.A. et al. (1998) Mol. Cell. Biol. 18, 2949–2956.
[56] de Santa-Barbara, P. et al. (1998) Mol. Cell. Biol. 18, 6653–6665.
[57] Ito, M. et al. (1997) Mol. Cell. Biol. 17, 1476–1483.
[58] Tremblay, J.J. et al. (1998) Mol. Endocrinol. 12, 428–441.
[59] Tremblay, J.J. et al. (1999) EMBO J. 18, 3431–3441.
[60] Tremblay, J.J. et al. (1999) Mol. Cell. Biol. 19, 2567–2576.
[61] Crawford, P.A. et al. (1997) Mol. Endocrinol. 11, 1626–1635.
[62] Ito, M. et al. (1998) Mol. Endocrinol. 12, 290–301.
[63] Zhou, D. et al. (2000) Mol. Endocrinol. 14, 986–998.
[64] Li, F.Q. et al. (1994) Mol. Cell. Biol. 14, 3013–3021.
[65] Takemaru, K.I. et al. (1997) Proc. Natl Acad. Sci. USA 94, 7251–7256.
[66] Kabe, Y. et al. (1999) J. Biol. Chem. 274, 34196–34202.
[67] Parker, K.L. et al. (1999) Annu. Rev. Physiol. 61, 417–433.
[68] Hatano, O. et al. (1994) Development 120, 2787–2797.
[69] Ikeda, Y. et al. (1994) Mol. Endocrinol. 8, 654–662.
[70] Ikeda, Y. et al. (1995) Mol. Endocrinol. 9, 475–486.
[71] Ikeda, Y. et al. (1996) Mol. Endocrinol. 10, 1261–1272.
[72] Ramayya, M.S. et al. (1997) J. Clin. Invest. Metab. 82, 1799–1806.
[73] Crawford, P.A. et al. (1997) Mol. Cell. Biol. 17, 3997–4006.
[74] Hanley, N.A. et al. (1999) Mech. Dev. 87, 175–180.
[75] Rausa, F.M. et al. (1999) Mech. Dev. 89, 185–188.
[76] Woodward, C.T. et al. (1994) Cell 79, 607–615.
[77] Broadus, J. et al. (1999) Mol. Cell 3, 143–149.
[78] Huet, F. et al. (1995) Development 121, 1195–1204.
[79] Horner, M.A. et al. (1995) Dev. Biol. 168, 490–502.
[80] Liu, D. et al. (2000) J. Biol. Chem. 275, 16758–16766.

81 Ito, M. et al. (2000) J. Biol. Chem. 275, 31708–31714.

82 Li, L.A. et al. (1998) Biochem. Biophys. Res. Commun. 250, 318–320.

83 Taketo, M. et al. (1995) Genomics 25, 565–567.

84 Swift, S. and Ashworth, A. (1995) Genomics 28, 609–610.

85 Hammer, G.D. et al. (1999) Mol. Cell 3, 521–526.

86 Luo, X. et al. (1994) Cell 77, 481–490.

87 Ingraham, H.A. et al. (1994) Genes Dev. 8, 2302–2312.

88 Sadovsky, Y. et al. (1995) Proc. Natl Acad. Sci. USA 92, 10939–10943.

89 Luo, X. et al. (1995) Mol. Endocrinol. 9, 1233–1239.

90 Luo, X. et al. (1999) J. Steroid Biochem. Mol. Biol. 69, 13–18.

91 Sugawara, T. et al. (1996) Biochemistry 35, 9052–9059.

92 Carlone, D.L. and Richards, J.A. (1997) Mol. Endocrinol. 11, 292–304.

93 Zhang, P. and Mellon, S.H. (1996) Mol. Endocrinol. 10, 147–158.

94 Tsai, C. and Gergen, P. (1995) Development 121, 453–462.

95 Han, W. et al. (1998) Mol. Cell. Biol. 18, 3384–3394.

96 Murata, T. et al. (1996) Mol. Cell. Biol. 16, 6509–6515.

97 Gilbert, S. et al. (2000) J. Virol. 74, 5032–5039.

98 del Castillo-Olivares A. et al. (2000) J. Biol. Chem. 275, 17793–17799.

99 Parker, K.L. and Schimmer, B.P. (1996) Trends Endocrinol. Metab. 7, 203–207.

100 Morohashi, K.I. (1999) Trends Endocrinol. Metab. 10, 169–173.

101 Jacob, A.L. and Lund, J. (1998) J. Biol. Chem. 273, 13391–13394.

102 Morohashi, K.I. et al. (1993) Mol. Endocrinol. 7, 1196–1204.

103 Leers-Sucheta, S. et al. (1997) J. Biol. Chem. 272, 7960–7967.

104 Tsukiyama, T. and Niwa, O. (1992) Nucl. Acids Res. 20, 1477–1482.

105 Hu, Z.Z. et al. (1997) J. Biol. Chem. 272, 14263–14271.

106 Hu, Z.Z. et al. (1998) J. Biol. Chem. 273, 26225–26235.

107 Halvorson, L.M. (1996) J. Biol. Chem. 271, 6645–6650.

108 Dorn, C. et al. (1999) J. Biol. Chem. 274, 13870–13876.

109 Wolfe, M.W. (1999) J. Biol. Chem. 13, 1497–1510.

110 Corley, D.R. et al. (2000) Mol. Hum. Reprod. 6, 671–676.

111 Duval, D.L. et al. (1997) Biol. Reprod. 56, 160–168.

112 Ngan, E.S. et al. (1999) Endocrinology 140, 2452–2462.

113 Reinhart, A.J. et al. (1999) Mol. Endocrinol. 13, 729–741.

114 Wooton-Kee, C.R. and Clark, B.J. (2000) Endocrinology 141, 1345–1355.

115 Le Dréan, Y. et al. (1996) Mol. Endocrinol. 10, 217–229.

116 Giuili, G. et al. (1997) Development 124, 1799–1807.

117 Shen, W.H. et al. (1994) Cell 77, 651–661.

118 Teixeira, J. et al. (1999) Proc. Natl Acad. Sci. USA 96, 13831–13838.

119 Cammas, F.M. et al. (1997) Mol. Endocrinol. 11, 867–876.

120 Zimmermann, S. et al. (1998) Mol. Endocrinol. 12, 706–713.

121 Nomura, M. et al. (1996) J. Biol. Chem. 271, 8243–8249.

122 Lee, Y.K. et al. (1999) J. Biol. Chem. 274, 20869–20873.

123 Kawabe, K. et al. (1999) Mol. Endocrinol. 13, 1267–1284.

124 Lam, G.T. et al. (1997) Development 124, 1757–1769.

125 White, K.P. et al. (1997) Science 276, 114–117.

126 Lin, W.W. et al. (2000) Biochem. J. 348, 439–446.

127 Woodson, K.G. et al. (1997) Mol. Endocrinol. 11, 117–126.

128 Haisenleder, D.J. et al. (1996) Endocrinology 137, 5719–5722.

[129] Harris, A.N. and Mellon, P.L. (1998) Mol. Endocrinol. 12, 714–726.

[130] Daggett, M.A. et al. (2000) Biol. Reprod. 62, 670–679.

[131] Nomura, M. et al. (1995) J. Biol. Chem. 270, 7453–7461.

[132] Bland, M.L. et al. (2000) Proc. Natl Acad. Sci. USA 97, 14488–14493.

[133] Chawla, A. et al. (2000) Cell 103, 1–4.

[134] Lu, T.T. et al. (2000) Mol. Cell 6, 507–515.

[135] Goodwin, B. et al. (2000) Mol. Cell 6, 517–526.

GCNF

Names

Only one gene encoding the GCNF (NR6A1) orphan receptor is known in this group[1]. This gene has also been described as RTR and NCNF[2,3]. No other paralogous gene in vertebrates is known. Recently, a homologue of this gene in insects has been described and called GRF[4].

Species	Other names	Accession number	References
Official name: NR6A1			
Human	GCNF	U64876	5–7
Mouse	GCNF, RTR, NCNF	U14666	1–3
Xenopus	xGCNF	U58683	8
Zebrafish	zfGCNF	AJ007703	9
Official name: NR6A2			
Bombyx	GRF	AF124981	23
Tenebrio	THR4, TmGRF	AJ005685	4

Isolation

NR6A1 (GCNF) was first isolated from a mouse heart cDNA library by Bert O'Malley's laboratory by low-stringency screening using a fragment of the orphan receptor PPARβ (NR1C2) as a probe. The gene was called GCNF (for 'germ cell nuclear factor') because of its restricted expression pattern in germ cells[1]. The same gene, called RTR (retinoid receptor-related testis-associated receptor) was isolated by PCR with degenerated oligonucleotides corresponding to highly conserved sequences of the DBD[2]. The human homologue was later found by several research groups[5-7]. The Xenopus homologue was found independently in Christine Dreyer's laboratory using a PCR-assisted cDNA cloning strategy with degenerated oligonucleotides[8]. More recently a homologue has also been described in zebrafish[9].

Homologues of GCNF were recently identified in insects such as *Bombyx*[23] and *Tenebrio*[4]. These genes were called either THR-4 or GRF (for GCNF-related factor) and corresponds to NR6A2 in the nuclear receptors nomenclature. A genomic clone corresponding to this gene has been found in *Drosophila* (sequence EG:133E12.2 in FlyBase: http://fly.ebi.ac.uk).

DNA binding

GCNF has been shown to bind to DNA on extended half-site sequences identical to the SF-1 response element (SFRE: TCA AGGTCA)[1,10,11]. Binding to DR0 element (AGGTCA AGGTCA; which also encompasses the SFRE) was also observed, apparently with a better affinity[1,8,10,11]. PCR-coupled EMSA selection of specific binding sites indicated that the preferential target sequence is TCA AG(G/T)TCA[11]. Interestingly, the A-box of GCNF is almost identical to that of SF-1, which also recognizes the SFRE[1]. It has been shown by mixing experiments on gel-shift assays as well as by using specific antibodies that GCNF binds to DNA as a homodimer to the DR0 element[10,11]. It has been shown that the C-terminal extension of the DBD (i.e. the T- and A-boxes) is important for DNA binding[10,12]. This suggests that the specific binding to DNA is similar to that of monomer-binding orphan receptors such as SF-1 or NGFIB. The requirement of the T- and A-boxes for efficient binding is consistent with the fact that GCNF is unable to bind to half-site motifs containing different 5' bases. For example, GCNF does not recognize the NBRE sequence (AA AGGTCA)[1]. In addition, it has been shown recently that GCNF binds as a dimer to half-site sequences of the SFRE type[13]. Interestingly, dimer formation requires both the DBD and the LBD (helices 3 and 12). Modeling and mutation studies reveal that the GCNF LBD has the potential to adopt different conformations with distinct dimerization properties. The helix 12 region regulates the switch between these dimerization conformations and dictates the DNA-binding behavior of the protein[13].

The *Tenebrio* and *Bombyx* GRF proteins bind specifically to the TCA AGGTCA motif[4,23]. The *Bombyx* GRF was shown to bind to the DR0 element[23]. It was shown by mixing experiments in gel-shift assay that, in contrast to its vertebrate homologue, *Bombyx* GRF binds to the SFRE as a monomer[23].

Partners

It has been shown that GCNF is unable to heterodimerize with RXR[10], USP or EcR[23].

Ligands

No ligand or activator has been described for GCNF.

Three-dimensional structure

No 3D structure has been determined for GCNF. Nevertheless, modeling of the LBD has been carried out. This, in conjunction with careful mutation studies, has revealed the presence of a new homodimerization interface in the LBD that encompasses helices 3 and 12[13].

Expression

In mice, two mRNA species (7.5 and 2.4 kb) were observed[1,2]. These differ according to their use of alternative polyadenylation sites[14]. In humans, the size of the smaller transcript is 2.2 kb[5,6]. Recent data suggest a regulation by retinoic acid in human cells[25].

The first expression studies carried out in mice have revealed a very narrow expression pattern in testis with very low levels in ovary, liver and kidney[1,2]. *In situ* hybridization studies have shown a specific expression in both male and female germ cells, both in mice and humans[1,14,15]. This has provided the basis for the names GCNF and RTR, and has led to the suggestion that it may be implicated in meiosis[1,2]. Northern blot analysis has allowed the expression stage of GCNF during male germ cell differentiation to be identified as the round spermatid stage, i.e. suggesting a role in post-meiotic cells[2]. More recently, it has been shown at both the

RNA and protein levels that, in adult mouse testis, GCNF is expressed in late pachytene spermatocytes and round spermatids[16,17]. This suggests a role for GCNF in regulating both meiotic and post-meiotic events.

The cloning and characterization of the homologue of GCNF in *Xenopus laevis* has given the first indication that this gene is also expressed during early embryogenesis[8]. Transcripts of xGCNF are found in oocytes and in much smaller amounts in testis. In zebrafish, zfGCNF is expressed predominantly in pre-vitellogenic oocytes as well as in spermatocytes[9]. Only one of the several transcripts found in ovary was detected in testis. Taken together, these data suggest that the testis specific expression of GCNF is not conserved in all vertebrates.

During *Xenopus* embryogenesis, GCNF expression peaks in mid-neurula, mostly in neural plate and neural crest[8]. Interestingly, from the late gastrula to mid-neurula stages, an anterior to posterior gradient of expression was observed. The significance of this graded expression was nicely shown by overexpression of a full-length or a dominant negative GCNF during embryogenesis[18]. Since GCNF is either transcriptionally inactive or is a repressor, the dominant negative mutant was constructed by deleting the N-terminal part of the protein, including most of the DBD. The resultant protein is believed to act as a dominant negative by heterodimer formation with the intact protein. Early events of embryogenesis were not affected by the overexpression of either the wild-type or the dominant negative form of GCNF. In contrast, and in accordance with the expression pattern, ectopic posterior overexpression of the wild-type GCNF caused posterior defects and disturbed somite formation. Conversely, overexpression of the dominant negative form interfered with differentiation of the neural tube and affected the differentiation of anterior structures such as the eyes. These experiments suggest that GCNF has an essential function in anteroposterior differentiation during organogenesis[18]. Recently, GCNF was cloned in a differential display experiment devoted to the identification of genes regulated by bone morphogenetic proteins (BMP) during neural induction in *Xenopus*[24]. It was shown in this study that GCNF plays a role in the formation of the midbrain–hindbrain boundary.

An expression of GCNF in embryos and in developing brain was also observed in the mouse[3,5,19]. GCNF was cloned from a library of neuronal derivatives of retinoic acid-induced embryonic carcinoma cells and was called NCNF for neuronal cell nuclear factor[3,7,19]. In fact, GCNF is highly expressed in embryonic stem cells but its expression is repressed on retinoic acid-induced differentiation[3,5,20]. Northern blot analysis has also demonstrated an expression of GCNF during mouse gastrulation and early organogenesis[19]. *In situ* hybridization revealed that the early expression is localized in ectodermal cells as well as in the primitive streak. During further development, the expression appears restricted more to the cells of the developing nervous system[19].

In the coleopteran insect *Tenebrio molitor*, GCNF is expressed during metamorphosis[4]. In epidermal cells, its expression reaches a maximum both during the last larval instar stage and at the pupal stage, just after the ecdysteroid peak when the epidermal cells begin to synthesize cuticular

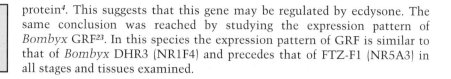

protein[4]. This suggests that this gene may be regulated by ecdysone. The same conclusion was reached by studying the expression pattern of *Bombyx* GRF[23]. In this species the expression pattern of GRF is similar to that of *Bombyx* DHR3 (NR1F4) and precedes that of FTZ-F1 (NR5A3) in all stages and tissues examined.

Activity and target genes

A DR0 element found in the promoter of the mouse protamine-2 gene is recognized by GCNF[11,21]. The protamine-2 gene is expressed in the same stage of spermatogenesis (i.e. round spermatids) as GCNF, suggesting that it could be a bona fide target gene *in vivo*. Strikingly, the first trials to show an up-regulation of the activity of this promoter with GCNF in transient transfection assays were unsuccessful[11]. This lack of transactivation is confirmed by the observation that on synthetic DR0 elements GCNF behaves in transient transfection assays as a transcriptional repressor[12,13]. This is consistent with the apparent lack of a bona fide AF-2-AD core region observed in GCNF[1,10,13].

Knockout

Details of GCNF null mice have not yet been published.

Associated disease

No disease has yet been associated with GCNF.

Gene structure, promoter and isoforms

Several isoforms differing in the A/B region have been described in humans[5-7,15,22]. These isoforms all contain the same four amino acids (MERDE), but differ on the presence or absence of two short stretches of 22 and 4 amino acids, respectively. This suggests that these isoforms all start on the same promoter but are submitted to alternative splicing events.

The mouse GCNF contains 11 exons with the first four exons coding for the A/B region and exon 4 coding for the DBD[26]. Thus, in contrast to other nuclear receptors, there is no intron lying between the two zinc fingers. Exons 7–11 encode the LBD.

Chromosomal location

The human GCNF gene was mapped by FISH analysis with a BAC clone containing the gene on chromosome 9 on q33–34.1[15]. No etiology of the inherited diseases that have been located in that region fits with the known biology of GCNF[5,15].

The *Drosophila* GRF gene was found at position 2C1–2 on chromosome X in the sequence of the complete *Drosophila* genome (see information on sequence EG:133E12.2 in FlyBase).

Amino acid sequence for mouse GCNF (NR6A1)

Accession numbers: U14666, U09563

```
  1 MERDERPPSGGGGGGSAGFLEPPAALPPPPRNGFCQDELAELDPGTNGETDSLTLGQGHIPVSVPDDRA
 71 EQRTCLICGDRATGLHYGIISCEGCKGFFKRSICNKRVYRCSRDKNCVMSRKQRNRCQYCRLLKCLQMGM
141 NRKAIREDGMPGGRNKSIGPVQISEEEIERIMSGQEFEEEANHWSNHGDSDHSSPGNRASESNQPSPGST
211 LSSSRSVELNGFMAFRDQYMGMSVPPHYQYIPHLFSYSGHSPLLPPQQARSLDPQSYSLIHQLMSAEDLEP
281 LGTPMLIEDGYAVTQAELFALLCRLADELLFRQIAWIKKLPFFCELSIKDYTCLLSSTWQELILLSSLTV
351 YSKQIFGELADVTAKYSPSDEELHRFSDEGMEVIERLIYLYHKFHQLKVSNEEYACMKAINFLNQDIRGL
421 TSASQLEQLNKRYWYICQDFTEYKYTHQPNRFPDLMMCLPEIRYIAGKMVNVPLEQLPLLFKVVLHSCKT
491 STVKE 495
```

Amino acid sequence for *Tenebrio* GRF (NR6A2)

Accession number: AJ005685

```
  1 MTLTRAPCELDKMSLFQDLKLKRRKVDSRCSSDGESVADTSTSSPDLVSPSSPKMSEAVLNPPSPDSTPI
 71 QIHPPEPTLDKISSRLEDSGVFDGGGTASVIRSLPGVRSQSPPLRPAPTSSPCQPPRPHSSPGRPANTSP
141 VIIHNPTIAQPTSSRAKLSTMVTCQEPTSSSPSAVSKSRGVIISHPTSLTQSQLWMKHSRINGVKPELIG
211 GNFSGALGHYSELKSPTPAGAMQRPSSNPPVRQTPTVIMGEAGGVRTMIWSQPTLGASPTSPVESPHHAT
281 TSWASGANMSNTEESAAQMLLNLGQDRLRSPVSRTLVSPQSPSTARFTTTPLNMERLWAGDLRQLPVNQQ
351 TQALNLSSPTPGPPGVYCGSVSDVKISIMNESSTSESQEATEEEEQPMICMICEDKATGLHYGIITCEGC
421 KGFFKRTVQNRRVYTCVADGNCEITKAQRKRCPYCRFKKCIEQGMVLQAVREDRMPGGRNSGAVYNLYKV
491 KYKKHKKPACKQPQKAAEKNILSQQFKVEQPSSIPANLVNGTILKTALTNPSEVVRLRQRLDSAVSSSRD
561 RNFSIEYSLSMIKTLIDCDEFQDIATLQNLDDLLDHNTDLSEKLCHIGDSIVYKLVQWTKRLPFYLELPV
631 EVHTRLLTHKWHELLVLTTSAYQAIHKAGDQLTTVIKTDFNHEVETNLCTLQSCLTSMMGREITIEQLRQ
701 DVGLMIEKITHVTLMFRQIKLTMEEYVCLKVITMLNQAKPASSSGNSELESIHERYMTCLRVYTQHMYPQ
771 QTTRFQDLLGRLPEIQSAAFLLLESKMFYVPFLLNSAIQR  811
```

References

1. Chen, F. et al. (1994) Mol. Endocrinol. 8, 1434–1444.
2. Hirose, T. et al. (1995) Gene 152, 247–251.
3. Bauer, U.M. et al. (1997) Eur. J. Biochem. 249, 826–837.
4. Mouillet, J-F. et al. (1999) Eur. J. Biochem. in press.
5. Lei, W.T. et al. (1997) J. Mol. Endocrinol. 18, 167–176.
6. Süsens, U. and Borgmeyer, U. (1996) Biochim. Biophys. Acta 1309, 179–182.
7. Schneider-Hirsch, S. et al. (1998) J. Recept. Signal Transduct. Res. 18, 1–13.
8. Joos, T.O. et al. (1996) Mech. Dev. 60, 45–57.
9. Braat, A.K. et al. (1999) Mol. Reprod. Dev. 53, 369–375.
10. Borgmeyer, U. (1997) Eur. J. Biochem. 244, 120–127.
11. Yan, Z.H. et al. (1997) J. Biol Chem. 272, 10565–10572.
12. Cooney, A.J. et al. (1998) Biochem. Biophys. Res. Commun. 245, 94–100.
13. Greschik, H. et al. (1999) Mol. Cell. Biol. 19, 690–703.
14. Katz, D. et al. (1997) Endocrinology 138, 4364–4372.
15. Agoulnik, I.Y. et al. (1998) FEBS Lett. 424, 73–78.
16. Bauer, U.M. et al. (1998) FEBS Lett. 208, 208–214.
17. Zhang, Y.L. et al. (1998) Mol. Reprod. Dev. 50, 93–102.
18. David, R. et al. (1998) Mech. Dev. 79, 137–152.

[19] Süsens, U. et al. (1997) Dev. Neurosci. 19, 410–420.
[20] Heinzer, C. et al. (1998) Biol. Chem. 379, 349–359.
[21] Hummelke, G.C. et al. (1998) Mol. Reprod. Dev. 50, 396–405.
[22] Kapelle, M. et al. (1997) Biochim. Biophys. Acta 1352, 13–17.
[23] Charles, J.P. et al. (1999) Eur. J. Biochem. 266, 181–190.
[24] Song, K. et al. (1999) Dev. Biol. 213, 170–179.
[25] Scmitz, T.P. et al. (1999) Biochim. Biophys. Acta 1446, 173–180.
[26] Süsens, U. and Borgmeyer, U. (2000) Genome Biol. 1(3): research006.

Knirps

Names

Names

The Knirps group of genes contains atypical members of the superfamily that are found exclusively in insects and among insects only in dipterans. These genes encode transcription factors that contain a DBD similar to the C domain of nuclear receptors but their C-terminal parts do not bear any sequence identity with the LBD of nuclear hormone receptors. The two first members of this group to be cloned were called knirps (as it corresponds to a known *Drosophila* mutant having this name[1]) and Knirps-related[2,3]. The third gene was called EGON (for embryonic gonad) when it was identified because of its expression pattern but was shown later to be identical to the gene mutated in the eagle mutant of *Drosophila* harboring an adult phenotype of wings held out at right angles to the body described by Thomas Hunt Morgan in 1930.

Species	Other names	Accession number	References
Official name: NR0A1			
D. melanogaster	Knirps, KNI	X13331	2
D. virilis	Knirps, KNI	L36177	6
Musca domestica	Knirps, KNI	Not available	7
Official name: NR0A2			
D. melanogaster	Knirps-related, KNRL	X14153	3
Official name: NR0A3			
D. melanogaster	embryonic gonad, EGON, eagle, eg	X16631	4, 5

Isolation

KNI was cloned by Herbert Jäckle's laboratory in a search for the gene responsible for the Knirps mutation[2]. Since this mutant was known to be located on the third chromosome in region 77E, this chromosome region was microcloned and used to initiate a chromosome walk by comparison with the chromosomal breakpoint of a small deletion inducing the Knirps phenotype. KNI was found to complement the Knirps mutation, proving that a mutation or deletion of this gene was responsible for the phenotype.

KNRL was cloned by Ron Evans's laboratory by the screening of a *Drosophila* genomic library with a human retinoic acid receptor DBD probe[3]. It was shown to be close to the KNI gene and to harbor the same atypical structure with a DBD but no recognizable LBD.

EGON was first identified by Herbert Jäckle's laboratory during a search for KNI-related genes[4]. Six years later it was independently reisolated using an enhancer trap strategy[5]. The LacZ expression pattern was observed in the enhancer trap line. Since recessive mutations of the isolated gene failed to complement the eagle mutation, the cloned gene was identified as being responsible for the mutation and called EAGLE.

The three genes harbor the same structure with a DBD harboring 92% identity between KNI and KNRL, 86% between KNI and EGON, and 80% between KNRL and EGON[4]. Their respective C-terminal parts, which are large (full size proteins are 429, 647 and 373 amino-acids long, respectively,

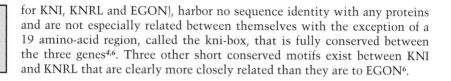

for KNI, KNRL and EGON), harbor no sequence identity with any proteins and are not especially related between themselves with the exception of a 19 amino-acid region, called the kni-box, that is fully conserved between the three genes[4,6]. Three other short conserved motifs exist between KNI and KNRL that are clearly more closely related than they are to EGON[6].

DNA binding

Soon after its discovery it has been shown by *in vivo* footprinting that KNI binds to DNA in a sequence-specific manner[8–10]. Further experiments identified a consensus target sequence for KNI (AA/TCTAA/GATC) that is very different from that of other nuclear receptors despite the fact that KNI harbors a classical P-box sequence (CEGCKS)[11]. The DNA binding of KNI strictly requires the DBD[6]. It has been shown that KNI binds DNA as a monomer, an observation consistent with the fact that KNI target sequences are not organized in palindrome or direct repeats in the regulatory regions of KNI target genes.

A nuclear localization signal has been mapped inside the KNI protein[6]. Interestingly, this NLS overlaps partially with the strictly conserved 19 amino-acid kni-box. The sequences sufficient for KNI nuclear localization start within the second zinc finger and extend into the kni-box (amino acids 52–93 of *D. melanogaster* KNI)[6]. This region contains basic amino acids and has sequence similarities to certain NLS motifs. The kni-box itself is not the NLS, since it was not sufficient alone when fused to β-gal to target the fusion protein in the nucleus. The kni-box nevertheless plays a role in the nuclear localization, since its deletion results in a less efficient nuclear localization of KNI. Its precise role remains unclear since there are no alleles bearing kni mutation in this region.

It has been shown that KNRL binds *in vitro* to the same sequence as KNI[11]. There is no functional data concerning EGON, but the high sequence identity argues for their ability to bind the same DNA sequence and that the NLS is located in the same region.

Partners

KNI is unable to bind to the RXR homologue USP and did not homodimerize[6]. Nevertheless, recent data demonstrated that KNI can form heterodimers with the product of the segmentation gene Krüppel[12]. In fact, KNI binds the DNA-bound form of Krüppel and this heterodimer plays a repressive role in gene regulation. The association with KNI is thus able to modify the transcriptional potential of Krüppel from an activator to a repressor. Interestingly, previous experiments have shown that KNI competes with the Tailless (TLL, NR2E2) gene product for the regulation of the Krüppel gene itself[11]. This suggests that a complex regulatory network exists between these genes.

Ligands

There are no known ligands for knirps. Given the absence of sequence identity between the C-terminal ends of KNI group members and other nuclear receptors, KNI group members are considered as real orphan receptors.

Three-dimensional structure

No 3D structure data are available for Knirps group of genes.

Expression

KNI is expressed as two 2.2 and 2.5 kb mRNA transcripts, owing to alternative polyadenylation[2]. The smaller transcript is expressed only transiently at the blastoderm stage, although the larger one continues to be expressed after gastrulation. The developmental profile of the 2.2 kb transcript is therefore consistent with the expected expression period of KNI for a role in segmentation. *In situ* hybridization reveals that KNI expression can be first detected after the 11th nuclear division forming a broad band in the posterior region of the embryo[4]. Soon afterwards, KNI transcripts are also observed at the anterior tip of the embryo. During cellularization, a second anterior KNI domain forms posteriorly adjacent to the previous one.

Northern blots show that KNRL is expressed as a 3.8 kb transcript[3,4]. The KNRL expression pattern coincides precisely with that of KNI in both early and late stages of embryogenesis[3,4]. One difference between them is that KNRL is a maternal mRNA, whereas KNI expression is exclusively zygotic. Another difference lies in the respective intensity of the anterior and posterior expression territories for KNI and KNRL. Both genes are also expressed later during embryogenesis, consistent with a late function (see e.g. reference 17).

The expression of EGON was first described as restricted to the embryonic gonad[4]. Nevertheless, when the gene was identified by an enhancer trap strategy as EAGLE, its expression was shown to be totally different and no expression in the embryonic gonad was observed[5]. The reason for this discrepancy is unknown. By *in situ* hybridization, transcripts first became discernible in a subset of lateral central nervous system cells at stages 10–12. These cells were identified as neuroblasts, giving rise to serotonin neurons[5,18]. Expression disappeared at stage 13 except for some thoracic cells.

Activity and target genes

It has been clearly demonstrated that KNI acts as a transcriptional repressor. For example, it has been shown that both KNI and Tailless compete with the homeodomain-containing Bicoid protein for the binding to a specific 16 bp regulatory element of the gap gene Krüppel[11]. In such a system, KNI acts as a competitive repressor of the Bicoid-mediated activation of Krüppel. A functional analysis of KNI protein has suggested that a 47-amino-acid stretch partially conserved in KNRL and located in the middle of the C-terminal moiety of the protein exerts a strong repressive activity[6]. This repressor domain is able to act when fused to an heterologous DBD such as GAL4. Another study has shown that, in addition to competing with activators for the occupancy of DNA target sequence, KNI can quench or locally inhibit upstream activators within a heterologous promoter, and can also act as a dominant repressor of transcription and block multiple enhancers in a modular promoter[21]. This data suggests that the repressive properties of KNI are of several types and that their precise role is far from being understood. Recently, it has been shown that KNI as well as two other repressors, Krüppel and Snail, bind to a common corepressor, called dCtBP, which was first discovered in mammals as an attenuator of transcriptional activation by the adenovirus E1A protein[24]. Interestingly, the three repressors interact with dCtBP by a common motif harboring the conserved sequence PXDLSXK/H, which is present in the 47 AA C-terminal repressor domain of KNI[24].

A number of target genes, consistent with its role as a gap gene, has been proposed for KNI and the precise role of the protein among the regulatory cascade controlling segmentation in *Drosophila* has been studied extensively (see reference 10 for a review). Among the proposed targets for KNI are the genes Krüppel[11], Hairy[9,20,23] or Even-skipped[22].

Knockout

KNI is essential for the development of the abdominal region of the embryo. In the absence of KNI activity, in lack-of-function mutants, the establishment of abdominal segments is abnormal. The head, the thorax and the tail region of these embryos develop normally but, of the eight abdominal segments (A1–A8), only A8 is formed properly[1,2,19]. In place of the other abdominal segments, a single large denticle field that contains pattern elements for both A1 and A7 is found (reviewed in references 10 and 19).

Late functions during embryogenesis were described for both KNI and KNRL by experiments involving ectopic expression. It has, for example, been shown that they control cell migration and branch morphogenesis during tracheal development in *Drosophila*[17].

The function of EGON remained mysterious until it was identified as the EAGLE gene[5]. This gene is expressed in a subset of neuroblasts and regulates the fate of their progeny in the *Drosophila* central nervous system. More recently, the study of loss of function mutations revealed that this gene is required for the specification of serotonin neurons[18].

Associated disease

There is no disease associated with knirps. There are apparently no vertebrate homologues for these genes[13].

Gene structure, promoter and isoforms

The KNI gene contains three exons interrupted by two small introns of 733 and 214 bp, respectively[2]. The position of the intron in the DBD is conserved in other members of the KNI group[3,4].

Although having a genomic structure closely related to that of KNI, KNRL exhibits a marked difference from it. KNRL is a very large gene spaning 23 kbp. By comparison, the size of the KNI transcriptional unit is 3 kbp. This difference in size is simply due to the length of their introns. Interestingly, elegant experiments have shown that this size difference has important functional consequences[14]. It has been shown that KNRL cannot substitute for KNI in early *Drosophila* embryogenesis except if an intronless (e.g. short) version of KNRL is used. This effect is explained by the speed of the first mitotic cycle of *Drosophila* that acts as a physiological barrier for the action of the large KNRL gene. This is direct proof of the importance that the size of a gene may have for development (see reference 15 for a specific review).

The reading frame of EGON is interrupted by a unique intron of 1421 bp suggesting that, as KNI, EGON is a relatively small gene[6]. Nevertheless, the recloning of the gene as EAGLE reveals that it contains two alternative upstream non-coding exons giving rise to two alternative transcripts that encode the same protein[5]. The EGON transcriptional unit spans *c*. 10 kbp.

The promoter of KNI was studied because, as it is a gap gene, it was important to understand the elements that control its expression. Genetic studies have indicated that KNI was a target for the posterior determinant gene Nanos but the precise regulatory cascade that controls KNI was not known. It was shown that a fragment 4.4 kbp upstream of the KNI transcriptional start site directs the expression of a reporter gene with a pattern that fits with that of the endogenous KNI transcript. Further experiments demonstrate that this region contains binding sites for Hunchback and Tailless that are responsible for spatially restricting the KNI expression from the anterior and posterior side of the embryo, respectively[16].

Chromosomal location

The three genes are close to each other on the chromosome 3L of *D. melanogaster*[2-4]. The location of KNI and KNRL is not distinguishable by *in situ* hybridization to polytene chromosome. These two genes are located on region 77E1–2. EGON, which is more divergent at the sequence level, is also slightly more distant to KNI and KNRL, since it is located in region 79B[4].

Amino acid sequence for *Drosophila melanogaster* KNI (NR0B1)

Accession number: X13331

```
  1  MNQTCKVCGEPAAGFHFGAFTCEGCKSFFGRSYNNISTISECKNEGKCIIDKKNRTTCKACRLRKCYNVG
 71  MSKGGSRYGRRSNWFKIHCLLQEHEQAAAAAGKAPPLAGGVSVGGAPSASSPVGSPHTPGFGDMAAHLHH
141  HHQQQQQQQVPRHPHMPLLGYPSYLSDPSAALPFFSMMGGVPHQSPFQLPPHLLFPGYHASAAAAAASAA
211  DAAYRQEMYKHRQSVDSVESQNRFSPASQPPVVQPTSSARQSPIDVCLEEDVHSVHSHQSSASLLHPIAI
281  RATPTTPTSSSPLSFAAKMQSLSPVSVCSIGGETTSVVPVHPPTVSAQEGPMDLSMKTSRSSVHSFNDSG
351  SEDQEVEVAPRRKFYQLEAECLTTSSSSSSHSAAHSPNTTTAHAEVKRQKLGGAEATHFGGFAVAHNAAS
421  AMRGIFVCV   429
```

Note: the KNI box has been underlined and italicized.

Amino acid sequence for *Drosophila melanogaster* KNRL (NR0B2)

Accession number: X14153

```
  1  MMNQDNPYAMNQTCKVCGEPAAGFHFGAFTCEGCKSFFGRSYNNLSSISDCKNNGECIINKKNRTACKAC
 71  RLKKCLMVGMSKSGSRYGRRSNWFKIHCLLQEQQQQAVAAMAAHHNSQQAGGGSSGGSGGGQGMPNGVKG
141  MSGVPPPAAAAAALGMLGHPGGYPGLYAVANAGGSSRSKEELMMLGLDGSVEYGSHKHPVVASPSVSSPD
211  SHNSDSSVEVSSVRGNPLLHLGGKSNSGGSSSGADGSHSGGGGGGGGGVTPGRPPQMRKDLSPFLPLPFP
281  GLASMPVMPPPAFLPPSHLLFPGYHPALYSHHQGLLKPTPEQQQAAVAAAAVQHLFNSSGAGQRFAPGTS
351  PFANHQQHHKEEDQPAPARSPSTHANNNHLLTNGGAADELTKRFYLDAVLKSQQQSPPPTTKLPPHSKQD
421  YSISALVTPNSESGRERVKSRQNEEDDEARADGIIDGAEHDDEEEDLVVSMTPPHSPAQQEERTPAGEDP
491  RPSPGQDNPIDLSMKTTGSSLSSKSSSPEIEPETEISSDVEKNDTDDDDEDLKVTPEEEISVRETADPEI
561  EEDHSSTTETAKTSIENTHNNNNSISNNNNNNNNNNNSILSDSEASETIKRKLDELIEASSENGKRLRLE
631  APVKVATSNALDLTTKV
```

Note: the KNI box has been underlined and italicized.

Amino acid sequence for *Drosophila melanogaster* EGON (NR0B3)

Accession number: X16631

```
  1  MNQLCKVCGEPAAGFHFGAFTCEGCKSFFGRTYNNIAAIAGCKHNGDCVINKKNRTACKACRLRKCLLVG
 71  MSKSGSRYGRRSNWFKIHCLLQEQQTTSGLGGGSVGSGSGGGVSSASLEQLARLQQASNQARQTYQDKT
141  NPCIKSATATTSPRIEGAAVGTGIGGGASPSFLQAAKLHHQRQLKLDSRLSNTPSDSGASSAGDPNEDGV
211  TSVLGGQIATPSSTNATSLPKLDLRHPNFPATSEPDADMQRQRHQELLEIFRSHSEPLYSSFAPFSHLPP
281  VLLAAGVPQLPIFKDQFKAELLFPTTSSPELEEPIDLSFRSRADHASPMAHNSNSPSLSEPAAASHCLGE
351  STNFVRKSTPLDLTLVRSQTLTG  373
```

Note: the KNI box has been underlined and italicized.

References

[1] Nüsslein-Volhard, C. and Wieschaus, E. (1980) Nature 287, 795–801.
[2] Nauber, U. et al. (1988) Nature 336, 489–492.
[3] Oro, A.E. et al. (1988) Nature 336, 493–496.
[4] Rothe, M. et al. (1989) EMBO J. 8, 3087–3094.
[5] Higashijima, S. et al. (1996) Development 122, 527–536.
[6] Gerwin, N. et al. (1994) Mol. Cell. Biol. 14, 7899–7908.
[7] Sommer, R. and Tautz, D. (1991) Development 113, 419–430.
[8] Pankratz, M.J. et al. (1989) Nature 341, 337–340.
[9] Pankratz, M.J. et al. (1990) Cell 61, 309–317.
[10] Pankratz, M.J. and Jäckle, H. (1990) Trends Genet. 6, 287–292.
[11] Hoch, M. et al. (1992) Science 256, 94–97.
[12] Sauer, F. and Jäckle, H. (1995) EMBO J. 14, 4773–4780.
[13] Laudet, V. (1997) J. Mol. Endocrinol. 19, 207–226.
[14] Rothe, M. et al. (1992) Nature 359, 156–159.
[15] O'Farrell, P.H. (1992) Nature 359, 366–367.
[16] Pankratz, M.J. et al. (1992) Science 255, 986–989.
[17] Chen, C-K. et al. (1998) Development 125, 4959–4968.
[18] Lundell, M.J. and Hirsh, J. (1998) Development 125, 463–472.
[19] Lehmann, R. (1988) Development (Suppl.) 104, 17–27.
[20] Hartmann, C. et al. (1994) Mech. Dev. 45, 3–13.

[21] Arnosti, D.N. et al. (1996) EMBO J. 15, 3659–3666.

[22] Kosman, D. and Small, S. (1997) Development 124, 1343–1354.

[23] Häder, T. et al. (1998) Mech. Dev. 71, 177–186.

[24] Nibu, Y. et al. (1998) EMBO J. 17, 7009–7020.

Odr-7

Names

To date this gene is known in only one species, the nematode *Caenorhabditis elegans*. In the nomenclature, it is in subfamily '0', which groups all the unusual receptors that contain only the C (group A) or E (group B) domain.

Species	Other names	Accession number	References
Official name: NR0A4			
C. elegans	Odr-7, T18D3.2	U16708	1

Isolation

This gene was isolated during a screen for mutants that failed to respond to odorant molecules[1]. Two different mutants that failed to complement each other, odr-7(ky4) and odr-7(ky55), were identified. These mutants are defective for the function of two specific neurons implicated in olfaction in *C. elegans*, the AWA neurons. The Odr-7 gene was mapped genetically and cloned by rescue with cosmids and cosmid fragments. A 7.3 kb fragment was the smaller one to rescue the mutant phenotype of both odr-7(ky4) and odr-7(ky55). This fragment harbors two genes, the leftmost being able to rescue the phenotype.

Odr-7 exhibits unusual properties for a member of the nuclear hormone receptor superfamily, since it does not contain LBD. The gene encodes a 457-amino-acid protein with a large N-terminal domain (330 amino acids) rich in hydrophobic residues that is not conserved, a C domain with a divergent P-box (see below) and 48 divergent amino acids in the C-terminal. The last 41 amino acids did not appear to be necessary for the function of the protein, since they are not present in the shorter clone (7.3 kb cosmid fragment) that rescues the odr-7 mutations.

Overall, the C domain of Odr-7 exhibits 40% identity with the C domain of Tailless (TLL) and 38% with RARβ or RXRα[1]. In accordance with these values in a phylogenetical tree, Odr-7 appears to be a member of the subfamily II connected to TLL[2].

DNA binding

No functional assays using Odr-7 have been published. Thus the DNA target of this unusual receptor is unknown. Of note, the P-box of Odr-7 is unusual because most other nuclear receptors contain at least two hydrophilic residues among the four variable residues of the P-box sequence; Odr-7 harbors the sequence Cys-Ala-Ala-Cys-Ala-Ala[1]. Identical or closely related sequences were also found in other unusual NHRs found in nematode[3,4]. Odr-7 also contains four extra amino acids in the second zinc finger that are not found in other NHRs. Lastly, Odr-7 gene products may contain a third helix analogous to that present in the T- and A-boxes in the N-terminal part of the D domain of RXR and other superfamily members, since four of the eight residues in this region are identical between Odr-7 and RXRα.

Partners

There are no known partners for Odr-7.

Ligands

There are no known ligands for Odr-7. Given that the E domain is absent (i.e. no LBD), Odr-7 should be considered as an orphan receptor.

Three-dimensional structure

No 3D structure is available for Odr-7.

Expression

Odr-7 is specifically expressed in the AWA neurons of *C. elegans* in the fourth larval stage and in the adult[1]. The expression was also detected in embryos by RT-PCR experiments.

Activity and target genes

No functional data regarding the transcriptional activity of the Odr-7 gene product are available. Genetic data suggest nevertheless that Odr-7 controls the expression of Odr-10, a seven transmembrane domain olfactory receptor required for the response to diacetyl[5]. The Odr-10 transcript was detected in AWA neurons in wild-type animals but not in odr-7(ky4) null mutants. In addition, a fragment of 1 kb containing the promoter of Odr-10 linked to GFP was expressed in AWA neurons in wild-type animals but not in odr-7(ky4) mutants. This suggests that Odr-10 is a direct target of Odr-7.

Recent results also suggest that Odr-7 may repress the expression of another seven-transmembrane olfactory receptor, Str-2[6]. A 4 kb upstream fragment of this gene linked to GFP directs expression to the AWC neuron. Interestingly, in the odr-7(ky4) mutant an ectopic expression of Str-2 is observed in the AWA neurons. This effect was not observed in the mild odr-7(ky55) mutant. Taken together, these data suggest that Odr-7 may behave either as a transcriptional activator (on Odr-10) or as a transcriptional repressor (on Str-2), depending on the target gene. There is no information on the sequence of the Odr-10 and Str-2 promoters that indicates how Odr-7 regulates these promoters.

Knockout

Two different mutants obtained by an EMS mutagenesis and a screen for mutants with a defect in odorant response were characterized for the Odr-7 gene. The phenotype of the odr-7(ky4) mutants is very similar to a loss of AWA function, whereas odr-7(ky55) animals have defects in only a subset of AWA function. For example, odr-7(ky55) animals failed to respond by chemotaxis (attraction or repulsion) to some compounds (diacetyl, whereas it responded normally to others (benzaldehyde, butanone, isoamyl alcohol). The odr-7(ky4) mutants do not respond to any of these compounds and are, in addition, defective for their response to pyrazine. These two mutants respond normally to other odorants that implicate other neurons (such as AWB or AWC) for their detection. In both mutants, the AWA neurons are indistinguishable from those of wild-type animals, suggesting that the function but not the presence of these neurons is impaired in the mutants.

Interestingly, the odr-7(ky55) mutation (a G to A transition) causes the substitution of a glutamic acid residue (position 340) for a glycine in the C domain of Odr-7. This glutamic acid is located in the tip of the first zinc finger and is conserved in most members of the NHR superfamily. This suggests that the C domain of Odr-7 is important for the function of the protein. The odr-7(ky4) mutation is located in the N-terminal part of the protein. It is a C to T mutation that induces the formation of a stop codon at position 148 and results in the truncation of the protein that completely lacks the putative DBD.

Associated disease

This is not relevant for Odr-7.

Gene structure, promoter and isoforms

The Odr-7 gene contains six exons and five introns. The C domain is encoded by the last three exons. In this domain the first intron lies just after the P-box (in the sequence CAACAAFFR*R, where the asterisk indicates the position of the intron). The same position was found for the vitamin D receptor. The second intron inside the C domain is located in the putative third helix, i.e. in the putative T/A boxes. Most other NHR-encoding genes harbor introns in this region, although the precise location of this intron in Odr-7 appears to be further downstream than for the other NHRs.

Chromosomal location

Odr-7 was mapped on the chromosome X of the nematode *C. elegans*, to the right of the *lin*-2 gene[1].

Amino acid sequence for *C. elegans* Odr-7 (NR0A4)

Accession number: U16708

```
  1  MIVPDTEGLLIYSYGLMYGSYCMACQMLIPHFQCIPGIFPNFRISTELIKTMTDKLEQPNNNVPQQPWGP
 71  FPPAFGGRPSGEQTDGNPGEFDNDAAHQQTAPFMTHFFPRIGLQFPDFTEYQRFNGFQRNAFFPNPFGSQ
141  FTGQAFAQSFPLHNSMTTMDGFNLTHAPHPFSTNTNSTKPKDIENTVQSTIKHSSENIQDKPPVLSVEYP
211  VKYDSELKFDANVDFTAVPKQESSDDSTLKNLKKSDQQLQQPQQFTFPPPLLAEKSFEQPRMREDVLPFH
281  PQFYPAPLDMGTNFKQEMRTPPIDGHIDYRKFDASGKRMEFQPPGAL HDCQVC LSTHANGLHFGART CAA
351  CAAFFRRTISDDKRYV C KRNQR C NNASRDGTGYRKICRS C RMKR C LEIGML PENVQHKRNRRDSGSPPRK
421  TPFDTFFNGFYPSFQPSGSAAQPITVSSSESPRHTTN  457
```

References

1   Sengupta, P. et al. (1994) Cell 79, 971–980.
2   Laudet, V. (1997) J. Mol. Endocrinol. 19, 207–226.
3   Sluder, A.E. et al. (1999) Genome Res. 9, 103–120.
4   Miyabayashi, T. et al. (1999) Dev. Biol. 215, 314–331.
5   Sengupta, P. et al. (1996) Cell 84, 899–909.
6   Sagasti, A. et al. (1999) Genes Dev. 13, 1794–1806.

DAX1 and SHP

Names

These two genes are closely related both at the sequence and structure level. They are unique in the nuclear receptor superfamily in that they do not contain a DBD. At the sequence level they are distantly related to the TLL group of orphan receptors[4]. In the nomenclature of nuclear receptors, they were grouped in the subfamily '0' that contains unusual receptors[5].

Species	Other names	Accession number	References
Official name: NR0B1			
Human	DAX1, AHCH	S74720	1
Mouse	DAX1	U41568	2
Rat	DAX1	X99470	Unpub.
Pig	DAX1	AF019044	16
Macropus eugenii	DAX1	U96075	38
Chicken	DAX1	AF202991	37
Alligator	DAX1	AF180295	35
Official name: NR0B2			
Human	SHP	L76571	3
Mouse	SHP	L76567	3
Rat	SHP	D86580	8

Isolation

Human DAX1 was found by a consortium of laboratories searching for the gene responsible for a human developmental disorder, X-linked adrenal hypoplasia congenita (AHC)[1]. It has been suggested in the same study that the gene may also be the X-linked locus involved in sex determination (DSS, for dosage-sensitive sex reversal). The name DAX1 refers to this double relationship, since DAX1 stands for DSS–AHC critical region on the X chromosome, gene 1. Several mutations inside the DAX1 gene were linked to AHC as well as to hypogonadotropic hypogonadism, which is frequently associated with AHC[6]. The mouse gene was found by screening a mouse genomic library with a human DAX1 probe[2]. Both the human and mouse DAX1 genes contain 66.5% amino-acid identity and exhibit the same structure with a N-terminal domain of 53 amino acids organized in three and a half repeats of a 67–68-amino-acid motif and a putative LBD. The 67–68-amino-acid motif is not related to the regular DBD of nuclear receptors, although it contains conserved cysteines. The presence of the last and incomplete repeat was not confirmed by modeling studies, suggesting that, in fact, DAX1 contains only three repeats[9]. Interestingly, the DAX1 gene is located on an autosome in marsupials as well as in chicken, suggesting that it was originally not involved in X-linked dose-dependent sex determination[37,38]. Nevertheless, it was found expressed in the gonads in both males and females in alligator and chicken, which indicates that it is implicated in gonadal development in vertebrates long before its recruitment in the X-chromosome[35–38].

SHP was identified by a two-hybrid screen as an heterodimeric partner of the orphan receptor CARβ (NR1I4)[3]. It was shown to interact with numerous members of the superfamily and exhibits a structure closely

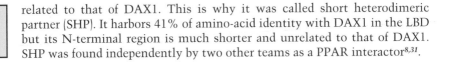

related to that of DAX1. This is why it was called short heterodimeric partner (SHP). It harbors 41% of amino-acid identity with DAX1 in the LBD but its N-terminal region is much shorter and unrelated to that of DAX1. SHP was found independently by two other teams as a PPAR interactor[8,31].

DNA binding

Although its N-terminal domain is unrelated to the DBD of nuclear receptors, human DAX1 was first shown to be able to bind to a retinoic acid response element (RARE) from the RARβ2 promoter[1]. This finding was apparently not confirmed in subsequent studies and it was reported that DAX1 bound to DNA hairpin structure with no strong sequence specificity[7]. These experiments also reveal that DAX1 interacts with the minor groove of the double helix of DNA. These characteristics are reminiscent of the HMG family proteins such as Sry and SOX9, which are involved in mammalian sex determination. Nevertheless, the role played by the DNA binding of DAX1 is unclear because there is no known mutations in the DBD of DAX1 in AHC. Interestingly, it has recently been found that DAX1 is an RNA-binding protein associated with polyribosomes that shuttle between the cytoplasm and the nucleus[27]. The N-terminal repeats as well as, curiously, the LBD are directly implicated in the RNA recognition. In fact, the same study demonstrates that the LBDs of RARα and γ are also able to bind efficiently to RNA[27]. Of note, mutations in DAX1 found in AHC patients significantly impair DAX1 RNA binding, suggesting that it is a functionally relevant interaction.

SHP is apparently unable to bind to RARE or to other DNA elements containing the AGGTCA core motif[3]. Its binding ability on the hairpin structure has not been described.

Partners

DAX1 was shown to interact physically with SF-1 (NR5A1), another orphan receptor implicated in sex determination in mammals, and to inhibit SF-1-mediated transactivation[11]. This inhibition of SF-1 function by DAX1 was surprising given that the phenotype of the SF-1 knockout mice closely resembles the phenotype of the AHC patients. The interaction between SF-1 and DAX1 occurs through a repressive domain within the carboxy-terminus of SF-1[12]. The interaction between DAX1 and SF-1 also antagonizes the synergy existing between SF-1 and WT1, the Wilm's tumor gene, which is also implicated in sex determination[13]. It was thus proposed that WT1 and DAX1 functionally oppose each other in testis development by modulating SF-1-mediated transactivation. The repressive effect of DAX1 on SF-1 was demonstrated to be linked to the recruitment of the corepressor NCoR to SF-1[12]. Interestingly, mutations of DAX1 found in AHC patients allow the SF-1-DAX1 interaction but markedly diminish NCoR recruitment. The interaction of DAX1 with NCoR is similar to that existing between NCoR and Rev-erb because it does not occur with the related SMRT corepressor. Of note, both SF-1 and WT1 regulate the DAX1 promoter (see below). It has also been shown recently that DAX1 interacts with the new corepressor Alien and that this interaction is abrogated by mutations in DAX1 that cause AHC[28].

SHP was found as an heterodimeric partner and is able to bind to a variety of nuclear receptors among which are CARβ, RAR, RXR, PPAR, HNF4, ER or TR[3,8,34]. When it interacts with classical receptors, the interaction is much stronger in the presence rather than in the absence of ligand[3]. This interaction results in an inhibition of the transcriptional activity of these receptors. This interaction has been studied in more detail in the case of estrogen receptors, HNF4 and RXR[29-31]. Consistent with its apparent ligand dependence, the interaction occurs through the AF-2 domain of the receptors. This regions recognize LXXLL motifs found in the central region of SHP. Thus, SHP is believed to mediate its inhibitory activity by a competition mechanism with coactivators for the binding of the AF-2 domain of nuclear receptors[10,29-31].

In accordance with its role in repressing other nuclear receptors, SHP has recently been shown to impair the ability of LRH-1 to up-regulate the cholesterol 7α-hydroxylase CYP7A1 gene (reviewed in reference 39). This is due to the direct physical interaction between SHP and LRH-1[40,41]. There is an interesting parallel between the SHP/LRH-1 and the DAX-1/SF-1 complexes that involves both couples of paralogous genes: SHP and DAX-1 on the one hand, and LRH-1 and SF-1 on the other. This may suggest that this interaction is evolutionary ancient and has been conserved after duplication of the respective genes.

Ligands

There are no known ligands for either DAX1 or SHP.

Three-dimensional structure

No 3D structure is yet available for DAX1 or SHP. Nevertheless, the structure of the LBD of DAX1 was modeled based on apo-RXRα and holo-RARγ structures[9]. The domains corresponding to helices 1–12 were identified in human and mouse DAX1 sequences. Surprisingly, it was found that helix 1 encompasses the region previously defined as the last and incomplete repeat, suggesting that this repeat is a part of the LBD.

Expression

DAX1 has a narrow expression pattern restricted to the three layers of the adrenal cortex, ovarian granulosa and theca cells, testicular Leydig and Sertoli cells, anterior pituitary gonadotrope cells and the neurons of the ventromedial nucleus of the hypothalamus[1,2,14]. This pattern is reminiscent of that of SF-1[14] and is consistent with a function of DAX1 in sex determination as well as in the control of the hypothalamus–pituitary–adrenal axis (see reference 15 for a review). In mice, DAX1 expression was first detected in the gonadal urogenital ridge at E10.5 and in the adrenal primordium at E12.5. In the pituitary, the expression starts at E14.5, whereas in the diencephalon it starts at E11.5. The first sexually divergent expression of DAX1 is seen in the gonad. In both cases, DAX1 expression is high at E12 but in the male (and not in the female) a rapid decline of expression follows. Similar findings have been observed in another mammal, suggesting that this expression pattern is conserved[16].

DAX1 expression is regulated negatively by FSH in Sertoli cells[17]. The same study has demonstrated that DAX1 transcript levels vary at different stages of spermatogenesis and that transcript peak levels apparently coincide with low FSH receptor expression. This study suggests that, in addition to a role in sex determination *per se*, DAX1 may play a role in spermatogenesis.

SHP is encoded by a 1.3 kb transcript expressed in liver and at lower levels in heart, adrenal gland, spleen and pancreas[3,33]. It has been shown recently that the SHP promoter is activated by SF-1 (NR5A1) and its paralogue LRH1 (NR5A2)[32]. The mouse SHP promoter was shown to contain five SFREs which are all required for the activation by LRH1. In addition, it was shown that SHP is coexpressed with SF-1 in adrenal glands as well as with LRH1 in liver.

Activity and target genes

DAX1 acts as a powerful transcriptional repressor via binding to DNA hairpin structures[7]. The C-terminus of DAX1 contains a potent transcriptional silencing activity, which can be transferred to a heterologous DBD[9]. The modeling studies described above have hinted that two domains of DAX1 cooperate for the silencing activity, one located within helix H3 (called silencing domain A) and the other within H12 (silencing domain B). This has been verified by mutational analysis. This has also been confirmed by the presence of AHC mutations in these domains that led to transcriptionally inactive DAX1 variants[9,12]. The repression of SF-1–WT1 synergism by DAX1 required the LBD[13]. These domains are important for the interaction with NCoR and mutations in AHC patient impair DAX1–NCoR interactions, suggesting that it explains most of the transcriptional inhibition by DAX1[12]. It is also this domain that mediates the interaction with the corepressor Alien[28].

DAX1 itself and the StAR (steroidogenic acute regulatory protein) gene are target genes of the DAX1 protein[7]. The repression of StAR by DAX1 led to a drastic decrease in steroid production, consistent with a role of DAX1 in steroidogenesis.

By its interaction with many nuclear receptors, and more specifically with the classical receptor in the holo form, SHP inhibits their transcriptional activity[3,8,10,34]. This repression results from two mechanisms: SHP inhibits the DNA-binding activity of its partners and it directly represses transcription via a repressor function located in the N-terminal part of the LBD[10]. SHP interacts with its partners through LXXLL motifs that recognize the AF-2 domain. Thus, it is clear that SHP competes for the binding of coactivators to the receptors[29–31]. In addition,

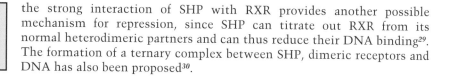

the strong interaction of SHP with RXR provides another possible mechanism for repression, since SHP can titrate out RXR from its normal heterodimeric partners and can thus reduce their DNA binding[29]. The formation of a ternary complex between SHP, dimeric receptors and DNA has also been proposed[30].

Knockout

As it has been demonstrated to be the gene responsible for the dosage-sensitive sex reversal[1,2], which occurred when the DSS locus of chromosome Xp21 is duplicated, DAX1 has been used in transgenic experiments in order to test the effects of modifying gene dosages[20]. It was shown that XY mice carrying extra copies of mouse DAX1 as a transgene show delayed testis development when the gene is expressed at high levels but do not normally show sex reversal except when the transgene was introduced in mice strains carrying weak Sry alleles. This has confirmed the notion that DAX1 is responsible for the DSS syndrome.

In order to examine the function of DAX1 in both males and females, a strain of mice carrying inactivated DAX1 alleles has been generated[18]. Strikingly, although DAX1 was first postulated to function in ovarian determination, the female mice do not exhibit abnormalities of ovarian development or fertility. In contrast, in males, the lack of DAX1 causes progressive degeneration of the testicular germinal epithelium, suggesting that DAX1 is essential for spermatogenesis. In addition, these animals exhibit abnormalities in gonadotropin and testosterone production, further stressing the role of DAX1 in steroidogenesis and hypothalamus–pituitary–adrenal axis regulation. All these data highlight the male-specific function of DAX1 and suggest that its product should be replaced in the male sex determination cascade[19,21].

No knockout data are available yet for SHP.

Associated disease

DAX1 was cloned as implicated in X-linked AHC as well as in hypogonadotropic hypogonadism (HHG) and was later demonstrated to be the gene responsible for the DSS syndrome[1,2]. The gene was found mutated in AHC and HHG patients (see reference 15 for a review and references for mutations). In a first screen, DAX1 was found deleted in 14 patients and mutated in the coding region in 12 unrelated individuals, confirming that the gene was responsible for both AHC and HHG[6]. All types of missense mutations in DAX1 resulting in AHC localize in the LBD, suggesting that the DNA-binding activity of DAX1 is not critical for its biological function. Many mutations are frameshift or nonsense mutations that lead to truncated DAX1 protein[15,22,23]. It has been shown that several mutations impair the transcriptional repression activity of DAX1, further confirming the importance of this inhibitory activity[9,12].

No disease is known to be associated with SHP.

Gene structure, promoter and isoforms

The gene structure of DAX1 is very simple with a unique intron that lies in the region coding the LBD[1]. Interestingly, this intron is located in a conserved position in many other nuclear receptors, further suggesting that the LBD of DAX1 really originates from a bona fide nuclear receptor through gene duplication[4].

The mouse DAX1 promoter was cloned and was shown to be regulated by several factors among which are DAX1 itself and SF-1[7,24]. The delineation of the SF-1 binding site inside this promoter has led to conflicting results[24,25]. It was proposed that SF-1 binds a complex element containing two SFRE imperfect repeats (TCA AGGTCA) and that this element can be a target for repression by COUP-TF, suggesting that a balance between SF-1 and COUP-TF regulates DAX1 expression[24]. These findings have been challenged recently and a new SF-1 target sequence has been found[25]. Interestingly, this study reinforces the importance of SF-1 for DAX1 expression, since DAX1 expression is significantly impaired in SF-1 knockout mice[25].

A recent study has also identified WT1 as a regulator of DAX1 expression[26]. WT1 is able to bind to a GC-rich tract found upstream of the DAX1 TATA box and activates DAX1 transcription. It is striking that SF-1 and WT1, two proteins that are able to interact and synergize together and whose positive action on transcription is inhibited by DAX1[13], are both positive regulators of the DAX1 gene transcription.

The genomic organization of SHP is known[33]. The gene is composed of two exons interrupted by one small intron lying in the same location as the DAX1 intron. The mouse and human SHP promoters were cloned[33]. It was shown that the mouse promoter contains five SFRE sequences that are the target for activation of this promoter by SF1 and LRH1[32]. This promoter is also regulated by FXR, which explains why SHP gene expression can be activated by bile acids, the natural ligand of FXR[40,41].

Chromosomal location

The DAX1 gene is located on the chromosome Xp21, the region containing the locus for AHC, HHG and DSS[1]. SHP is located in chromosome 1p36.1 in humans[33].

Amino acid sequence for human DAX1 (NR0B1)

Accession number: S74720.
The three sequence repeats discussed in the text were successively italicized, underlined and italicized again.

```
  1  MAGENHQWQGSILYNMLMSAKQTRAAPEAPETRLVDQCWGCSCGDEPGVGREGLLGGRNVALLYRCCFCG

 71  KDHPRQGSILYSMLTSAKQTYAAPKAPEATLGPCWGCSCGSDPGVGRAGLPGGRPVALLYRCCFCGEDHP

141  RQGSILYSLLTSSKQTHVAPAAPEARPGGAWWDRSYFAQRPGGKEALPGGRATALLYRCCFCGEDHPQQG

211  STLYCVPTSTNQAQAAPEERPRAPWWDTSSGALRPVALKSPQVVCEAASAGLLKTLRFVKYLPCFQVLPL

281  DQQLVLVRNCWASLLMLELAQDRLQFETVEVSEPSMLQKILTTRRRETGGNEPLPVPTLQHHLAPPAEAR

351  KVPSASQVQAIKCFLSKCWSLNISTKEYAYLKGTVLFNPDVPGLQCVKYIQGLQWGTQQILSEHTRMTHQ

421  GPHDRFIELNSTLFLLRFINANVIAELFFRPIIGTVSMDDMMLEMLCTKI   470
```

Amino acid sequence for mouse SHP (NR0B2)

Accession number: L76567.

The region non-homologous to LBD of nuclear receptors is underlined.

```
  1 MSSGQSGVCPCQGSAGRPTILYALLSPSPRTRPVAPASHSHCLCQQQRPVRLCAPHRTCREALDVLAKTV
 71 AFLRNLPSFCHLPHEDQRRLLECCWGPLFLLGLAQDAVTFEVAEAPVPSILKKILLEEASSGTQGAQPSD
141 RPQPSLAAVQWLQRCLESFWSLELGPKEYAYLKGTILFNPDVPGLRASCHIAHLQQEAHWALCEVLEPWY
211 PASQGRLARILLMASTLKNIPGTLLVDLFFRPIMGDVDITELLEDMLLLR  260
```

References

1   Zanaria, E. et al. (1994) Nature 372, 635–641.
2   Swain, A. et al. (1996) Nature Genet. 12, 404–409.
3   Seol, W. et al. (1996) Science 272, 1336–1339.
4   Laudet, V. et al. (1997) J. Mol. Endocrinol. 19, 207–226.
5   Nuclear Receptor Nomenclature Committee (1999) Cell 97, 161–163.
6   Muscatelli, F. et al. (1994) Nature 372, 672–676.
7   Zazopoulos, E. et al. (1997) Nature 390, 311–315.
8   Masuda, N. et al. (1997) Biochim. Biophys. Acta 1350, 27–32.
9   Lalli, E. et al. (1997) Mol. Endocrinol. 11, 1950–1960.
10 Seol, W. et al. (1997) Mol. Cell. Biol. 17, 7126–7131.
11 Ito, M. et al. (1997) Mol. Cell. Biol. 17, 1476–1483.
12 Crawford, P.A. et al. (1998) Mol. Cell. Biol. 18, 2949–2956.
13 Nachtigal, M.W. et al. (1998) Cell 93, 445–454.
14 Ikeda, Y. et al. (1996) Mol. Endocrinol. 10, 1261–1272.
15 Yu, R.N. et al. (1998) Trends Endocrinol. Metab. 9, 169–175.
16 Parma, P. et al. (1997) Mol. Cell. Endocrinol. 135, 49–58.
17 Tamai, K.T. et al. (1996) Mol. Endocrinol. 10, 1561–1569.
18 Yu, R.N. et al. (1998) Nature Genet. 20, 353–357.
19 Parker, K.L. and Schimmer, B.P. (1998) Nature Genet. 20, 318–319.
20 Swain, A. et al. (1998) Nature 391, 761–767.
21 Jimenez, R. and Burgos, M. (1998) BioEssays 20, 696–699.
22 Zhang, Y.H. et al. (1998) Am. J. Hum. Genet. 62, 855–864.
23 Habiby, R.L. et al. (1996) J. Clin. Invest. 98, 1055–1062.
24 Yu, R.N. et al. (1998) Mol. Endocrinol. 12, 1010–1022.
25 Kawabe, K. et al. (1999) Mol. Endocrinol. 13, 1267–1284.
26 Kim, J. et al. (1999) Mol. Cell. Biol. 19, 2289–2299.
27 Lalli, E. et al. (2000) Mol. Cell. Biol. 20, 4910–4921.
28 Altincicek, B. et al. (2000) J. Biol. Chem. 275, 7662–7667.
29 Lee, Y.K. et al. (2000) Mol. Cell. Biol. 20, 187–195.
30 Johansson, L. et al. (2000) Mol. Cell. Biol. 20, 1124–1133.
31 Johansson, L. et al. (1999) J. Biol. Chem. 274, 345–353.
32 Lee, Y.K. et al. (1999) J. Biol. Chem. 274, 20869–20873.
33 Lee, H.K. et al. (1998) J. Biol. Chem. 273, 14398–14402.
34 Seol, W. et al. (1998) Mol. Endocrinol. 12, 1551–1557.
35 Western, P.S. et al. (2000) Gene 24, 223–232.
36 Smith, C.A. et al. (1999) Gene 234, 395–402.
37 Smith, C.A. et al. (2000) J. Mol. Endocrinol. 24, 23–32.
38 Pask, A. et al. (1997) Genomics 41, 422–426.
39 Chawla, A. et al. (2000) Cell 103, 1–4.

40 Lu, T.T. et al. (2000) Mol. Cell 6, 507–515.
41 Goodwin, B. et al. (2000) Mol. Cell 6, 517–526.

Index